D1459580

FUNDAMENTALS OF BIOMECHANICS

Equilibrium, Motion, and Deformation

Nihat Özkaya, Ph.D.
Margareta Nordin, Med.Dr.Sci.

VNR VAN NOSTRAND REINHOLD
New York

Library of Congress Catalog Number 91-14305
ISBN 0-442-00313-7

Manufactured in the United States of America.

Published by Van Nostrand Reinhold
115 Fifth Avenue
New York, New York 10003

Chapman and Hall
2-6 Boundary Row
London, SE1 8HN, England

Thomas Nelson Australia
102 Dodds Street
South Melbourne 3205
Victoria, Australia

Nelson Canada
1120 Birchmount Road
Scarborough, Ontario MIK 5G4, Canada

16 15 14 13 12 11 10 9 8 7 6 5 4 3 2 1

Library of Congress Cataloging-in-Publication Data

Özkaya, Nihat, 1956-
Fundamentals of biomechanics : equilibrium, motion, and deformation / Nihat Özkaya, Margareta Nordin ; edited by Dawn Leger.
p. cm.
Includes index.
ISBN 0-442-00313-7
1. Biomechanics. I. Nordin, Margareta. II. Leger, Dawn. III. Title.
QP303.O95 1991
612.7'6–dc20

91-14305
CIP

To My Parents.

"Sevgili Anneme ve Babama."

Nihat Özkaya

CONTENTS

* Sections which can be omitted with little or no loss in continuity.

FOREWORD

Bioengineering is a relatively recent field which has developed worldwide from the recognition that the theories and methods developed in conventional engineering are often useful for understanding and solving problems in physiology and medicine as well. Biomechanics is an important part of bioengineering, particularly for the field of orthopaedics. Biomechanics considers the applications of classical mechanics, including statics, dynamics, solid mechanics, and fluid mechanics to biological problems.

In the theoretical and practical advances that have been made to date, there has usually been a cooperation of a biological specialist or physician with an engineer or physicist who is thoroughly grounded in mathematics and physical sciences. Such a dialogue requires a certain amount of common vocabulary. The engineer must learn some anatomy and physiology to be useful. The medical personnel need to learn the basic concepts and vocabulary of the physical science and mathematics involved. This is often a difficult task and an impediment to bonafide collaboration because most textbooks are written for their own specialists and are not easily accessible to others.

This is where the present book will be useful. It provides a sound introduction to applied mechanics concepts that are useful in biomechanics. The basics given here will allow understanding of the results of more advanced theories than are presented here. It will also serve as an introduction for engineers who know mechanics, but wish to explore other aspects of interest in biomechanics.

On a personal level, it is a pleasure to see this book come to fruition because it represents a second and successful stage in the development of bioengineering. Dr. Nihat Özkaya was a doctoral student at the Columbia University School of Engineering and Applied Sciences while I was a Professor there, and I can vouch for the rigor, accuracy, and care he has brought to all his work, including this book. It is a second stage in that most books and papers in bioengineering tend to be written by bioengineers for engineers or by biologists for biologists. But here is a genuine effort by an engineer to explain the foundations of mechanics in a manner suitable for study by biologists and medical personnel. This dialogue is enhanced by Dr. Margareta Nordin as a coauthor which also insures the accuracy of the medical content.

Finally, it is a pleasure to see a recent student publish a useful book. It makes the effort of being involved in education seem worthwhile and lends assurance that we are collectively on a useful track.

Richard Skalak, Ph.D., M.D. (Hon.)
Professor of Bioengineering
University of California, San Diego

FOREWORD

Biomechanics is a discipline utilized by different groups of professionals. It is a required basic science for orthopaedic surgeons, physiatrists, rheumatologists, physical and occupational therapists, and athletic trainers. These medical and paramedical specialists usually do not have a strong mathematics and physics background. Biomechanics must be presented to them in a rather non-mathematical way so that they may learn the concepts of mechanics without a rigorous mathematical approach.

On the other hand, many engineers work in fields in which biomechanics plays a significant role. Human factors engineering, ergonomics, biomechanics research, and prosthetic research and development all require that the engineers working in the field have a strong knowledge of biomechanics. They are equipped to learn biomechanics through a rigorous mathematical approach. Classical textbooks in the engineering fields do not approach the biological side of biomechanics.

The *Fundamentals of Biomechanics* by Drs. Nihat Özkaya and Margareta Nordin approaches biomechanics through a rigorous mathematical standpoint whilst emphasizing the biological side. This book will be very useful for engineers studying biomechanics and for medical specialists enrolled in courses who desire a more intensive study of biomechanics and are equipped through previous study of mathematics to develop a deeper comprehension of engineering as it applies to the human body.

This work was prepared in a combined clinical setting at the Hospital for Joint Diseases Orthopaedic Institute and teaching setting within the Program of Ergonomics and Biomechanics at New York University. The authors of this volume have the unique experience of teaching biomechanics in a clinical setting to professionals with diverse backgrounds. This work reflects their many years of classroom teaching, rehabilitation treatment, and practical and research experience.

Victor H. Frankel, M.D., Ph.D.
President and Chairman
Hospital for Joint Diseases Orthopaedic Institute
Professor of Orthopaedic Surgery
New York University School of Medicine

PREFACE

Courses in biomechanics are taught within a wide variety of academic programs to students with quite different backgrounds. We believe that this text is a self-sufficient teaching and learning tool for health care professionals who are seeking a graduate degree in biomechanics but have limited backgrounds in calculus, physics and engineering mechanics. It is also a useful resource for undergraduate biomechanical and bioengineering programs. It is an intermediate level text in biomechanics aimed to bridge the gap between descriptive texts which avoid mathematics and those which require advanced background in mathematics and engineering principles.

This text is divided into three parts. The first part (Chapters 1 through 6, and Appendices A and B) introduces the basic concepts of mechanics, provides the mathematical tools necessary to explain these concepts, outlines the procedure for analyzing systems in "equilibrium," and applies this procedure to relatively simple mechanical systems and to the human musculoskeletal system. The second part of the text (Chapters 7 through 12, and Appendix C) is devoted to the analyses of "moving" systems with applications to human motion analyses and sports mechanics. The last part of the text (Chapters 13 through 17) provides the techniques for analyzing the "deformation" characteristics of materials with applications to orthopaedic biomechanics.

While preparing this text we paid particular attention to the applications of the concepts introduced, methods explained and procedures outlined by providing many solved example problems. Most of these examples are constructed to illustrate the relevance of engineering knowledge to human physiology. To provide proper visual aids, special attention is given to the quality and quantity of illustrations. To avoid overwhelming the reader with extensive lists of references, we direct the interested reader to sources which contain more complete literature surveys.

It was our purpose to illustrate how biological phenomena can be described in terms of mechanical concepts. We believe that the knowledge of the biomechanical aspects and structural behavior of the human musculoskeletal system is an essential prerequisite for any experimental, theoretical, or analytical approach to analyze its physiological function in the body. By preparing this text, we hope to contribute to an improved dialogue between those professionals who are primarily interested in the biological and physiological aspects of the human body and those who are interested in the structural behavior of the human body, through an engineering approach.

Nihat Özkaya
Margareta Nordin

ACKNOWLEDGMENTS

I would like to express my thanks to those individuals who followed the biomechanics courses I have been teaching within the Graduate Program of Ergonomics and Biomechanics at New York University, and who, one way or another contributed to the development of this book. In this regard, I extend my thanks to Ann Barr, Lino Chuang, Eva Frykman, Jane Miller, Lorna Ramos, Ellen Ross, Sydney Schachne, and Anthony Thomas for their suggestions and editorial comments.

Special thanks to Bob Esposito of Van Nostrand Reinhold for initiating this project; to Athena Aaron and Bernardus Willems for checking the solutions of the example problems; to my colleagues Drs. Paul-M. Brisson, Heinrich O. Hofer, Patricia Hogan, and Mohamad Parnianpour for their helpful suggestions and contributions; to Dr. Victor H. Frankel for his support and encouragement; and to Dr. Dawn L. Leger for her patience, support, and for editing the text.

Nihat Özkaya

Chapter 1

INTRODUCTION

1.1 Mechanics

Mechanics is a branch of physics that is concerned with the motion and deformation of bodies that are acted on by mechanical disturbances called forces. Mechanics is the oldest of all physical sciences, dating back to the times of Archimedes (287–212 B.C.). Galileo (1564–1642) and Newton (1642–1727) were the most prominent contributors to this field. Galileo made the first fundamental analyses and experiments in dynamics, and Newton formulated the laws of motion and gravity.

Engineering mechanics or *applied mechanics* is the science of applying the principles of mechanics. Applied mechanics is concerned with both the analysis and design of mechanical systems. The broad field of applied mechanics can be divided into three main parts, as illustrated in Table 1.1.

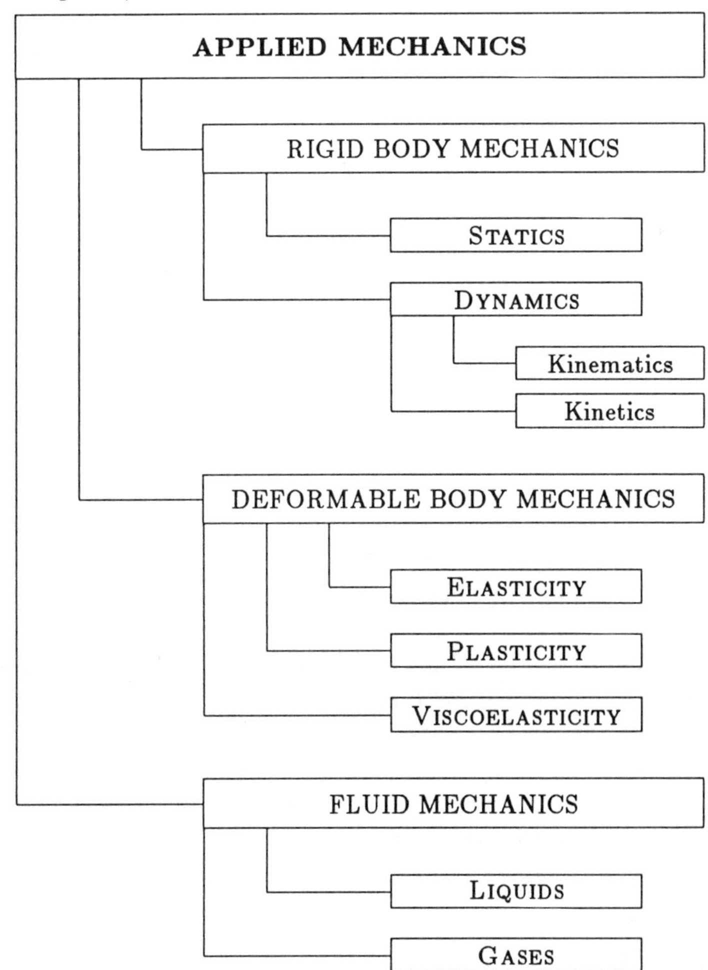

Table 1.1 *Classification of applied mechanics.*

In general, a material can be categorized as either a solid or fluid. Solid materials can be rigid or deformable. A *rigid body* is one that cannot be deformed. In reality, every object or material does undergo deformation to some extent when acted upon by external forces. In some cases the amount of deformation is so small that it does not affect the desired analysis. In such cases, it is preferable to consider the body as rigid and carry out the analysis with relatively simple computations.

Statics is the study of forces on rigid bodies at rest or moving with a constant velocity. *Dynamics* deals with bodies in motion. *Kinematics* is a branch of dynamics that deals with the geometry and time-dependent aspects of motion without considering the forces causing the motion. *Kinetics* is based on kinematics, and it includes the effects of forces and masses in the analysis.

Statics and dynamics are devoted primarily to the study of the external effects of forces on rigid bodies, bodies for which the deformation (change in shape) can be neglected. On the other hand, the *mechanics of deformable bodies* deals with the relations between externally applied loads and their internal effects on bodies. This field of applied mechanics does not assume that the bodies of interest are rigid, but considers the true nature of their material properties. The mechanics of deformable bodies has strong ties with the field of material science which deals with the atomic and molecular structure of materials. The principles of deformable body mechanics have important applications in the design of structures and machine elements. In general, analyses in deformable body mechanics are more complex as compared to the analyses required in rigid body mechanics.

The mechanics of deformable bodies is the field that is concerned with the deformability of objects. An *elastic* body is defined as one in which all deformations are recoverable upon removal of external forces. This feature of some materials can easily be visualized by observing a spring or a rubber band. If you gently stretch (deform) a spring and then release it (remove the applied force), it will resume its original (undeformed) size and shape. A *plastic* body, on the other hand, undergoes permanent (unrecoverable) deformations. One can observe this behavior again by using a spring. Apply a large force on a spring so as to stretch the spring extensively, and then release it. The spring will bounce back, but there may be an increase in its length. This increase illustrates the extent of plastic deformation in the spring. Note that depending on the extent and duration of applied forces, a material may exhibit elastic or elasto-plastic behavior as in the case of the spring.

To explain viscoelasticity, we must first define what is known as a *fluid*. In general, materials are classified as either solid or fluid. When an external force is applied to a solid body, the body will deform to a certain extent. The continuous application of the same force will not necessarily deform the solid body continuously. On the other hand, a continuously applied force on a fluid body will cause a continuous deformation (flow). *Viscosity* is a fluid property which is a quantitative measure of resistance to flow. In nature there are some materials that have both fluid and solid properties. The term *viscoelastic* is used to refer to the mechanical properties of such materials. Many biological materials exhibit viscoelastic properties.

Note that the distinctions between the various areas of applied mechanics are not sharp. For example, viscoelasticity simultaneously utilizes the principles of fluid and solid mechanics.

1.2 Biomechanics

Biomechanics combines the field of engineering mechanics with the fields of biology and physiology. Biomechanics is concerned with the human body. In biomechanics, the principles of mechanics are applied to the conception, design, development, and analysis of equipment and systems in biology and medicine. Although biomechanics is a relatively young and dynamic field, its history can be traced back to the fifteenth century, when Leonardo da Vinci (1452-1519) noted the significance of mechanics in his biological studies. As a result of contributions of researchers in the fields of biology, medicine, basic sciences, and engineering, the interdisciplinary field of biomechanics has been growing steadily in the last two decades.

The development of the field of biomechanics has improved our understanding of many things, including normal and pathological situations, mechanics of neuromuscular control, mechanics of blood flow in the microcirculation, mechanics of air flow in the lung, and mechanics of growth and form. It has contributed to the development of medical diagnostic and treatment procedures. It has provided the means for designing and manufacturing medical instruments, devices for the handicapped, artificial replacements and implants. It has suggested the means for improving human performance in the workplace and in athletic competition.

Different aspects of biomechanics utilize different components of applied mechanics. For example, the principles of statics are applied to determine the magnitude and nature of forces involved in various joints and muscles of the musculoskeletal system. The principles of dynamics are utilized for motion description and have many applications in sports mechanics. The principles of the mechanics of deformable bodies provide the necessary tools for developing the field and constitutive equations for biological materials and systems, which in turn are used to evaluate their functional behavior under different conditions. The principles of fluid mechanics are used to investigate the blood flow in the human circulatory system and air flow in the lung.

It is the aim of this textbook to expose the reader to the principles and applications of biomechanics. For this purpose, the basic tools and principles will first be introduced. Next, systematic and comprehensive applications of these principles will be carried out with many solved example problems. Attention will be focused on the applications of statics, dynamics, and the mechanics of deformable bodies (i.e., solid mechanics). A limited study of fluid mechanics and its applications in biomechanics will be provided.

1.3 Basic Concepts

Engineering mechanics is based on Newtonian mechanics in which the basic concepts are length, time, and mass. These are absolute concepts because they are independent of each other. *Length* is a concept for describing size quantitatively. *Time* is a concept for ordering the flow of events. *Mass* is the property of all matter and is the quantitative measure of inertia. *Inertia* is the resistance to the change in motion of matter.

Other important concepts in mechanics are not absolute but derived from the basic concepts. These include force, moment or torque, velocity, acceleration, work, energy, power, impulse, momentum, stress, and strain. *Force* can be defined in many ways, such as mechanical disturbance or load. Force is the action of one body on another. It is the force applied on a body which causes the body to move and/or deform. *Moment* or *torque* is the quantitative measure of the rotational, bending or twisting action of a force applied on a body. *Velocity* is defined as the time rate of change of position. The time rate of increase of velocity, on the other hand, is termed *acceleration*. Detailed descriptions of these and other concepts will be provided in subsequent chapters.

1.4 Newton's Laws

The entire field of mechanics rests on a few basic laws. Among these, the laws of mechanics introduced by Sir Isaac Newton form the basis for analyses in statics and dynamics.

Newton's first law states that a body that is originally at rest will remain at rest, or a body in motion will move in a straight line with constant velocity, if the net force acting upon it is zero. In analyzing this law, we must pay extra attention to a number of key words. The term "rest" implies no motion. For example, a book lying on a desk is said to be at rest. To be able to explain the concept of "net force" fully, we need to introduce vector algebra (see Chapter 2). The net force simply refers to the combined effect of all forces acting on a body. If the net force acting on a body is zero, it does not necessarily mean that there are no forces acting on the body. For example, there may be two equal and opposite forces applied on a body so that the combined effect of the two forces on the body is zero, assuming that the body is rigid. Note that if a body is either at rest or moving in a straight line with a constant velocity, then it is said to be in *equilibrium*.

Therefore, the first law states that if the net force acting on a body is zero, then the body is in equilibrium.

Newton's second law states that a body with a net force acting on it will accelerate in the direction of that force, and that the magnitude of the acceleration will be proportional to the magnitude of the net force. The important terms in the statement of the second law are "magnitude" and "direction," and they will be explained in detail in Chapter 2, within the context of vector algebra.

Newton's third law states that to every action there is always an equal reaction, and that the forces of action and reaction between interacting bodies are equal in magnitude, opposite in direction, and have the same line of action. This law can be simplified by saying that if you push a body, the body will push you back. This law has important applications in constructing *free-body diagrams* of components constituting large systems. The free-body diagram of a component of a structure is one in which the surrounding parts of the structure are replaced by equivalent forces. It is an effective aid to study the forces involved in the structure.

Newton's laws will be explained in detail in subsequent chapters, and they will be utilized extensively throughout this text.

1.5 Dimensional Analysis

The term "dimension" has several uses in mechanics. It is used to describe space, as for example while referring to one-dimensional, two-dimensional, or three-dimensional situations. Dimension is also used to denote the nature of quantities. Every measurable quantity has a *dimension* and a *unit* associated with it. Dimension is a general description of a quantity, whereas unit is associated with a system of units (see Section 1.6). Whether a distance is measured in meters or feet, it is a distance. We say that its dimension is "length." There are two sets of dimensions. *Primary*, or *basic, dimensions* are those associated with the basic concepts of mechanics. In this text, we shall use L, T, and M to specify the primary dimensions length, time, and mass, respectively. We shall use square brackets to denote the dimensions of a physical quantity. The basic dimensions are:

$$\begin{aligned}[\mathsf{LENGTH}] &= \mathsf{L}\\ [\mathsf{TIME}] &= \mathsf{T}\\ [\mathsf{MASS}] &= \mathsf{M}\end{aligned}$$

Secondary dimensions are associated with dependent concepts which are derived from basic concepts. For example, the area of a rectangle can be calculated by multiplying its width and length, both of which have the dimension of length. Therefore, the dimension of area is:

$$[\text{AREA}] = [\text{LENGTH}][\text{LENGTH}] = \text{LL} = \text{L}^2$$

The secondary dimensional quantities are established as a consequence of certain natural laws. If we know the definition of a physical quantity we can easily determine the dimension of that quantity in terms of basic dimensions. If the dimension of a physical quantity is known, then the units of that quantity in different systems of units can easily be determined. Furthermore, the validity of an equation relating a number of physical quantities can be verified by analyzing the dimensions of terms forming the equation or formula. In this regard, the *law of dimensional homogeneity* imposes restrictions on the formulation of such relations. To explain this law, consider the following arbitrary equation:

$$Z = aX + bY + c$$

For this equation to be dimensionally homogeneous, every grouping in the equation must have the same dimensional representation. In other words, if Z refers to a quantity whose dimension is length, then products aX and bY, and quantity c must all have the dimension of length. The numerical equality between both sides of the equation must also be maintained for all systems of units.

1.6 Systems of Units

There have been a number of different systems of units adopted in different parts of the world. For example, there is the British gravitational or foot-pound-second system, the Gausian (metric absolute) or centimeter-gram-second (cgs) system, and the metric gravitational or meter-kilogram-second (mks) system. The lack of a universal standard in units of measure often causes confusion.

In 1960, an International Conference on Weights and Measures was held to bring an order to the confusion surrounding the units of measure. Based on the metric system, this conference adopted a system called *Le Système International d'Unités* in French, which is abbreviated as SI. In English it is known as the International System of Units. Today, nearly the entire world is either using this modernized metric system or committed to its adoption. In the International System of Units, the units of length, time, and mass are meter (m), second (s), and kilogram (kg), respectively. The units of measure of these fundamental concepts in three different systems of units are listed in Table 1.2. Throughout this text, we shall use the International System of Units. Other units will be defined for informational purposes.

SYSTEM	LENGTH	MASS	TIME
SI	meter (m)	kilogram (kg)	second (s)
c-g-s	centimeter (cm)	gram (gm)	second (s)
British	foot (ft)	slug (slug)	second (s)

Table 1.2 *Units of fundamental quantities of mechanics.*

Once the units of measure for the primary concepts are agreed upon, the units of measure for the derived concepts can easily be determined provided that the dimensional relationship between the basic and derived quantities is known. All that is required is replacing the dimensional representation of length, mass, and time with their appropriate units. For example, the dimension of force is ML/T^2. Therefore, according to the International System of Units, force has the unit of kg-m/s^2 which is also known as Newton (N). Similarly, the unit of force is lb-ft/s^2 in the British system of units, and is gm-cm/s^2 or dyne (dyn) in the metric absolute or cgs system. Table 1.3 lists the dimensional representations of some of the derived quantities and their units in the International System of Units.

QUANTITY	DIMENSION	SI UNIT	SPECIAL NAME
Area	L^2	m^2	
Volume	L^3	m^3	
Velocity	L/T	m/s	
Acceleration	L/T^2	m/s^2	
Force	ML/T^2	kg-m/s^2	Newton (N)
Pressure & Stress	M/LT^2	N/m^2	Pascal (Pa)
Moment (Torque)	ML^2/T^2	N-m	
Work & Energy	ML^2/T^2	N-m	Joule (J)
Power	ML^2/T^3	J/s	Watt (W)

Table 1.3 *Dimensions and units of selected quantities in SI.*

Note that "kilogram" is the unit of mass in SI. For example, consider a 60 kg object. The weight of the same object in SI is (60 kg)×(9.8 m/s^2)=588 N, the factor 9.8 m/s^2 being the magnitude of the gravitational acceleration.

In addition to the primary and secondary units that are associated with the basic and derived concepts in mechanics, there are ***supplementary units*** such as plane angle and temperature. The common measure of an angle is degree (°). Three hundred sixty degrees is equal to one revolution (rev) or 2π radians (rad), where π=3.1416. The SI unit of temperature is Kelvin (°K). However, degree Celsius (°C) is more commonly used. The British unit of temperature is degree Fahrenheit (°F).

It should be noted that in most cases, a number has a meaning only if the correct unit is displayed with it. In performing calculations, the ideal method is to show the correct units with each number throughout the solution of equations. This approach helps in detecting conceptual errors and eliminates the need for determining the unit of the calculated quantity separately. Another important aspect of using units is consistency. One must not use the units of one system for some quantities and the units of another system for other quantities while carrying out calculations.

1.7 Conversion of Units

The International System of Units is a revised version of the metric system which is based on the decimal system. Table 1.4 lists the SI multiplication factors and corresponding prefixes. Table 1.5 lists factors needed to convert quantities expressed in British and metric systems to corresponding units in SI.

MULTIPLICATION FACTOR	SI PREFIX	SI SYMBOL
1 000 000 000 = 10^9	giga	G
1 000 000 = 10^6	mega	M
1 000 = 10^3	kilo	k
100 = 10^2	hecto	h
10 = 10^1	deka	da
.1 = 10^{-1}	deci	d
.01 = 10^{-2}	centi	c
.001 = 10^{-3}	milli	m
.000 001 = 10^{-6}	micro	μ
.000 000 001 = 10^{-9}	nano	n
.000 000 000 001 = 10^{-12}	pico	p

Table 1.4 *SI multiplication factors and prefixes.*

LENGTH	1 centimeter (cm) = 0.01 meter (m)
	1 inch (in) = 0.0254 m
	1 foot (ft) = 0.3048 m
	1 yard (yd) = 0.9144 m
	1 mile = 1609 m
	1 angstrom (Å) = 10^{-10} m
TIME	1 minute (min) = 60 second (s)
	1 hour (h) = 3600 s
	1 day (d) = 86400 s
MASS	1 pound mass (lbm) = 0.4536 kilogram (kg)
	1 slug = 14.59 kg
FORCE	1 kilogram force (kgf) = 9.807 Newton (N)
	1 pound force (lbf) = 4.448 N
	1 dyne (dyn) = 10^{-5} N
PRESSURE & STRESS	1 kg/m-s^2 = 1 N/m^2 = 1 Pascal (Pa)
	1 lbf/in^2 (psi) = 6896 Pa
	1 lbf/ft^2 (psf) = 992966 Pa
	1 dyn/cm^2 = 0.1 Pa
MOMENT (TORQUE)	1 dyn-cm = 10^{-7} N-m
	1 lbf-ft = 1.356 N-m
WORK & ENERGY	1 kg-m^2/s^2= 1 N-m = 1 Joule (J)
	1 dyn-cm = 1 erg = 10^{-7} J
	1 lbf-ft = 1.356 J
POWER	1 kg-m^2/s^3 = 1 J/s = 1 Watt (W)
	1 horsepower (hp) = 550 lbf-ft/s = 746 W
PLANE ANGLE	1 degree (°) = π/180 radian (rad)
	1 revolution (rev) = 360°
	1 rev = 2π rad = 6.283 rad
TEMPERATURE	°C = °K − 273.2
	°C = 5(°F − 32)/9

Table 1.5 *Conversion of units.*

1.8 Mathematics

The applications of biomechanics require certain knowledge of mathematics. These include simple geometry, properties of the right-triangle, basic algebra, differentiation, and integration. The appendices that follow the last chapter contain a summary of the mathematical techniques needed to carry out the calculations in this book. The reader may find it useful to examine them now, and review them later when those concepts are needed. In subsequent chapters throughout the text, the mathematics required will be reviewed and the corresponding appendix will be indicated.

During the formulation of the problems, we shall use Greek letters as well as the letters of the Latin alphabet. Greek letters will be used for example to refer to angles. The Greek alphabet is provided in Table 1.6 for quick reference.

alpha	A α	iota	I ι	rho	P ρ
beta	B β	kappa	K κ	sigma	Σ σ
gamma	Γ γ	lambda	Λ λ	tau	T τ
delta	Δ δ	mu	M μ	upsilon	Υ υ
epsilon	E ϵ	nu	N ν	phi	Φ ϕ
zeta	Z ζ	xi	Ξ ξ	chi	X χ
eta	H η	omicron	O o	psi	Ψ ψ
theta	Θ θ	pi	Π π	omega	Ω ω

Table 1.6 *Greek alphabet.*

1.9 Scalars and Vectors

In mechanics, two kinds of quantities are distinguished. A *scalar* quantity, such as mass, temperature, work, and energy, has magnitude. A *vector* quantity, such as force, velocity, and acceleration, has both a magnitude and a direction. Unlike scalars, vector quantities add according to a rule called the *parallelogram law.* Vector algebra will be covered in detail in Chapter 2.

1.10 Modeling and Approximations

One needs to make certain assumptions to simplify complex systems and problems so as to achieve analytical solutions. The complete model is the one that includes the effects of all parts constituting a system. However, the more detailed the model, the more difficult the formulation and solution of the problem. It is not always possible and in some cases it may not be necessary to include every detail in the analysis. For example, during most human activities, there is more than one muscle group activated at a time. If the task is to analyze the forces involved in the joints and muscles during a particular human activity, the best approach is to predict which muscle group is the most active and set up a model that neglects all other muscle groups. As we shall see in the following chapters, bone is a deformable body. If the forces involved are relatively small, then the bone can be treated as a rigid body. This approach may reduce the complexity of the problem under consideration.

In general, it is always best to begin with a simple basic model that represents the system. Gradually, the model can be expanded on the basis of experience gained and the results obtained from simpler models. The guiding principle is to make simplifications that are consistent with the required accuracy of the results. In this way, the researcher can set up a model that is simple enough to analyze and exhibit satisfactorily the phenomena under consideration. The more we learn, the more detailed our analysis can become.

1.11 Generalized Procedure

The general method of solving problems in biomechanics may be outlined as follows:

1. Select the system of interest.
2. Postulate the characteristics of the system.
3. Simplify the system by making proper approximations. Explicitly state important assumptions.
4. Form an analogy between the human body parts and basic mechanical elements.
5. Construct a mechanical model of the system.
6. Apply principles of mechanics to formulate the problem.
7. Solve the problem for the unknowns.
8. Compare the results with the behavior of the actual system. This may involve tests and experiments.
9. If satisfactory agreement is not achieved, steps 3 through 7 must be repeated by considering different assumptions and a new model of the system.

1.12 Scope of the Text

Courses in biomechanics are taught within a wide variety of academic programs to students with quite different backgrounds. This text is prepared to provide a teaching and learning tool to health care professionals who are seeking a graduate degree in biomechanics but have limited backgrounds in calculus, physics, and engineering mechanics. This text can also be a useful reference for undergraduate biomedical, biomechanical, or bioengineering programs.

This text is divided into three parts. The first part (Chapters 1 through 6, and Appendices A and B) will introduce the basic concepts of mechanics including force and moment vectors, provide the mathematical tools (geometry, algebra, and vector algebra) so that complete definitions of these concepts can be given, explain the procedure for analyzing the systems at "static equilibrium," and apply this procedure to analyze simple mechanical systems and the forces involved at various muscles and joints of the human musculoskeletal system. The second part of the text (Chapters 7 through 12, and Appendix C) is devoted to "dynamic" analyses. The concepts introduced in the second part are position, velocity and acceleration vectors, work, energy, power, impulse, and momentum. Also provided in the second part are the techniques for kinetic and kinematic analyses of systems undergoing translational and rotational motions. These techniques are applied for human motion analyses of various sports activities. The last part of the text (Chapters 13 through 17) provides the techniques for analyzing the "deformation" characteristics of materials under different load conditions. For this purpose, the concepts of stress and strain are defined. Classifications of materials based on their stress-strain diagrams are given. The concepts of elasticity, plasticity, and viscoelasticity are also introduced and explained. Topics such as torsion, bending, fatigue, endurance, and factors affecting the strength of materials are provided. The emphasis is placed upon applications to orthopaedic biomechanics.

It should be noted here that the topics covered in the first part of this text are prerequisites for both parts two and three. On the other hand, topics covered in the third part are relatively independent of the concepts introduced in the second part.

1.13 Notation

While preparing this text, special attention was given to the consistent use of notation. Important terms are italicized where they are defined or described (such as, *force* is defined as load or mechanical disturbance). Symbols for quantities are also italicized (for example, m for mass). Units are not italicized (for example, kg for kilogram). Underlined letters are used to refer to vector quantities (for example, force vector $\underline{F}$). Sections and subsections marked with a star (*) are considered optional. In other words, the reader can omit a section or subsection marked with a star without losing the continuity of the topics covered in the text.

1.14 References and Suggested Reading

We believe that this text is a self-sufficient teaching and learning tool. While preparing it, we utilized the information provided in a number of sources, some of which are listed below. Note, however, that it is not our intention to promote these titles, or to suggest that these are the only texts available on the subject matter. The field of biomechanics has been growing very rapidly. There are many other sources of information available, including scientific journals presenting peer-reviewed research articles in biomechanics.

Black, J. 1988. *Orthopaedic Biomaterials in Research and Practice*. New York: Churchill Livingstone.

Chaffin, D.B., and Andersson, G.B.J. 1991. *Occupational Biomechanics*. 2nd ed. New York: John Wiley & Sons.

Frankel, V.H., and Burnstein, A.H. 1970. *Orthopaedic Biomechanics*. Philadelphia: Lea & Febiger.

Hay, J. 1978. *The Biomechanics of Sports Techniques*. Englewood Cliffs, NJ: Prentice-Hall.

Hay, J.G., and Reid, J.G. 1988. *Anatomy, Mechanics and Human Motion*. 2nd ed. Englewood Cliffs, NJ: Prentice-Hall.

Kelly, D.L. 1971. *Kinesiology: Fundamentals of Motion Description*. Englewood Cliffs, NJ: Prentice-Hall.

Kirby, R., and Roberts, J.A. 1985. *Introductory Biomechanics*. Ithaca, NY: Mouvement Publications.

Nahum, A.M., and Melvin, J. (Eds.) 1985. *The Biomechanics of Trauma*. Norwalk, CT: Appleton-Century-Crofts.

Nordin, M., and Frankel, V.H. 1989. *Basic Biomechanics of the Musculoskeletal System*. 2nd ed. Philadelphia: Lea & Febiger.

Simon, S.R., Riggins, R.S., Wirth, C.R., and Fox, M.L. (Eds.) 1986. *Orthopaedic Science*. Illinois: American Academy of Orthopaedic Surgeons.

Thompson, C.W. 1989. *Manual of Structural Kinesiology*. 11th ed. St. Louis, MO: Times Mirror/Mosby.

Wictorin, C.H., and Nordin, M. 1982. *Introduction to Problem Solving in Biomechanics*. Philadelphia: Lea & Febiger.

Williams, M., and Lissner, H.R. 1977. *Biomechanics of Human Motion*. 2nd ed. (B. LeVeau, ed.) Philadelphia: Saunders.

Winter, D.A. 1990. *Biomechanics and Motor Control of Human Behavior*. 2nd ed. New York: John Wiley & Sons.

The following sources contain advanced topics and up-to-date information in the fields of biomechanics and bioengineering.

Fung, Y.C. 1981. *Biomechanics: Mechanical Properties of Living Tissues*. New York: Springer-Verlag.

Fung, Y.C. 1990. *Biomechanics: Motion, Flow, Stress, and Growth*. New York: Springer-Verlag.

Mow, V.C., Ratcliff, A., and Woo, S.L-Y. (Eds.) 1990. *Biomechanics of Diarthrodial Joints*. New York: Springer-Verlag.

Schmid-Schönbein, G.W., Woo, S.L-Y., and Zweifach, B.W. (Eds.) 1985. *Frontiers in Biomechanics.* New York: Springer-Verlag.

Skalak, R., and Chien, S. (Eds.) 1987. *Handbook of Bioengineering.* New York: McGraw-Hill.

Winters, J.M., and Woo, S.L-Y. (Eds.) 1990. *Multiple Muscle Systems.* New York: Springer-Verlag.

In addition to the biomechanics related books listed above, there are many physics and engineering mechanics books dedicated to undergraduate engineering education. These books may be useful for reviewing the basic concepts of mechanics. A few examples of such titles are listed below.

Sandor, B.I. 1983. *Engineering Mechanics: Statics and Dynamics.* 2nd ed. Englewood Cliffs, NJ: Prentice-Hall.

Servay, R.A. 1982. *Physics: For Scientists and Engineers.* Philadelphia: Saunders.

Shames, I.H. 1980. *Engineering Mechanics: Statics and Dynamics.* 3rd ed. Englewood Cliffs, NJ: Prentice-Hall.

The last part of this text will introduce the principles of mechanics of deformable bodies and use these principles to explain certain aspects of biomechanics. The following books can be reviewed to gain more detailed information on these principles.

Crandall, S.H., Dahl, N.C., and Lardner, T.J. 1978. *An Introduction to the Mechanics of Solids.* 2nd ed. New York: McGraw-Hill.

Popov, E.P. 1978. *Mechanics of Materials.* 2nd ed. Englewood Cliffs, NJ: Prentice-Hall.

Pytel, A., and Singer, F.L. 1987. *Strength of Materials.* 4th ed. New York: Harper & Row.

Chapter 2

VECTOR ALGEBRA

2.1 Definitions

Most of the concepts in mechanics are either scalar or vector. A *scalar* quantity has a magnitude only, as opposed to a *vector* quantity which has both a magnitude and a direction associated with it. Concepts such as mass, energy, power, work and temperature are scalar quantities. Saying that the room temperature is 20°C sufficiently describes the temperature in the room. Force, velocity, and acceleration are samples of vector quantities. To describe an applied force fully, one must state how much force is applied and in which direction it is applied.

It should be noted that both scalars and vectors are special forms of a more general category of all quantities called *tensors.* Scalar quantities are also known as "zero order tensors," whereas vectors are "first order tensors." Stress and strain, on the other hand, are examples of "second order tensors."

2.2 Notation

There are various notations used to refer to vector quantities. In this text, we shall use letters with underbars. For example, $\underline{A}$ will refer to a vector quantity, whereas A without the underbar will be used to refer to a scalar quantity. In graphical solutions, vectors are commonly represented by arrows, as illustrated in Figure 2.1. If there is a need for showing more than one vector in a single drawing, then the length of each arrow must be proportional to the magnitude of the vector it is representing. There are two ways of referring to the magnitude of a vector quantity: either by dropping the underbar (A), or by enclosing the vector quantity with a set of vertical lines known as the absolute sign ($|\underline{A}|$). The *magnitude* of a vector is a scalar quantity. The magnitude of a quantity is always a positive number of units whose value corresponds to the numerical measure of that quantity. The orientation of the arrow indicates the *line of action* of the vector. The arrowhead denotes the direction of the vector by defining its *sense* along its line of action. If the vector represents for example an applied force, then the base (tail) of the arrow corresponds to the *point of application* of the force vector.

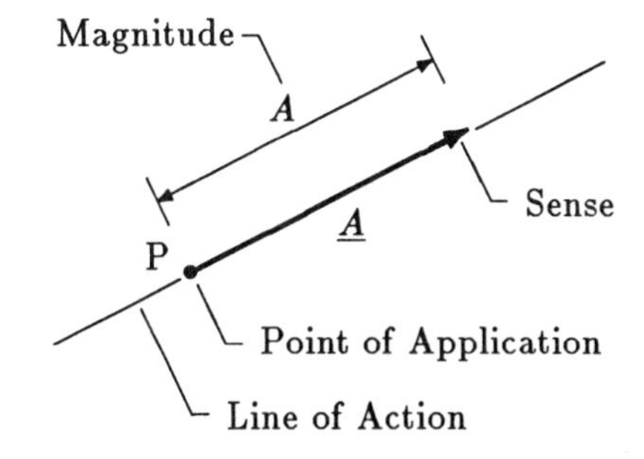

Figure 2.1 *Graphical representation of a vector $\underline{A}$.*

2.3 Multiplication of a Vector by a Scalar

Let $\underline{A}$ be a vector quantity with magnitude A, and m be a scalar quantity. The product $m\underline{A}$ is equal to a new vector, $\underline{B}=m\underline{A}$, such that it has the same direction as vector $\underline{A}$ but a magnitude equal to m times the magnitude of $\underline{A}$ ($B=mA$). For example, if $m=2$ then the magnitude of the product vector $\underline{B}$ is twice as large as the magnitude of vector $\underline{A}$.

2.4 Negative Vector

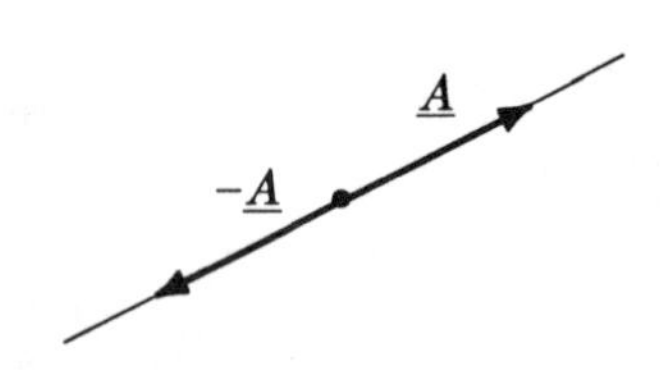

Figure 2.2 *Negative vector.*

Let $\underline{A}$ be a vector quantity with magnitude A. $-\underline{A}$ is called a ***negative vector*** and it differs from vector $\underline{A}$ in that $-\underline{A}$ has a direction opposite to that of vector $\underline{A}$ (Figure 2.2). Since magnitudes of vector quantities are positive scalar quantities, both $-\underline{A}$ and $\underline{A}$ have the same magnitude equal to A. Therefore, for vector quantities a negative sign implies change in direction and it has nothing to do with the magnitude of the vector.

2.5 Addition of Vectors: Graphical Methods

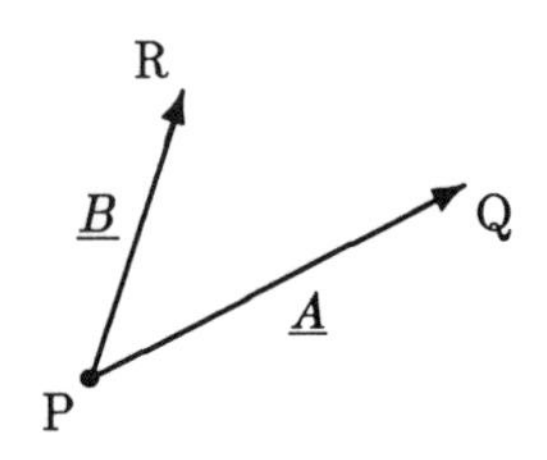

Figure 2.3 *Vectors $\underline{A}$ and $\underline{B}$.*

There are two ways of adding two or more vectors graphically: parallelogram and tail-to-tip methods. Consider the two vectors $\underline{A}$ and $\underline{B}$ shown in Figure 2.3. Vector $\underline{A}$ is pointing towards point Q and vector $\underline{B}$ is pointing towards point R. The ***parallelogram*** method of adding two vectors involves construction of a parallelogram by drawing a line at the tip of one of the vectors parallel to the other vector, and repeating the same thing for the second vector. If S corresponds to the point of intersection of these parallel lines, then an arrow drawn from point P towards S will represent a vector that is equal to the sum of vectors $\underline{A}$ and $\underline{B}$ (Figure 2.4). Referring to this new vector as $\underline{C}$, which can be called the ***resultant***, ***sum***, or the ***net*** vector, then:

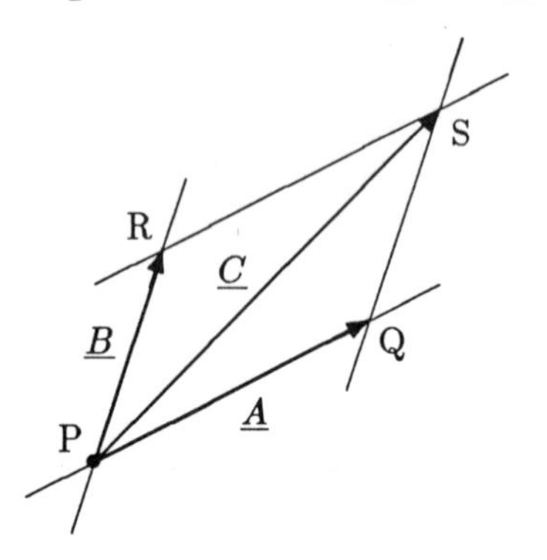

Figure 2.4 *Parallelogram.*

$$\underline{C} = \underline{A} + \underline{B} \tag{2.1}$$

The ***tail-to-tip*** method of adding two vectors graphically is illustrated in Figure 2.5. In this case, without changing its orientation, one of the vectors to be added is translated to the tip of the other vector in such a way that the tip of one of the vectors coincides with the tail of the other. An arrow drawn from the tail of the first vector towards the tip of the second vector represents the resultant vector.

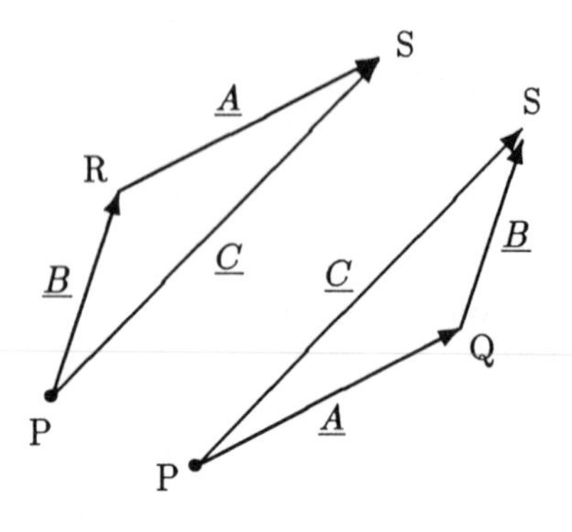

Figure 2.5 *Tail-to-tip method.*

Notice that while performing vector addition, the order of appearance of vectors is arbitrary. That is, the addition of vectors is a ***commutative*** operation. The sum of $\underline{A}$ and $\underline{B}$, and the sum of $\underline{B}$ and $\underline{A}$ result in the same vector $\underline{C}$:

$$\underline{C} = \underline{A} + \underline{B} = \underline{B} + \underline{A} \tag{2.2}$$

2.6 Subtraction of Vectors

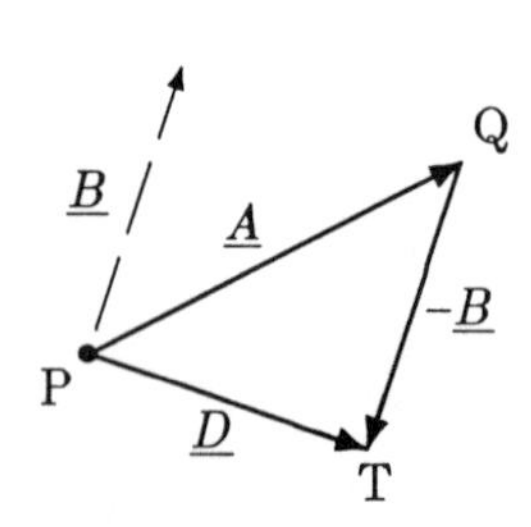

Figure 2.6 $\underline{D} = \underline{A} - \underline{B}$

The subtraction of one vector from another can easily be done by noting that the negative of any vector is a vector of the same magnitude pointing in the opposite direction. For example, the subtraction of vector $\underline{B}$ from vector $\underline{A}$ can be expressed as $\underline{A}-\underline{B}=\underline{A}+(-\underline{B})$. Therefore, as illustrated in Figure 2.6, instead of subtracting $\underline{B}$ from $\underline{A}$, $-\underline{B}$ can be added to $\underline{A}$ to determine the resultant vector.

Note that subtraction of vector $\underline{A}$ from $\underline{B}$ follows a similar approach. If $\underline{D}$ and $\underline{E}$ are two vectors such that

$$\underline{D} = \underline{A} - \underline{B}$$
$$\underline{E} = \underline{B} - \underline{A}$$

then since $\underline{A}$–$\underline{B}$=–($\underline{B}$–$\underline{A}$), vectors $\underline{D}$ and $\underline{E}$ have an equal magnitude (D=E) but opposite directions ($\underline{D}$=–$\underline{E}$).

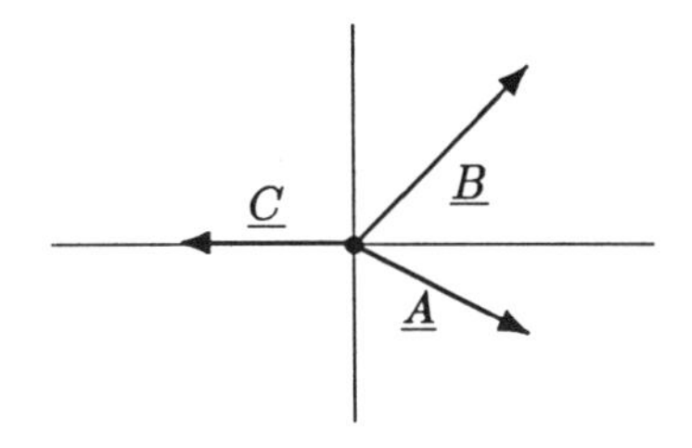

Figure 2.7 *Vectors $\underline{A}$, $\underline{B}$, $\underline{C}$.*

2.7 Addition of More Than Two Vectors

Let $\underline{A}$, $\underline{B}$ and $\underline{C}$ be three vectors such that all three vectors lie on a plane (two-dimensional) surface, as illustrated in Figure 2.7. Such a system of vectors is called *coplanar*. Notice that any two vectors always form a coplanar system. There are a number of different ways to add three vectors together. Since the addition of vectors is a commutative operation, the order of appearance of vectors to be added does not influence the resultant vector.

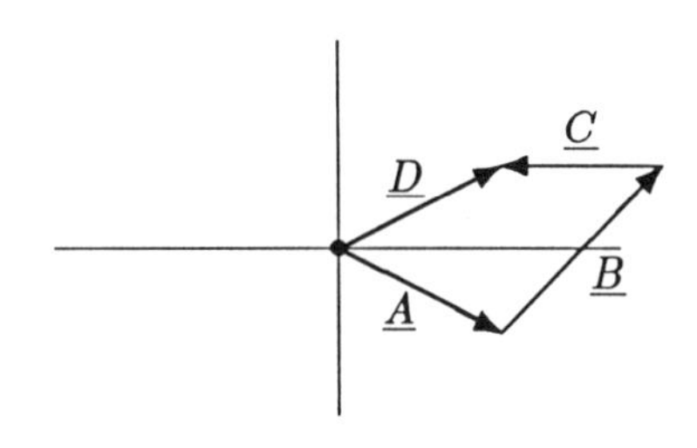

Figure 2.8 *Alternative ways of adding three vectors.*

The three vectors shown in Figure 2.7 are graphically added together in Figure 2.8 by using the tail-to-tip method. In the case illustrated, vector $\underline{C}$ is added to vector $\underline{B}$ which is added to vector $\underline{A}$ to obtain the resultant vector $\underline{D}$. That is:

$$\underline{A} + \underline{B} + \underline{C} = \underline{D}$$

The parallelogram method of adding the same vectors is illustrated in Figure 2.9. In this case, two ($\underline{B}$ and $\underline{C}$) of the three vectors are added together first. The resultant (vector $\underline{E}$) of this operation is then added to vector $\underline{A}$ and the overall resultant (vector $\underline{D}$) is determined. The sequence of additions illustrated in Figure 2.9 can be mathematically expressed as:

$$\underline{A} + (\underline{B} + \underline{C}) = \underline{A} + \underline{E} = \underline{D}$$

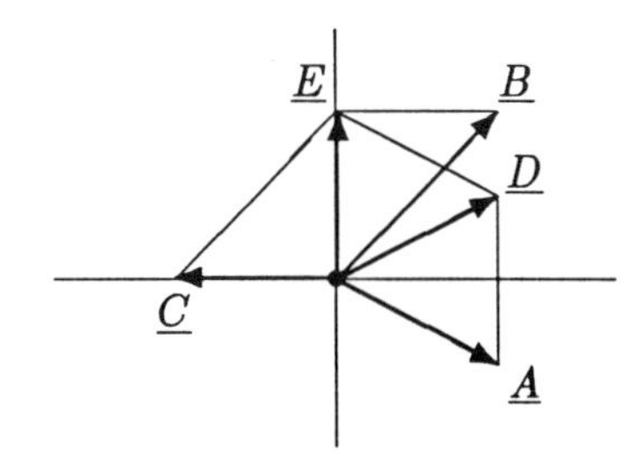

Figure 2.9 *Parallelogram method of adding three vectors.*

2.8 Projection of Vectors

Let s represent a line and $\underline{A}$ be a vector whose line of action makes an angle θ with s, as shown in Figure 2.10. To determine the *projection* or the *component* of vector $\underline{A}$ on s, we drop a straight line from the tip (point Q) of the vector which cuts the line defined by s at right angles. If R denotes the point of intersection of these two lines and A_s is the length of the line segment between points P and R, then A_s is equal to the projection of vector $\underline{A}$ on s. Points P, Q and R define a right-triangle (see Appendix A), and therefore:

$$A_s = A \cos\theta \tag{2.3}$$

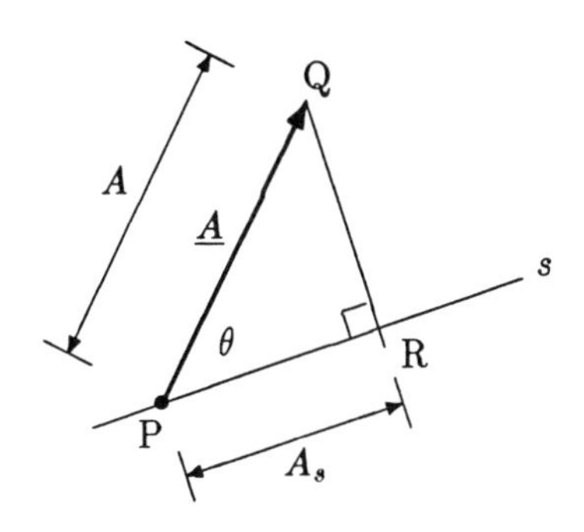

Figure 2.10 *Projection of a vector on a given direction.*

2.9 Resolution of Vectors

Resolution of a vector into its components is the reverse action of adding two vectors. For most applications (for example, while analyzing two-dimensional problems) it is useful to resolve a vector into its components along two mutually perpendicular directions. In Figure 2.11, x and y indicate the horizontal and vertical, respectively. $\underline{A}$ is a vector acting on the plane defined by x and y. To determine the component of $\underline{A}$ along the x, we drop a vertical line from the tip of the vector that cuts the horizontal line defined by x at right angles. Similarly, to determine the component of $\underline{A}$ along the y, we draw a horizontal line passing through the tip of the vector that cuts the vertical line defined by y at right angles. What we obtain are two similar right-triangles forming a rectangle (a parallelogram). Lengths of the sides of this rectangle are the *scalar components* of vector $\underline{A}$ along the x and y, which can be determined by utilizing the properties of right-triangles:

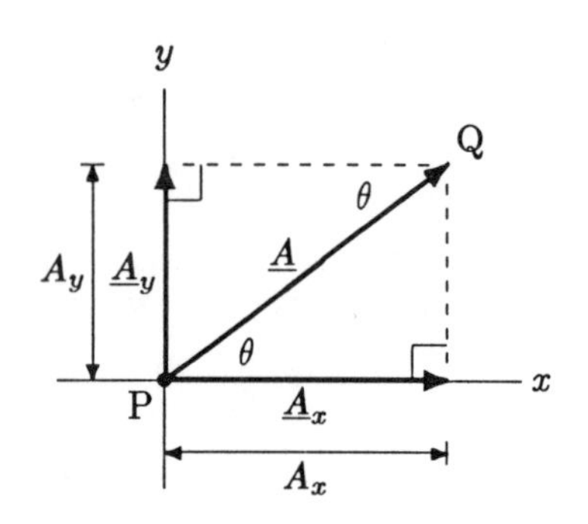

Figure 2.11 *Components of a vector along two mutually perpendicular directions.*

$$\begin{aligned} A_x &= A \cos\theta \\ A_y &= A \sin\theta \end{aligned} \tag{2.4}$$

In Eq. (2.4), θ is the angle that vector $\underline{A}$ makes with the horizontal direction. Vector $\underline{A}$ can also be represented as the sum of its *vector components* along the x and y:

$$\underline{A} = \underline{A}_x + \underline{A}_y \tag{2.5}$$

2.10 Unit Vectors

Usually it is convenient to express a vector $\underline{A}$ as a product of its magnitude A and a vector $\underline{a}$ of unit magnitude which has the same direction as vector $\underline{A}$ (Figure 2.12). $\underline{a}$ is called the ***unit vector***. For a given vector, its unit vector can be obtained by dividing that vector with its magnitude:

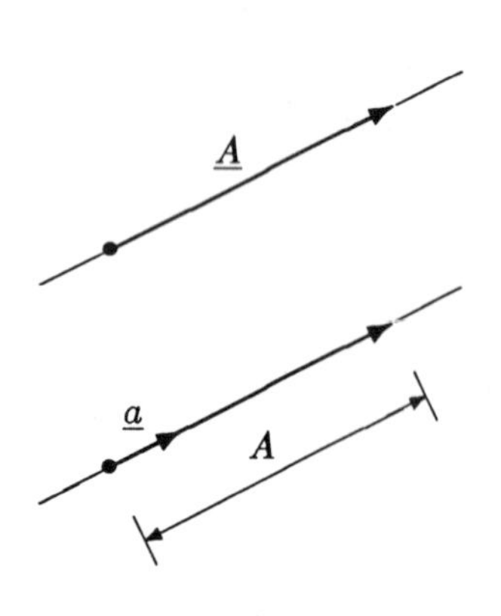

Figure 2.12 $\underline{A} = A\underline{a}$.

$$\underline{a} = \frac{\underline{A}}{A} \tag{2.6}$$

The original vector $\underline{A}$ can now be expressed as:

$$\underline{A} = A\,\underline{a} \tag{2.7}$$

Since the magnitude of $\underline{a}$ is equal to the magnitude of $\underline{A}$ (which is A) divided by A, the magnitude of vector $\underline{a}$ is always equal to one. Unit vectors have no physical significance, and they are simply used as a convenient way of describing directions.

2.11 Rectangular Coordinates

To be able to define the position of an object in space and to be able to analyze changes in position, we have to make our measurements based on a reference frame. In other words, we need a coordinate system. There are a number of widely used coordinate systems. Among these, the *Cartesian*, or *rectangular, coordinate* system is the one most commonly used. The Cartesian coordinate system consists of three mutually perpendicular axes, as shown in Figure 2.13.

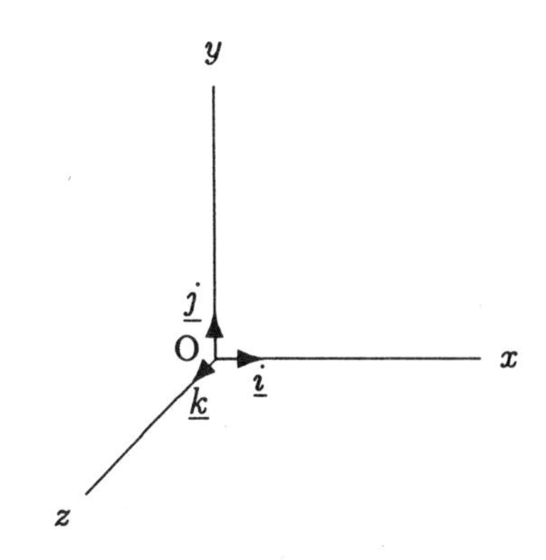

Figure 2.13 *Cartesian or rectangular coordinates.*

The concept of the unit vector has an important application in the construction of coordinate systems. The unit vectors along the Cartesian coordinate axes are so frequently used that they have widely accepted names. Symbols $\underline{i}$, $\underline{j}$, and $\underline{k}$ are commonly used to refer to *unit coordinate vectors* indicating positive x, y and z directions, respectively.

2.12 Addition of Vectors: Trigonometric Method

We have seen the addition and subtraction of vectors, and the resolution of vectors into their components by means of graphical methods. Adding and resolving vectors by measuring the length of arrows on graphs is time-consuming and is not accurate. Faster and more precise results can be obtained using trigonometric identities.

To apply the trigonometric method of addition and subtraction of vectors, one must first resolve each vector into its components. For example, consider the two vectors shown in Figure 2.14. Vectors $\underline{A}$ and $\underline{B}$ have magnitudes equal to A and B, and they make angles α and β with the horizontal (x axis), respectively. The scalar components of $\underline{A}$ and $\underline{B}$ can be determined by utilizing the properties of right-triangles (see Appendix A). For vector $\underline{A}$:

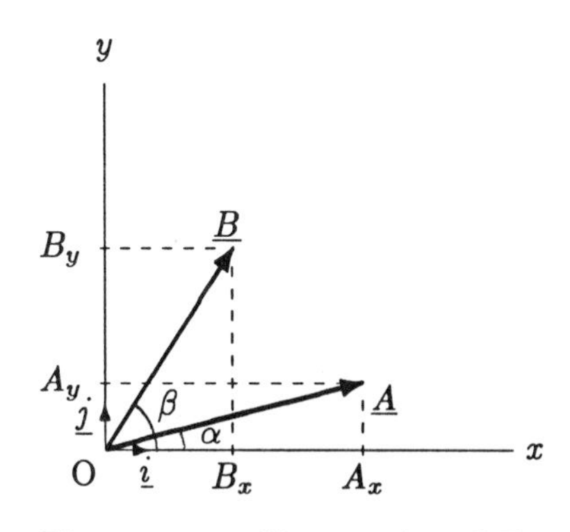

Figure 2.14 *Vectors $\underline{A}$ and $\underline{B}$.*

$$\begin{aligned} A_x &= A \cos\alpha \\ A_y &= A \sin\alpha \end{aligned} \tag{2.8}$$

A_x and A_y are the scalar components of $\underline{A}$ in the x and y directions, respectively. Making use of the unit vectors $\underline{i}$ and $\underline{j}$ which show positive x and y directions, the vector components of $\underline{A}$ in rectangular coordinates can be determined easily from its scalar components:

$$\begin{aligned} \underline{A}_x &= A_x\,\underline{i} \\ \underline{A}_y &= A_y\,\underline{j} \end{aligned} \tag{2.9}$$

Now, vector $\underline{A}$ can alternately be expressed as follows:

$$\begin{aligned} \underline{A} &= \underline{A}_x + \underline{A}_y \\ &= A_x\,\underline{i} + A_y\,\underline{j} \\ &= A\cos\alpha\,\underline{i} + A\sin\alpha\,\underline{j} \end{aligned} \tag{2.10}$$

Similarly for vector $\underline{B}$:

$$\begin{aligned}\underline{B} &= \underline{B}_x + \underline{B}_y \\ &= B_x\,\underline{i} + B_y\,\underline{j} \\ &= B\,\cos\beta\,\underline{i} + B\,\sin\beta\,\underline{j}\end{aligned} \tag{2.11}$$

Note that A_x, A_y, and A correspond to the sides of a right-triangle with A being the hypotenuse. Therefore we must have:

$$A = \sqrt{(A_x)^2 + (A_y)^2} \tag{2.12}$$

Once the vectors are expressed in terms of their components, the next step is to add (or subtract) the x components of all vectors together. This will yield the x component of the resultant vector. Similarly, adding the y components of all vectors will give the y component of the resultant vector. For example, the addition of vectors $\underline{A}$ and $\underline{B}$ can be performed as follows:

$$\begin{aligned}\underline{A} + \underline{B} &= (A_x\,\underline{i} + A_y\,\underline{j}) + (B_x\,\underline{i} + B_y\,\underline{j}) \\ &= (A_x + B_x)\underline{i} + (A_y + B_y)\underline{j} \\ &= (A\,\cos\alpha + B\,\cos\beta)\underline{i} + (A\,\sin\alpha + B\,\sin\beta)\underline{j}\end{aligned} \tag{2.13}$$

If $\underline{C}$ refers to the sum of $\underline{A}$ and $\underline{B}$, then:

$$\begin{aligned}\underline{C} &= \underline{A} + \underline{B} \\ &= \underline{C}_x + \underline{C}_y \\ &= C_x\,\underline{i} + C_y\,\underline{j}\end{aligned} \tag{2.14}$$

Comparing Eqs. (2.13) and (2.14) we can conclude that:

$$\begin{aligned}C_x &= A_x + B_x = A\,\cos\alpha + B\,\cos\beta \\ C_y &= A_y + B_y = A\,\sin\alpha + B\,\sin\beta\end{aligned} \tag{2.15}$$

If A, B, α, and β are given, then we can determine C_x and C_y from Eqs. (2.15). C_x, C_y, and C again form a right-triangle, with C being the hypotenuse (Figure 2.15). Therefore:

$$C = \sqrt{(C_x)^2 + (C_y)^2} \tag{2.16}$$

If γ represents the angle vector $\underline{C}$ makes with the x axis, then:

$$\tan\gamma = \frac{C_y}{C_x} \tag{2.17}$$

Therefore, angle γ is:

$$\gamma = \tan^{-1}\left(\frac{C_y}{C_x}\right) \tag{2.18}$$

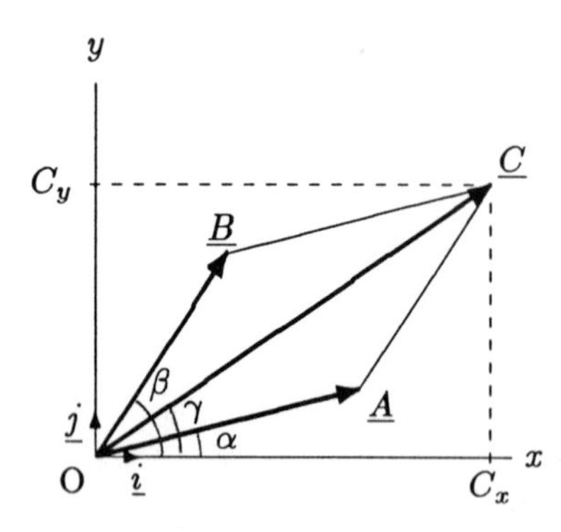

Figure 2.15 *Sum of $\underline{A}$ and $\underline{B}$ is equal to $\underline{C}$.*

$\tan^{-1}$ is called the *inverse tangent* which is also known as the *arctangent* (abbreviated as arctan).

Subtraction of one vector from the other is as straightforward as adding the two. For example:

$$\begin{aligned} \underline{A} - \underline{B} &= (A_x\, \underline{i} + A_y\, \underline{j}) - (B_x\, \underline{i} + B_y\, \underline{j}) \\ &= (A_x - B_x)\underline{i} + (A_y - B_y)\underline{j} \\ &= (A\, \cos\alpha - B\, \cos\beta)\underline{i} + (A\, \sin\alpha - B\, \sin\beta)\underline{j} \end{aligned} \tag{2.19}$$

Example 2.1 A vector $\underline{A}$ is given such that its magnitude is A=5 units and it makes an angle α=36.87° with the horizontal, as shown in Figure 2.16.

(a) Determine the components of $\underline{A}$ in the horizontal and vertical directions.
(b) Represent $\underline{A}$ in terms of its components.

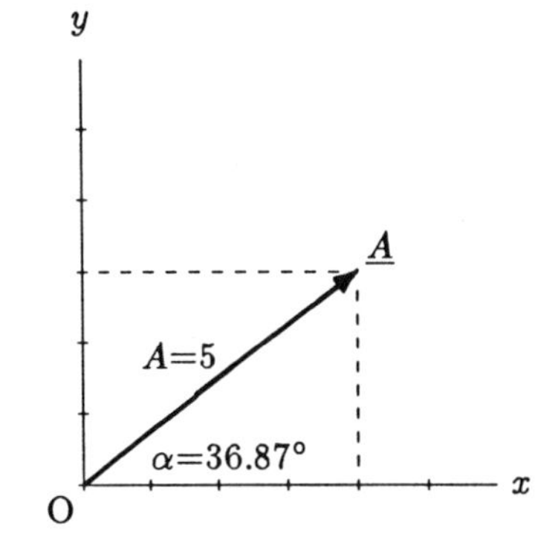

Figure 2.16 *Example 2.1.*

Solution:
(a) Using Eqs. (2.8):

$$A_x = A\, \cos\alpha = 5\, \cos(36.87°) = 4$$
$$A_y = A\, \sin\alpha = 5\, \sin(36.87°) = 3$$

(b) From Eq. (2.10):

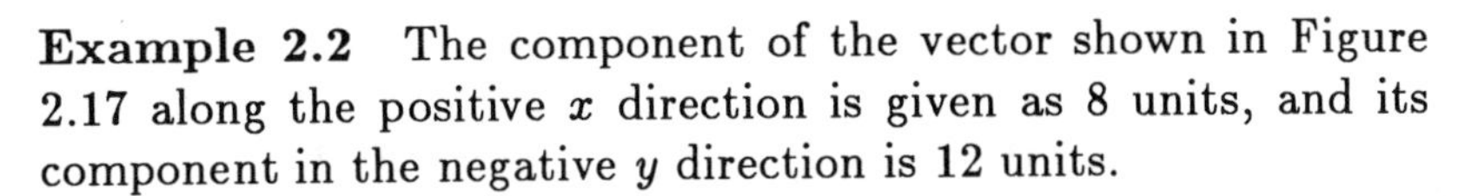

$$\underline{A} = A_x\, \underline{i} + A_y\, \underline{j} = 4\, \underline{i} + 3\, \underline{j}$$

Example 2.2 The component of the vector shown in Figure 2.17 along the positive x direction is given as 8 units, and its component in the negative y direction is 12 units.

(a) Write vector $\underline{B}$ in terms of its components.
(b) Determine the magnitude of vector $\underline{B}$, and the angle β it makes with the positive x direction.

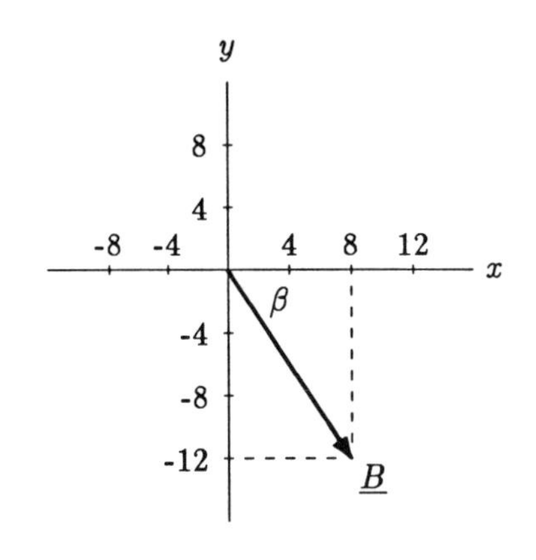

Figure 2.17 *Example 2.2.*

Solution:
(a) While expressing a vector in terms of its components, we have to pay particular attention to the correct directions of the vector components. In this example, the horizontal component of vector $\underline{B}$ is acting in the positive x direction and the vertical component is acting in the negative y direction. Therefore:

$$\underline{B} = 8\, \underline{i} - 12\, \underline{j}$$

(b) Note that B_x=8 and B_y=12. Hence:

$$B = \sqrt{(B_x)^2 + (B_y)^2} = \sqrt{(8)^2 + (12)^2} = \sqrt{208} = 14.42$$

$$\beta = \tan^{-1}\left(\frac{B_y}{B_x}\right) = \tan^{-1}\left(\frac{12}{8}\right) = 56.31^\circ$$

Example 2.3 Consider the two vectors shown in Figure 2.18. Components of these vectors are given as follows:

$$A_x = 15 \qquad B_x = 5$$
$$A_y = 10 \qquad B_y = 10$$

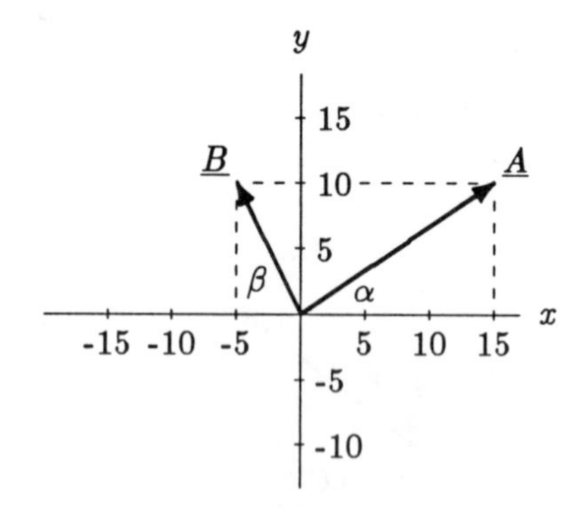

Figure 2.18 *Example 2.3.*

(a) Write $\underline{A}$ and $\underline{B}$ in terms of their components.
(b) Determine magnitudes of $\underline{A}$ and $\underline{B}$, and angles α and β.
(c) Determine $\underline{A} + \underline{B}$.
(d) Determine $\underline{A} - \underline{B}$.
(e) Determine $\underline{B} - \underline{A}$.
(f) Determine $-\underline{A} - \underline{B}$.

Solution:
(a) In this example, both components of vector $\underline{A}$ and the vertical component of vector $\underline{B}$ are acting in the positive directions. However, the horizontal component of vector $\underline{B}$ is pointing the negative x direction. Therefore:

$$\underline{A} = 15\,\underline{i} + 10\,\underline{j}$$
$$\underline{B} = -5\,\underline{i} + 10\,\underline{j}$$

(b) Magnitudes of $\underline{A}$ and $\underline{B}$:

$$A = \sqrt{(A_x)^2 + (A_y)^2} = \sqrt{15^2 + 10^2} = 18.03$$
$$B = \sqrt{(B_x)^2 + (B_y)^2} = \sqrt{5^2 + 10^2} = 11.18$$

Angles α and β:

$$\alpha = \tan^{-1}\left(\frac{10}{15}\right) = 33.69^\circ$$
$$\beta = \tan^{-1}\left(\frac{10}{5}\right) = 63.43^\circ$$

(c) Let $\underline{C} = \underline{A} + \underline{B}$. Then

$$\begin{aligned}\underline{C} &= (15\underline{i} + 10\underline{j}) + (-5\underline{i} + 10\underline{j})\\ &= (15-5)\underline{i} + (10+10)\underline{j}\\ &= 10\underline{i} + 20\underline{j}\end{aligned}$$

(d) Let $\underline{D} = \underline{A} - \underline{B}$. Then

$$\begin{aligned}\underline{D} &= (15\underline{i} + 10\underline{j}) - (-5\underline{i} + 10\underline{j}) \\ &= (15 + 5)\underline{i} + (10 - 10)\underline{j} \\ &= 20\underline{i}\end{aligned}$$

(e) Let $\underline{E} = \underline{B} - \underline{A}$. Then

$$\begin{aligned}\underline{E} &= (-5\underline{i} + 10\underline{j}) - (15\underline{i} + 10\underline{j}) \\ &= (-5 - 15)\underline{i} + (10 - 10)\underline{j} \\ &= -20\underline{i}\end{aligned}$$

(f) Let $\underline{F} = -\underline{A} - \underline{B}$. Then

$$\begin{aligned}\underline{F} &= -(15\underline{i} + 10\underline{j}) - (-5\underline{i} + 10\underline{j}) \\ &= (-15 + 5)\underline{i} + (-10 - 10)\underline{j} \\ &= -10\underline{i} - 20\underline{j}\end{aligned}$$

The resultant vectors $\underline{C}$, $\underline{D}$, $\underline{E}$, and $\underline{F}$ are illustrated in Figure 2.19. A careful examination of these results shows that vectors $\underline{C}$ and $\underline{F}$ form a pair of negative vectors with equal magnitude and opposite directions, and so are vectors $\underline{D}$ and $\underline{E}$. Also note that vectors $\underline{D}$ and $\underline{E}$ have no components along the y direction.

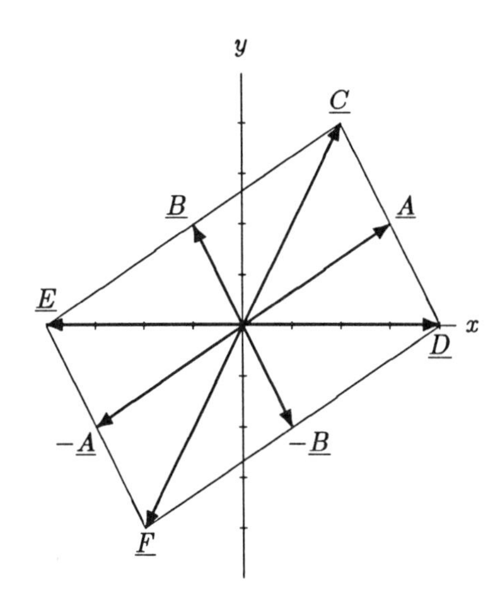

Figure 2.19 *Addition and subtraction of vectors.*

2.13* Three-Dimensional Components of Vectors

For most of our applications we shall use coplanar systems for which each vector quantity involved will have at most two components. However, the trigonometric addition and subtraction of vectors with components in the x, y, and z directions are simply the extension of the principles introduced for two-dimensional systems.

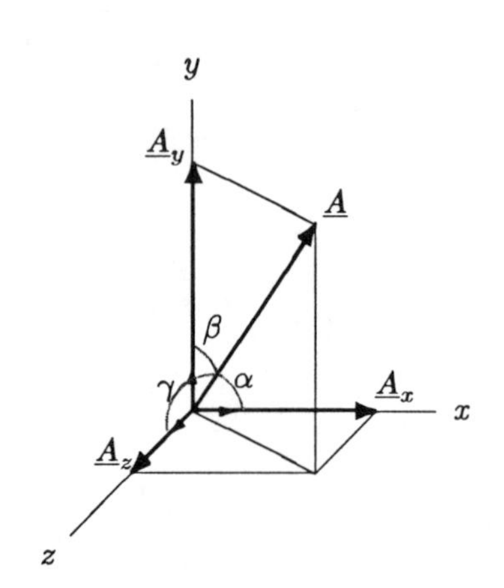

Figure 2.20 *Components of a three-dimensional vector.*

In Cartesian coordinate system, to define a vector quantity such as $\underline{A}$ uniquely, either all three components A_x, A_y, and A_z of vector $\underline{A}$, or the magnitude A of the vector along with the angles the vector makes with the x, y, and z directions, must be provided (Figure 2.20). We can alternatively express $\underline{A}$ as:

$$\begin{aligned} \underline{A} &= \underline{A}_x + \underline{A}_y + \underline{A}_z \\ &= A_x\,\underline{i} + A_y\,\underline{j} + A_z\,\underline{k} \end{aligned} \tag{2.20}$$

If α, β, and γ refer to the angles that vector $\underline{A}$ makes with the x, y, and z directions, respectively, then:

$$\begin{aligned} A_x &= A\,\cos\alpha \\ A_y &= A\,\cos\beta \\ A_z &= A\,\cos\gamma \end{aligned} \tag{2.21}$$

Consider that we have a second vector, $\underline{B}$, that we want to add to vector $\underline{A}$. Calling the resultant vector $\underline{C}$:

$$\begin{aligned} \underline{C} &= \underline{A} + \underline{B} \\ &= (A_x\,\underline{i} + A_y\,\underline{j} + A_z\,\underline{k}) + (B_x\,\underline{i} + B_y\,\underline{j} + B_z\,\underline{k}) \\ &= (A_x + B_x)\underline{i} + (A_y + B_y)\underline{j} + (A_z + B_z)\underline{k} \end{aligned} \tag{2.22}$$

Notice that the components of the resultant vector $\underline{C}$ in the x, y, and z directions are:

$$\begin{aligned} C_x &= A_x + B_x \\ C_y &= A_y + B_y \\ C_z &= A_z + B_z \end{aligned} \tag{2.23}$$

As is true for coplanar systems, the addition of two or more vector quantities having components in all three directions requires addition of the x components of all vectors together. The same procedure must be repeated for the y and z components. Subtraction of two or more three-dimensional vector quantities follows a similar approach.

2.14* Dot (Scalar) Product of Vectors

Two of the most commonly encountered vector quantities are the force and the displacement vectors. In physics, *work done* by a force is defined as the product of the force component in the direction of displacement times the magnitude of displacement. These concepts will be covered in detail in later chapters. Although force and displacement are vector quantities, their product – work – is a scalar quantity. In other physical situations, vectors are associated in a similar manner as to result in scalar quantities. An operation that represents such situations concisely is called the *dot*, or *scalar, product*.

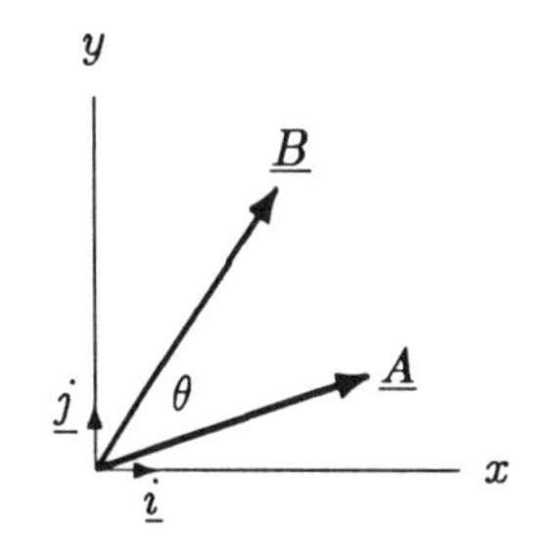

Figure 2.21 *Dot (scalar) product of two vectors.*

The dot product of any two vectors is defined as a scalar quantity equal to the product of magnitudes of the two vectors multiplied by the cosine of the smaller angle between them. For example, consider vectors $\underline{A}$ and $\underline{B}$ shown in Figure 2.21. The dot product of these vectors is:

$$\underline{A} \cdot \underline{B} = A\ B\ \cos\theta \tag{2.24}$$

where A and B are the magnitudes of the two vectors, and θ is the smaller angle between them. The operation given by Eq. (2.24) is representative of first projecting vector $\underline{A}$ onto the line of action of vector $\underline{B}$ (or vice versa) and then multiplying the magnitudes of the projected component and the other vector.

Note that the dot product may result in positive or negative quantities depending on whether the smaller angle between the vectors is less or greater than 90°. Since $\cos 90°=0$, the dot product of two vectors is equal to zero if the vectors are perpendicular to each other. Unit vectors $\underline{i}$ and $\underline{j}$, which define the positive x and y directions, respectively, are also vector quantities with magnitudes equal to unity. The concept of the dot product is applicable to unit vectors as well:

$$\begin{aligned} \underline{i} \cdot \underline{i} = \underline{j} \cdot \underline{j} = \underline{k} \cdot \underline{k} = 1\ 1\ \cos 0° = 1 \\ \underline{i} \cdot \underline{j} = \underline{j} \cdot \underline{k} = \underline{k} \cdot \underline{i} = 1\ 1\ \cos 90° = 0 \end{aligned} \tag{2.25}$$

To take the dot product of the vectors shown in Figure 2.21, we can first represent the vectors in terms of their components along the x and y directions, and then apply the dot product to the unit vectors:

$$\begin{aligned} \underline{A} \cdot \underline{B} &= (A_x\ \underline{i} + A_y\ \underline{j}) \cdot (B_x\ \underline{i} + B_y\ \underline{j}) \\ &= A_x B_x (\underline{i} \cdot \underline{i}) + A_x B_y (\underline{i} \cdot \underline{j}) + \\ &\quad A_y B_x (\underline{j} \cdot \underline{i}) + A_y B_y (\underline{j} \cdot \underline{j}) \\ &= A_x B_x + A_y B_y \end{aligned} \tag{2.26}$$

The following are some of the properties of the dot product.

- The dot product of two vectors is a commutative operation:

$$\underline{A} \cdot \underline{B} = \underline{B} \cdot \underline{A}$$

- The dot product is a distributive operation:

$$\underline{A} \cdot (\underline{B} + \underline{C}) = \underline{A} \cdot \underline{B} + \underline{A} \cdot \underline{C}$$

- A vector multiplied by itself as a dot product is equal to the square of the magnitude of the vector:

$$\underline{A} \cdot \underline{A} = A^2$$

- The dot product of two three-dimensional vector quantities $\underline{A}$ and $\underline{B}$ expressed in terms of their Cartesian components is:

$$\underline{A} \cdot \underline{B} = A_x B_x + A_y B_y + A_z B_z$$

- The scalar component of a vector along a given direction is equal to the dot product of the vector times the unit vector along that direction. For example, the x and y components of a vector $\underline{A}$ are:

$$A_x = \underline{A} \cdot \underline{i} \qquad A_y = \underline{A} \cdot \underline{j}$$

2.15* Cross (Vector) Product of Vectors

There are other interactions between vector quantities which result in other vector quantities. An example of such interactions is the moment or torque generated by an applied force (see Chapter 4). The mathematical tool developed to define these interactions is called the *cross*, or *vector, product.*

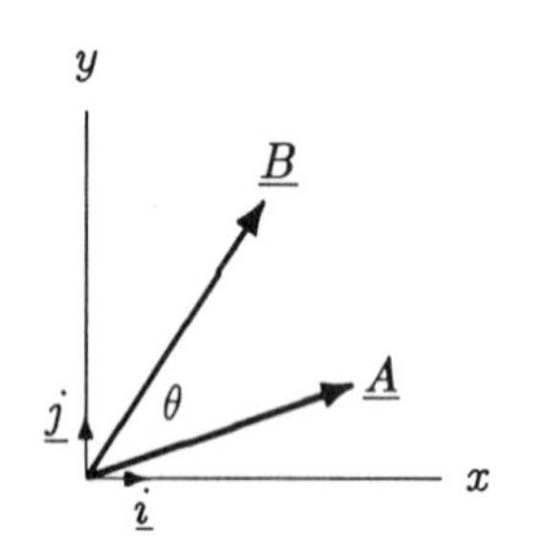

Figure 2.22 *Two vectors* $\underline{A}$ *and* $\underline{B}$.

Consider the vectors $\underline{A}$ and $\underline{B}$ shown in Figure 2.22. The cross product of these two vectors is equal to a third vector, $\underline{C}$. The commonly used mathematical notation for the cross product is:

$$\underline{A} \times \underline{B} = \underline{C} \tag{2.27}$$

The vector $\underline{C}$ has a magnitude equal to the product of the magnitudes of vectors $\underline{A}$ and $\underline{B}$ times the sine of the smaller angle (angle θ in Figure 2.22) between the two:

$$C = A\ B\ \sin\theta \tag{2.28}$$

Vector $\underline{C}$ has a direction perpendicular to the plane defined by vectors $\underline{A}$ and $\underline{B}$. For example, if both $\underline{A}$ and $\underline{B}$ are in the xy plane, then $\underline{C}$ acts in the z direction.

The sense of vector $\underline{C}$ can be determined by the ***right-hand-rule.*** To apply this rule, first the fingers of the right hand are pointed in the direction of vector $\underline{A}$ (the first vector), and then they are curled toward vector $\underline{B}$ (the second vector) in such a way as to cover the smaller angle between $\underline{A}$ and $\underline{B}$. The extended thumb points in the direction of the product vector $\underline{C}$. To specify the direction of the product vector, it may also be sufficient to say that the vector is either clockwise (cw) or counterclockwise (ccw).

Note again that the concept of the cross product is applicable to the Cartesian unit vectors $\underline{i}$, $\underline{j}$, and $\underline{k}$, which are mutually perpendicular. Since $\sin 0°=0$ and $\sin 90°=1$, we can write:

$$\begin{aligned} &\underline{i} \times \underline{i} = \underline{j} \times \underline{j} = \underline{k} \times \underline{k} = 0 \\ &\underline{i} \times \underline{j} = \underline{k} \qquad \underline{j} \times \underline{i} = -\underline{k} \\ &\underline{j} \times \underline{k} = \underline{i} \qquad \underline{k} \times \underline{j} = -\underline{i} \\ &\underline{k} \times \underline{i} = \underline{j} \qquad \underline{i} \times \underline{k} = -\underline{j} \end{aligned} \tag{2.29}$$

Using the above relations between the unit vectors, we can determine mathematical relations to evaluate the vector products which would include both the magnitude and the direction of the product vector. For example, if $\underline{A}$ and $\underline{B}$ are two vectors in the xy plane (Figure 2.23), then:

$$\begin{aligned} \underline{C} &= \underline{A} \times \underline{B} \\ &= (A_x\,\underline{i} + A_y\,\underline{j}) \times (B_x\,\underline{i} + B_y\,\underline{j}) \\ &= A_xB_x(\underline{i} \times \underline{i}) + A_xB_y(\underline{i} \times \underline{j}) + \\ &\qquad A_yB_x(\underline{j} \times \underline{i}) + A_yB_y(\underline{j} \times \underline{j}) \\ &= A_xB_x(0) + A_xB_y(\underline{k}) + A_yB_x(-\underline{k}) + A_yB_y(0) \\ &= (A_xB_y - A_yB_x)\underline{k} \end{aligned} \tag{2.30}$$

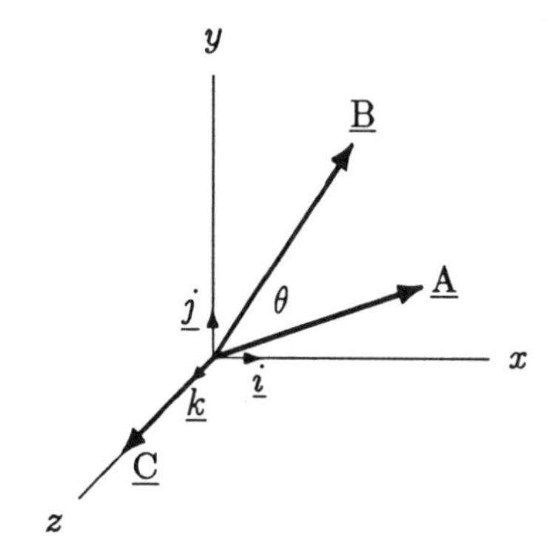

Figure 2.23 $\underline{C} = \underline{A} \times \underline{B}$

The following are some of the properties of the cross product.

- The cross product is not a commutative operation:

$$\underline{A} \times \underline{B} \neq \underline{B} \times \underline{A}$$

But

$$\underline{A} \times \underline{B} = -\underline{B} \times \underline{A}$$

- The cross product is a distributive operation:

$$\underline{A} \times (\underline{B} + \underline{C}) = \underline{A} \times \underline{B} + \underline{A} \times \underline{C}$$

- The vector product of two three-dimensional vector quantities $\underline{A}$ and $\underline{B}$ expressed in terms of their Cartesian components is:

$$\underline{A} \times \underline{B} = (A_yB_z - A_zB_y)\underline{i} + (A_zB_x - A_xB_z)\underline{j} + (A_xB_y - A_yB_x)\underline{k}$$

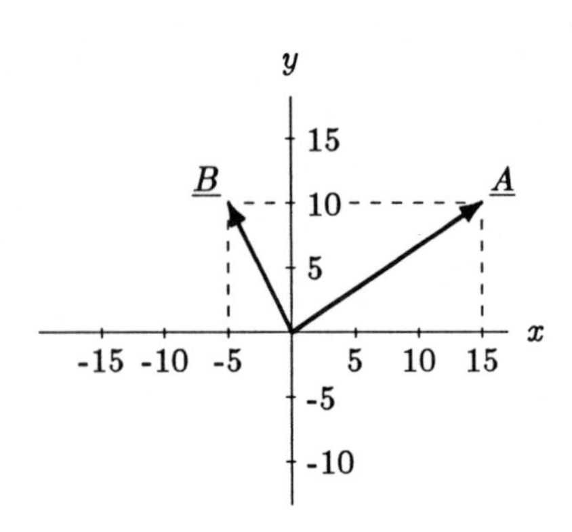

Figure 2.24 *Example 2.4.*

Example 2.4 Vectors $\underline{A}$ and $\underline{B}$ shown in Figure 2.24 are given in terms of their Cartesian components:

$$\underline{A} = 15\,\underline{i} + 10\,\underline{j}$$
$$\underline{B} = -5\,\underline{i} + 10\,\underline{j}$$

Determine
(a) The scalar product $c = \underline{A} \cdot \underline{B}$
(b) The vector product $\underline{D} = \underline{A} \times \underline{B}$
(c) The vector product $\underline{E} = \underline{B} \times \underline{A}$

Solution:
(a)

$$\begin{aligned} c &= \underline{A} \cdot \underline{B} \\ &= A_x B_x + A_y B_y \\ &= (15)(-5) + (10)(10) \\ &= 25 \end{aligned}$$

(b)

$$\begin{aligned} \underline{D} &= \underline{A} \times \underline{B} \\ &= (A_x B_y - A_y B_x)\,\underline{k} \\ &= \left[(15)(10) - (10)(-5)\right]\underline{k} \\ &= 200\,\underline{k} \end{aligned}$$

(c)

$$\begin{aligned} \underline{E} &= \underline{B} \times \underline{A} = -D \\ &= (B_x A_y - B_y A_x)\,\underline{k} \\ &= \left[(-5)(10) - (10)(15)\right]\underline{k} \\ &= -200\,\underline{k} \end{aligned}$$

Note that the product vectors $\underline{D}$ and $\underline{E}$ have equal magnitude of 200 units. $\underline{D}$ has a counterclockwise direction, whereas vector $\underline{E}$ is clockwise. Both vectors are acting in the direction perpendicular to the surface of the page, $\underline{D}$ out of the page and $\underline{E}$ into the page.

Chapter 3

THE FORCE VECTOR

3.1 Definition of Force

Force can be defined as mechanical disturbance or load. When you pull or push an object you apply a force to it. You also exert a force when you throw or kick a ball, in which case the force is associated with the result of muscular activity. A force acting on a body can deform the body, change its state of motion, or both. Although forces cause motion, it does not necessarily follow that force is always associated with motion. For example, a person sitting on a chair applies a force equal to his/her weight on the chair, and yet the chair remains stationary.

Force is a vector quantity and the principles of vector algebra must be applied to analyze problems involving forces. To describe a force fully, its magnitude and direction must be specified.

There are relatively few basic laws which govern the relationship between force and motion. The entire structure of mechanics is based on these laws, that were established by Sir Isaac Newton.

3.2 Newton's First and Second Laws

Newton's first law states that a body that is originally at rest will remain at rest, or a body moving with a constant velocity in a straight line will maintain its motion unless an external resultant force acts on the body. The tendency of a body to maintain its state of rest or uniform motion in a straight line is called *inertia*. Newton's first law must be considered in conjunction with his second law.

Newton's second law states that if the net or the resultant force acting on a body is not zero then the body will accelerate in the direction of the resultant force. Furthermore, the acceleration of the body is directly proportional to the net force acting on the body and inversely proportional to its mass. Newton's second law of motion can be formulated as:

$$\underline{F} = m\,\underline{a} \tag{3.1}$$

Eq. (3.1) is also known as the *equation of motion*. In Eq. (3.1), $\underline{F}$ is the net or the resultant force (vector sum of all forces) acting on the body, m is the mass of the body and $\underline{a}$ is its acceleration. Notice that both force and acceleration are vector quantities while mass is a scalar quantity.

According to Newton's second law of motion, an object will accelerate if the net force acting on the object is not equal to zero. If the net force is zero, then the acceleration of the body is zero, and consequently the velocity of the body is either constant or zero. When the acceleration of the body is zero, the body is said to be in *equilibrium*. If the velocity of the body is zero as well, then the body is in *static equilibrium* or at *rest*. From Eq. (3.1)

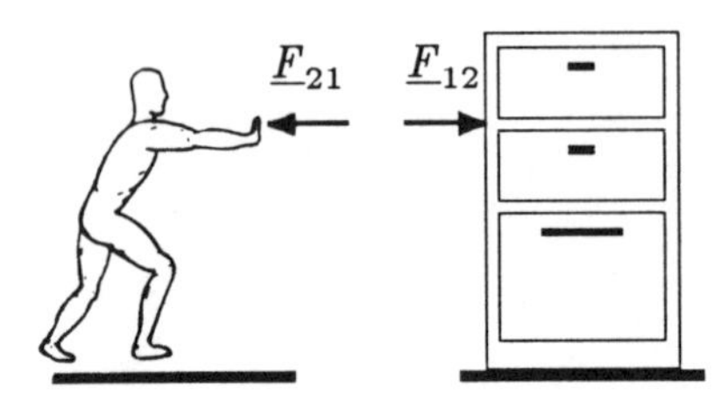

Figure 3.1 *Action and reaction.*

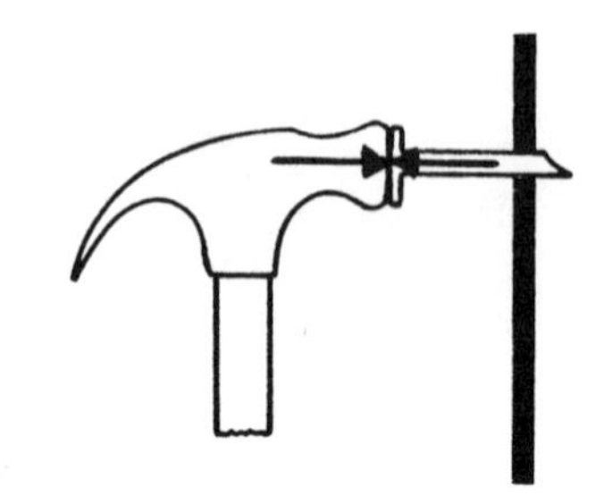

Figure 3.2 *Hammering.*

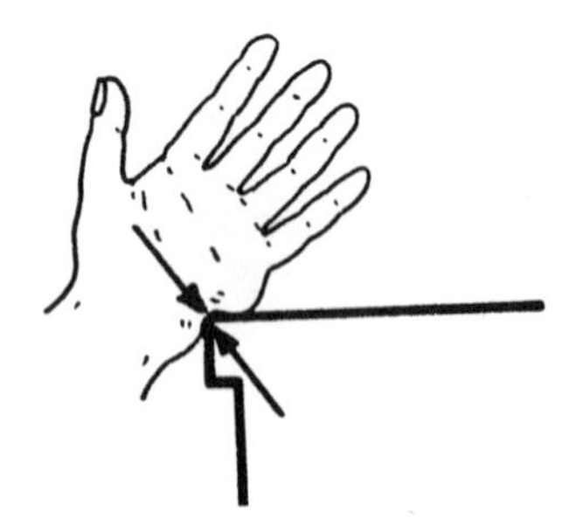

Figure 3.3 *The harder you push, the harder you will be pushed.*

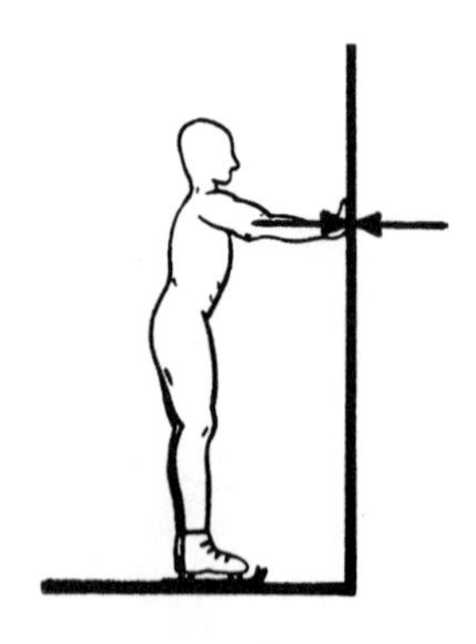

Figure 3.4 *An ice skater.*

it follows that acceleration and mass are inversely proportional. Mass is a measure of the inertia of an object. The more inertia an object has, the harder it is to start moving it from rest, to change its motion, or to change its direction of motion.

3.3 Newton's Third Law

This law is based on the observation that there are always two sides when it comes to forces. A force applied to an object is always applied by another object. When a worker pushes a cart, a child pulls a wagon, and a hammer hits a nail, force is applied by one body onto another. Newton's third law states that, if two bodies are in contact and body 1 is exerting a force on body 2, then body 2 will apply a force on body 1 in such a way that the two forces will have an equal magnitude but opposite directions.

Consider the person in Figure 3.1 who is trying to push a file cabinet towards the right. Let $\underline{F}_{12}$ be the horizontal force applied by the person on the file cabinet. $\underline{F}_{12}$ is acting towards the right. In return, the cabinet is applying a force, say $\underline{F}_{21}$, on the person such that the magnitude of $\underline{F}_{21}$ is equal to the magnitude of $\underline{F}_{12}$ ($F_{21}=F_{12}$). However, the direction of $\underline{F}_{21}$ is towards the left, opposite to that of $\underline{F}_{12}$.

In the case of a hammer pushing a nail illustrated in Figure 3.2, it is the force applied by the hammer on the nail that causes the nail to advance. However, it is the force of the nail applied back on the hammer that causes the hammer to stop after every impact upon the nail.

If you press your hand against the edge of a desk, you can see the shape of your hand change, and also feel the force exerted by the desk on your hand (Figure 3.3). The harder you press your hand against the desk, the harder the desk will push your hand back.

Perhaps the best example which can help us understand Newton's third law is a skater applying a force on a wall (Figure 3.4). By pushing against the wall the skater can move backwards. It is the force exerted by the wall on the skater that causes the motion of the skater.

Newton's third law of motion can be summarized as "to every action there is an equal and opposite reaction." This law is particularly useful in analyzing complex problems in mechanics in which there are several interacting bodies.

3.4 Dimension and Units of Force

From Newton's second law of motion, force is equal to mass times acceleration. By definition, acceleration is the time rate of change of velocity, while velocity is defined as the time rate of change of relative position. The change in position is measured in terms of length units. Therefore, velocity has a dimension of length divided by time, and acceleration has a dimension of velocity divided by time:

$$[\text{VELOCITY}] = \frac{[\text{POSITION}]}{[\text{TIME}]} = \frac{\text{L}}{\text{T}}$$

$$[\text{ACCELERATION}] = \frac{[\text{VELOCITY}]}{[\text{TIME}]} = \frac{\text{L/T}}{\text{T}} = \frac{\text{L}}{\text{T}^2}$$

$$[\text{FORCE}] = [\text{MASS}]\,[\text{ACCELERATION}] = \frac{\text{M L}}{\text{T}^2}$$

Units of force in different systems of units are provided in Table 3.1.

SYSTEM	UNITS OF FORCE	SPECIAL NAME
SI	kilogram-meter/(second)^2	Newton (N)
c-g-s	gram-centimeter/(second)^2	dyne (dyn)
British	slug-foot/(second)^2	pound (lb)

Table 3.1 *Units of force* (1 N = 10^5 dyn, and 1 N = 0.225 lb).

3.5 Properties of Force as a Vector Quantity

Newton's second law of motion states that the acceleration of an object is proportional to the net or the resultant force acting on the object. The net force is the vector sum of all forces acting on the object. The force vectors must be added according to the principles introduced in Chapter 2. While analyzing problems involving two or more forces, one can also utilize the concept of resolving vectors into their components along specified directions, such as the Cartesian coordinate axes. To determine the resultant force for a coplanar force system, we can first express individual forces in terms of their components along the x and y directions. Adding x components of all forces will yield the x component of the resultant force, and the sum of y components of all forces will be the y component of the resultant force.

The following examples will utilize some of the concepts introduced in Chapter 2 to analyze problems involving forces.

Example 3.1 Consider the block shown in Figure 3.5. Two workers are pulling the block with forces $\underline{F}_1$ and $\underline{F}_2$ applied in directions perpendicular to one another. If $F_1=F_2=100$ N are the magnitudes of these forces, determine the magnitude and direction of the net force acting on the block.

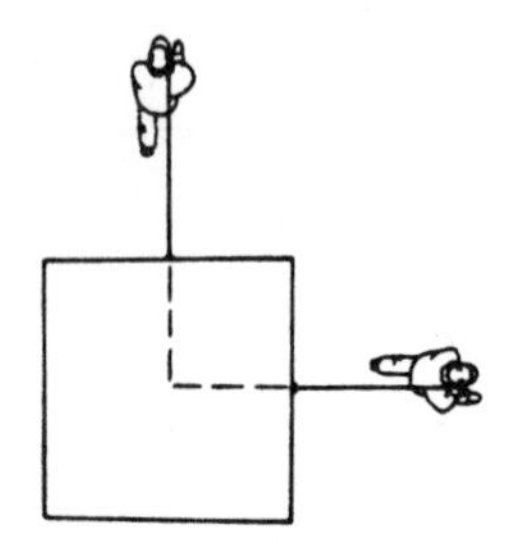

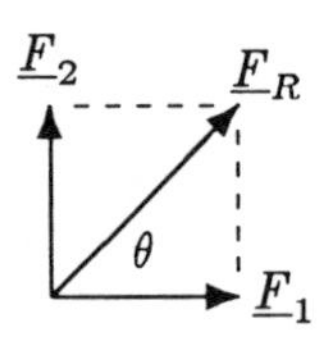

Figure 3.5 *Forces applied by the workers are perpendicular to one another.*

Solution: We have a system of two forces ($\underline{F}_1$ and $\underline{F}_2$) that are mutually perpendicular. As illustrated in Figure 3.5, the resultant $\underline{F}_R$ of these forces can be obtained by applying the parallelogram method of adding two vectors. Since $\underline{F}_1$ and $\underline{F}_2$ are perpendicular and have equal magnitudes, they form the sides of a square consisting of two similar right-triangles. The magnitude of the resultant force is the length of the hypotenuse of these triangles:

$$F_R = \sqrt{(F_1)^2 + (F_2)^2} = \sqrt{(100)^2 + (100)^2} = 141 \text{ N}$$

The angle θ that the resultant force makes with the horizontal can be determined by utilizing the properties of right-triangles:

$$\theta = \tan^{-1}\left(\frac{F_2}{F_1}\right) = \tan^{-1}\left(\frac{100}{100}\right) = \tan^{-1}(1) = 45^\circ$$

Therefore, the combined effort of the workers is such that they can move the block in a direction which makes 45° with the horizontal with a net force of 141 N.

Assume that the block has a mass m=60 kg. Knowing the mass of the block and the net force acting on it, we can determine the acceleration of the block. From Newton's second law of motion, the direction of acceleration is the same as the direction of the resultant force. The magnitude of the acceleration of the block can be calculated by rewriting Eq. (3.1) as:

$$a = \frac{F_R}{m} = \frac{141}{60} = 2.4 \text{ m/s}^2$$

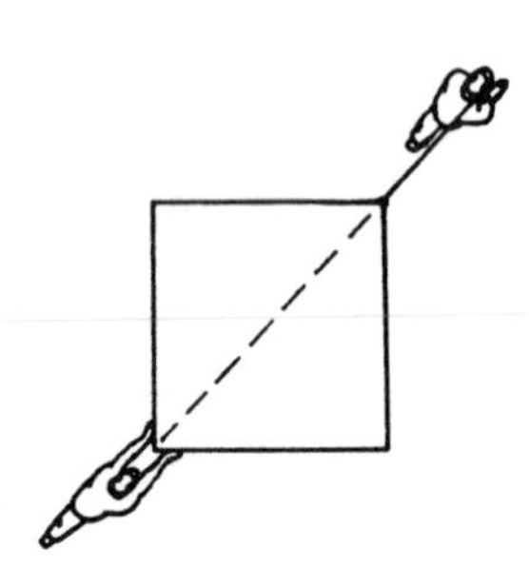

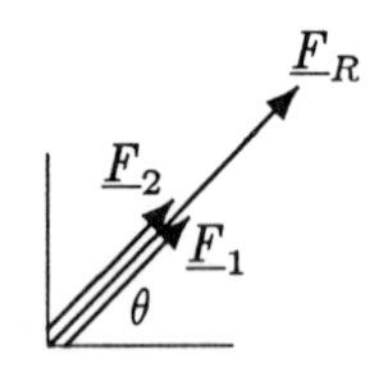

Figure 3.6 *Forces applied in the same direction can improve efficiency.*

Note that if the purpose of the task is to pull the block towards the direction of the resultant force, then the workers are not utilizing their energy efficiently. If they pull or push the block in the direction that makes 45° with the horizontal (Figure 3.6), then the net force applied on the block has a magnitude equal to the arithmetic sum (100 N+100 N=200 N) of the magnitudes of the forces they apply. The improvement in efficiency can be determined by:

$$\frac{200 - 141}{141} \times 100 = 42$$

Therefore, a minor adjustment of the direction of applied forces increases the efficiency by 42%.

Example 3.2 Consider the system of forces acting on the object shown in Figure 3.7. Assume that the magnitudes of these forces are given as F_1=31.6 N, F_2=36 N, and F_3=30 N. The angles that $\underline{F}_1$ and $\underline{F}_2$ make with the x axis are α=18.4° and β=56.3°, respectively. Force $\underline{F}_3$ is acting in the negative x direction.

(a) Determine the components of the applied forces in the x and y directions.
(b) Express these forces in terms of their components.
(c) Determine the magnitude and direction of the net force acting on the object.

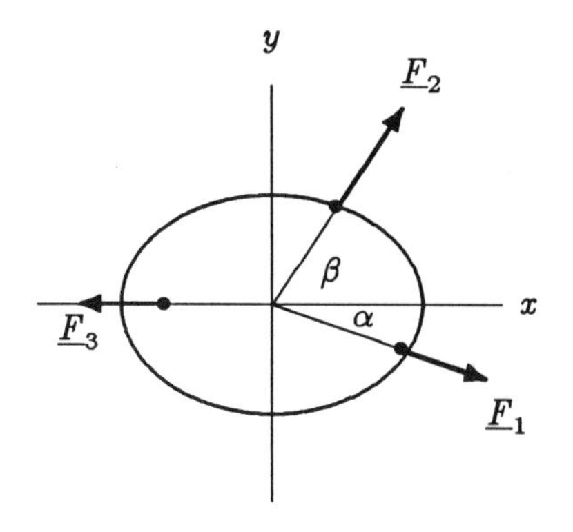

Figure 3.7 *Coplanar and concurrent system of forces.*

Solution:
(a) The Cartesian components of the applied forces:

$$F_{1x} = F_1 \cos\alpha = (31.6)(\cos 18.4°) = 30 \text{ N} \quad (\rightarrow)$$
$$F_{1y} = F_1 \sin\alpha = (31.6)(\sin 18.4°) = 10 \text{ N} \quad (\downarrow)$$

$$F_{2x} = F_2 \cos\beta = (36)(\cos 56.3°) = 20 \text{ N} \quad (\rightarrow)$$
$$F_{2y} = F_2 \sin\beta = (36)(\sin 56.3°) = 30 \text{ N} \quad (\uparrow)$$

$$F_{3x} = F_3 = 30 \text{ N} \quad (\leftarrow)$$
$$F_{3y} = 0$$

(b) The applied forces in terms of their components:

$$\underline{F}_1 = 30\,\underline{i} - 10\,\underline{j}$$
$$\underline{F}_2 = 20\,\underline{i} + 30\,\underline{j}$$
$$\underline{F}_3 = -30\,\underline{i}$$

(c) The resultant force vector:

$$\begin{aligned}\underline{F}_R &= \underline{F}_1 + \underline{F}_2 + \underline{F}_3 \\ &= (30\,\underline{i} - 10\,\underline{j}) + (20\,\underline{i} + 30\,\underline{j}) + (-30\,\underline{i}) \\ &= (20\,\underline{i} + 20\,\underline{j})\end{aligned}$$

The magnitude and direction of the resultant force vector:

$$F_R = \sqrt{(20)^2 + (20)^2} = 28 \text{ N}$$

$$\theta = \tan^{-1}(20/20) = 45°$$

3.6 Classification of Forces

Any two or more forces acting on a single body form a *force system*. There are a number of different ways of classifying forces. Forces can be classified according to the way they affect the body they are acting on, or according to their orientation as compared to one another.

3.7 External and Internal Forces

Forces can be broadly classified as *external* and *internal*. Almost all commonly known forces are external forces. For example, when you push a cart, hammer a nail, sit on a chair, kick a football, or shoot a basketball, you apply an external force on the cart, nail, chair, football or basketball. Internal forces, on the other hand, are the ones which hold a body together when the body is under the effect of externally applied forces. For example, a piece of string does not necessarily break when it is pulled from both ends. When a piece of rubber band is stretched, the band elongates to a certain extent. What holds any material together under externally applied forces, including the rope or the band, is the internal forces generated within that material. If we consider the human body as a whole, then the forces generated by muscle contractions are also internal forces. The significance and details of internal forces will be studied by introducing the concept of "stress" in later chapters.

3.8 Normal and Tangential Forces

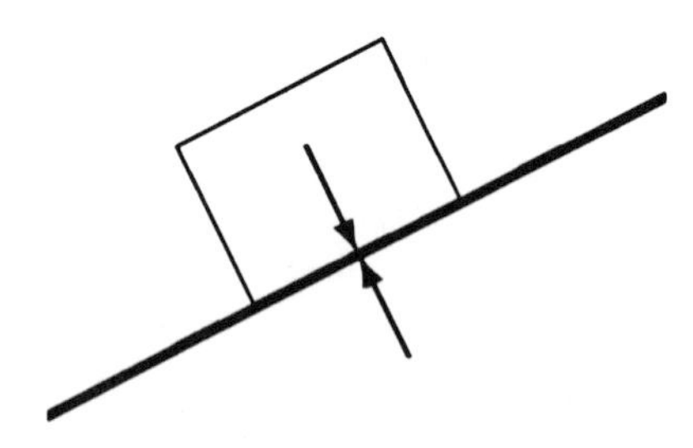

Figure 3.8 *Forces normal to the surfaces in contact.*

If a force acting on a surface is applied in a direction perpendicular (normal) to that surface, then the force is called a *normal* force. For example, a book on a flat horizontal desk applies a normal force on the desk, the magnitude of which is equal to the weight of the book. A block resting on an inclined surface applies a normal force on the surface in the direction perpendicular to the incline (Figure 3.8).

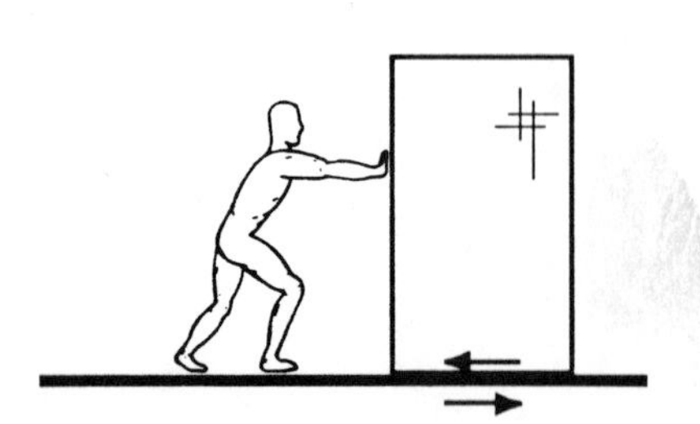

Figure 3.9 *Frictional forces are tangential forces.*

A *tangential* force is that applied on a surface in the direction parallel to the surface. A good example of tangential forces is the frictional force. As illustrated in Figure 3.9, pushing or pulling a block will cause a frictional force to occur between the bottom surface of the block and the floor. The line of action of the frictional force is always tangent to the surfaces in contact (see Section 3.16).

3.9 Tensile and Compressive Forces

A *tensile* force applied on a body will tend to stretch or elongate the body, whereas a *compressive* force will tend to shrink the body in the direction that it is applied (Figure 3.10). For example, a tensile force applied on a rubber band will stretch the band. Poking into an inflated balloon will produce a compressive force on the balloon. It must be noted that there are certain materials upon which only tensile forces can be applied. For example, a rope, a cable or a string cannot withstand compressive forces. The shapes of these materials will be completely distorted under compressive forces. Similarly, muscles contract to produce tensile forces that pull together the bones that they are attached to. Muscles can neither produce compressive forces nor exert a push.

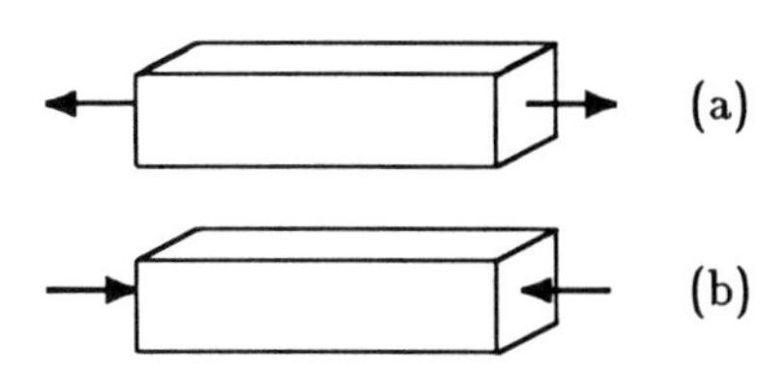

Figure 3.10 *(a) tensile and (b) compressive forces.*

3.10 Coplanar Forces

A system of forces is said to be *coplanar* if all the forces are acting on a two-dimensional (plane) surface. Forces forming a coplanar system have at most two non-zero components. Therefore, with respect to the Cartesian coordinate frame, it is sufficient to analyze coplanar force systems by considering the x and y components of the forces involved.

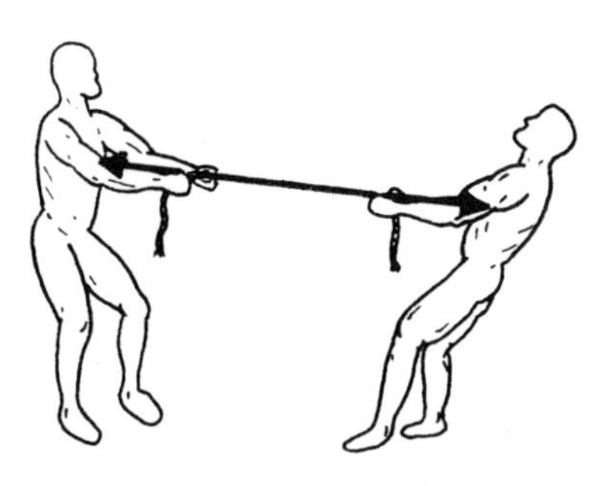

Figure 3.11 *Collinear forces.*

3.11 Collinear Forces

A system of forces is *collinear* if all the forces have a common line of action. For example, forces applied on a rope in a rope-pulling contest as illustrated in Figure 3.11, or the forces applied by the workers in Figure 3.6, form collinear force systems. To determine the magnitude of the resultant force vector for a collinear force system, all we have to do is arithmetically add and/or subtract the magnitudes of forces forming the force system.

3.12 Concurrent Forces

A system of forces is *concurrent* if the lines of action of the forces have a common point of intersection. Examples of concurrent systems of forces can be seen in various traction devices, as illustrated in Figure 3.12. Due to the weight in the weight pan, the cables stretch and forces are applied on the pulleys and the leg. The force applied on the leg holds the leg in place. If we know the weight in the weight pan and the angles that the cables make with the horizontal or vertical, then we can determine the tensile forces in the cables. We can resolve these forces into their components along the horizontal and vertical directions, easily determine the force resultant, and undertake an in-depth analysis of the entire system (see Chapter 5).

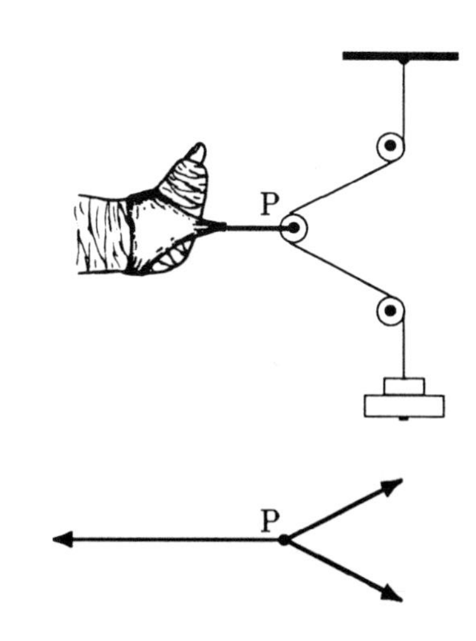

Figure 3.12 *Concurrent forces.*

3.13 Parallel Forces

A set of forces form a *parallel* force system if the lines of action of the forces are parallel to each other. An example of parallel force systems is illustrated in Figure 3.13 by a human arm flexed

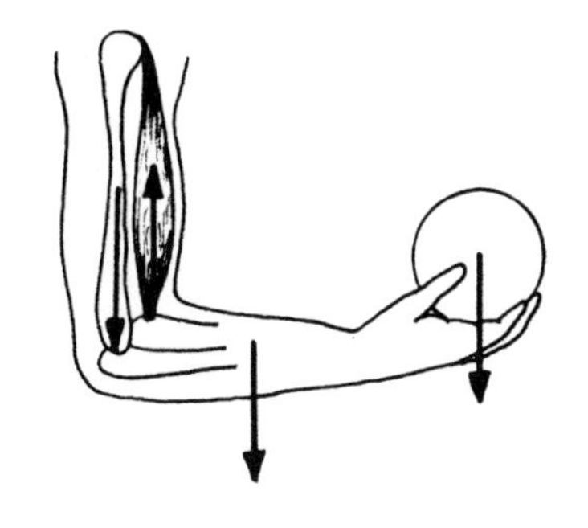

Figure 3.13 *Parallel forces.*

at 90° and holding an object. The forces on the forearm are the weight of the object, the weight of the arm itself, the tension in the biceps muscle, and the joint reaction force at the elbow. A detailed analysis of this and other anatomical systems will be presented in Chapter 6.

3.14 Gravitational Forces (Weight)

The force exerted by Earth on an object is called the *gravitational force*, or *weight*, of the object. The weight $\underline{W}$ of an object is equal to the mass m of the object times the gravitational acceleration $\underline{g}$:

$$\underline{W} = m\,\underline{g} \tag{3.2}$$

If we note that weight is a special form of force, and gravity is a special form of acceleration, then Eq. (3.2) is a special form of Eq. (3.1).

The magnitude W of weight is:

$$W = m\,g$$

The magnitude of the gravitational acceleration can vary slightly with altitude. For our applications, we shall assume g to be a constant. The magnitude of gravitational acceleration for different unit systems is listed in Table 3.2. These values are valid only on the surface of Earth. For example, on the moon, g is about one-sixth of what it is on Earth. Therefore, a 10 kg mass on Earth weighs about 98 Newtons on Earth, while it weighs about 17 Newtons on the moon.

SYSTEM	GRAVITATIONAL ACCELERATION
SI	9.81 m/s^2
c-g-s	981 cm/s^2
British	32.2 ft/s^2

Table 3.2 *Magnitudes of gravitational acceleration.*

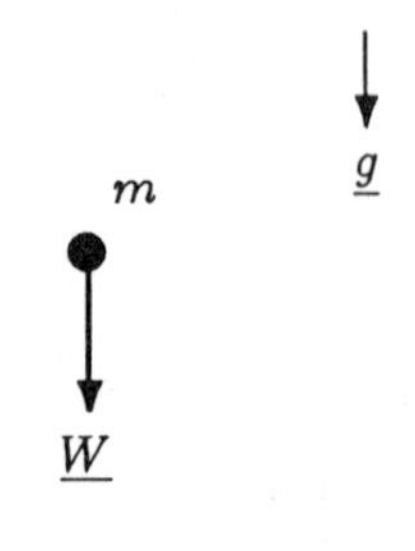

Figure 3.14 *Free-fall is due to the force of gravity.*

The direction of gravitational acceleration and gravitational force vectors is always towards the center of Earth, or always vertically downward. The force of gravity acts on an object at all times. If we drop an object from a height, it is the force of gravity which will make the object fall to the ground (Figure 3.14). When an object is at rest on the ground, the gravitational force does not disappear. An object at rest or in equilibrium simply means that the net force acting on the object is zero. For the net force on the block shown in Figure 3.15 to be equal to zero, there has to be another force acting on the block in the vertical direction to balance the force of gravity. Newton's third law of motion states that for every action there is an equal and opposite reaction. Therefore, the forces acting on the block shown in Figure 3.15 are the downward force of gravity ($\underline{W}$) and the upward normal force ($\underline{N}$) applied by the ground on the block. These two forces have the same magnitude equal to the weight of the block ($N=W=m\,g$).

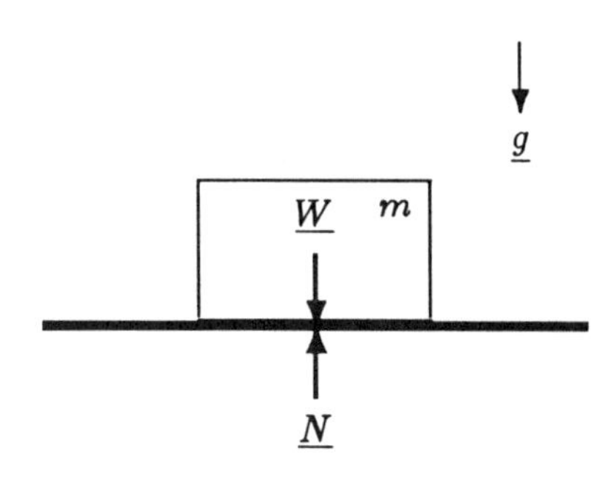

Figure 3.15 *The net force on an object at rest is zero.*

The terms mass and weight are often confused with one another. Mass is a property of a body. Weight is the force of gravity acting on the body. A body has the same mass on Earth and on the moon. However, the weight of a body is about six times as much on Earth as on the moon, because of the difference between the forces of gravity.

3.15 Distributed Force Systems and Pressure

Consider a pile of sand lying on a horizontal surface, as illustrated in Figure 3.16(a). The sand exerts force or load on the surface, which is distributed over the area under the sand. The load is not uniformly distributed over this area. The marginal regions under the pile are loaded less as compared to the central regions (Figure 3.16b). For practical purposes, the distributed load applied by the sand can be represented by a single force, called the *equivalent force* or *concentrated load.* The magnitude of the equivalent force is equal to the total weight of the sand (Figure 3.16c). The line of action of this force passes through a point, called the *center of gravity* or the *center of mass.* For some applications, we can assume that the entire weight of the pile is concentrated at the center of gravity of the load. For uniformly distributed loads, such as the load applied by the rectangular block on the horizontal surface shown in Figure 3.17, the center of gravity coincides with the geometric center of the load. For nonuniformly distributed loads, the center of gravity can be determined by experimentation (see Chapter 5).

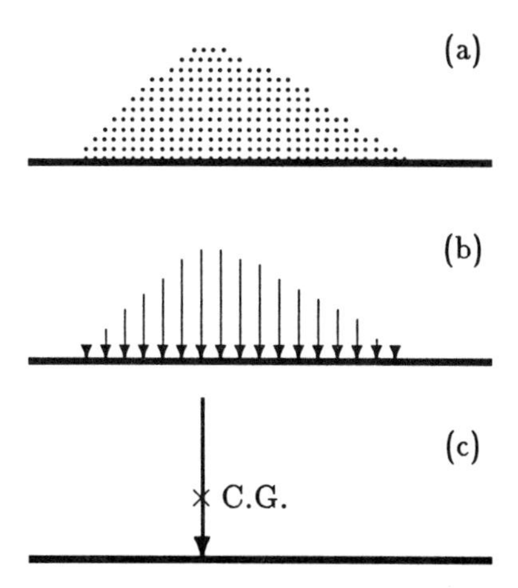

Figure 3.16 *A pile of sand (a), distributed load on the ground (b), and an equivalent force (c).*

Another important concept is *pressure*, which is a measure of the intensity of distributed loads. By definition, pressure is equal to total applied force divided by the area over which the force is applied. For example, if the bottom surface area of the rectangular block in Figure 3.17 is A and the total weight of the block is W, then the magnitude of the pressure exerted by the block on the horizontal surface is:

$$p = \frac{W}{A} \tag{3.3}$$

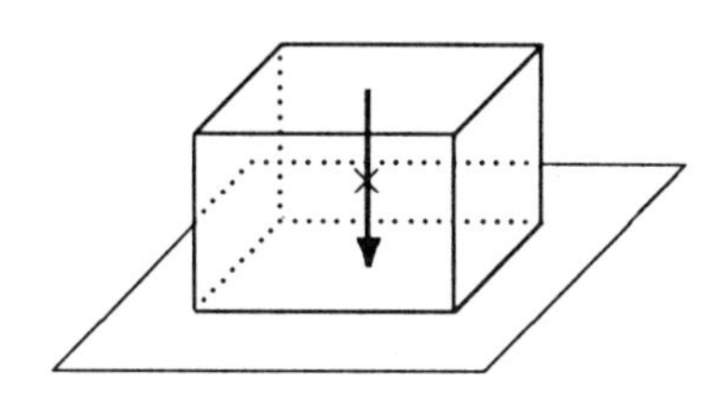

Figure 3.17 *Rectangular block.*

It follows that the dimension of pressure has the dimension of force ($\mathrm{ML/T^2}$) divided by the dimension of area ($\mathrm{L^2}$):

$$[\mathrm{PRESSURE}] = \frac{[\mathrm{FORCE}]}{[\mathrm{AREA}]} = \frac{\mathrm{M}}{\mathrm{LT^2}}$$

Units of pressure in different unit systems are listed in Table 3.3.

The principles behind the concept of pressure have many applications. Note that the larger the area over which a force is applied, the lower the magnitude of pressure. If we observe two people standing on soft snow, one wearing a pair of shoes and the other wearing skis, we can easily notice that the person wearing shoes stands deeper in the snow than the skier. This is simply because the weight of the person wearing shoes is distributed over a smaller area on the snow, and therefore applies a larger force per unit area of snow (Figure 3.18).

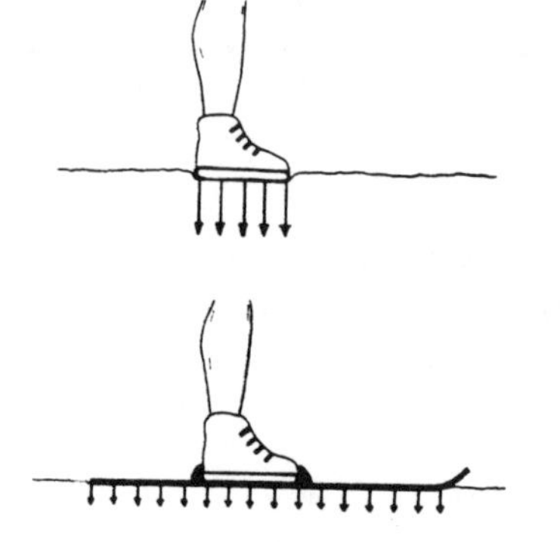

Figure 3.18 *Intensity of force (pressure) applied on the snow by a pair of shoes is higher than that applied by a pair of skis.*

SYSTEM	UNITS OF PRESSURE	SPECIAL NAME
SI	$kg/m\text{-}s^2$ or N/m^2	Pascal (Pa)
c-g-s	$gm/cm\text{-}s^2$ or dyn/cm^2	
British	lb/ft^2 or lb/in^2	psf or psi

Table 3.3 *Units of pressure.*

In general, to avoid pressure sore to the skin, a force must be applied over a larger skin area. It is obvious that the sensation and pain induced by a sharp object is much more severe than that produced by a force that is applied by a dull object. A prosthesis that fits the amputated limb, or a set of dentures that fits the gum and the bony structure properly, would feel and function better than an improperly fitted implant or replacement device. The idea is to distribute the forces involved as uniformly as possible over a large area.

3.16 Frictional Forces

Frictional forces occur between two surfaces in contact when one surface slides or tends to slide over the other. When a body is in motion on a rough surface, or when an object moves in a fluid (a viscous medium such as water), there is resistance to motion because of the interaction of the body with its surroundings. In some applications friction may be desirable, while in others it may have to be reduced to a minimum. For example, walking and running require sufficient friction between the sole of the foot and the floor. In the absence of frictional forces it would be impossible to start walking. Automobile, bicycle, and wheelchair brakes utilize the principles of friction. While setting up a traction device, if the concern is patient comfort, then the patient must be positioned on the bed so as to minimize the frictional forces developed between the patient and the bed. On the other hand, the traction device and the patient must be configured to generate sufficient opposing (frictional) forces in order to achieve the intended effect (see Chapter 5).

Friction can cause heat to be generated between the surfaces in contact. In the operation of machinery, friction dissipates energy as heat. Excess heat can cause early, unexpected failure of machine parts. Friction may also cause wear. For mechanical efficiency and to increase the life expectancy of machine parts, it is necessary to keep the energy loss and wear at a minimum. The effects of friction and wear may be reduced by the use of lubricants. Lubricants placed between the moving parts absorb the frictional effects, and reduce wear by reducing direct contact between the moving parts. In the case of the human body, the joints are lubricated by synovial fluid.

Consider the block resting on the floor, as shown in Figure 3.19. The block is applying its weight $\underline{W}$ on the floor. In return the floor is applying a force $\underline{N}$ on the block, such that $N=W$. Now consider that a horizontal force with magnitude F is applied on the block to move it towards the right. This will cause a frictional force, $\underline{f}$, to develop between the block and the floor. As long as the block remains stationary (in static equilibrium), the magnitude of the frictional force would be equal to the magnitude of the applied force. This frictional force is called the *static friction* ($\underline{f}_s$).

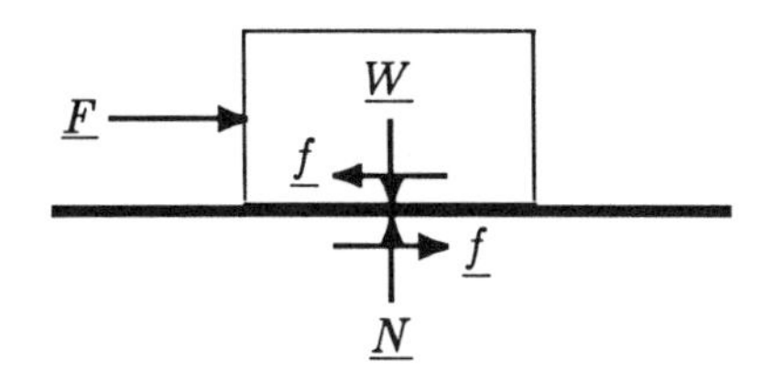

Figure 3.19 *Friction occurs on surfaces when one surface slides or tends to slide over the other.*

If the magnitude of the applied force is increased, the block will eventually slip or begin sliding over the floor. When the block is on the verge of sliding, the magnitude of the static friction is maximum (f_{max}). When F exceeds f_{max} the block moves towards the right. When the block is in motion, the resistance to motion at the surfaces of contact is called the *kinetic* or *dynamic friction*, $\underline{f}_k$. In general, the magnitude of the force of kinetic friction is lower than the maximum static friction ($f_k<f_{max}$) and the magnitude of the applied force ($f_k<F$). The unbalanced force in the horizontal direction ($F-f_k$) produces an acceleration towards the right. If $F=f_k$, the block moves with constant speed. If the applied force is removed, then the frictional force acting towards the left would decelerate the block and eventually bring it to rest.

There are several factors that influence frictional forces. Friction depends on the nature of the two sliding surfaces. For example, if all other conditions are the same, the friction between two metal surfaces would be different than the friction between two wood surfaces in contact. Friction is generally higher for like materials sliding on one another. Friction is larger for materials that strongly interact. Friction depends on the surface quality and surface finish. A good surface finish reduces frictional effects. The frictional force does not depend on the total surface area of contact.

It has been experimentally determined that the magnitudes of both static and kinetic friction are proportional to the normal force acting on the surfaces in contact. In Figure 3.19, the normal force is the one whose magnitude is given as N. The constant of proportionality is commonly referred to with μ (mu) and is called the *coefficient of friction*, which depends on such factors as the material properties, quality of surface finish, and the conditions of the surfaces in contact. The coefficient of friction also varies depending on whether the bodies in contact are stationary or sliding over each other. It is important to note that it takes more force to set an object in motion than to keep it moving. To be able to distinguish the frictional forces involved at static and dynamic conditions, two different friction coefficients are defined. The *coefficient of static friction* (μ_s) is associated with static friction, and the *coefficient of kinetic friction* (μ_k) is associated with friction on bodies in motion. The magnitude of the static frictional force is:

$$f_s \leq \mu_s N \tag{3.4}$$

The equality, $f_s = \mu_s N = f_{max}$, occurs when the block is on the verge of sliding. The inequality, $f_s < \mu_s N$, holds when the magnitude of the applied force is less than the maximum frictional force, in which case the force of static friction is equal in magnitude to the applied force ($f_s = F$). The formula relating the kinetic friction and the normal force is:

$$f_k = \mu_k N \tag{3.5}$$

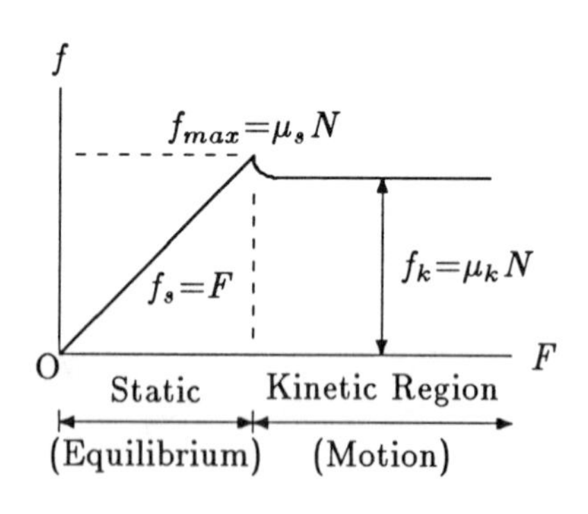

Figure 3.20 *The variation of frictional force as a function of applied force.*

The variations of friction coefficients with respect to the force applied in a direction parallel (tangential) to the surfaces in contact are shown in Figure 3.20. For any given pair of materials, the coefficient of kinetic friction is usually lower than the coefficient of static friction. The coefficient of kinetic friction is approximately constant at moderate sliding speeds. At higher speeds, μ_k may decrease because of the heat generated by friction.

Frictional forces always act in a direction parallel to the surfaces in contact. If one of the two bodies in contact is moving, then the frictional force acting on that body will have a direction opposite to the direction of motion. For example, under the action of applied force $\underline{F}$, the block in Figure 3.20 moves (or tends to move) towards the right. The direction of the frictional force acting on the block is towards the left, trying to stop the motion of the block. The frictional forces occur in pairs because there are always two surfaces in contact for friction to occur. Newton's third law states that action and reaction must have an equal magnitude. Therefore, a frictional force is also acting on the floor, whose magnitude is equal to the magnitude of the frictional force acting on the block. Newton's third law also states that the action and reaction must have opposite directions. Therefore, the frictional force on the floor is acting towards the right.

Chapter 4

THE MOMENT VECTOR

4.1 Definition of Moment
4.2 Magnitude of Moment
4.3 Direction of Moment
4.4 Dimension and Units of Moment
4.5 Some Fine Points About the Moment Vector
4.6 The Net (Resultant) Moment
4.7 The Couple and Couple-Moment
4.8 Translation of Forces
4.9* Moment as a Vector Product

4.1 Definition of Moment

A force applied to an object can translate, deform, and/or rotate the object, depending on how the force is applied and the object is situated. For example, if you push an open door, the door will swing about the edge along which it is hinged to the wall. What causes the door to swing is the moment or torque generated by the force you apply on the door. If you stand at the free-end of a diving board, the board will bend. What bends the diving board is the moment of your body weight about the fixed end of the board. It is easy to explain and understand the concept of force. It is relatively more challenging to explain the concept of moment and its physical significance.

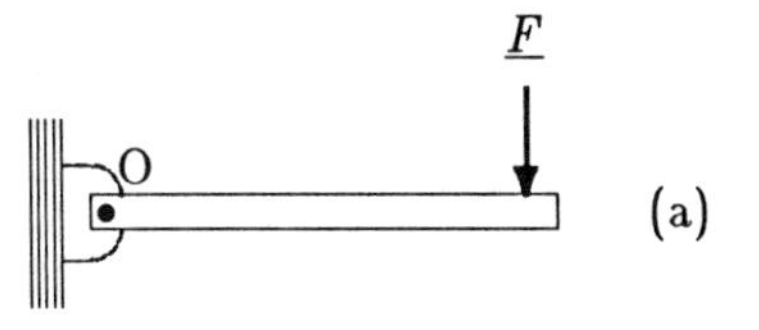

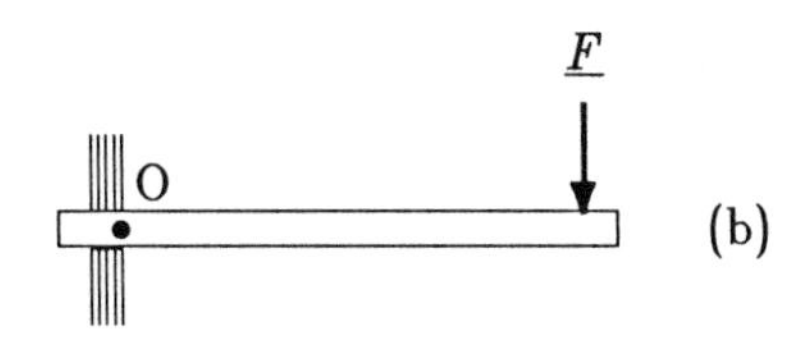

Figure 4.1 *Force $\underline{F}$ is causing (a) rotation and (b) bending action about point O.*

Note that when a body is free to rotate about a point, as for the hinged beam in Figure 4.1(a) about point O, the applied force will cause the body to rotate about that point. If the body is held fixed at one end, as for the cantilever beam in Figure 4.1(b), the force will tend to bend the body. The mathematical definition of moment and torque are the same. However, torque is related to the rotating and twisting action, while moment is associated with the bending action of an applied force.

4.2 Magnitude of Moment

In Figure 4.2, $\underline{F}$ is a force vector with magnitude F, applied at point P on a body. Let AB be the line of action of $\underline{F}$, and O refer to another point on the same body which is r distance away from point P. The magnitude of *moment* or *torque* of force $\underline{F}$ about point O is equal to the magnitude of force times the length of the shortest (perpendicular) distance between point O and the line of action of the force.

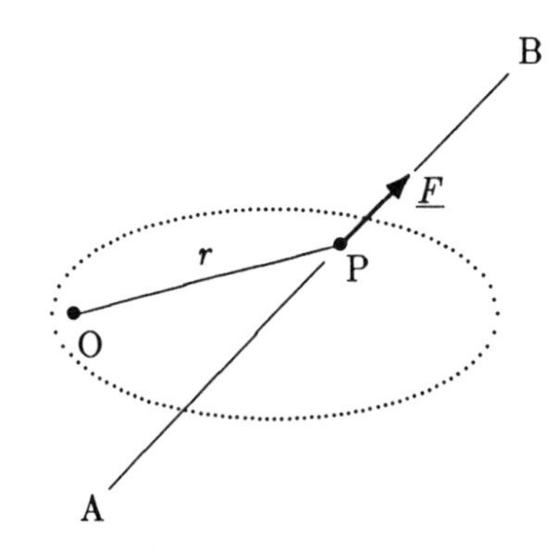

Figure 4.2 *What is the moment of force $\underline{F}$ about point O?*

To determine the magnitude of the moment of $\underline{F}$ about point O, first draw a straight line from point O which cuts line AB at right angles, as shown in Figure 4.3. Let R be the point of intersection of these perpendicular lines. The line joining points O and R is called the *moment arm* or *lever arm*, and it represents the shortest distance between point O and the line of action of $\underline{F}$. If d refers to the length of the moment arm and θ is the smaller angle between lines OP and AB, then the magnitude of moment vector $\underline{M}$ of force $\underline{F}$ about point O is:

$$M = d\,F \tag{4.1}$$

Note that points O, P, and R form the corners of a right-triangle, and $d = r \sin\theta$. Therefore, in terms of length r and angle θ:

$$M = r\;F\;\sin\theta \tag{4.2}$$

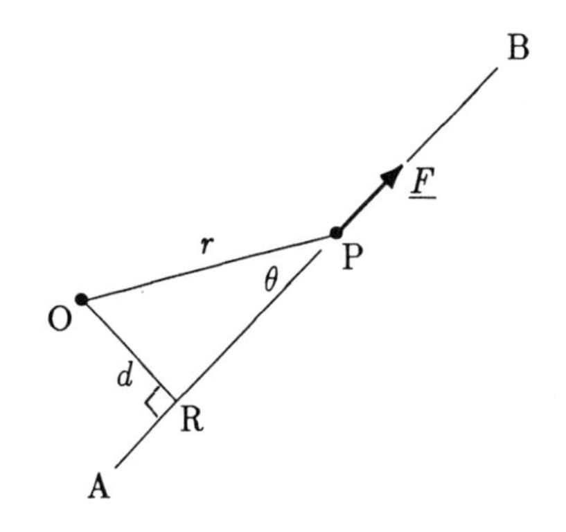

Figure 4.3 *Moment is equal to force times the moment arm.*

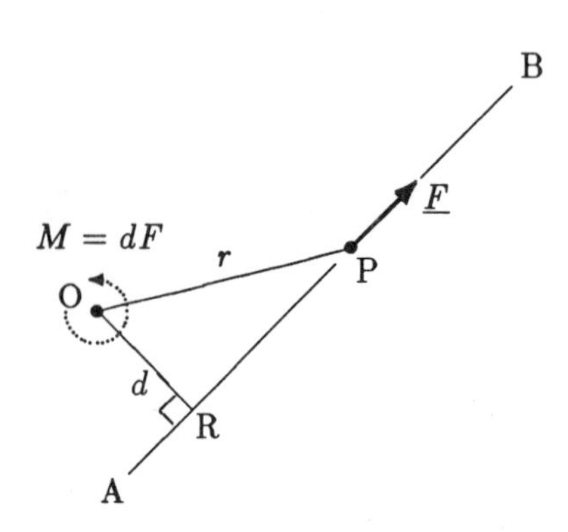

Figure 4.4 *The magnitude and direction of moment of $\underline{F}$ about O.*

Figure 4.5 *The right-hand-rule.*

4.3 Direction of Moment

Moment is a vector quantity. It has a direction as well as a magnitude. The moment of a force about a point acts in a direction perpendicular to the plane upon which the point and the force lie. For example, in Figure 4.4, if we assume that lines OP and AB lie on the surface of the page, then the line of action of moment $\underline{M}$ is perpendicular to the page. The direction or sense of $\underline{M}$ can be determined using the ***right-hand-rule***, as illustrated in Figure 4.5. When the fingers of the right hand curl in the direction that the applied force tends to rotate the body about point O, the right hand thumb points in the direction of the moment vector.

Note that to refer to the direction of moments of coplanar force systems, it can be sufficient to say that a particular moment is either clockwise (cw) or counterclockwise (ccw). Again for coplanar force systems, the direction of a moment vector can be specified graphically by using a dot (•) or a cross (×). The dot implies that the direction of the moment is out of the page (i.e., the tip of the arrow head is towards the observer), and the cross implies that the direction of the moment vector is into the page (i.e., observing the tail of the arrow).

4.4 Dimension and Units of Moment

By definition, moment is equal to the product of applied force and the length of the moment arm. Therefore, the dimension of moment is equal to the dimension of force (ML/T^2) times the dimension of length (L):

$$[\text{MOMENT}] = [\text{FORCE}]\,[\text{MOMENT ARM}] = \frac{\text{ML}}{\text{T}^2}\,\text{L} = \frac{\text{ML}^2}{\text{T}^2}$$

The units of moment in different systems are listed in Table 4.1.

SYSTEM	UNITS OF MOMENT AND TORQUE
SI	kilogram-meter^2/(second)^2 or N-m
c-g-s	gram-centimeter^2/(second)^2 or dyn-cm
British	slug-foot^2/(second)^2 or lb-ft

Table 4.1 *Units of moment and torque. (1 lb-ft=1.3573 N-m)*

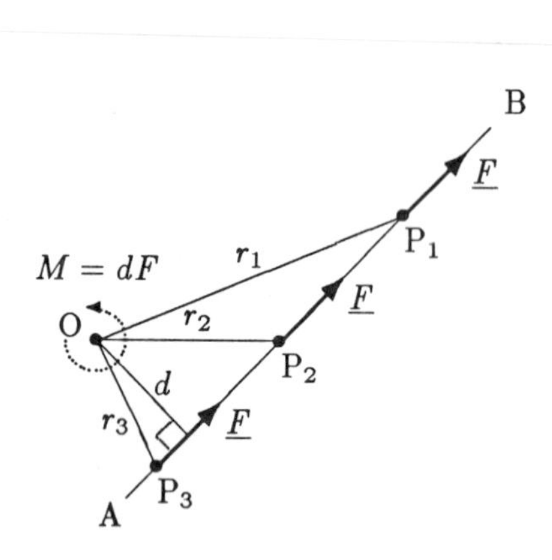

Figure 4.6 *Moment is invariant under the operation of sliding the force along its line of action.*

4.5 Some Fine Points About the Moment Vector

- The moment of a force is invariant under the operation of sliding the force vector along its line of action. As illustrated in Figure 4.6, the point of application of a force can be moved without influencing the magnitude and direction of the resultant moment vector, as long as the point of application of the force is kept along the line of action of the force. For all cases illustrated, the moment of force $\underline{F}$ about point O is:

$$M = d\,F \qquad \text{(ccw)}$$

where length d is always the shortest distance between point O and the line of action of $\underline{F}$. Again for all three cases shown in Figure 4.6, the forces generate a moment in the counterclockwise direction.

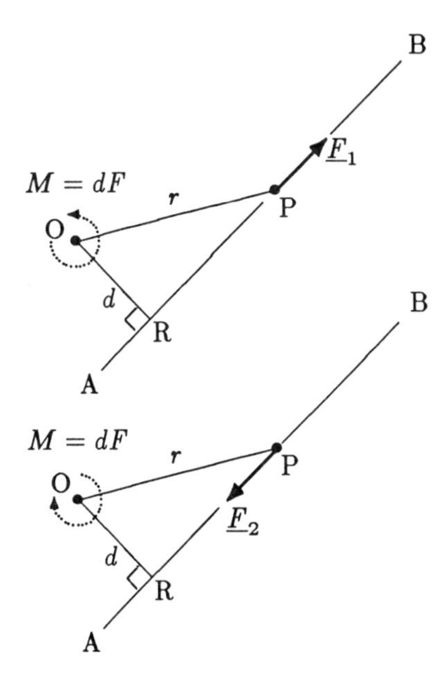

Figure 4.7 *Opposite moments.*

- Let $\underline{F}_1$ and $\underline{F}_2$ shown in Figure 4.7 be two forces with equal magnitude ($F_1{=}F_2{=}F$) and same line of action, but acting in opposite directions. The moment $\underline{M}_1$ of force $\underline{F}_1$ and the moment $\underline{M}_2$ of force $\underline{F}_2$ about point O have an equal magnitude ($M_1{=}M_2{=}M{=}dF$), but opposite directions ($\underline{M}_1{=}{-}\underline{M}_2$).

- The magnitude of the moment of an applied force increases with an increase in the length of the moment arm.

- The moment of a force about one point which lies on the line of action of the force is zero, because the length of the moment arm is zero (Figure 4.8).

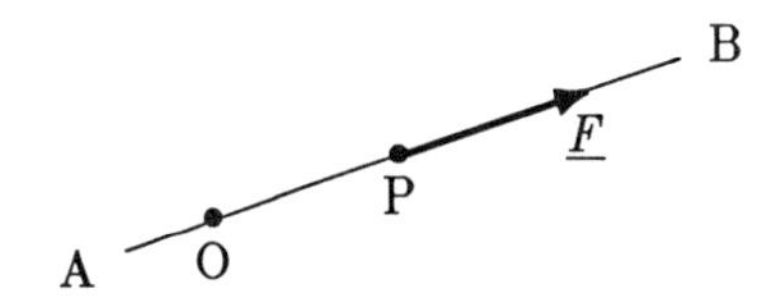

Figure 4.8 *The moment of $\underline{F}$ about point O is zero.*

- A force applied to a body may tend to rotate or bend the body in one direction with respect to one point and in the opposite direction with respect to another point in the same plane.

- The principles of resolution of forces into their components along appropriate directions can be utilized to simplify the calculation of moments. For example, in Figure 4.9, line OP coincides with the x axis. Applied force $\underline{F}$ can be resolved into its components $\underline{F}_x$ and $\underline{F}_y$ along the x and y directions, such that:

$$F_x = F\,\cos\theta$$
$$F_y = F\,\sin\theta$$

Since point O lies on the line of action of $\underline{F}_x$, the moment arm of $\underline{F}_x$ is zero. Therefore, the moment of $\underline{F}_x$ about point O is zero. The line of action of force $\underline{F}_y$ is perpendicular to line OP, and r is the length of the moment arm. Therefore, the moment of force $\underline{F}_y$ about point O is:

$$M = r\,F_y = r\,F\,\sin\theta \qquad \text{(cw)}$$

which is also the moment generated by the total force vector $\underline{F}$ about point O.

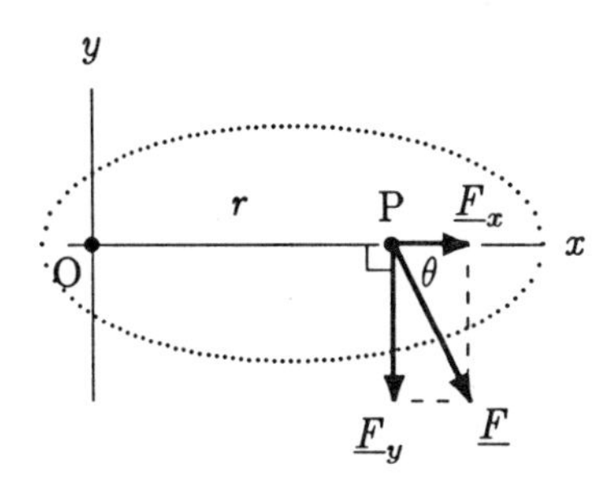

Figure 4.9 *Components of $\underline{F}$.*

4.6 The Net (Resultant) Moment

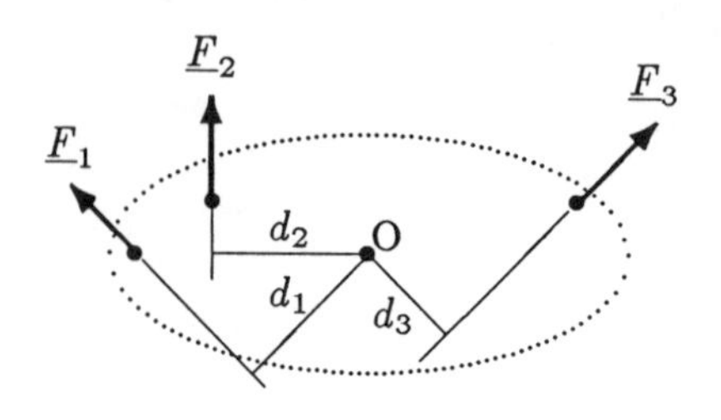

Figure 4.10 *Three-force system.*

When there is more than one force acting on a body, the net or the resultant moment about a point in the body can be calculated by considering the vector sum of the moments of all forces acting on that body about that point. Consider the three-force system shown in Figure 4.10. Assume that $\underline{F}_1$, $\underline{F}_2$, and $\underline{F}_3$ form a coplanar system of forces. These forces produce moments $\underline{M}_1$, $\underline{M}_2$, and $\underline{M}_3$ about point O, respectively. Let d_1, d_2, and d_3 be the lengths of the moment arms of $\underline{F}_1$, $\underline{F}_2$, and $\underline{F}_3$ as measured from point O. To determine the net moment about point O, first calculate the magnitudes M_1, M_2 and M_3 of moments about point O, and indicate the correct direction of each moment vector:

$$M_1 = d_1\, F_1 \qquad \text{(cw)}$$
$$M_2 = d_2\, F_2 \qquad \text{(cw)}$$
$$M_3 = d_3\, F_3 \qquad \text{(ccw)}$$

The net moment $\underline{M}_{net}$ is equal to the vector sum of all moments:

$$\underline{M}_{net} = \underline{M}_1 + \underline{M}_2 + \underline{M}_3 \tag{4.3}$$

In the case shown in Figure 4.10, the moments $\underline{M}_1$, $\underline{M}_2$, and $\underline{M}_3$ have either clockwise or counterclockwise directions.

Next choose either the clockwise or the counter-clockwise direction to be the positive direction for the resultant moment vector. Assuming that we have chosen the clockwise direction to be positive, the moments $\underline{M}_1$ and $\underline{M}_2$ are positive vectors, while moment $\underline{M}_3$ is a negative vector. The magnitude of the net moment can now be determined by simply adding the magnitudes of the positive moments and subtracting the negatives:

$$\begin{aligned} M_{net} &= M_1 + M_2 - M_3 \\ &= d_1 F_1 + d_2 F_2 - d_3 F_3 \end{aligned}$$

Depending on the magnitudes of forces and the lengths of the moment arms, either a positive, negative, or a zero value for the net moment will be calculated from the above equation. If the computed value for M_{net} is positive, then the direction chosen (in this case the clockwise direction) was correct and the net moment has a clockwise direction. In other words, with respect to point O, the applied forces would tend to rotate the body in the clockwise direction. If the net moment is negative, then the applied forces tend to rotate the body about point O in the counterclockwise direction. If the net moment is equal to zero, the body is said to be in *rotational equilibrium*.

Example 4.1 A 60 kg person is getting ready to dive into a pool, as illustrated in Figure 4.11. The diving board has a mass of 50 kg and is mounted to the ground at O. The weight of the person can be assumed as a concentrated force applied on the diving board at A. The distance between O and A is ℓ=4 m.

Compute the moments generated about point O due to the weight of the person, the weight of the board, and the net moment about point O.

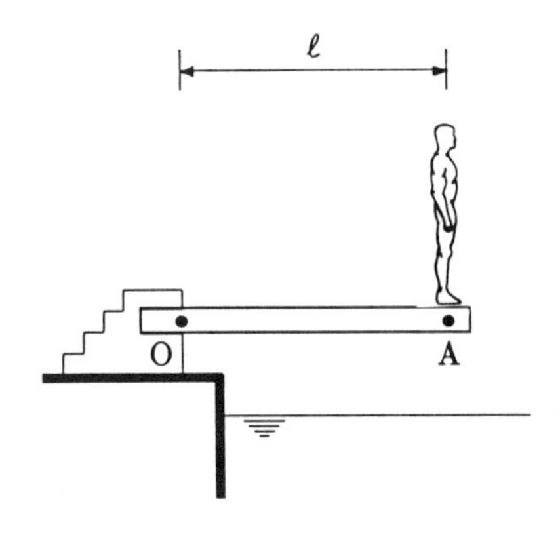

Figure 4.11 *A person at point A (free-end of the diving board) is preparing to dive.*

Solution: The magnitudes W_1 and W_2 of weights of the person and diving board can be determined, because their masses are given:

$$W_1 = m_1\, g = (60)(9.8) = 588 \text{ N}$$
$$W_2 = m_2\, g = (50)(9.8) = 490 \text{ N}$$

The weight of the person is applied at the free-end (A) of the diving board which is ℓ=4 m away from point O. The mass of the diving board produces a load system distributed over the length of the board. We can assume that the weight of the board is a concentrated force acting at a point (B) where the center of gravity of the board is located. Since the board has a uniform thickness, point B is located at a distance equal from both ends of the board (Figure 4.12).

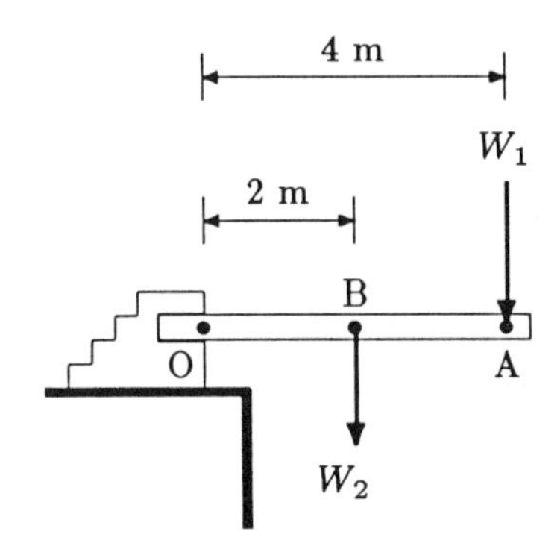

Figure 4.12 *Forces acting on the diving board.*

We have a two force ($\underline{W}_1$ and $\underline{W}_2$) system. These forces are acting in the vertical direction perpendicular to the length of the board. Let $\underline{M}_1$ and $\underline{M}_2$ be the moments of forces $\underline{W}_1$ and $\underline{W}_2$ about point O. The length of the lever arm for force $\underline{W}_1$ is ℓ=4 m, and it is $\ell/2$=2 m for $\underline{W}_2$. Therefore, the moments due to $\underline{W}_1$ and $\underline{W}_2$ about point O are:

$$M_1 = \ell\, W_1 = (4)(588) = 2352 \text{ N-m} \quad \text{(cw)}$$
$$M_2 = \frac{\ell}{2}\, W_2 = (2)(490) = 980 \text{ N-m} \quad \text{(cw)}$$

Since both moments have a clockwise direction, the net moment has a clockwise direction as well. The magnitude of the net moment about point O is:

$$M_{net} = M_1 + M_2 = 3332 \text{ N-m} \quad \text{(cw)}$$

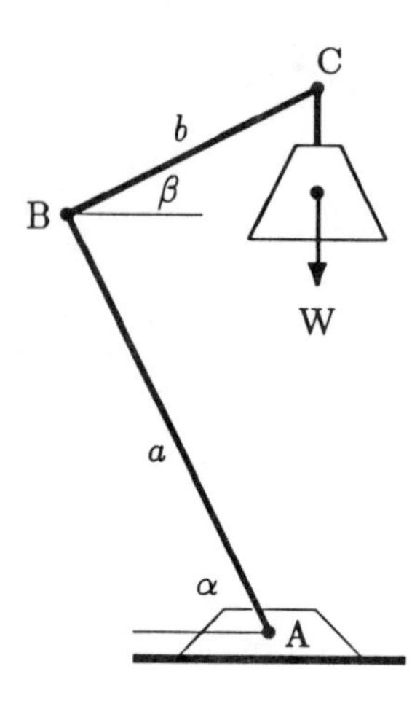

Example 4.2 The fixture (head) of the adjustable lamp shown in Figure 4.13 weighs W=10 N. a=50 cm and b=30 cm are the lengths of arms AB and BC, respectively. The angles that the arms make with the horizontal are α=63.4° and β=26.6°.

Neglecting the weights of the arms, determine the moments generated about points A, B, and C due to the weight of the fixture of the lamp.

Solution: In this example, the task is to determine appropriate moment arms by considering the geometry of the problem. The length of the moment arm for point C is zero, because point C lies on the line of action of $\underline{W}$. Therefore, the moment generated by the weight of the lamp about point C is zero:

$$M_C = 0$$

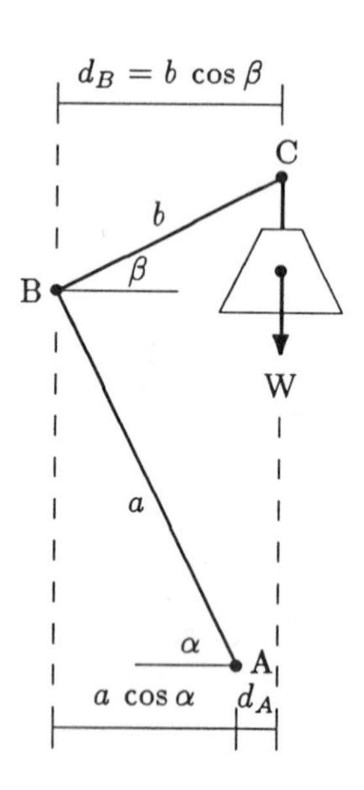

Figure 4.13 *An adjustable lamp.*

The length of the moment arm for point B is d_B=$b\cos\beta$. Therefore, the magnitude of the moment generated about point B is:

$$\begin{aligned} M_B &= d_B\, W = (b\, \cos\beta)\, W \\ &= (0.30)(\cos 26.6°)(10) = 2.68 \text{ N-m} \quad (\text{cw}) \end{aligned}$$

Again from the geometry of the problem, the length of the moment arm for point A is d_A=$b\cos\beta$–$a\cos\alpha$. Hence, the magnitude of the moment generated by the weight of the lamp about point A is:

$$M_A = d_A\, W = (b\, \cos\beta - a\, \cos\alpha)\, W = 0.44 \text{ N-m} \quad (\text{cw})$$

Note that under the action of the weight of the lamp, the arms AB and BC of the adjustable lamp should rotate in the clockwise direction. What maintains this system in the configuration shown are the frictional forces generated at A, B, and C, which are applied by nuts and bolts.

Example 4.3 The simple structure shown in Figure 4.14 is called a *cantilever beam* and is one of the fundamental mechanical elements in engineering. A cantilever beam is one that is fixed at one end and free at the other. Consider that the beam shown has a weight W=100 N and a length ℓ=1 m. A force with magnitude F=150 N is applied at the free-end (C) of the beam in a direction that makes an angle θ=45° with the horizontal.

Determine the magnitude and direction of the net moment developed at the fixed-end (A) of the beam.

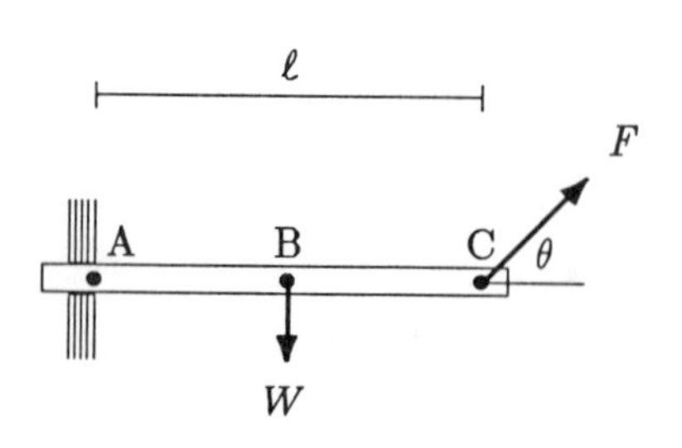

Figure 4.14 *A cantilever beam.*

Solution 1: Let x be the horizontal direction and y show the vertical direction. As shown in Figure 4.15, we can resolve the applied force $\underline{F}$ into its components along the x and y directions, such that:

$$F_x = F\ \cos\theta = (150)(\cos 45^\circ) = 106 \text{ N-m} \quad (\rightarrow)$$
$$F_y = F\ \sin\theta = (150)(\sin 45^\circ) = 106 \text{ N-m} \quad (\uparrow)$$

Arrows in parentheses indicate the directions of applied forces.

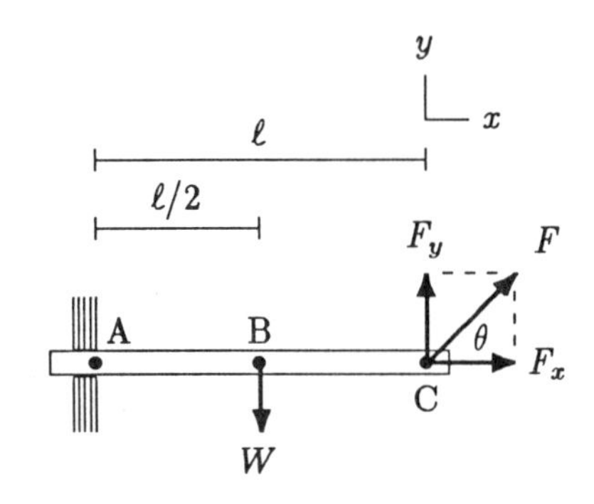

Figure 4.15 *Solution 1.*

Note that we now have a three-force system ($\underline{F}_x$, $\underline{F}_y$, and $\underline{W}$). To determine the moment generated about point A, we need to know the lengths of the moment arms for $\underline{F}_x$, $\underline{F}_y$, and $\underline{W}$. The line of action of $\underline{F}_x$ is passing through point A, and therefore the moment of $\underline{F}_x$ about A is zero. Since the beam has a uniform thickness, the weight of the beam can be assumed to concentrate at the middle (B) of the beam. The lines of action of $\underline{F}_y$ and $\underline{W}$ are perpendicular to line AB. Therefore, ℓ and $\ell/2$ are the lengths of the moment arms for $\underline{F}_y$ and $\underline{W}$ with respect to point A. $\underline{F}_y$ causes a counterclockwise moment and $\underline{W}$ a clockwise moment about point A. If we choose the clockwise direction to be positive, then the moment about point A is:

$$M_A = \frac{\ell}{2} W - \ell\, F_y = \frac{1}{2}(100) - (1)(106) = -56$$

The negative sign in front of 56 is significant, and it shows that we have chosen an incorrect positive direction for the moment. Now we can correct it and write:

$$M_A = 56 \text{ N-m} \quad (\text{ccw})$$

Solution 2: A second approach to the solution of the same problem is illustrated in Figure 4.16. In this case, we remain with the two-force system ($\underline{F}$ and $\underline{W}$). By extending the line of action of $\underline{F}$ and dropping a perpendicular to it, we obtain the right-triangle ACD. The length d of one of the sides of the triangle is the length of the moment arm for $\underline{F}$ about point A. From the right-triangle ACD:

$$d = \ell\ \sin\theta = (1)(\sin 45^\circ) = 0.707 \text{ m}$$

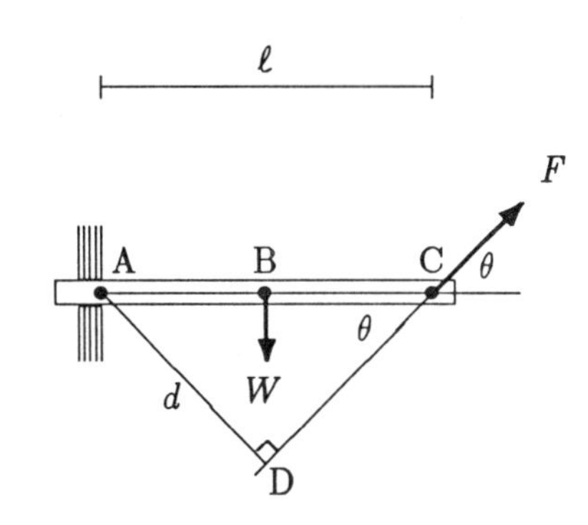

Figure 4.16 *Solution 2.*

This time, we can choose the counterclockwise direction to be the positive direction and calculate the moment about point A as:

$$M_A = d\,F - \frac{\ell}{2} W = (0.707)(150) - \frac{1}{2}(100) = 56 \text{ N-m} \quad (\text{ccw})$$

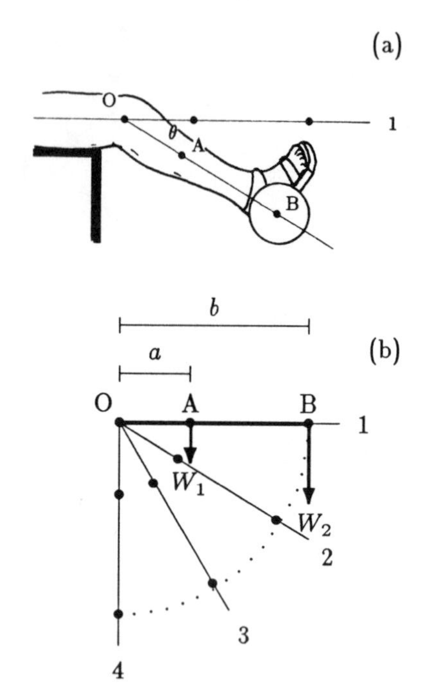

Figure 4.17 *Example 4.4.*

Example 4.4 As illustrated in Figure 4.17(a), consider an athlete wearing a weight boot, and from a sitting position, doing lower leg flexion/extension exercises to strengthen quadriceps muscles. The weight of the athlete's lower leg is W_1=50 N and the weight of the boot is W_2=100 N. As measured from the knee joint at O, the center of gravity (A) of the lower leg is at a distance a=20 cm, and the center of gravity (B) of the weight boot is at a distance b=50 cm.

Determine the net moment generated about the knee joint when the lower leg is extended horizontally (position 1), and when the lower leg makes an angle of 30° (position 2), 60° (position 3), and 90° (position 4) with the horizontal (Figure 4.17b).

Solution: At position 1, the lower leg is extended horizontally and the long axis of the leg is perpendicular to the lines of action of $\underline{W}_1$ and $\underline{W}_2$. Therefore, a and b are the lengths of the moment arms for $\underline{W}_1$ and $\underline{W}_2$, respectively. Both $\underline{W}_1$ and $\underline{W}_2$ apply clockwise moments about the knee joint. The moment about the knee joint when the lower leg is at position 1 is:

$$
\begin{aligned}
M_O &= a\,W_1 + b\,W_2 \\
&= (0.20)(50) + (0.50)(100) \\
&= 60 \text{ N-m} \qquad \text{(cw)}
\end{aligned}
$$

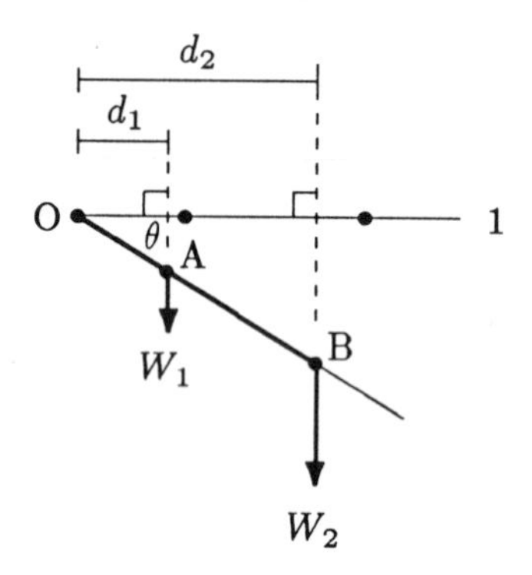

Figure 4.18 *Forces and moment arms when the lower leg makes an angle θ with the horizontal.*

Figure 4.18 illustrates the external forces acting on the lower leg and their moment arms (d_1 and d_2) when the lower leg makes an angle θ with the horizontal. From the geometry of the problem:

$$
\begin{aligned}
d_1 &= a \cos\theta \\
d_2 &= b \cos\theta
\end{aligned}
$$

Therefore, the moment about O is:

$$
\begin{aligned}
M_O &= d_1\,W_1 + d_2\,W_2 \\
&= a \cos\theta\,W_1 + b \cos\theta\,W_2 \\
&= (a\,W_1 + b\,W_2)\cos\theta
\end{aligned}
$$

The term in the parentheses has already been calculated as 60 N-m. Therefore we can write M_O as:

$$M_O = 60 \cos\theta$$

For position 1:	$\theta = 0°$	$M_O = 60$ N-m	(cw)
For position 2:	$\theta = 30°$	$M_O = 52$ N-m	(cw)
For position 3:	$\theta = 60°$	$M_O = 30$ N-m	(cw)
For position 4:	$\theta = 90°$	$M_O = 0$	(cw)

In Figure 4.19, the moment generated about the knee joint is plotted as a function of angle θ.

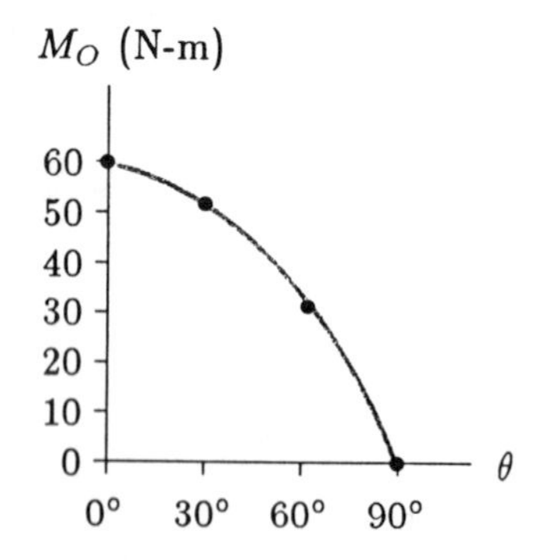

Figure 4.19 *Variation of moment with angle θ.*

Example 4.5 Figure 4.20(a) illustrates an athlete doing shoulder muscle strengthening exercises by lowering and raising a barbell with arms straight. The position of the arms, when they make an angle θ with the vertical, is simplified in Figure 4.20(b). O is where the shoulder joint is located, A is the center of gravity of the arm, and B is a point along the axis of the barbell. The distance between O and A is a=24 cm and between O and B is b=60 cm. The arm weighs W_1=50 N and the total weight of the barbell is W_2=300 N.

Determine the net moment generated about the shoulder joint as a function of θ, which is the angle the arm makes with the vertical. Calculate the moments for $\theta = 0°$, 15°, 30°, 45°, and 60°.

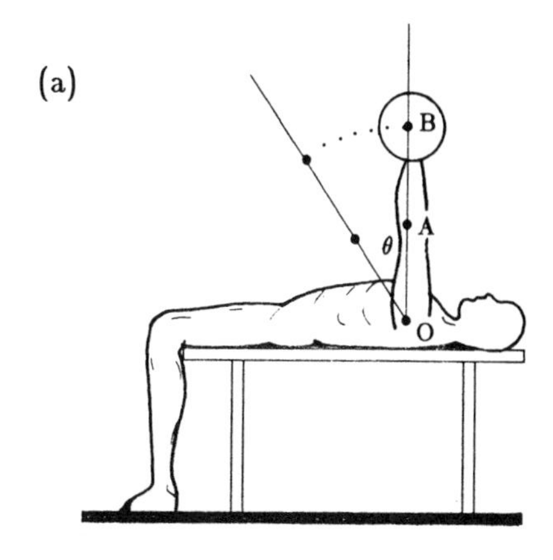

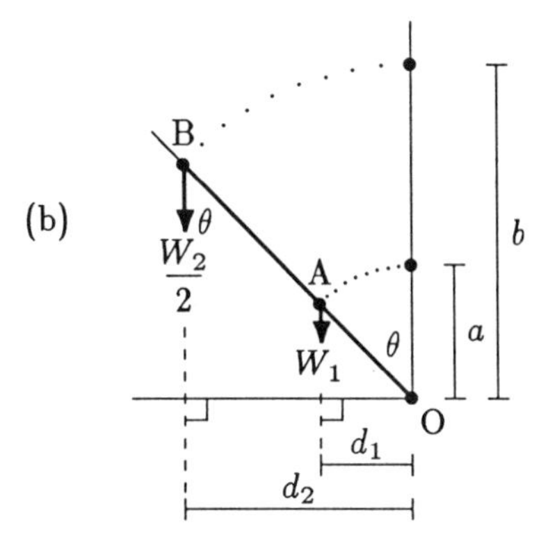

Figure 4.20 *An exercise to strengthen the shoulder muscles, and a simple model of the arm.*

Solution: The external forces acting on the arm are the weight of the arm and the weight of the barbell. To determine the moment generated about the shoulder joint, we need to know the moment arms. From the geometry of the problem (Figure 4.20b), the lengths of the moment arms are:

$$d_1 = a \sin\theta$$
$$d_2 = b \sin\theta$$

Since the athlete is using both arms, the total weight of the barbell is assumed to be equally shared by each arm. Also note that both the weight of the arm and the weight of the barbell are trying to rotate the arm in the counterclockwise direction. Hence, the moment generated about the shoulder joint as a function of angle θ is:

$$\begin{aligned} M_O &= d_1 W_1 + d_2 \frac{W_2}{2} \\ &= \left(a W_1 + b \frac{W_2}{2}\right) \sin\theta \\ &= \left[(0.24)(50) + (0.60)\frac{(300)}{2}\right] \sin\theta \\ &= 102 \sin\theta \ \text{N-m} \quad (\text{ccw}) \end{aligned}$$

θ	$\sin\theta$	M_O (N-m)
0°	0.000	0.0
15°	0.259	26.4
30°	0.500	51.0
45°	0.707	72.1
60°	0.866	88.3

Table 4.2 *Moment about the shoulder joint (Example 4.5).*

To determine the magnitude of moment about O, for $\theta = 0°$, 15°, 30°, 45°, and 60°, all we need to do is evaluate the sines and carry out the multiplications. The results are listed in Table 4.2.

In-depth analysis of this and similar problems will be provided in Chapter 6.

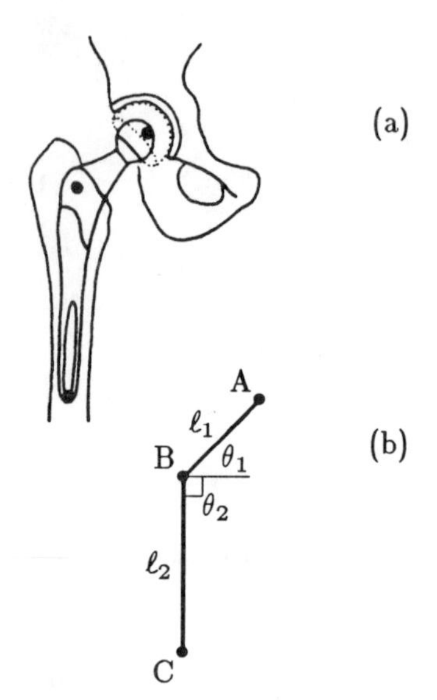

Figure 4.21 *Total hip joint prosthesis.*

Example 4.6 Consider the total hip joint prosthesis shown in Figure 4.21. The geometric parameters of the prosthesis are such that ℓ_1=50 mm, ℓ_2=100 mm, θ_1=45°, and θ_2=90°. Assume that, when standing symmetrically on both feet, a joint reaction force of F=400 N is acting at the femoral head due to the body weight of the patient. For the sake of illustration, consider three different lines of action for the applied force, which are illustrated in Figure 4.22.

Determine the moments generated about points B and C on the prosthesis for all cases shown.

Solution: For each case shown in Figure 4.22, the line of action of the joint reaction force is different, and therefore the lengths of the moment arms are different.

From the geometry of the problem in Figure 4.22(a), we can see that the moment arm of force F about points B and C are the same:

$$d_1 = \ell_1 \cos\theta_1 = (50)(\cos 45°) = 35 \text{ mm}$$

Therefore, the moments generated about points B and C are:

$$M_B = M_C = d_1\, F = (0.035)(400) = 14 \text{ N-m} \quad \text{(cw)}$$

For the case shown in Figure 4.22(b), point B lies on the line of action of the joint reaction force. Therefore, the length of the moment arm for point B is zero, and:

$$M_B = 0$$

For the same case, the length of the moment arm and the moment about point C are:

$$d_2 = \ell_2 \cos\theta_1 = (100)(\cos 45°) = 71 \text{ mm}$$

$$M_C = d_2\, F = (0.071)(400) = 28 \text{ N-m} \quad \text{(ccw)}$$

For the case shown in Figure 4.22(c), the moment arms for B and C are:

$$d_3 = \ell_1 \sin\theta_1 = (50)(\sin 45°) = 35 \text{ mm}$$
$$d_4 = d_3 + \ell_2 = (35) + (100) = 135 \text{ mm}$$

Therefore, the moments generated about points B and C are:

$$M_B = d_3\, F = (0.035)(400) = 14 \text{ N-m} \quad \text{(ccw)}$$

$$M_C = d_4\, F = (0.135)(400) = 54 \text{ N-m} \quad \text{(ccw)}$$

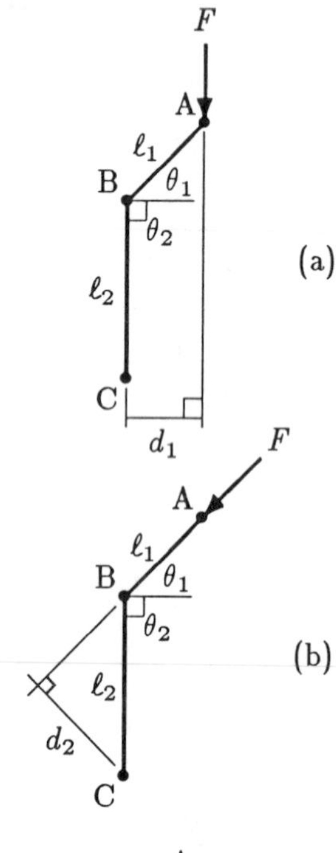

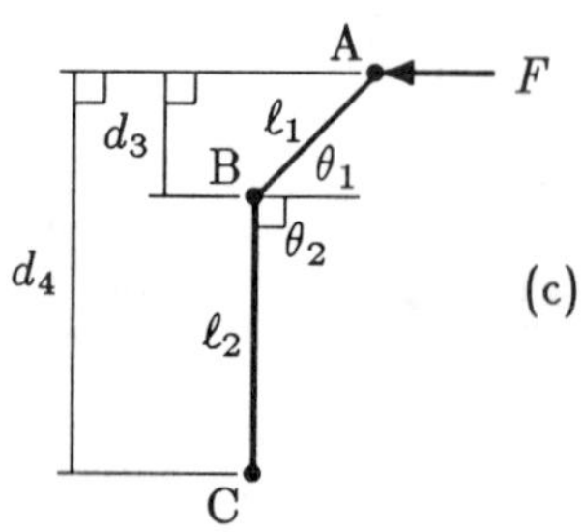

Figure 4.22 *Example 4.6.*

4.7 The Couple and Couple-Moment

A special arrangement of forces that is of importance is called *couple*. A couple is formed by any two parallel forces with equal magnitude and opposite directions. On a rigid body, the couple has only one effect, namely, the twisting action. The twisting action of a couple is quantitatively called the *couple-moment*. The moment of a couple is a vector whose line of action is perpendicular to the plane of the couple and whose direction can be determined by the right-hand-rule.

Figure 4.23 helps illustrate the concepts of couple and couple-moment. The pole shown is mounted to the ground, and a body is attached to it at points A and B which are d distance apart. On the basis of intuition, the attached body would apply its weight on the pole as to rotate the pole in the clockwise direction. Also on the basis of intuition, the attached body would pull the pole towards the right at A and push it towards the left at B. Furthermore, it can be shown that the magnitude of these forces applied on the vertical pole would be equal, and therefore, the net force applied on the pole in the horizontal direction would be zero. (Note that in addition to these horizontal forces, there are forces acting in the vertical direction due to the friction between the pole and the attachment. These are the forces holding the attachment in place. Here we are only interested in the "couple.")

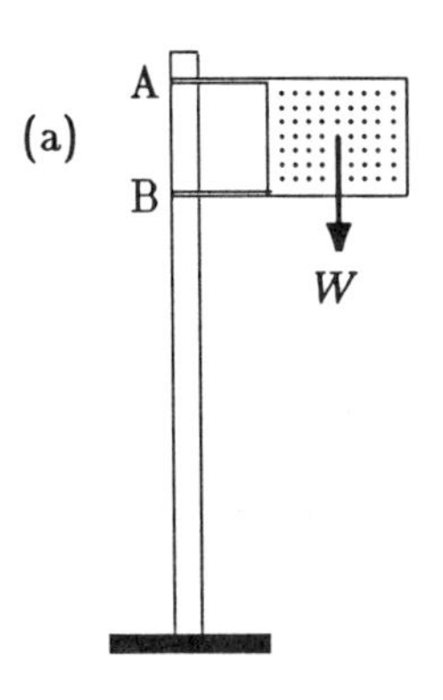

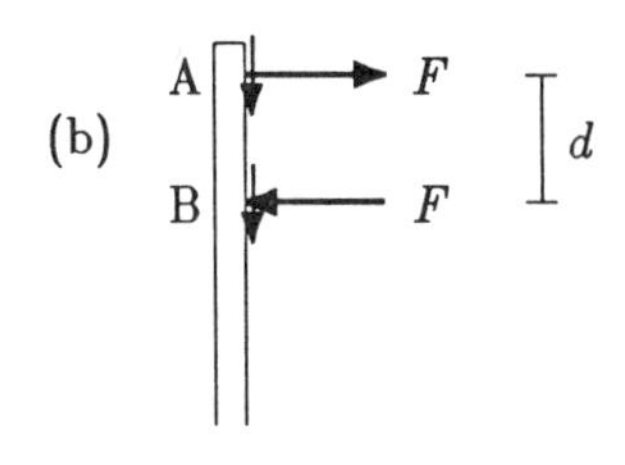

Figure 4.23 *Forces on a pole due to the weight of the attachment.*

If F is the magnitude of the forces forming the couple and d is the perpendicular distance between the lines of actions of the forces, then the magnitude of the couple-moment is:

$$M = d\,F \tag{4.4}$$

To prove the validity of Eq. (4.4), consider the couple in Figure 4.24. The net moment about point A is equal to $M{=}dF$ (cw), which is due to the force applied at B. The net moment about point B is also equal to $M{=}dF$ (cw), and it is due to the force applied at A. Consider a third point, C, located at b distance away from B. The net moment about C is equal to the sum of clockwise moments of forces applied at A and B, whose moment arms are $(d-b)$ and b, respectively. Therefore:

$$M = (d-b)\,F + b\,F = d\,F \qquad \text{(cw)}$$

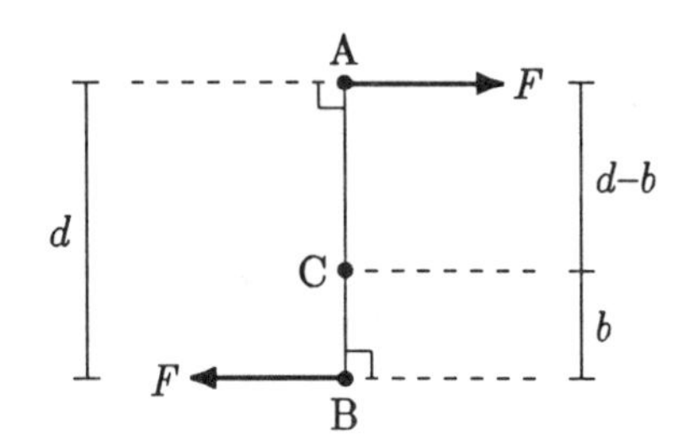

Figure 4.24 *A couple.*

We can conclude, without further proof, that the couple has the same moment about every point in space. The couple-moment is known as a *free vector*, meaning that it can be moved anywhere in space without changing its meaning, provided that its magnitude and direction are kept intact. For example, the couple-moment generated by the body attached to the pole in Figure 4.23 acts everywhere along the pole.

4.8 Translation of Forces

We have seen in Section 4.5 that a force vector may be moved along its line of action without changing its effect. In other words, forces are transmissible vectors. Also, the net effect of a pair of forces applied on a rigid body is zero, provided that the forces have an equal magnitude and the same line of action, but are acting in opposite directions. We have seen in the previous section that the only effect a couple has on a rigid body is the couple-moment it produces. These important features about the force vector can be utilized to translate a force vector to a position parallel to its original position.

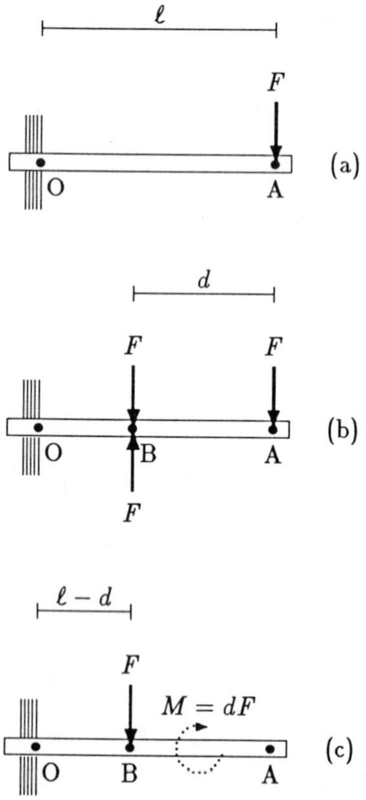

Consider the rigid cantilever beam shown in Figure 4.25(a). A downward force of F is applied at the free-end (A) of the beam. As shown in Figure 4.25(b), we can place a pair of forces with equal magnitude (F) and the same line of action, but which are acting in opposite directions at a point B on the beam, without disturbing the original equilibrium of the beam. Note that the downward force at A and the upward force at B form a couple. This couple produces a clockwise moment whose magnitude is $M{=}dF$. Therefore, the couple can be replaced by the couple-moment, as shown in Figure 4.25(c).

Figure 4.25 *Translation of a force from position A to B.*

If the length of the beam is ℓ, then the moment of force F in Figure 4.25(a) about point O is:

$$M_O = \ell\, F \qquad \text{(cw)}$$

The moment about point O due to the three forces shown in Figure 4.25(b) is:

$$M_O = (\ell - d)\, F - (\ell - d)\, F + \ell\, F = \ell\, F \qquad \text{(cw)}$$

The clockwise couple-moment with magnitude $M{=}dF$ in Figure 4.25(c) is acting everywhere along the length of the beam. Therefore, the net moment about point O is equal to the sum of the moment $(\ell - d)F$ due to force F applied at point B and the couple-moment dF:

$$M_O = (\ell - d)\, F + dF = \ell\, F \qquad \text{(cw)}$$

In other words, the one-force system in Figure 4.25(a), the three-force system in Figure 4.25(b), and the one-force and one couple-moment system in Figure 4.25(c) are mechanically equivalent.

4.9* Moment as a Vector Product

We have been applying the scalar method of determining the moment of a force about a point. The scalar method is satisfactory to analyze relatively simple coplanar force systems and systems in which the perpendicular distance between the point and the line of action of the applied force are easy to compute. The analysis of more complex problems can be simplified by utilizing additional mathematical tools.

The concept of *vector (cross) product* of two vectors was introduced in Chapter 2 and will be reviewed here. Consider vectors $\underline{A}$ and $\underline{B}$, shown in Figure 4.26. The cross product of $\underline{A}$ and $\underline{B}$ is equal to a third vector, $\underline{C}$:

$$\underline{C} = \underline{A} \times \underline{B} \tag{4.5}$$

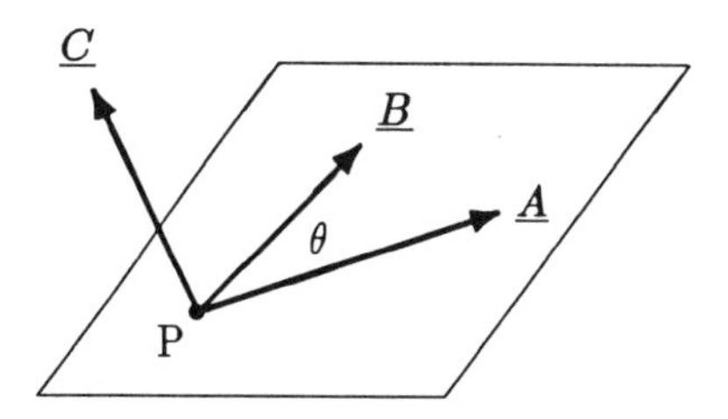

Figure 4.26 *$\underline{C}$ is the vector product of $\underline{A}$ and $\underline{B}$.*

The product vector $\underline{C}$ has the following properties:

- The magnitude of $\underline{C}$ is equal to the product of the magnitudes of $\underline{A}$, $\underline{B}$, and $\sin\theta$, where θ is the smaller angle between $\underline{A}$ and $\underline{B}$.

$$C = A\,B\,\sin\theta \tag{4.6}$$

- The line of action of $\underline{C}$ is perpendicular to the plane formed by vectors $\underline{A}$ and $\underline{B}$.
- The direction (sense) of $\underline{C}$ obeys the right-hand-rule.

The principle of vector or cross product can be applied to determine moments of forces. The moment of a force about a point is defined as the vector product of position and force vectors. The *position vector* of a point P with respect to another point O is defined by a straight line drawn from point O to point P. To help understand the definition of moment as a vector product, consider Figure 4.27. Force $\underline{F}$ acts in the xy-plane, and has a point of application at P. Force $\underline{F}$ can be expressed in terms of its components F_x and F_y along the x and y directions:

$$\underline{F} = F_x\,\underline{i} + F_y\,\underline{j} \tag{4.7}$$

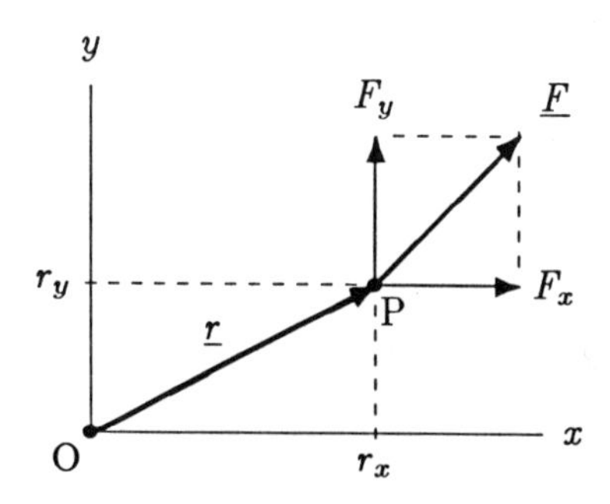

Figure 4.27 *Moment about O is $\underline{M} = \underline{r} \times \underline{F}$.*

The position vector of point P with respect to point O is represented by vector $\underline{r}$, which can be written in terms of its components:

$$\underline{r} = r_x\,\underline{i} + r_y\,\underline{j} \tag{4.8}$$

The components r_x and r_y of the position vector are simply the x and y coordinates of point P as measured from point O.

The moment of force $\underline{F}$ about point O is equal to the vector product of position vector $\underline{r}$ and force vector $\underline{F}$:

$$\underline{M} = \underline{r} \times \underline{F} \tag{4.9}$$

Using Eqs. (4.7) and (4.8), Eq. (4.9) can alternatively be written as follows:

$$\begin{aligned}\underline{M} &= (r_x\,\underline{i} + r_y\,\underline{j}) \times (F_x\,\underline{i} + F_y\,\underline{j}) \\ &= r_x F_x\,(\underline{i}\times\underline{i}) + r_x F_y\,(\underline{i}\times\underline{j}) + \\ &\quad r_y F_x\,(\underline{j}\times\underline{i}) + r_y F_y\,(\underline{j}\times\underline{j})\end{aligned} \tag{4.10}$$

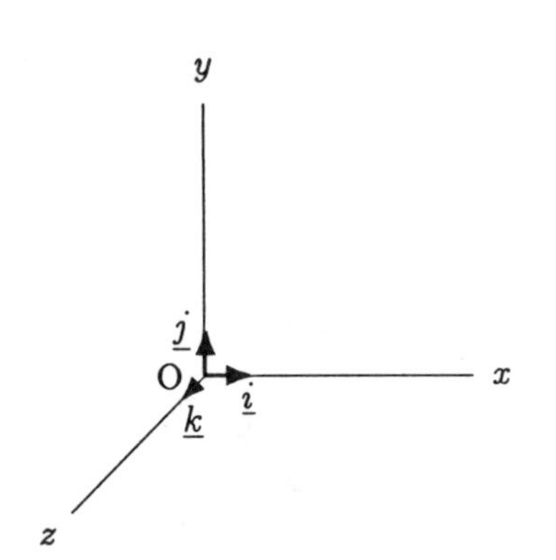

Figure 4.28 *The right-hand-rule applies to Cartesian coordinate directions, as well.*

Recall that $\underline{i}\times\underline{i} = \underline{j}\times\underline{j} = 0$ since the angle that a unit vector makes with itself is zero, and $\sin 0° = 0$. $\underline{i}\times\underline{j} = \underline{k}$ because the angle covered while going from the positive x axis to the positive y axis is 90° ($\sin 90° = 1$). On the other hand, $\underline{j}\times\underline{i} = -\underline{k}$. For the last two cases, the product is either in the positive z (counter-clockwise or out of the page) or negative z (clockwise or into the page) direction. z and unit vector $\underline{k}$ designate the direction perpendicular to the xy-plane (Figure 4.28). Now, Eq. (4.10) can be simplified as:

$$\underline{M} = (r_x F_y - r_y F_x)\,\underline{k} \tag{4.11}$$

To show that the definition of moment as the vector product of position and force vectors is consistent with the scalar method of finding the moment, consider the simple case illustrated in Figure 4.29. The force vector $\underline{F}$ is acting in the positive y direction and its line of action is d distance away from point O. Applying the scalar method, the moment about O is:

$$M = d\,F \qquad \text{(ccw)} \tag{4.12}$$

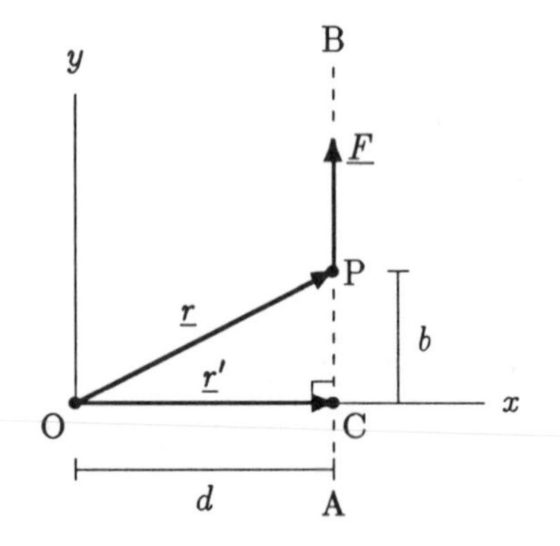

Figure 4.29 $M = d\,F$ *(ccw).*

The force vector is acting in the positive y direction. Therefore:

$$\underline{F} = F\,\underline{j} \tag{4.13}$$

If b is the y coordinate of the point of application of the force, then the position vector of point P is:

$$\underline{r} = d\,\underline{i} + b\,\underline{j} \tag{4.14}$$

Therefore, the moment of $\underline{F}$ about point O is:

$$\begin{aligned}\underline{M} &= \underline{r}\times\underline{F} \\ &= (d\,\underline{i} + b\,\underline{j}) \times (F\,\underline{j}) \\ &= d\,F\,(\underline{i}\times\underline{j}) + b\,F\,(\underline{j}\times\underline{j}) \\ &= d\,F\,\underline{k}\end{aligned} \tag{4.15}$$

Eqs. (4.12) and (4.15) carry exactly the same information, in two different ways. Furthermore, the y coordinate (b) of point P does not appear in the solution. This is consistent with the definition of the moment, which is the vector product of the position vector of any point on the line of action of the force and the force itself.

In Figure 4.29, if C is the point of intersection of the line of action of force $\underline{F}$ and the x axis, then the position vector $\underline{r}'$ of point C with respect to O is:

$$\underline{r}' = d\,\underline{i} \tag{4.16}$$

Therefore, the moment of $\underline{F}$ about point O can alternatively be determined as:

$$\begin{aligned} \underline{M} &= \underline{r}'\,\underline{F} \\ &= (d\,\underline{i}) \times (F\,\underline{j}) \\ &= d\,F\,\underline{k} \end{aligned} \tag{4.17}$$

For any two-dimensional problem composed of a system of coplanar forces in the xy-plane, the resultant moment vector has only one component. The resultant moment vector has a direction perpendicular to the xy-plane, acting in the positive or negative z direction.

The method we have outlined to study coplanar force systems using the concept of the vector (cross) product can easily be expanded to analyze three-dimensional situations. In general, the force vector $\underline{F}$ and the position vector $\underline{r}$ of a point on the line of action of $\underline{F}$ about a point O would have up to three components. With respect to the Cartesian coordinate frame:

$$\underline{F} = F_x\,\underline{i} + F_y\,\underline{j} + F_z\,\underline{k} \tag{4.18}$$

$$\underline{r} = r_x\,\underline{i} + r_y\,\underline{j} + r_z\,\underline{k} \tag{4.19}$$

The moment of $\underline{F}$ about point O can be determined as:

$$\begin{aligned} \underline{M} &= \underline{r} \times \underline{F} \\ &= (r_x\,\underline{i} + r_y\,\underline{j} + r_z\,\underline{k}) \times (F_x\,\underline{i} + F_y\,\underline{j} + F_z\,\underline{k}) \\ &= (r_y F_z - r_z F_y)\,\underline{i} + (r_z F_x - r_x F_z)\,\underline{j} + \\ &\quad (r_x F_y - r_y F_x)\,\underline{k} \end{aligned} \tag{4.20}$$

The moment vector can be expressed in terms of its components along the x, y, and z directions:

$$\underline{M} = M_x\,\underline{i} + M_y\,\underline{j} + M_z\,\underline{k} \tag{4.21}$$

By comparing Eqs. (4.20) and (4.21), we can conclude that:

$$\begin{aligned} M_x &= r_y F_z - r_z F_y \\ M_y &= r_z F_x - r_x F_z \\ M_z &= r_x F_y - r_y F_x \end{aligned} \tag{4.22}$$

The following example provides an application of the analysis outlined in the last three sections of this chapter.

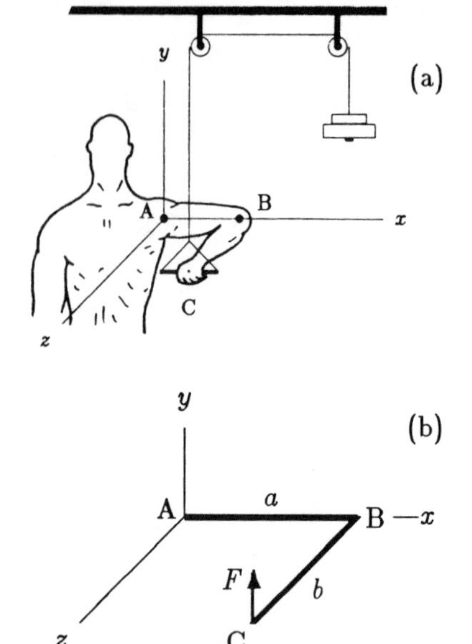

Figure 4.30 *Example 4.7*

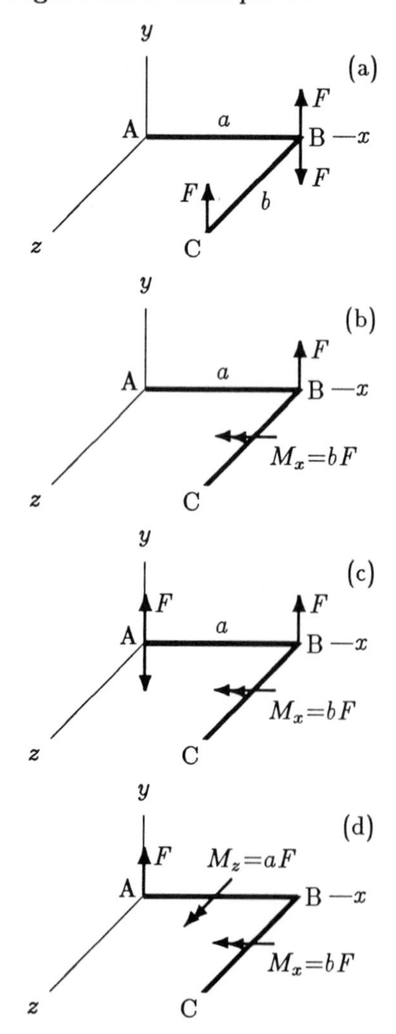

Figure 4.31 *Scalar method.*

Example 4.7 Figure 4.30(a) illustrates a person on an exercise machine. The "L" shaped beam shown in Figure 4.30(b) represents the left arm of the person. Points A and B correspond to the shoulder and elbow joints, respectively. The upper arm (AB) is extended towards the left (x direction) and the forearm (BC) is extended forward (z direction). At this instant the person is grasping a handle that is connected by a cable to a suspending weight. The weight applies an upward force (in the y direction) of F on the arm at point C. The lengths of the upper arm and forearm are a=25 cm and b=30 cm, respectively, and the magnitude of the applied force is F=200 N.

Explain how force F can be translated to the shoulder joint at A, and determine the magnitudes and directions of moments developed at the forearm and upper arm by F.

Solution 1: Scalar method.
The scalar method of finding the moments generated on the forearm (BC) and upper arm (AB) is illustrated in Figure 4.31, and it utilizes the concepts of couple and couple-moment.

The first step is placing a pair of forces at B with equal magnitude (F) and opposite directions, both being parallel to the original force at C (Figure 4.31a). The upward force at C and the downward force at B form a couple. Therefore, they can be replaced by a couple-moment (shown by a double-headed arrow in Figure 4.31b), whose magnitude is equal to bF. Applying the right-hand-rule, we can see that the couple-moment acts in the negative x direction. If M_x refers to the magnitude of this couple-moment, then:

$$M_x = b\,F \qquad (-x \text{ direction})$$

The next step is to place another pair of forces at A where the shoulder joint is located (Figure 4.31c). This time, the upward force at B and the downward force at A form a couple, and again, they can be replaced by a couple-moment (Figure 4.31d). The magnitude of this couple-moment is aF, and it has a direction perpendicular to the xy-plane, or, it is acting in the positive z direction. Referring to the magnitude of this moment as M_z, then:

$$M_z = a\,F \qquad (+z \text{ direction})$$

To evaluate the magnitudes of the couple-moments, the numerical values of a, b, and F must be substituted in the above equations:

$$M_x = (0.30)(200) = 60 \text{ N-m}$$
$$M_z = (0.25)(200) = 50 \text{ N-m}$$

The effect of the force applied at C is such that at the elbow joint, the person feels an upward force with magnitude F, and a moment with magnitude M_x is trying to *rotate* the forearm in the yz-plane. At the shoulder joint, the feeling is such that there is an upward force of F, a torque with magnitude M_x trying to twist the upper arm in the yz-plane, and a moment M_z trying to rotate or bend the upper arm in the xy-plane. If the person is able to hold the arm in this position, then he/she is producing sufficient muscle forces to counter-balance these applied forces and moments.

Solution 2: Vector product method.
The definition of moment as the vector product of position and force vectors is more straightforward to apply. The position vector of point C (where the force is applied) with respect to point A (where the shoulder joint is located) and the force vector, shown in Figure 4.32, can be expressed as follows:

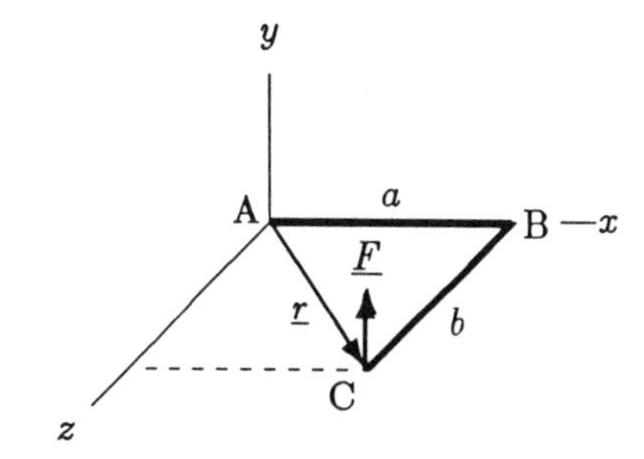

Figure 4.32 *Vector product method (Example 4.7).*

$$\underline{r} = a\,\underline{i} + b\,\underline{k}$$

$$\underline{F} = F\,\underline{j}$$

The cross product of $\underline{r}$ and $\underline{F}$ will yield the moment of force $\underline{F}$ about point A:

$$\begin{aligned}\underline{M} &= \underline{r} \times \underline{F} \\ &= (a\,\underline{i} + b\,\underline{k}) \times (F\,\underline{j}) \\ &= a\,F\,(\underline{i} \times \underline{j}) + b\,F\,(\underline{k} \times \underline{j}) \\ &= a\,F\,\underline{k} - b\,F\,\underline{i} \\ &= (0.25)(200)\,\underline{k} - (0.30)(200)\,\underline{i} \\ &= 50\,\underline{k} - 60\,\underline{i}\end{aligned}$$

The negative sign in front of $60\,\underline{i}$ indicates that the x component of the moment vector is acting in the negative x direction. Furthermore, there is no component associated with the unit vector $\underline{j}$, which implies that the y component of the moment vector is zero. Therefore, the moment about point A has the following components:

$$\begin{aligned}M_x &= 60 \text{ N-m} \qquad (-x \text{ direction}) \\ M_y &= 0 \\ M_z &= 50 \text{ N-m} \qquad (+z \text{ direction})\end{aligned}$$

These results are consistent with those obtained using the scalar method.

Chapter 5

STATICS: SYSTEMS IN EQUILIBRIUM

5.1 Definitions

Statics is an area within the field of applied mechanics, which is concerned with forces on rigid bodies in equilibrium. A *rigid body* is one that is assumed to undergo no deformation under the effect of externally applied forces. In mechanics, the term *equilibrium* implies that the body of concern is either at rest or moving with a constant velocity. Newton's second law of motion sets the relationship between the applied forces and motion:

$$\underline{F} = m\,\underline{a} \tag{5.1}$$

In Eq. (5.1), m stands for the mass of the body, $\underline{a}$ is the acceleration vector, and $\underline{F}$ is the net or resultant force vector. If there is more than one force acting on the body, then $\underline{F}$ is equal to the vector sum of all forces.

If a body is in equilibrium, then the net force acting on it is zero. It follows from Eq. (5.1) that if $\underline{F}$=0 then the acceleration of the body is zero as well, because the mass of a body cannot be zero. *Acceleration* is defined as the time rate of change of velocity. For the acceleration vector to be equal to zero, the velocity vector must either be equal to zero or to a constant. If the velocity of a body is equal to zero, then the body is said to be in *static equilibrium.*

In this chapter, the principles of statics will be introduced. Applications of these principles to relatively simple systems will be provided. In Chapter 6, these principles will be applied to analyze forces involved in the human musculoskeletal system.

5.2 Conditions for Static Equilibrium

For a rigid body to be in static equilibrium, it has to be in both translational and rotational equilibrium.

5.2.1 Translational equilibrium

A body is said to be in *translational equilibrium* if the net force acting on it is equal to zero. Consider the arbitrarily shaped body shown in Figure 5.1. There are three forces acting on the body. The vector sum of these forces is equal to the net or the resultant force acting on the body:

$$\sum \underline{F} = \underline{F}_1 + \underline{F}_2 + \underline{F}_3 \tag{5.2}$$

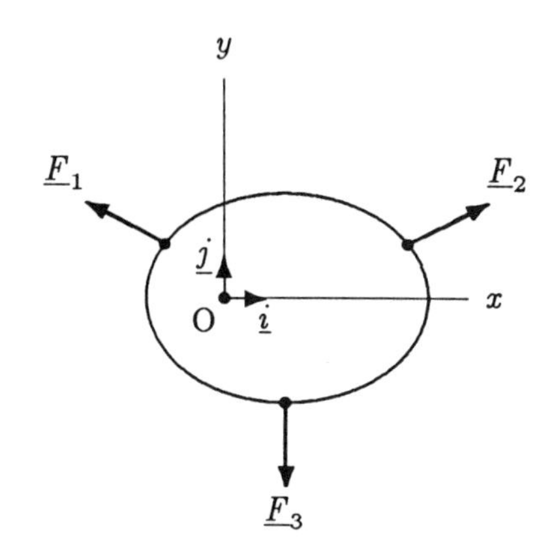

Figure 5.1 *For translational equilibrium, the net force on the body must be equal to zero.*

For the translational equilibrium of the body, the resultant force vector must be equal to zero:

$$\sum \underline{F} = 0 \tag{5.3}$$

Assuming that the forces form a coplanar force system in the xy plane, they can be represented in terms of their components along the x and y directions:

$$\begin{aligned} \underline{F}_1 &= F_{1x}\,\underline{i} + F_{1y}\,\underline{j} \\ \underline{F}_2 &= F_{2x}\,\underline{i} + F_{2y}\,\underline{j} \\ \underline{F}_3 &= F_{3x}\,\underline{i} + F_{3y}\,\underline{j} \end{aligned} \tag{5.4}$$

Eqs. (5.4) can be substituted into Eq. (5.2), and the x components of all forces and y components of all forces can be grouped together to write:

$$\sum \underline{F} = (F_{1x} + F_{2x} + F_{3x})\,\underline{i} + (F_{1y} + F_{2y} + F_{3y})\,\underline{j} \tag{5.5}$$

For Eq. (5.3) to be valid, both x and y components of the resultant force vector must be equal to zero. Therefore:

$$\begin{aligned} F_{1x} + F_{2x} + F_{3x} &= 0 \\ F_{1y} + F_{2y} + F_{3y} &= 0 \end{aligned} \tag{5.6}$$

These results can be generalized as:

$$\begin{aligned} \sum F_x &= 0 \\ \sum F_y &= 0 \end{aligned} \tag{5.7}$$

In other words, for a body to be in translational equilibrium, the sum of all forces acting in the x direction and the sum of all forces acting in the y direction must be equal to zero. This result is valid for coplanar force systems. For three-dimensional situations, this statement can be expanded such that the sum of all forces acting in the z direction must also be equal to zero ($\sum F_z = 0$).

Note that while applying the translational equilibrium condition, for example in the x direction, the forces acting in the positive x direction must be added together and the forces acting in the negative x direction must be subtracted.

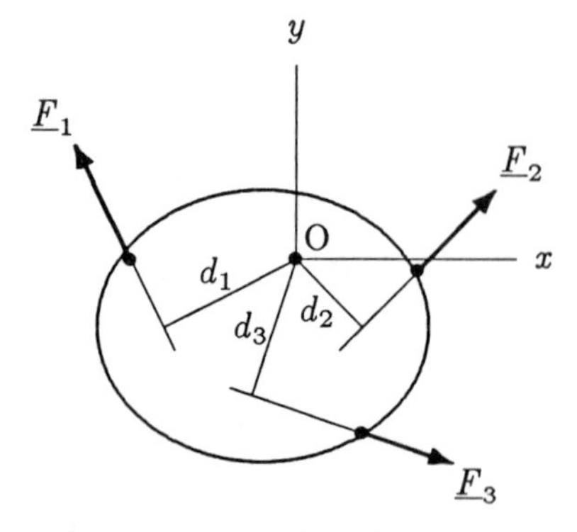

Figure 5.2 *For rotational equilibrium, the net moment about any point O must be zero.*

5.2.2 Rotational equilibrium

For a rigid body to be in *rotational equilibrium*, the net moment or torque about any point in the body due to the externally applied forces must be zero. Consider the system shown in Figure 5.2. If $\underline{M}_{O1}$, $\underline{M}_{O2}$, and $\underline{M}_{O3}$ correspond to moments about point O due to forces $\underline{F}_1$, $\underline{F}_2$, and $\underline{F}_3$, respectively, then the net moment about point O is:

$$\sum \underline{M}_O = \underline{M}_{O1} + \underline{M}_{O2} + \underline{M}_{O3} \tag{5.8}$$

For rotational equilibrium:

$$\sum \underline{M}_O = 0 \tag{5.9}$$

If the system of forces is coplanar in the xy plane, then the corresponding moments are acting in a direction perpendicular to the xy plane, in the z direction. For coplanar force systems, Eq. (5.9) can then be simplified by eliminating the vector sign:

$$\sum M_O = 0 \tag{5.10}$$

Some of the moments applied by external forces may act in the clockwise direction, while others act in the counterclockwise direction. For example, the moments of forces shown in Figure 5.2 about point O have the following magnitudes and directions:

$$\begin{aligned} M_{O1} &= d_1\, F_1 \qquad (\text{cw}) \\ M_{O2} &= d_2\, F_2 \qquad (\text{ccw}) \\ M_{O3} &= d_3\, F_3 \qquad (\text{ccw}) \end{aligned} \tag{5.11}$$

Depending on the choice of the positive direction as either clockwise or counterclockwise, some of the moments are positive and others are negative. If we consider the clockwise direction to be positive, then for the situation illustrated in Figure 5.2, the net moment about point O is:

$$\sum M_O = d_1\, F_1 - d_2\, F_2 - d_3\, F_3 \tag{5.12}$$

For rotational equilibrium, Eq. (5.12) must be equal to zero.

Eqs. (5.7) and (5.10) set the conditions for static equilibrium of a rigid body under the effect of a system of coplanar forces. In other words, if a rigid body is in static equilibrium, then the forces and moments acting on the body must satisfy Eqs. (5.7) and (5.10).

Note here that Eq. (5.9) is a special form of the equation of rotational motion:

$$\underline{M} = I\, \underline{\alpha} \tag{5.13}$$

where I is called the *mass moment of inertia*, $\underline{\alpha}$ is the *angular acceleration* vector, and $\underline{M}$ is the moment or torque vector. Eq. (5.13) is the rotational equivalent of Newton's second law of motion as given in Eq. (5.1). These concepts will be explained in detail in Part II of this text within the context of dynamic analysis.

5.3 Free-Body Diagrams

Free-body diagrams are constructed to help identify the forces and moments acting on individual parts of a system and to ensure the correct use of the equations of statics. For this purpose, the parts constituting a system are isolated from their surroundings, and the effects of the surroundings are replaced by corresponding forces. In a free-body diagram, all known and unknown forces and moments are shown. A force is unknown if its magnitude or direction is not known. For the known forces, we indicate the correct directions. If the direction of a force is not known, then we simply choose a direction for it. If this initial guess is not correct, it will appear in the solutions as a negative force.

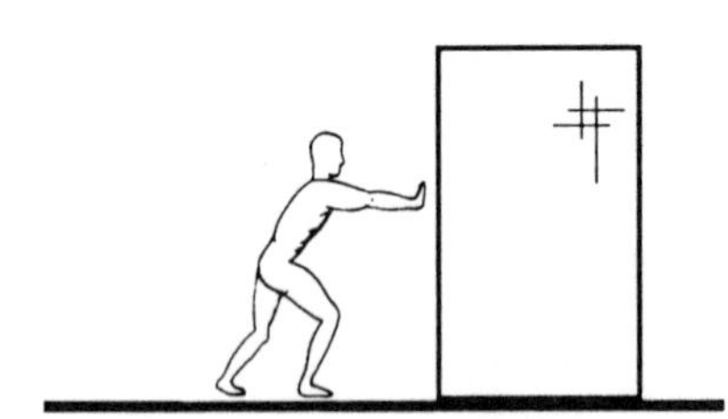

Figure 5.3 *A person pushing a block.*

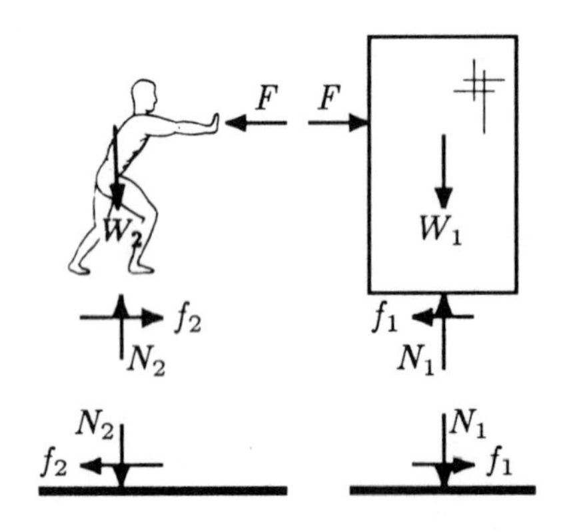

Figure 5.4 *Free-body diagrams.*

For example, consider a person trying to push a block to the right on a rough surface, as illustrated in Figure 5.3. There are three parts constituting this system: the person, the block, and the horizontal surface representing Earth. The free-body diagrams of these parts are shown in Figure 5.4. F is the magnitude of the horizontal force $\underline{F}$ applied by the person on the block to move the block to the right. Since action and reaction must have equal magnitudes (Newton's third law), F is also the magnitude of the force applied by the block on the person. Since action and reaction must have opposite directions, the force applied by the block on the person pushes the person to the left. W_1 and W_2 are the weights of the block and the person, respectively, and are always directed vertically downward. N_1 and N_2 are the magnitudes of forces on the horizontal surface applied by the block and the person, respectively. They are also the magnitudes of the forces applied by the horizontal surface on the block and the person. f_1 and f_2 are the magnitudes of the frictional forces $\underline{f}_1$ and $\underline{f}_2$ between the block, the person, and the horizontal surface, both acting in the horizontal direction parallel to the surface. Since the block tends to move towards the right, $\underline{f}_1$ on the block acts towards the left and $\underline{f}_2$ is the driving force for the person to push the block.

For many applications, it is sufficient to consider the free-body diagrams of only a few of the individual parts forming a system. In this example, the free-body diagram of the block provides sufficient detail.

5.4 Procedure to Analyze Systems in Equilibrium

The general procedure for analyzing the forces and moments acting on rigid bodies in equilibrium can be outlined as follows:

- Draw a simple, neat diagram of the system to be analyzed.

- Draw the free-body diagrams of the parts of the system. Show all the known and unknown forces on the free-body diagrams. Indicate the correct directions of the known forces. If the direction of a force is not known, simply choose a direction for it.

• Establish proper x and y directions for each free-body diagram, and resolve all external forces into their components along these mutually perpendicular directions. A good selection of the x and y directions can simplify the analyses considerably.

• For each free-body diagram, apply the necessary translational and rotational equilibrium conditions. For two-dimensional plane problems, the number of equations available is three:

$$\sum F_x = 0 \qquad \sum F_y = 0 \qquad \sum M_O = 0$$

Solve these equations simultaneously for the unknowns.

• Include the correct directions and the units of forces and moments in the solution.

5.5 Notes Concerning the Equilibrium Equations

• While applying the translational equilibrium conditions, pay special attention to the directions of forces. Let a force be positive if it is acting in the positive x (or y) direction, and let it be a negative force if it is acting in the negative x (or y) direction.

• While applying the rotational equilibrium condition, choose a proper point about which all moments are to be calculated. The choice of this point is arbitrary, but a good choice can simplify the computations.

• For the solution of a two-dimensional problem in statics, you may or may not need all three equations. For example, if there are no forces in the x direction, then the translational equilibrium condition in the x direction will be satisfied automatically.

• In some cases it may be more convenient to apply the rotational equilibrium condition more than once. For example, the rotational equilibrium condition may be applied twice by considering the moments of applied forces about two different points. In such cases, the third independent equation which might be used is the translational equilibrium condition either in the x or y direction. The rotational equilibrium condition may also be applied three times in a single problem. In this case, the moments about three points must be considered. However, the three points chosen must not form a straight line.

• If there are only two forces acting on a body, then the body is in equilibrium if the forces are collinear and have an equal magnitude but opposite directions. If there are three forces, then the body is in equilibrium if the forces form either a parallel or a concurrent force system.

• In general, the number of equations needed to solve a problem is equal to the number of unknowns. However, the number of unknowns can not exceed the total number of equations available. For two-dimensional problems, the maximum number of unknowns that can be computed by applying the equations of equilibrium is limited to three.

- Problems in statics require solving a number of equations simultaneously for the unknowns. The reader may find it helpful to study Appendix B, where the solutions of algebraic equations with one, two, or three unknowns are reviewed.

5.6 Constraints and Reactions

One way of classifying forces is by saying that they are either active or reactive. Active forces include applied loads and gravity forces. The reactive forces (and moments) are those supplied by the ground, by the supports such as rollers and knife-edges, and by the connecting members such as cables, pivots, and hinges. Knowing the common characteristics of these mechanical elements can facilitate drawing free-body diagrams. The properties of these elements will be reviewed within the context of solved example problems provided in the next section.

The reactive forces and moments applied on a body, which are simply known as *reactions*, act in such a way as to constrain the motion of the body. The reactions are usually unknown and must be determined by applying the equations of equilibrium.

5.7 Applications of Equilibrium Equations

The equations of equilibrium can be used to analyze a wide range of problems in mechanics. These equations can also be applied to study certain problems in biomechanics, as covered in the following chapter.

While solving problems in statics, it may be necessary to use other equations in addition to those derived from the conditions of static equilibrium. For example, if the mass of a body is known, then the weight of the body can be determined using $W=mg$. Another useful relationship is $f=\mu N$, which may be required to calculate the frictional forces acting on surfaces in contact. Also, the properties of the right-triangle are important.

Before we attempt to solve a problem, we must check whether certain assumptions could be made to simplify the problem. For example, if the problem states that a surface is "smooth," then frictional effects can be neglected. "Rough surface," on the other hand, implies that the frictional effects cannot be ignored. While considering the rope-pulley systems, we may ignore the weight of the pulley and assume that the pulley is frictionless, unless otherwise specified. If a body has a "uniform" geometry, then the weight of that body can be considered to be a concentrated force acting at the geometric center of the body.

5.8 Simply Supported Structures

Mechanical systems are composed of a number of elements linked together in various ways. "Beams" are the fundamental elements that form the building blocks in mechanics. A beam connected to the ground, or a number of beams connected together or to the ground by means of rollers, knife-edges, hinges, pivots, or cables, form a mechanical system. In this section, we shall analyze some of these cases.

Example 5.1 As illustrated in Figure 5.5, a 60 kg worker is standing at a distance d=3 m measured from the left end of a uniform, horizontal beam. The beam is resting on two frictionless knife-edge (wedge) supports. The length and weight of the beam are ℓ=5 m and W_1=900 N, respectively.

Calculate the forces applied on the supports at A and B.

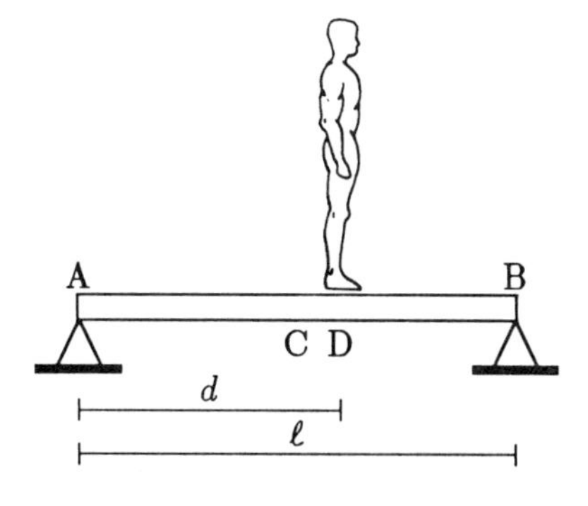

Figure 5.5 *A simply supported beam.*

Solution: The free-body diagram of the beam is shown in Figure 5.6. The forces applied on the beam form a parallel force system. W_1 and W_2 are the weights of the beam and the worker, respectively. The mass of the person is given as 60 kg. Therefore, $W_2=mg$=600 N. R_A and R_B are the magnitudes of unknown reaction forces applied on the beam by the supports. To a horizontal beam, frictionless knife-edge supports can only apply reaction forces in the vertical direction.

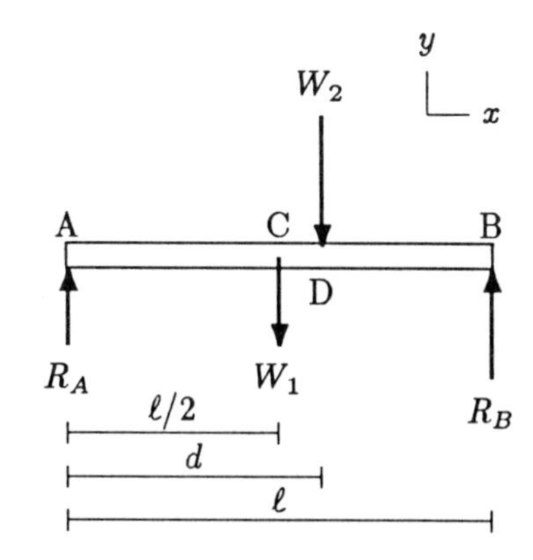

Figure 5.6 *Free-body diagram of the beam.*

There are two unknowns (R_A and R_B), and therefore we need two equations. For the solution of the problem, we can either apply the vertical equilibrium condition along with the rotational equilibrium condition about point A or B, or consider the rotational equilibrium of the beam about points A and B. Here, we shall use the latter approach, and check the results by considering the equilibrium in the vertical direction.

Assuming that the clockwise moments are positive, for the rotational equilibrium of the beam about point A:

$$\sum M_A = 0: \qquad \frac{\ell}{2} W_1 + d\,W_2 - \ell\,R_B = 0$$

$$R_B = \frac{1}{\ell}\left(\frac{\ell}{2} W_1 + d\,W_2\right)$$

$$= \frac{W_1}{2} + \frac{d}{\ell} W_2$$

$$= \frac{900}{2} + \frac{3}{5}\,600 = 810 \text{ N} \quad (\uparrow)$$

For the rotational equilibrium about point B (cw +):

$$\sum M_B = 0: \qquad \ell\, R_A - \frac{\ell}{2} W_1 - (\ell - d)\, W_2 = 0$$

$$R_A = \frac{1}{\ell}\left(\frac{\ell}{2} W_1 + (\ell - d)\, W_2\right)$$

$$= \frac{W_1}{2} + \left(1 - \frac{d}{\ell}\right) W_2$$

$$= \frac{900}{2} + \left(1 - \frac{3}{5}\right) 600 = 690 \text{ N} \quad (\uparrow)$$

Remarks:

• The arrows in parentheses show the correct directions of the forces. Since the results obtained are positive, the assumed directions for the reaction forces were correct.

• We can check these results by considering the equilibrium of the beam in the vertical (y) direction:

$$\sum F_y = 0: \qquad R_A - W_1 - W_2 + R_B \overset{?}{=} 0$$

$$690 - 900 - 600 + 810 \overset{\checkmark}{=} 0$$

• In the statement of the problem, we are asked to determine the forces applied by the beam on the supports. What we have determined are the forces applied by the supports on the beam. However, since action and reaction must be equal but opposite, the forces applied by the beam on the supports at A and B are R_A=690 N ($\downarrow$) and R_B=810 N ($\downarrow$), respectively.

• Because of the type of supports used, the beam analyzed in this example has a limited use. A frictionless knife-edge, a point support, a fulcrum, or a roller can provide a support only in the direction normal to the surfaces in contact. Such supports cannot provide reaction forces in the direction tangent to the contact surfaces. When there are forces applied on the beam with components in the long axis (x) of the beam (as in Example 5.2), these supports cannot provide the necessary horizontal reaction forces to maintain the horizontal equilibrium of the beam.

Example 5.2 The beam shown in Figure 5.7 is hinged to the ground at A and is supported by a frictionless roller at B. The distance between A and B is ℓ=8 m and the weight of the beam is W=1000 N. A force with magnitude P=2000 N, which makes an angle θ=45° with the horizontal, is applied on the beam at point D which is at a distance d=6 m from A.

Calculate the reaction forces at A and B of the beam.

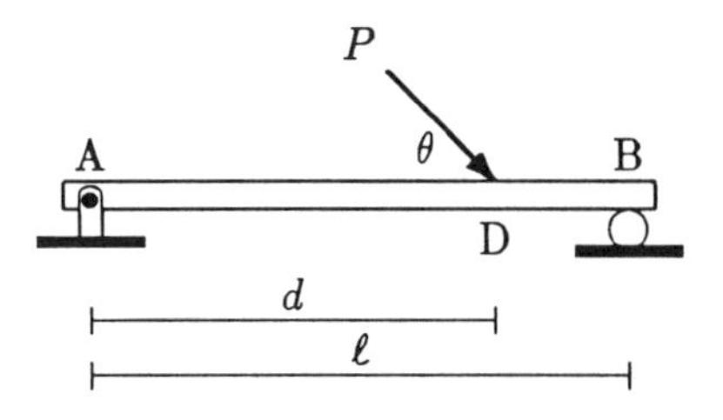

Figure 5.7 *A beam hinged at A and supported by a roller at B.*

Solution: Figure 5.8 illustrates the free-body diagram of the beam under consideration. The horizontal and vertical directions are designated by the x and y axes, respectively. P_x and P_y are the x and y components of the applied force such that:

$$P_x = P \cos\theta = (2000)(\cos 45°) = 1414 \text{ N} \quad (+x)$$
$$P_y = P \sin\theta = (2000)(\sin 45°) = 1414 \text{ N} \quad (-y)$$

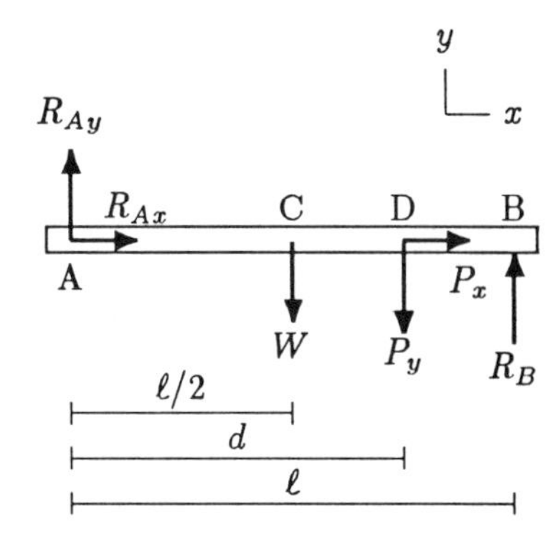

Figure 5.8 *Free-body diagram of the beam.*

The hinge joint at A constrains the beam both in the x and y directions, and permits no translational movement of the beam. In other words, there exists a reaction force at A that has components both in the x and y directions. Forces denoted by R_{Ax} and R_{Ay} in Figure 5.8 are the components of the reaction force applied by the ground on the beam through the hinge joint at A. At B, the beam is resting on a frictionless roller. A frictionless roller can only apply a force normal to the contact surfaces, in this case, in the vertical direction. If the frictional effects were also considered, then there would be an additional force tangent to the surfaces in contact. R_B is the reaction force applied by the ground on the beam through the frictionless roller.

There are three unknowns (R_{Ax}, R_{Ay}, R_B), and therefore we need three equations to solve this problem. Consider the equilibrium condition in the horizontal direction:

$$\sum F_x = 0: \qquad R_{Ax} + P_x = 0$$
$$R_{Ax} = -P_x = -1414$$

Note that R_{Ax} is determined to be a negative quantity. This is not permitted because R_{Ax} corresponds to a scalar quantity and scalar quantities cannot have negative values. The negative sign implies that while drawing the free-body diagram of the beam, we assumed the wrong direction ($+x$) for the horizontal component of the reaction force at A. Now we can correct the direction of force component $\underline{R}_{Ax}$ by writing:

$$R_{Ax} = 1414 \text{ N} \quad (-x)$$

The rest of the solution of this problem is similar to that carried out in Example 5.1. Consider the rotational equilibrium of the beam about point A (ccw +):

$$\sum M_A = 0: \qquad \ell\, R_B - \frac{\ell}{2} W - d\, P_y = 0$$

$$R_B = \frac{W}{2} + \frac{d}{\ell} P_y = 1560.5 \text{ N} \quad (+y)$$

Consider the rotational equilibrium of the beam about point B (cw +):

$$\sum M_B = 0: \qquad \ell\, R_{Ay} - \frac{\ell}{2} W - (\ell - d)\, P_y = 0$$

$$R_{Ay} = \frac{W}{2} + \left(1 - \frac{d}{\ell}\right) P_y = 853.5 \text{ N} \quad (+y)$$

Check the results by considering the vertical equilibrium of the beam:

$$\sum F_y = 0: \qquad R_{Ay} - W - P_y + R_B \stackrel{?}{=} 0$$

$$853.5 - 1000 - 1414 + 1560.5 \stackrel{\checkmark}{=} 0$$

The reaction forces acting on the beam at point A, where the beam is hinged to the ground, are shown in Figure 5.9. Note that R_{Ax} and R_{Ay} are the horizontal and vertical components of the reaction force $\underline{R}_A$ at A, whose magnitude is:

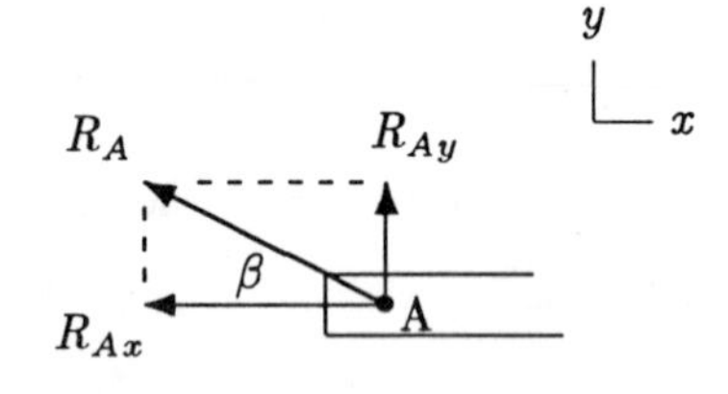

Figure 5.9 *Reaction force* $\underline{R}$ *at the hinge at A.*

$$R_A = \sqrt{(R_{Ax})^2 + (R_{Ay})^2} = 1651.6 \text{ N}$$

If β is the angle $\underline{R}_A$ makes with the horizontal, as shown in Figure 5.9, then:

$$\beta = \tan^{-1}\left(\frac{R_{Ay}}{R_{Ax}}\right) = 31^\circ$$

5.9 Structures Supported by Cables

Cables, wires, and ropes are used to connect various mechanical members together or to the ground, and they transmit forces from one member to the other. Cables, wires, and ropes are flexible members and can sustain only tensile forces, forces which tend to stretch them. They collapse under compressive forces. The tension in these members is uniform (constant) throughout the member. They are called *two-force members* because their free-body diagrams can have only two forces. Forces acting on two-force members must be collinear for the equilibrium of the member. Otherwise, there will be a net moment acting on the member, which will cause the member to rotate.

Example 5.3 Figure 5.10 illustrates a store sign. A uniform, horizontal beam which weighs W_1=100 N is hinged to the wall at point A. The other end (B) of the beam is supported by a cable which makes an angle θ=53° with the horizontal. The distance between points A and B is ℓ=1.5 m. The sign weighs W_2=20 N and is connected by two short, vertical cables to the beam at points D and E which are d=1 m apart.

Calculate the tensions in the cables, and the force at the hinge.

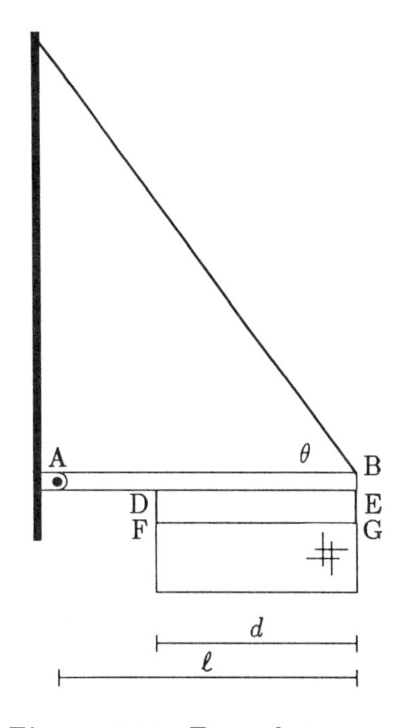

Figure 5.10 *Example 5.3.*

Solution: For the solution of this problem, we need to consider the free-body diagrams of both the beam and the sign, which are shown in Figure 5.11. In Figure 5.11(b), W_2 is the weight of the sign; T_D and T_E are the tensions in the short cables. Consider the rotational equilibrium of the sign about point F (ccw +):

$$\sum M_F = 0: \qquad d\,T_G - \frac{d}{2}W_2 = 0$$

$$T_G = \frac{W_2}{2} = \frac{20}{2} = 10 \text{ N} \quad (+y)$$

Consider the vertical equilibrium of the sign:

$$\sum F_y = 0: \qquad T_F - W_2 + T_G = 0$$

$$T_F = W_2 - T_G = 20 - 10 = 10 \text{ N} \quad (+y)$$

T_G and T_F are the forces applied by the short cables on the sign, and they are also the tensions in the cables, which are uniform throughout the cables. The forces applied by the cables to the horizontal beam are also equal to the tensions in the cables. Therefore:

$$T_D = T_F = 10 \text{ N} \quad (-y)$$

$$T_E = T_G = 10 \text{ N} \quad (-y)$$

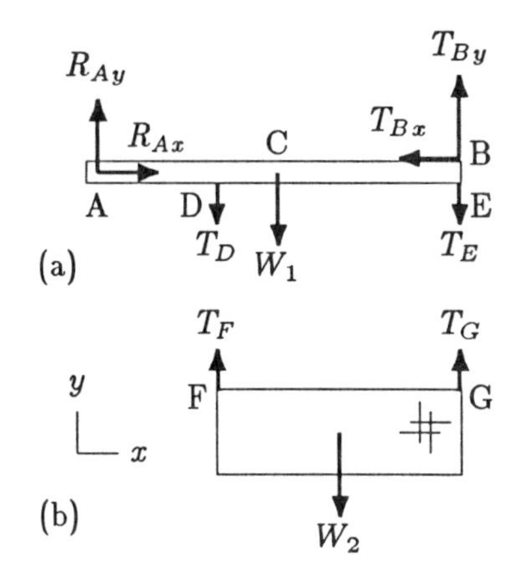

Figure 5.11 *Free-body diagrams of the beam and the sign.*

Consider the free-body diagram of the beam shown in Figure 5.11(a). W_1 is the weight of the beam, R_{Ax} and R_{Ay} are the components of the reaction force at A, and T_{Bx} and T_{By} are the components of the force $\underline{T}_B$ applied by the long cable on the beam. $\underline{T}_B$ is also the tensile force in the long cable that makes an angle $\theta = 53°$ with the horizontal. Therefore:

$$T_{Bx} = T_B \cos\theta \quad (-x) \qquad (i)$$

$$T_{By} = T_B \sin\theta \quad (+y) \qquad (ii)$$

We have a total of three unknowns: T_B, R_{Ax}, and R_{Ay}.

Consider the rotational equilibrium of the beam about point A (ccw +):

$$\sum M_A = 0: \; \ell T_{By} - (\ell - d)\, T_D - \frac{\ell}{2} W_1 - \ell T_E = 0$$

$$T_{By} = \left(1 - \frac{d}{\ell}\right) T_D + \frac{W_1}{2} + T_E = 63.3 \text{ N} \quad (+y)$$

Knowing T_{By}, we can determine T_B and T_{Bx} from (ii) and (i):

$$T_B = \frac{T_{By}}{\sin\theta} = \frac{63.3}{\sin 53} = 79.3 \text{ N} \quad (\nwarrow)$$

$$T_{Bx} = T_B \cos\theta = (79.3)(\cos 53) = 47.7 \text{ N} \quad (-x)$$

Consider the equilibrium in the x direction:

$$\sum F_x = 0: \quad R_{Ax} - T_{Bx} = 0$$

$$R_{Ax} = T_{Bx} = 47.7 \text{ N} \quad (+x)$$

Consider the equilibrium in the y direction:

$$\sum F_y = 0: \quad R_{Ay} - T_D - W_1 - T_E + T_{By} = 0$$

$$R_{Ay} = T_D + W_1 + T_E - T_{By} = 56.7 \text{ N} \quad (+y)$$

Note that the results can be checked by considering the rotational equilibrium condition of the beam about point B:

$$\sum M_B = 0: \qquad \ell R_{Ay} - d\, T_D - \frac{\ell}{2} W_1 \stackrel{?}{=} 0$$

$$(1.5)(56.7) - (1)(10) - (0.75)(100) \stackrel{\checkmark}{=} 0$$

5.10 Rope-Pulley Systems

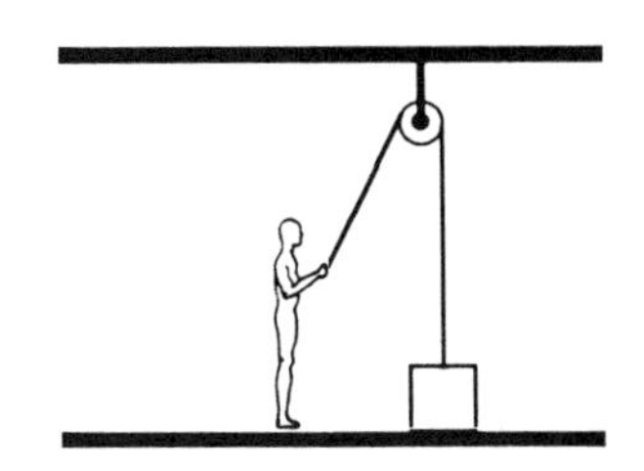

Figure 5.12 *A rope-pulley arrangement.*

Rope-pulley arrangements, such as the one shown in Figure 5.12, are commonly used to elevate weights and have practical applications in the design of traction devices used in patient rehabilitation. Figure 5.13(a) shows the free-body diagram of the pulley. r is the radius of the pulley, and O refers to its center (axle or shaft). R_x and R_y are the reactions at O. The rope is wrapped around the pulley between points A and B. T_A and T_B refer to the tensions in the rope at points A and B, respectively. Regardless of the way a rope is wrapped around a pulley, the rope would be tangent to the circumference of the pulley at the initial and final contact points A and B. This implies that straight lines drawn from O towards points A and B would be perpendicular to the rope at A and B. Hence, as measured from the center of the pulley, the moment arms of the tensile forces applied by the rope on the pulley are always equal to the radius of the pulley. Considering the moments of forces on the pulley about O:

$$\sum M_O = 0: \qquad r\,T_B - r\,T_A = 0$$

$$T_B = T_A$$

Therefore, the tension in the rope is the same on the two sides of the pulley. (This conclusion is valid under the assumption that the pulley is frictionless.)

The reaction forces R_x and R_y applied by the ground (through the ceiling) on the pulley at the axle of the pulley do not produce any moment about the center of the pulley. If needed, these forces can be determined in terms of the tension in the rope by considering the horizontal and vertical equilibrium of the pulley. Assuming that the weight of the pulley is negligible:

$$\sum F_x = 0: \qquad R_x = T\,\sin\theta$$

$$\sum F_y = 0: \qquad R_y = T + T\,\cos\theta$$

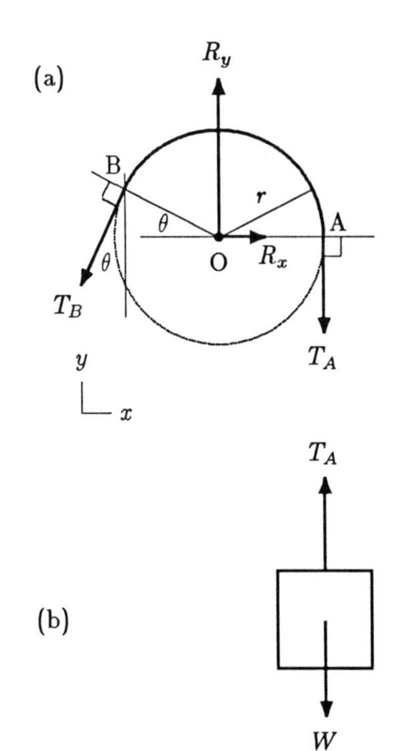

Figure 5.13 *Free-body diagrams of the pulley and the block.*

Here, T=T_A=T_B is the tension in the rope over the pulley. Note that the way the rope is wrapped around the pulley influences the magnitudes of these reaction forces through angle θ.

Figure 5.13(b) shows the free-body diagram of the block suspended by the rope. If W is the weight of the block, then for the vertical equilibrium of the block:

$$\sum F_y = 0: \qquad T_A = W$$

Therefore, the tension in the rope is equal to the weight of the block.

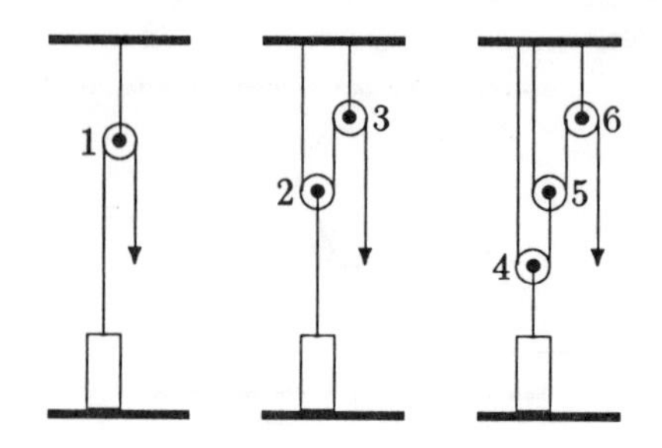

Figure 5.14 *Various rope-pulley arrangements.*

Example 5.4 A block of weight W has to be elevated to a certain height. Three people have proposed three different rope-pulley arrangements, each of which is believed to make the task easier. The first person suggested a single pulley system, the second person proposed a two pulley system, and the third person insisted on a three pulley system shown in Figure 5.14.

Determine how much force is required to initiate the lifting task by each proposed system.

Solutions:

System 1. We have already analyzed the single pulley system. For the vertical equilibrium of the block (Figure 5.15a):

$$T_1 = W$$

To be able to lift the block, the first person must apply a force on the rope at a magnitude equal to the weight of the block.

System 2. The analysis of multiple pulley systems should start at the pulley where the known force (in this case the weight of the block) is applied. For the vertical equilibrium of pulley 2 (Figure 5.15b):

$$T_2 = \frac{W}{2}$$

T_2 is the tension in the continuous rope around pulleys 2 and 3, and is held by the second person. To lift the block, the second person must apply half the force of the first person.

System 3. For the vertical equilibrium of pulley 4 (Figure 5.15c):

$$T_4 = \frac{W}{2}$$

This is the tension in the rope around pulley 4, which is connected to the axle of pulley 5. For the vertical equilibrium of pulley 5:

$$T_5 = \frac{T_4}{2} = \frac{W}{4}$$

T_5 is the tension in the rope held by the third person.

These results suggest that the best design is that proposed by the third person.

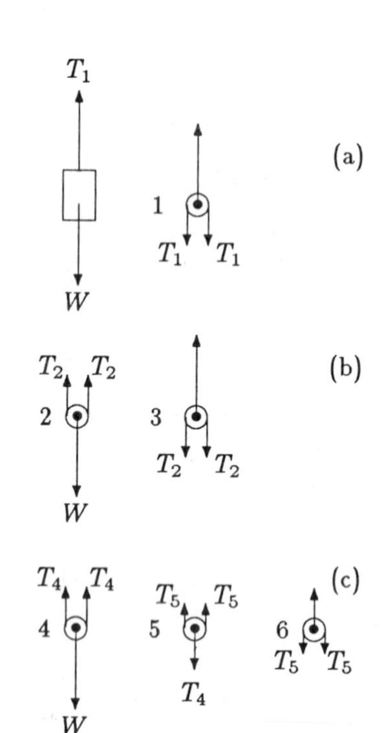

Figure 5.15 *Free-body diagrams.*

5.11 Traction Devices

Traction devices are designed to maintain parts of the human body in particular positions for healing purposes, and they provide typical applications of rope-pulley arrangements. The single pulley traction shown in Figure 5.16 pulls the leg towards the right by applying a horizontal force on the leg. The magnitude of this force is equal to the tension in the cable. For the vertical equilibrium of the weight pan, the tension in the cable must be equal to the weight in the weight pan. Therefore, this simple traction device pulls the leg with a force whose magnitude is equal to the weight in the weight pan. The three pulley traction device shown in Figure 5.17 exerts a horizontal force whose magnitude is twice that of the traction weight suspended on the cable. This result can be obtained by analyzing the forces applied on the middle pulley.

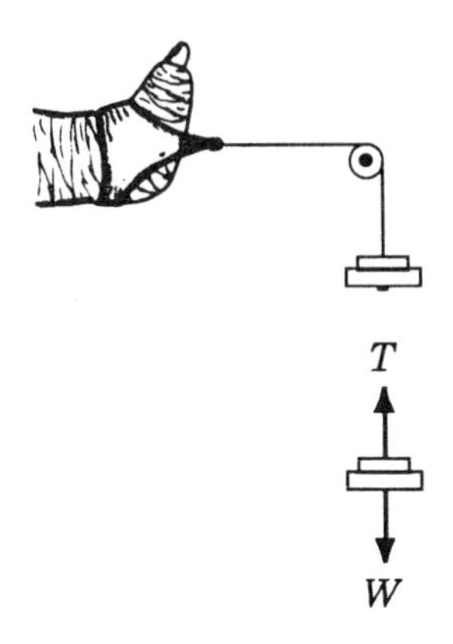

Figure 5.16 *A single pulley leg traction device.*

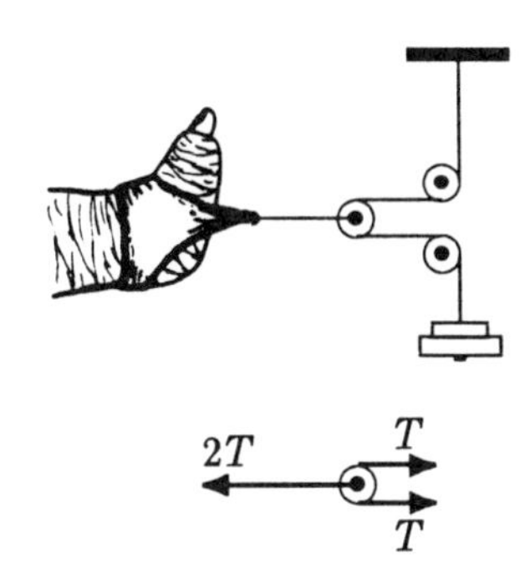

Figure 5.17 *A three pulley leg traction device.*

Example 5.5 Consider the split Russel traction device and a mechanical model of the leg shown in Figure 5.18. The leg is held in the position shown by two weights which are connected to the leg via two cables. The combined weight of the leg and the cast is W=300 N. ℓ is the horizontal distance between points A and B where the cables are attached to the leg. C is the center of gravity of the leg which is located at a distance two-thirds of ℓ as measured from A. The angle cable 2 makes with the horizontal is measured as β=45°.

Determine the tensions T_1 and T_2 in the cables, weights W_1 and W_2, and the angle α that cable 1 makes with the horizontal, so that the leg remains in equilibrium at the position shown.

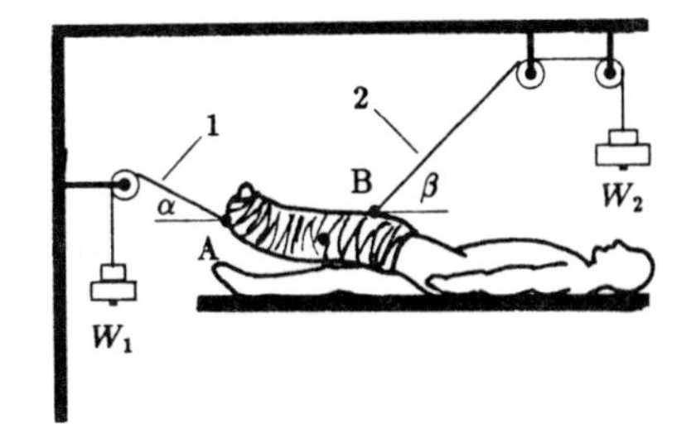

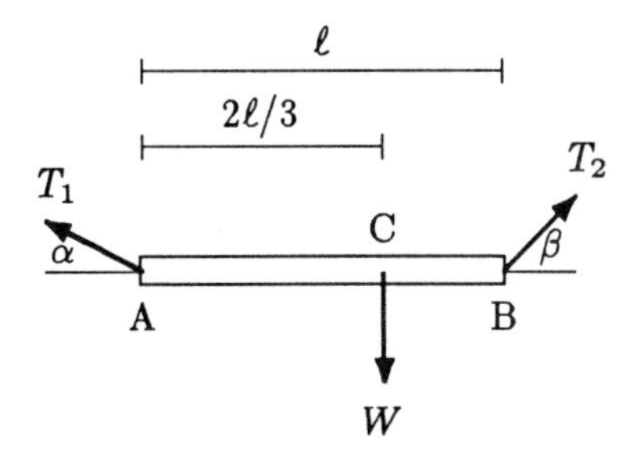

Figure 5.18 *Traction and a mechanical model of the leg.*

Solution: The free-body diagram of the mechanical model of the leg is shown in Figure 5.19(a). The components of T_1 and T_2 in the x and y directions can be expressed as follows:

$$T_{1x} = T_1 \cos\alpha \qquad (i)$$
$$T_{1y} = T_1 \sin\alpha \qquad (ii)$$
$$T_{2x} = T_2 \cos\beta \qquad (iii)$$
$$T_{2y} = T_2 \sin\beta \qquad (iv)$$

Consider the conditions for static equilibrium of the leg. For the rotational equilibrium of the leg about A:

$$\sum M_A = 0: \quad \ell T_{2y} - \frac{2\ell}{3} W = 0$$

$$T_{2y} = \frac{2}{3} W = \frac{2}{3} 300 = 200 \text{ N} \quad (\uparrow)$$

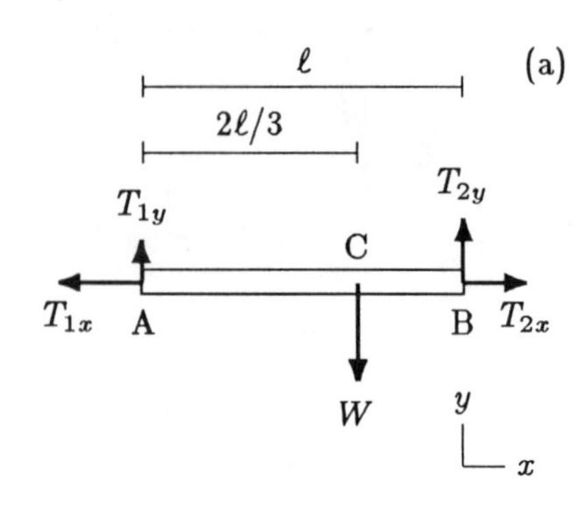

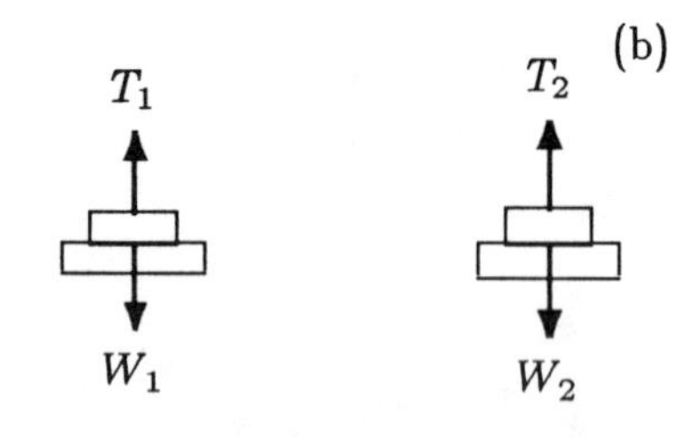

Figure 5.19 *Free-body diagrams.*

Using (*iv*):

$$T_2 = \frac{T_{2y}}{\sin\beta} = \frac{200}{\sin 45°} = 282.8 \text{ N} \quad (\nearrow)$$

Using (*iii*):

$$T_{2x} = T_2 \cos\beta = (282.8)(\cos 45°) = 200 \text{ N} \quad (\rightarrow)$$

For the horizontal equilibrium of the leg:

$$\sum F_x = 0: \quad T_{2x} - T_{1x} = 0$$

$$T_{1x} = T_{2x} = 200 \text{ N} \quad (\leftarrow)$$

For the vertical equilibrium of the leg:

$$\sum F_y = 0: \quad T_{1y} - W + T_{2y} = 0$$

$$T_{1y} = W - T_{2y} = 100 \text{ N} \quad (\uparrow)$$

The magnitude of the tensile force in cable 1:

$$T_1 = \sqrt{(T_{1x})^2 + (T_{1y})^2} = \sqrt{(200)^2 + (100)^2} = 223.6 \text{ N}$$

The angle $\underline{T}_1$ makes with the horizontal:

$$\alpha = \tan^{-1}\left(\frac{T_{1y}}{T_{1x}}\right) = \tan^{-1}\left(\frac{100}{200}\right) = 26.6°$$

Finally, to determine the weights of the blocks required to keep the traction in equilibrium, consider the vertical equilibrium of the blocks (Figure 5.19b):

$$W_1 = T_1 = 223.6 \text{ N}$$

$$W_2 = T_2 = 282.8 \text{ N}$$

Example 5.6 Consider the traction device shown in Figure 5.20. A weight pan is suspended on a long cable that passes over four pulleys. This cable is attached to the leg at B where the center of gravity of the leg is located. A second, relatively short cable is connected to the shaft of one of the pulleys and to the leg at point A, such that point A and the center of the pulley lie on a horizontal line. The distance between A and the center of gravity of the leg is ℓ=50 cm. The total weight of the leg and the cast is W_1=200 N, and angle α=75.5°.

Determine the tensions in the cables, angle β, and weight W_2 of the block required to maintain the leg in the position shown.

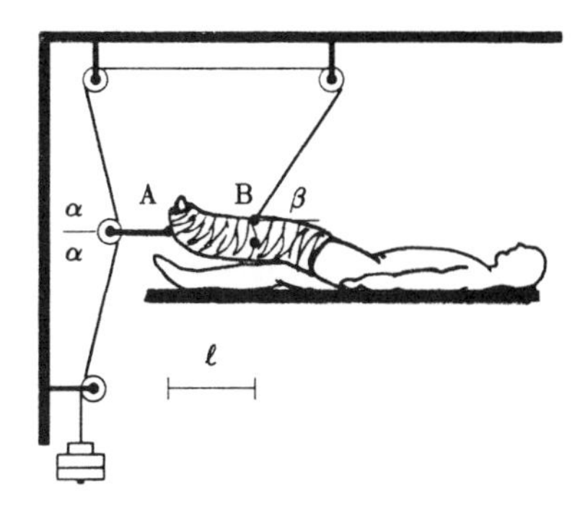

Figure 5.20 *Example 5.6.*

Solution: The tension T_1 is uniform throughout the long cable. It is equal to the magnitude of the force applied by the cable on the leg at point B, and it is also the force applied by the weight pan on the cable.

First consider the free-body diagram of the leg, shown in Figure 5.21(a). For the horizontal equilibrium:

$$\sum F_x = 0: \qquad T_{1x} - T_2 = 0$$
$$T_1 \cos\beta = T_2 \qquad (i)$$

For the vertical equilibrium:

$$\sum F_y = 0: \qquad T_{1y} - W_1 = 0$$
$$T_1 \sin\beta = W_1 \qquad (ii)$$

Now consider the forces acting on the pulley, closest to A. The free-body diagram of this pulley is shown in Figure 5.21(b). For the horizontal equilibrium:

$$\sum F_x = 0: \qquad T_2 - T_{1x} - T_{1x} = 0$$
$$T_2 = 2\,T_{1x} = 2\,T_1 \cos\alpha \qquad (iii)$$

Note that the vertical equilibrium condition for the pulley is automatically satisfied since we already assumed that the tension in the cable is uniform.

Next, consider the equilibrium of the weight pan (Figure 5.21c):

$$\sum F_y = 0: \qquad T_1 = W_2 \qquad (iv)$$

Substitute (iii) into (i) so as to eliminate T_2:

$$T_1 \cos\beta = 2\,T_1 \cos\alpha$$
$$\cos\beta = 2\cos\alpha = 2\,(\cos 75.5°) = 0.5$$
$$\beta = \cos^{-1}(0.5) = 60°$$

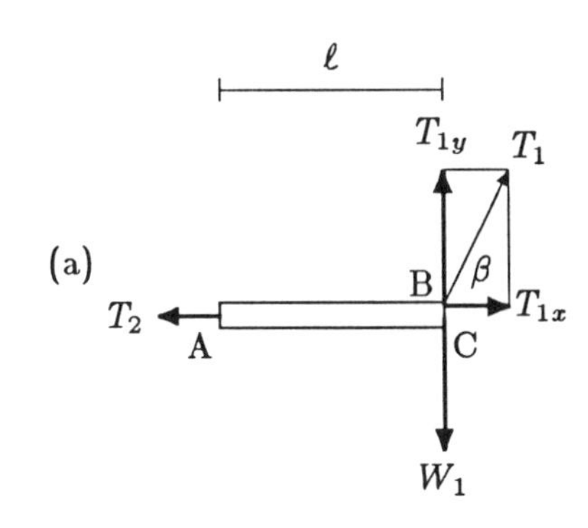

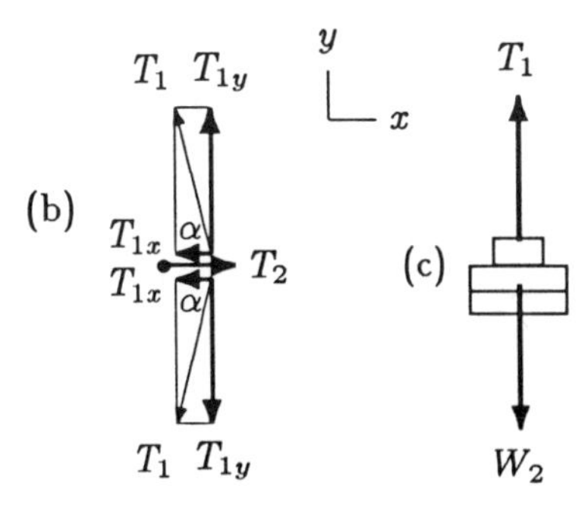

Figure 5.21 *Free-body diagrams.*

From (*ii*):

$$T_1 = \frac{W_1}{\sin\beta} = \frac{200}{\sin 60} = 230.9 \text{ N}$$

From (*iii*):

$$T_2 = 2\,T_1 \cos\alpha = 2(230.9)(\cos 75.5^\circ) = 115.6 \text{ N}$$

From (*iv*):

$$W_2 = T_1 = 230.9 \text{ N}$$

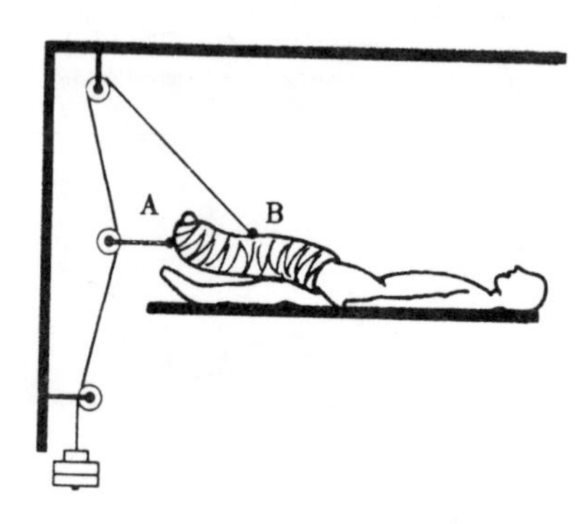

Figure 5.22 *A leg traction.*

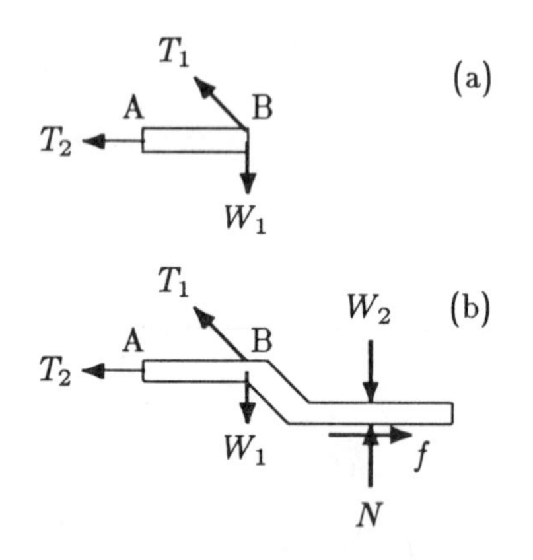

Figure 5.23 *Free-body diagrams.*

A remark: Consider the traction shown in Figure 5.22. An incorrect (incomplete) free-body diagram of the leg is illustrated in Figure 5.23(a). It is not complete because under the forces shown, the leg cannot be in equilibrium.

T_1 and T_2 are the tensions applied on the leg. $\underline{T}_2$ is acting towards the left, and $\underline{T}_1$ has components in the horizontal and vertical directions. The horizontal component of $\underline{T}_1$ is acting towards the left as well. Therefore, according to the free-body diagram in Figure 5.23(a), there is no force on the leg to balance the added effect of T_2 and T_{1x}, and therefore the leg is not under translational equilibrium in the horizontal direction. To be able to see the complete picture of what is happening, we must look at the free-body diagram of patient's body as a whole (Figure 5.23b). The patient can maintain his/her position in bed with the help of a frictional force generated between his/her back and the bed. However, if the concern is patient comfort, then the frictional forces exerted on the patient's body must be minimized. This can be achieved by rearranging the traction so as to counterbalance the applied forces, as in the previous examples.

5.12 Systems Involving Friction

Frictional forces are discussed in detail in Chapter 3.16. Here, we shall analyze a problem for which frictional forces play an important role.

Example 5.7 Figure 5.24 illustrates a person trying to push a block on an inclined surface by applying a force parallel to the incline. The weight of the block is W, the coefficient for maximum friction between the block and the incline is μ, and the incline makes an angle θ with the horizontal.

Determine the magnitude of the force (in terms of W, μ, and θ) that the person must apply in order to start moving the block.

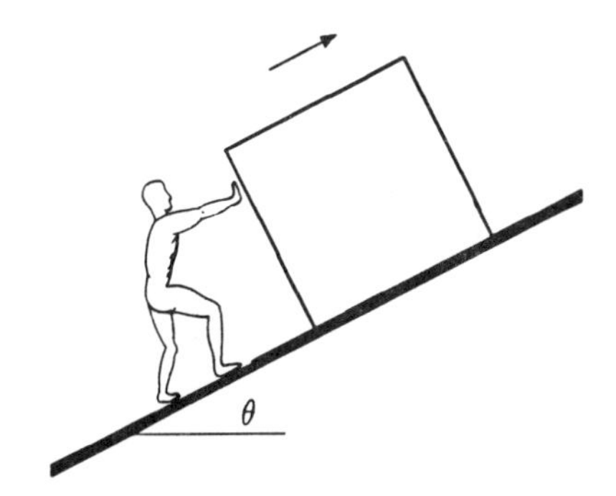

Figure 5.24 *Example 5.7.*

Solution: The free-body diagram of the block is shown in Figure 5.25. x and y correspond to the directions parallel and perpendicular to the incline, respectively. W is the weight of the block and has components along the x and y directions:

$$W_x = W \sin\theta$$

$$W_y = W \cos\theta$$

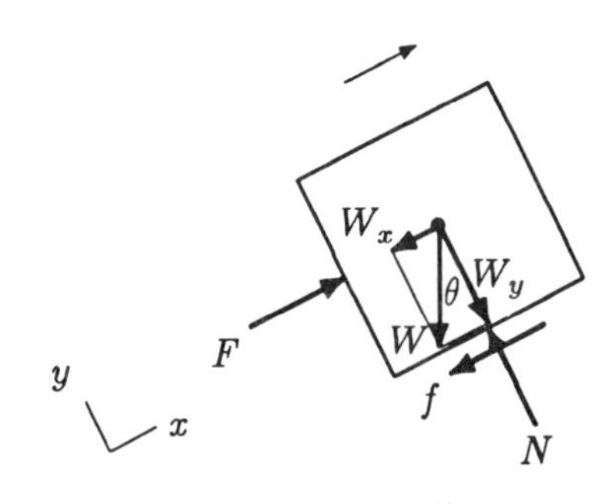

Figure 5.25 *Free-body diagram of the block.*

F is the magnitude of the force that the person must apply on the block to start moving the block in the positive x direction. f is the magnitude of the frictional force developed between the bottom surface of the block and the inclined surface. Since the intended movement of the block is in the positive x direction, the frictional force on the block is acting in the negative x direction trying to stop the block from moving. N is the magnitude of the normal force applied by the inclined surface on the block, whose line of action is perpendicular to the surfaces in contact (i.e., in the y direction). By definition, the magnitude of the frictional force is:

$$f = \mu N$$

Note that if the person pushes the block by applying a force closer to the top of the block, then he/she can cause the block to tilt (rotate in the clockwise direction) about its bottom right corner. Here, we are assuming that there is no such effect of the applied force but only sliding of the block on the inclined surface.

The unknowns in this example are F and N. Since we have an expression relating f and N, if N is known so is f. For the solution of the problem, we must consider equilibrium equations in the y and x directions:

$$\sum F_y = 0: \qquad N = W_y = W \cos\theta$$

$$\sum F_x = 0: \qquad F = f + W_x$$

$$F = \mu N + W \sin\theta$$

$$= \mu W \cos\theta + W \sin\theta$$

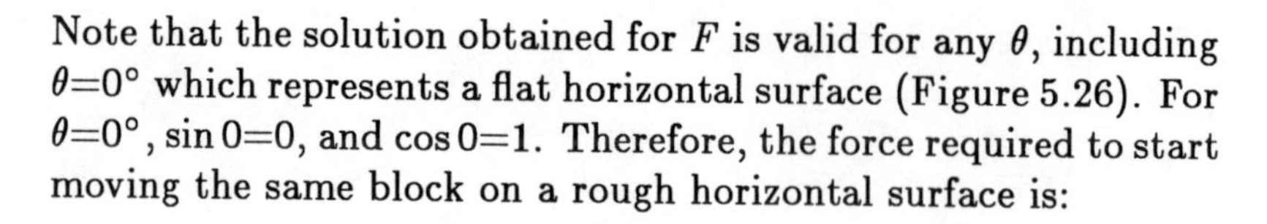

Note that the solution obtained for F is valid for any θ, including θ=0° which represents a flat horizontal surface (Figure 5.26). For θ=0°, sin 0=0, and cos 0=1. Therefore, the force required to start moving the same block on a rough horizontal surface is:

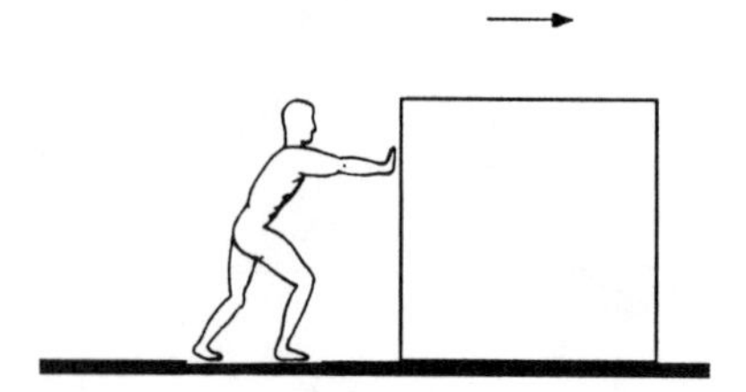

Figure 5.26 *Pushing a block on a horizontal surface.*

$$F = f = \mu W$$

Numerical example: Let W=1000 N, μ=0.3, and θ=15°. Then:

$$F = (0.3)(1000)(\cos 15°) + (1000)(\sin 15°) = 548.6 \text{ N} \quad (+x)$$

Therefore, to be able to start moving a 1000 N block up the 15° incline which has a surface friction coefficient of 0.3, the person must apply a force greater than 548.6 N in a direction parallel to the incline.

To start moving the same block on a horizontal surface with the same friction coefficient, the person must apply a horizontal force of:

$$F = (0.3)(1000) = 300 \text{ N}$$

As compared to a horizontal surface, the person must apply about 83% more force on the block to start moving the block on the 15° incline.

$$\left(\frac{548.6 - 300}{300} \times 100 = 82.9\right)$$

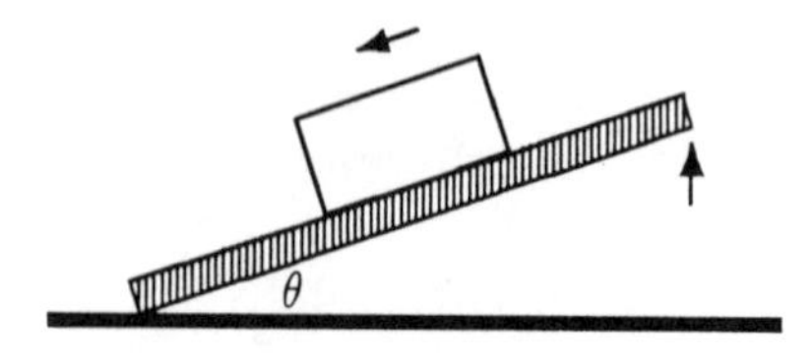

Figure 5.27 *Experimental determination of friction coefficient.*

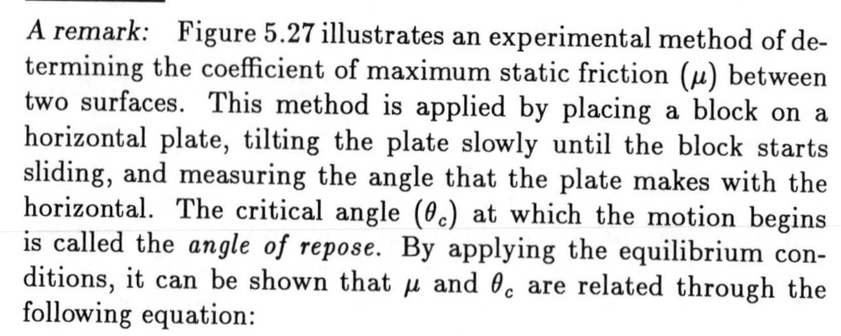

A remark: Figure 5.27 illustrates an experimental method of determining the coefficient of maximum static friction (μ) between two surfaces. This method is applied by placing a block on a horizontal plate, tilting the plate slowly until the block starts sliding, and measuring the angle that the plate makes with the horizontal. The critical angle (θ_c) at which the motion begins is called the ***angle of repose***. By applying the equilibrium conditions, it can be shown that μ and θ_c are related through the following equation:

$$\mu = \tan(\theta_c)$$

5.13 Built-in Structures

The beam shown in Figure 5.28 is built-in (fixed) to a wall and called a *cantilever beam.* Cantilever beams can withstand externally applied moments as well as forces.

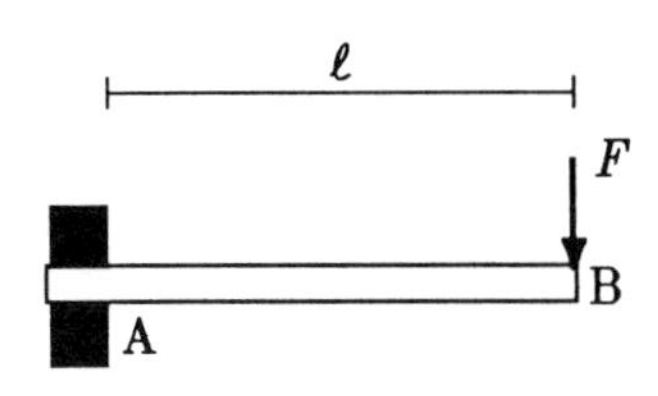

Figure 5.28 *Cantilever beam.*

For the sake of illustration, assume that the beam rests in a hole slightly larger than the beam itself. Due to the vertically applied force $\underline{F}$ (and the weight of the beam), the beam will tilt slightly and only two points at the built-in end of the beam will touch the wall, as illustrated in Figure 5.29(a). Let A and D refer to these points of contact which are d distance apart from each other. If R_A and R_D refer to the reaction forces applied by the wall on the beam at these contact points (Figure 5.29b), then the rotational equilibrium conditions about points A and D will yield:

$$\sum M_A = 0: \qquad R_D = \frac{\ell}{d} F$$

$$\sum M_D = 0: \qquad R_A = \frac{\ell}{d} F + F$$

Note that ℓ is the distance between points A and B, and that we ignored the weight of the beam for practical reasons. By comparing the expressions obtained for R_A and R_D, we can express R_A in terms of R_D and F:

$$R_A = R_D + F$$

Therefore, we can assume that R_A is the sum of two collinear forces: R_D and F, as illustrated in Figure 5.29(c). Note that R_D at D and R_D at A form a couple, and they can be replaced by a couple moment such that:

$$M = d\,R_D = d\left(\frac{\ell}{d} F\right) = \ell\, F \quad \text{(ccw)}$$

This moment acts everywhere along the beam, and the free-body diagram of the beam is shown in Figure 5.29(d).

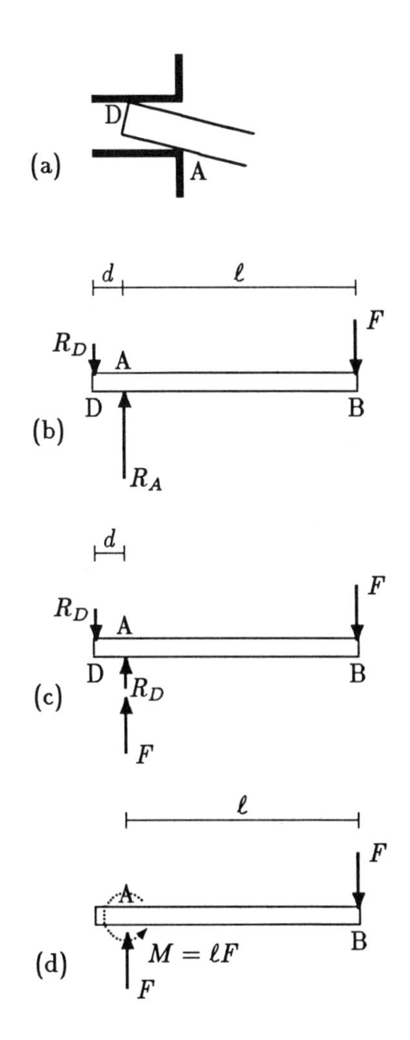

Figure 5.29 *Built-in beams can withstand applied moments as well as forces.*

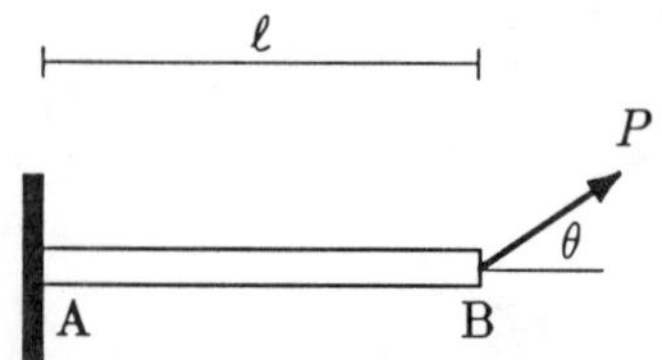

Figure 5.30 *Example 5.8.*

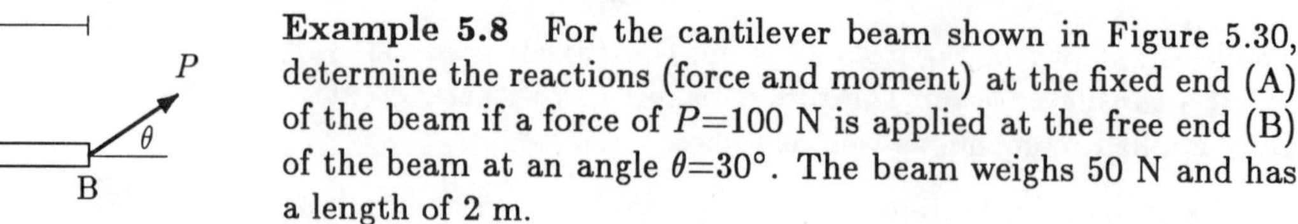

Example 5.8 For the cantilever beam shown in Figure 5.30, determine the reactions (force and moment) at the fixed end (A) of the beam if a force of P=100 N is applied at the free end (B) of the beam at an angle θ=30°. The beam weighs 50 N and has a length of 2 m.

Solution: The free-body diagram of the beam is shown in Figure 5.31. P_x and P_y are the x and y components of the applied force:

$$P_x = P\ \cos\theta = (100)(\cos 30°) = 86.6\ \text{N} \quad (+x)$$
$$P_y = P\ \sin\theta = (100)(\sin 30°) = 50\ \text{N} \quad (+y)$$

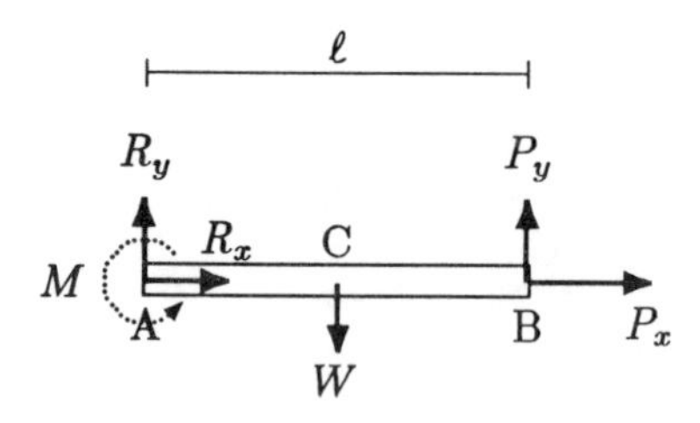

Figure 5.31 *Free-body diagram of the beam.*

R_x, R_y, and M are the reactions to be determined. To solve the problem, we need three equations because there are three unknowns. For equilibrium in the x direction:

$$\sum F_x = 0: \quad R_x + P_x = 0$$
$$R_x = -P_x = -86.6$$

The negative sign in front of 86.6 implies the wrong choice of direction:

$$R_x = 86.6\ \text{N} \quad (-x)$$

For the equilibrium in the y direction:

$$\sum F_y = 0: \quad R_y - W + P_y = 0$$
$$R_y = W - P_y = 50 - 50 = 0$$

In other words, there is no reaction at A in the vertical direction. Finally, for the rotational equilibrium of the beam about point A (ccw +):

$$\sum M_A = 0: \quad M - \frac{\ell}{2} W + \ell\, P_y = 0$$
$$M = \frac{\ell}{2} W - \ell\, P_y = \frac{2}{2}\, 50 - (2)(50) = -50$$

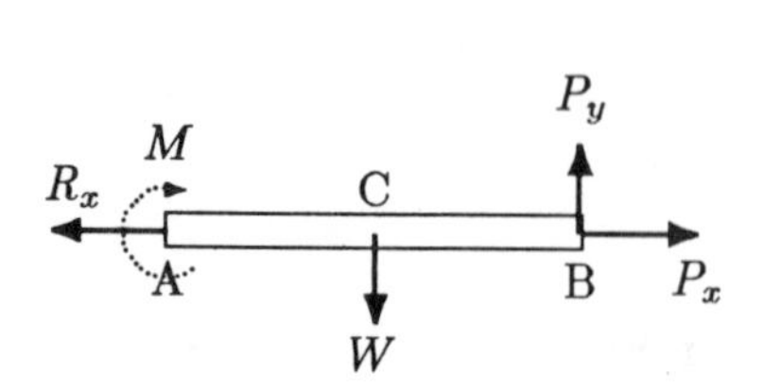

Figure 5.32 *Corrected free-body diagram of the beam.*

Again we have to correct the direction:

$$M = 50\ \text{N-m} \quad (\text{cw})$$

In the light of the solutions obtained, the free-body diagram of the beam is redrawn in Figure 5.32.

5.14 Center of Gravity Determinations

As discussed in Chapter 3.7, every object may be considered to consist of an infinite number of particles that are acted upon by the force of gravity, thus forming a distributed force system. The resultant of these tiny forces (i.e., individual weights of particles) is equal to the total weight of the object, which is a concentrated load acting at the *center of gravity* (cg) of the body. A concept related to the center of gravity is that of the *center of mass*, which is a point at which the entire mass of an object is assumed to be concentrated. In general, there is a difference between the centers of mass and gravity of an object. This may be worth considering if the object is large enough for the magnitude of the gravitational acceleration to vary at different parts of the object. For our applications, the centers of mass and gravity of an object refer to the same point.

The center of gravity of an object is located at a point about which the sum of moments of weights of all particles on one side of this point is equal to the sum of moments of weights of all particles on the other side. Therefore, if the object is balanced on a knife-edge at its center of gravity, the object will remain in rotational equilibrium because the net moment about the point of support will be zero. If the object has a symmetric, well-defined geometry (such as a sphere) and uniform composition, then its center of gravity is located at the geometric center of the object.

There are several methods of finding the centers of gravity of irregularly shaped objects. One method is by "suspending" the object. For the sake of illustration, consider the piece of paper shown in Figure 5.33 whose center of gravity is located at C. Let O and Q be two points on the paper. If the paper is pinned to the wall at O, there will be a net clockwise moment acting on the paper about point O because the center of gravity of the paper is located to the right of O. If the paper is released, the net moment about O will cause the paper to swing in the clockwise direction first, oscillate about O, and finally come to rest at a position in which C lies along a vertical line (*aa*) passing through point O (Figure 5.33b). At this position, the net moment about O is zero because the length of the moment arm is zero. If the paper is pinned at point Q, the paper will swing in the counterclockwise direction, and soon come to rest at a position in which C is located directly under Q, or along the vertical line (*bb*) passing through Q (Figure 5.33c). The point at which lines *aa* and *bb* intersect indicates the center of gravity of the paper.

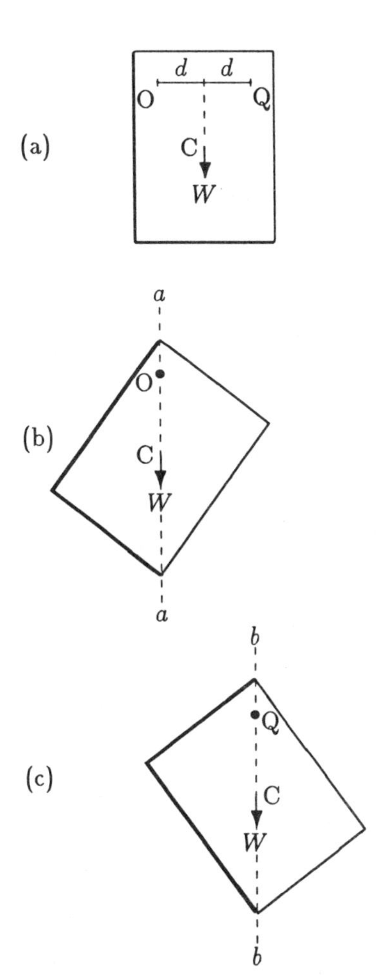

Figure 5.33 *Center of gravity of the paper is located at the intersection of lines aa and bb (suspension method).*

Note that a piece of paper is a two-dimensional object with negligible thickness, and that suspending the paper at two points is sufficient to locate its center of gravity. For a three-dimensional object, the object must be suspended at three points in two different planes.

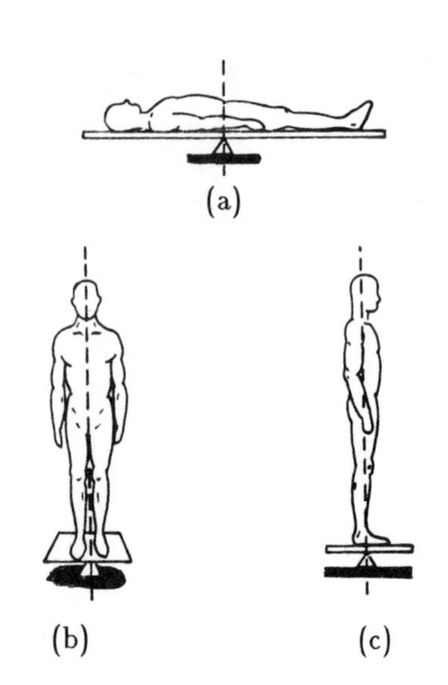

Figure 5.34 *Finding the center of gravity of a person (balance method).*

Another method of finding the center of gravity is by "balancing" the object on a knife-edge. As illustrated in Figure 5.34, to determine the center of gravity of a person, first balance a board on the knife-edge and then place the person supine on the board. Adjust the position of the person on the board until the board is again balanced (Figure 5.34a). The horizontal distance between the feet of the person and the point of contact of the knife-edge and the board is the height of the center of gravity of the person. Consider a plane which passes through the point of contact of the knife-edge with the board and which cuts the person into upper and lower portions. The center of gravity of the person lies somewhere on this plane. Note that the center of gravity of a three-dimensional object, such as a human being, has three coordinates. Therefore, the same method must be repeated in two other planes to establish the exact center of gravity. For this purpose, consider the anteroposterior balance of the person, as illustrated in Figures 5.34(b) and 5.34(c), which will yield two additional planes. The intersection of these planes corresponds to the center of gravity of the person.

The third method of finding the center of gravity of a body involves the use of a "reaction board" with two knife-edges fixed to its undersurface (Figure 5.35). Assume that the weight W_b of the board and the distance ℓ between the knife-edges are known. One of the two edges (A) rests on a platform and the other edge (B) rests on a scale, such that the board is horizontal. The location of the center of gravity of a person can be determined by placing the person on the board and recording the weight indicated on the scale, which is essentially the magnitude R_B of the reaction force on the board at B. Figure 5.35(b) illustrates the free-body diagram of the board. R_A is the magnitude of the reaction force at A, W_p is the known weight of the person. The weight W_b of the board is acting at the geometric center (C) of the board which is equidistant from both A and B. D is a point on the board directly under the center of gravity of the person. The unknown distance between A and D is designated by x_{cg} which can be determined by considering the rotational equilibrium of the board about point A (cw +):

$$\sum M_A = 0: \qquad \frac{\ell}{2} W_b + x_{cg} W_p - \ell R_B = 0$$

Solving this equation for x_{cg} yields:

$$x_{cg} = \frac{\ell}{W_p}\left(R_B - \frac{W_b}{2}\right)$$

Assuming that the person is placed on the board so that the feet are directly above the knife-edge at A, x_{cg} designates the height of the person's center of gravity as measured from the floor level.

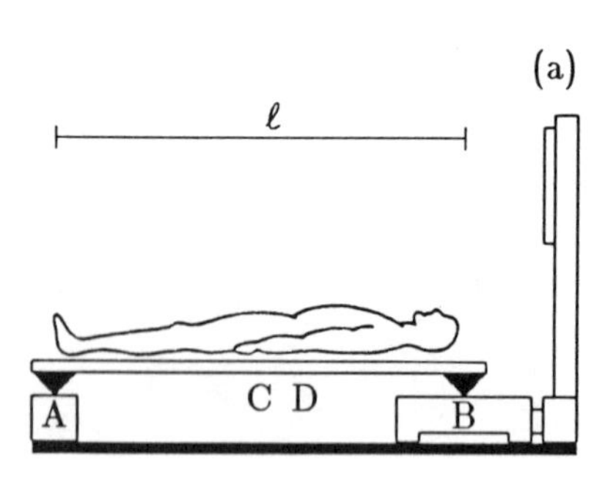

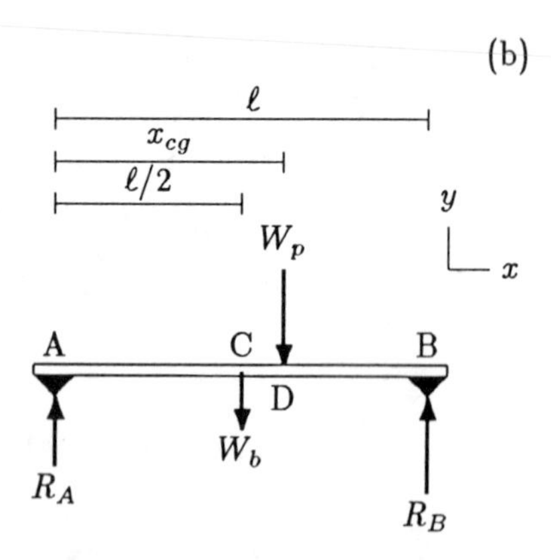

Figure 5.35 *Reaction board method.*

Sometimes, one must deal with a system made up of parts whose individual centers of gravity are known and the center of gravity of the system as a whole is required. This can be achieved simply by utilizing the definition of the center of gravity. Consider the system shown in Figure 5.36, which is composed of three spheres with weights W_1, W_2, and W_3 connected to one another through rods. Assume that the weights of the rods are negligible. Let x_1, x_2, and x_3 be the x coordinates of centers of gravity of each sphere. The net moment about point O due to the individual weights of the spheres is:

$$M_O = W_1\, x_1 + W_2\, x_2 + W_3\, x_3$$

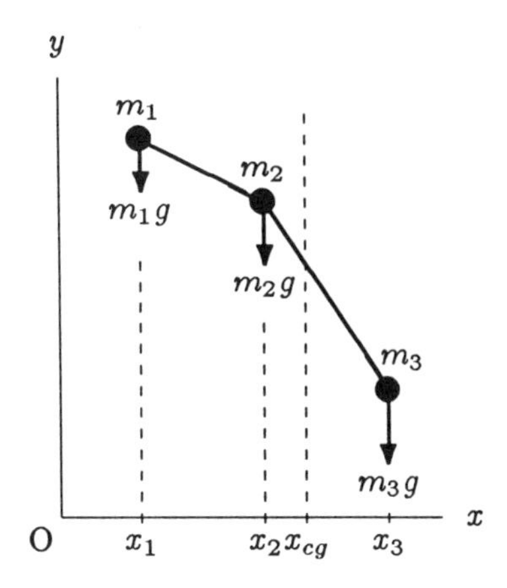

Figure 5.36 *x_{cg} is the x component of the center of gravity.*

The total weight of the system is $W_1+W_2+W_3$ which acts at the center of gravity of the system. If x_{cg} designates the x coordinate of the center of gravity of the entire system, then the moment of the total weight of the system about point O is:

$$M_O = (W_1 + W_2 + W_3)g\; x_{cg}$$

The above equations can be combined together so as to eliminate M_O and the magnitude g of gravitational acceleration. This will yield:

$$x_{cg} = \frac{W_1\, x_1 + W_2\, x_2 + W_3\, x_3}{W_1 + W_2 + W_3}$$

For any system composed of n parts, the above result can be generalized as:

$$x_{cg} = \frac{\sum_{i=1}^{n} W_i\, x_i}{\sum_{i=1}^{n} W_i} \tag{5.14}$$

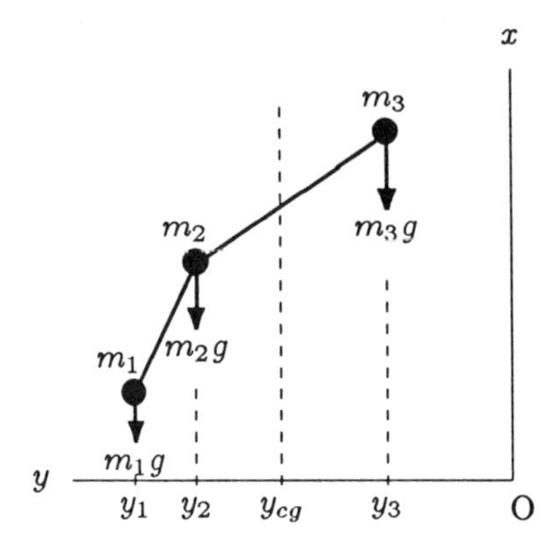

Figure 5.37 *y_{cg} is the y component of the center of gravity.*

Eq. (5.14) provides only the x coordinate of the center of gravity of the system. To determine the exact center, the y coordinate of the center of gravity must also be determined. For this purpose, the entire system must be rotated by an angle, preferably by 90°, as illustrated in Figure 5.37. If y_1, y_2, and y_3 correspond to the y coordinates of the centers of gravity of the spheres, then the y coordinate of the center of gravity of the system as a whole is:

$$y_{cg} = \frac{W_1\, y_1 + W_2\, y_2 + W_3\, y_3}{m_1 + m_2 + m_3}$$

For any system composed of n parts:

$$y_{cg} = \frac{\sum_{i=1}^{n} m_i\, y_i}{\sum_{i=1}^{n} m_i} \tag{5.15}$$

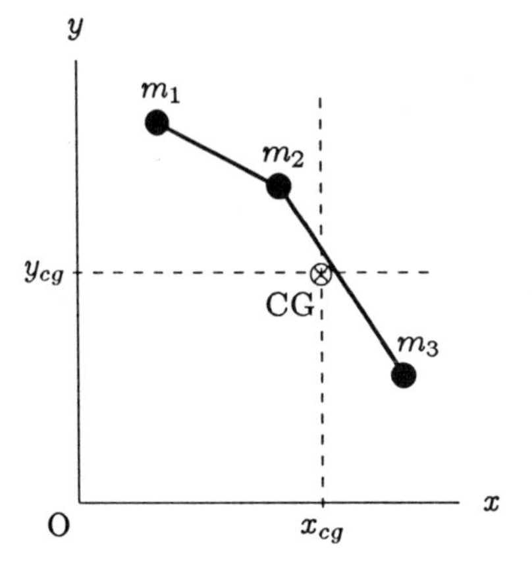

Figure 5.38 *⊗ is the center of gravity of the system.*

In Figure 5.38, the center of gravity of the entire system is located at the point of intersection of perpendicular lines passing through x_{cg} and y_{cg}.

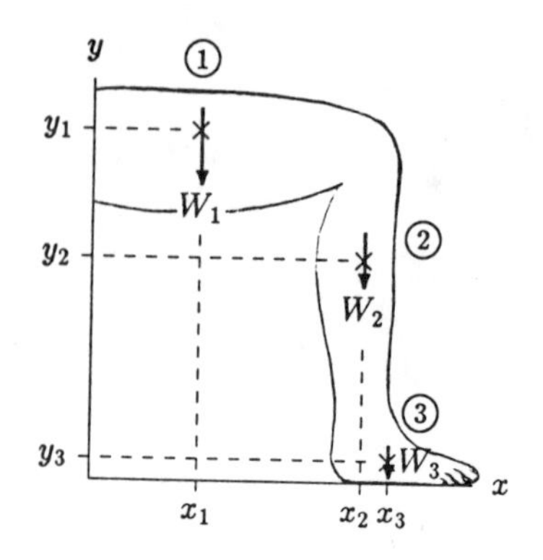

Figure 5.39 *Locating the center of gravity of a flexed leg.*

PART	x (cm)	y (cm)	$\%W$
①	17.3	51.3	10.6
②	42.5	32.8	4.6
③	45.0	3.3	1.7

Table 5.1 *Example 5.9.*

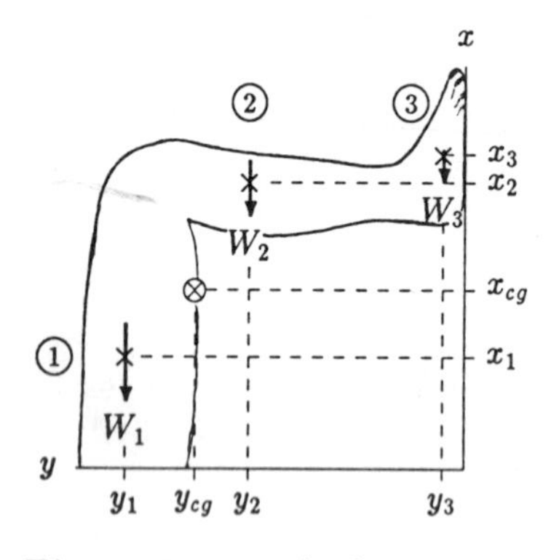

Figure 5.40 ⊗ *is the center of gravity of the leg.*

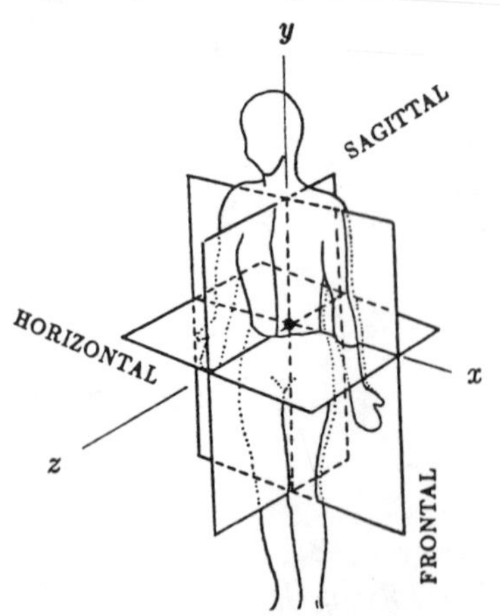

Figure 5.41 *Sagittal, frontal, and horizontal planes.*

Example 5.9 Consider the leg shown in Figure 5.39, which is flexed to a right angle. The coordinates of the centers of gravity of the upper leg, lower leg, and foot, as measured from the floor level directly in line with the hip joint, are given in Table 5.1. The weights of the segments of the leg as percentages of the total weight W of the person are also given in Table 5.1.

Determine the location of the center of gravity of the entire leg.

Solution: The coordinates (x_{cg}, y_{cg}) of the center of gravity of the entire leg can be determined by utilizing Eqs. (5.14) and (5.15). Using Eq. (5.14):

$$
\begin{aligned}
x_{cg} &= \frac{x_1 W_1 + x_2 W_2 + x_3 W_3}{W_1 + W_2 + W_3} \\
&= \frac{(17.3)(0.106W) + (42.5)(0.046W) + (45.0)(0.017W)}{0.106W + 0.046W + 0.017W} \\
&= 26.9 \text{ cm}
\end{aligned}
$$

To determine the y coordinate of the center of gravity of the leg, we must rotate the leg by 90°, as illustrated in Figure 5.40, and apply Eq. (5.15):

$$
\begin{aligned}
y_{cg} &= \frac{y_1 W_1 + y_2 W_2 + y_3 W_3}{W_1 + W_2 + W_3} \\
&= \frac{(51.3)(0.106W) + (32.8)(0.046W) + (3.3)(0.017W)}{0.106W + 0.046W + 0.017W} \\
&= 41.4 \text{ cm}
\end{aligned}
$$

Therefore, the center of gravity of the entire leg when bent at 90° is located at a horizontal distance of 26.9 centimeters from the hip joint, and at a height of 41.4 centimeters from the floor level.

Remarks:

- In an ordinary standing position with straight limbs, the center of gravity of a person lies within the pelvis anterior to the second sacral vertebra. If the origin of the rectangular coordinate system is placed at the center of gravity of the person, as illustrated in Figure 5.41, then the xy plane corresponds to the ***frontal, coronal,*** or ***longitudinal plane,*** the yz plane is the ***sagittal plane,*** and the xz plane is called the ***horizontal*** or ***transverse plane.*** The frontal plane divides the body into front and back portions, the sagittal plane divides the body into right and left portions, and the horizontal plane divides the body into upper and lower portions.
- The center of gravity of the entire body varies from person to person depending on build. For a given person, the position

of the center of gravity can shift depending on the changes in the relative alignment of the limbs during a particular physical activity. Locations of the centers of gravity of the upper and lower extremities can also vary. For example, the center of gravity of a lower limb shifts backwards as the knee is flexed. The center of gravity of the entire arm shifts forward as the elbow is flexed. These observations suggest that when the knee is flexed, the leg will tend to move forward bringing the center of gravity of the leg directly under the hip joint; and when the elbow is flexed, the arm will tend to move backward bringing its center of gravity under the shoulder joint.

- Knowing the center of gravity of the human body in various positions may facilitate our understanding of body mechanics. For example, high jumpers aim to extend their bodies in such a way as to locate the high bar between themselves and their center of gravities as illustrated in Figure 5.42.

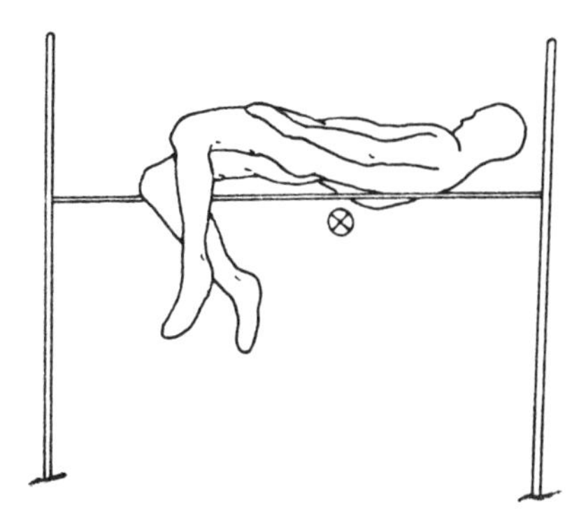

Figure 5.42 *A high jumper.*

5.15 Statically Determinate and Indeterminate Systems

A mechanical system in static equilibrium is one for which the conditions of equilibrium are satisfied for the system as a whole as well as for any part of the system. Analyses in statics are based on the assumption that the parts forming a system are not deformable.

Consider a horizontal beam hinged to a wall at one end and supported by a cable at the other end, as illustrated in Figure 5.43. Also shown in Figure 5.43 is the free-body diagram of the beam. W is the known weight of the beam. T is the tension in the cable, and R_x and R_y are reactions at the hinge, all of which are unknown forces. For the solution of this two-dimensional problem, we have three equilibrium equations. Therefore, the number of equations is equal to the number of unknowns. This implies that by solving the three equations simultaneously, we can determine a unique set of R_x, R_y, and T values which will satisfy the conditions of static equilibrium. Mechanical systems for which the equations of equilibrium are sufficient to determine the unknowns uniquely are called ***statically determinate***.

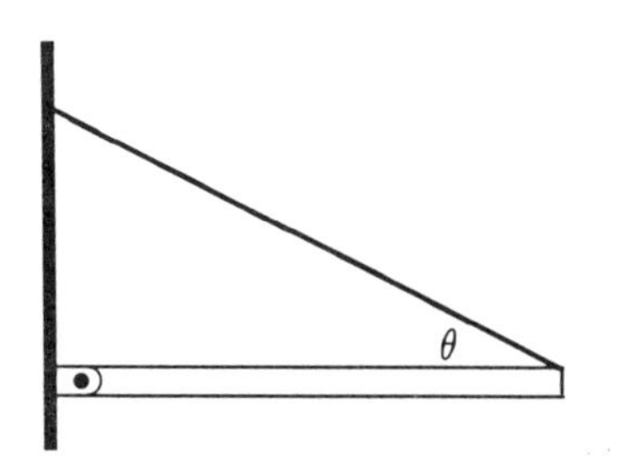

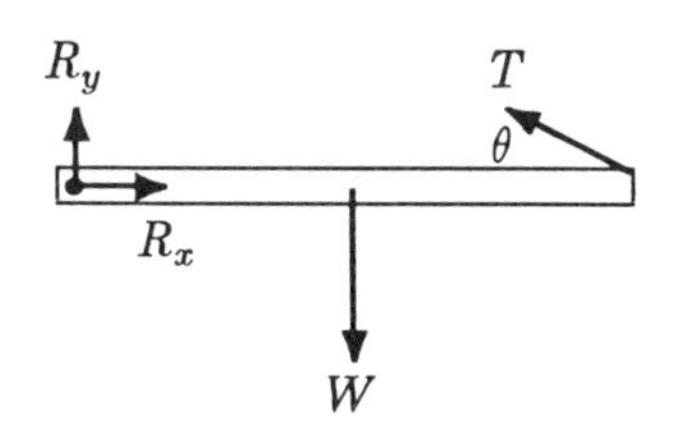

Figure 5.43 *A statically determinate system.*

Suppose now that the beam is additionally supported by a second cable, as shown in Figure 5.44. It is likely that tensions in the two cables are different, and we designate them T_1 and T_2. We again have the reactions R_x and R_y at the hinge. Therefore, we now have four unknowns. However, the number of equations available from statics is still three. Since the number of unknowns exceeds the number of equations, there exist an infinite number of possible sets of T_1, T_2, R_x, and R_y combinations, all of which may satisfy the equilibrium equations. It appears that there is

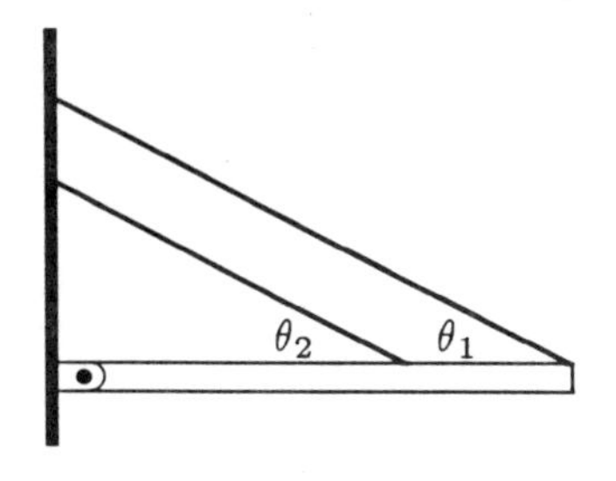

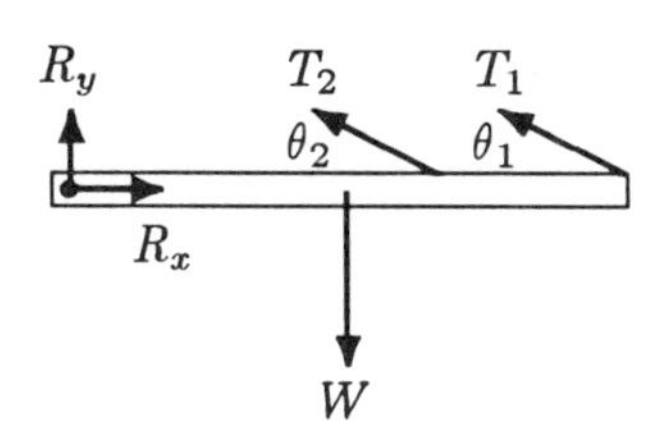

Figure 5.44 *A statically indeterminate system.*

no unique solution. To decide on the proper combination of supporting forces requires additional equations and computations. The additional equation(s) can be derived by considering the deformation properties of the beam and/or the cables. Systems for which the equations of equilibrium (statics) are not sufficient to determine the unknown forces are called ***statically indeterminate.***

As we shall investigate in the following chapter, the application of statics to analyze forces involved in the muscles and various joints of the human body requires certain assumptions. First of all, we must reduce the problem under consideration to one which is statically determinate. This can be achieved simply by assuming that during a particular activity only certain muscle groups are active. The complexity of the computations increases as more muscle groups are considered in the analysis.

5.16 Two- and Three-Force Members

Figure 5.45 *A collinear two-force system.*

It was demonstrated earlier in this chapter that if there are only two forces acting on a body, then the body will be in equilibrium if and only if the forces have an equal magnitude, opposite directions, and are collinear (Figure 5.45). Note that this condition must always be satisfied for members such as cables and ropes which are known as *two-force members.*

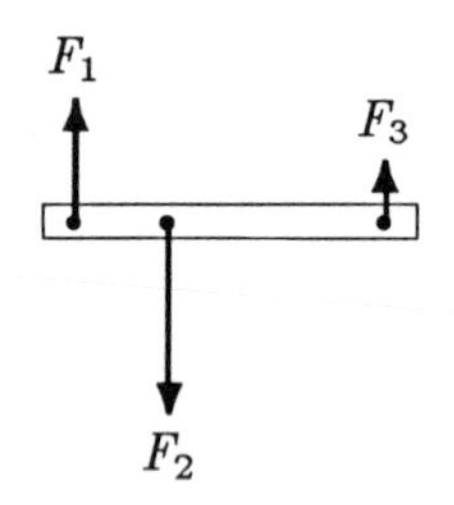

Figure 5.46 *A parallel three-force system.*

If there are three forces acting on a body and if the body is in equilibrium, then the forces must form either a parallel or a concurrent system of forces. A parallel three-force system is shown in Figure 5.46. Note that if two of the three forces are known to be parallel to one another, then the third force must also be parallel to the other two.

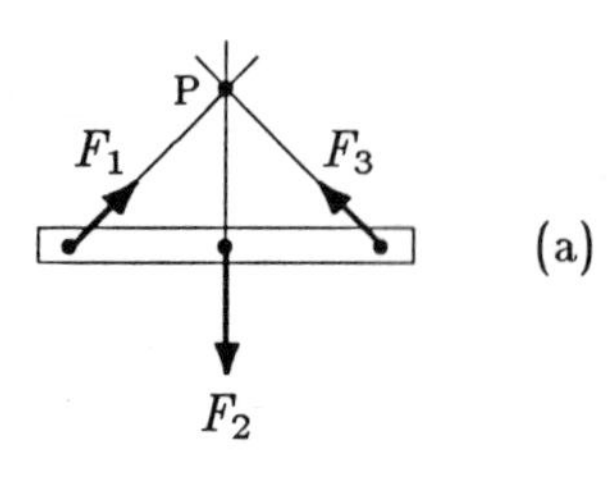

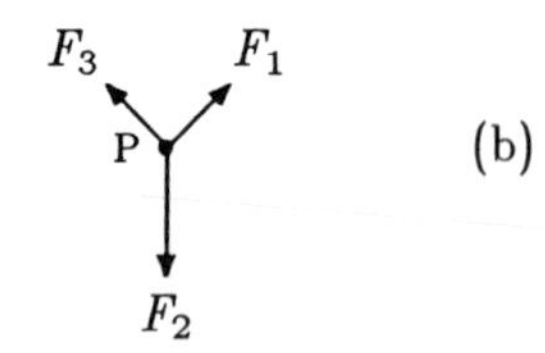

Figure 5.47 *A concurrent three-force system.*

A concurrent three-force system is shown in Figure 5.47. In this case, the lines of action of the forces have a common point of intersection (P in Figure 5.47a). Therefore, if the lines of action of two of the three forces are known, then the line of action of the third force can also be determined. For this purpose, we first extend the lines of action of the first two forces until they meet, label this point as P, and draw a straight line connecting point P and the point of application of the third force. The straight line thus obtained corresponds to the line of action of the third force. To analyze a concurrent force system, the forces forming the system can be moved to point P along their lines of action (Figure 5.47b). The forces can then be resolved into their components along two perpendicular directions, and the translational equilibrium conditions can be employed to determine two unknowns. Note in this case that we cannot utilize the rotational equilibrium condition, which would be satisfied automatically. However, by determining the point of intersection of the concurrent forces, we essentially determine the direction of the third force, and therefore determine one of the unknowns from the geometry of the problem.

Chapter 6

APPLICATIONS OF STATICS TO BIOMECHANICS

6.1 Skeletal Joints

The human body is rigid in the sense that it can maintain a posture, and flexible in the sense that it can change its posture and move. The flexibility of the human body is due primarily to the *joints*, or *articulations*, of the skeletal system. The primary function of joints is to provide mobility to the musculoskeletal system. In addition to providing mobility, a joint must also possess a degree of stability. Since different joints have different functions, they possess varying degrees of mobility and stability. Some joints are constructed so as to provide maximum mobility. For example, the construction of the shoulder joint (ball-and-socket) enables the arm to move in all three planes (triaxial motion). However, this high level of mobility is achieved at the expense of reduced stability, increasing the vulnerability of the joint to injuries, such as dislocations. On the other hand, the elbow joint provides movement only in one plane (uniaxial motion), but is more stable and less prone to injuries than the shoulder joint. The extreme case of increased stability is achieved at joints that permit no relative motion between the bones constituting the joint. The contacting surfaces of the bones in the skull are typical examples of such joints.

The joints of the human skeletal system may be classified based on their structure and/or function. *Synarthrodial joints*, such as those in the skull, are formed by two tightly fitting bones and do not allow any relative motion of the bones forming them. *Amphiarthrodial joints*, such as those between the vertebrae, allow slight relative motions, and feature an intervening substance (a cartilaginous or ligamentous tissue) whose presence eliminates direct bone-to-bone contact. The third and mechanically most significant type of articulations are called *diarthrodial joints* which permit varying degrees of relative motion and have articular cavities, ligamentous capsules, synovial membranes, and synovial fluids (Figure 6.1). The *articular cavity* is the space between the articulating bones. The *ligamentous capsule* holds the articulating bones together. The *synovial membrane* is the internal lining of the ligamentous capsule enclosing the *synovial fluid* which serves as a lubricant. The synovial fluid is a viscous material which functions to reduce friction, reduce wear and tear of the articulating surfaces by limiting direct contact between them, and nourish the articular cartilage lining the surfaces. The *articular cartilage*, on the other hand, is a cartilaginous material whose primary function is to absorb shock. Various diarthrodial joints can be further categorized as gliding (for example, vertebral facets), hinge (elbow and ankle), pivot (proximal radioulnar), condyloid (wrist), saddle (carpometacarpal of thumb), and ball-and-socket (shoulder and hip).

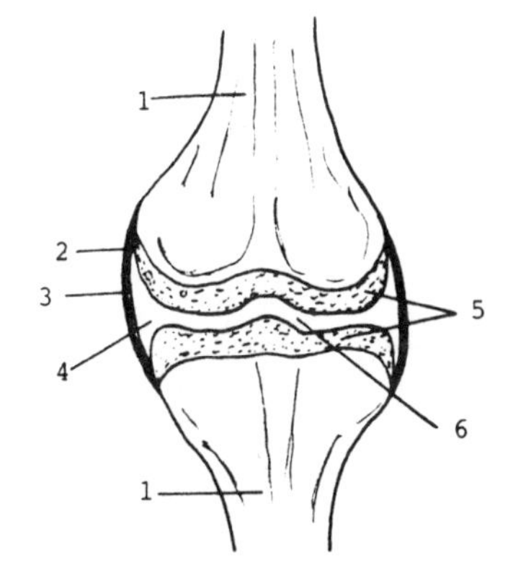

Figure 6.1 *A diarthrodial joint. (1) bone, (2) ligamentous capsule, (3, 4) synovial membrane & fluid, (5, 6) articular cartilage & cavity.*

The nature of motion about a diarthrodial joint and the stability of the joint are dependent upon many factors, including the manner in which the articulating surfaces fit together, the structure of the capsular ligament, the structure and length of the ligaments around the joint, and the degree to which the muscles that cross the joint can be stretched.

6.2 Skeletal Muscles

There are three types of muscles: skeletal, smooth, and cardiac. Movement of human body segments is achieved as a result of forces generated by ***skeletal muscles*** which convert chemical energy into mechanical work. The skeletal muscle is composed of muscle fibers and myofibrils. Myofibrils are made up of actin and myosin filaments. Muscles exhibit viscoelastic material behavior. That is, they have both solid and fluid like material properties. Muscles are elastic in the sense that when a muscle is stretched and released it will resume its original (unstretched) size and shape. Muscles are ***viscous*** in the sense that there is an internal resistance to motion.

A skeletal muscle is attached, via aponeuroses and/or tendons, to at least two different bones controlling the relative motion of one segment with respect to the other. When its fibers contract under the stimulation of a nerve, the muscle exerts a pulling effect on the bones to which it is attached. *Contraction* is a unique ability of the muscle tissue, which is defined as the development of tension in the muscle. In engineering mechanics, contraction implies shortening under compressive forces. In muscle mechanics, contraction can occur as a result of muscle shortening or muscle lengthening, or it can occur without any change in the muscle length. Furthermore, the result of a muscle contraction is always tension: a muscle can only exert a pull. Muscles can not exert a push.

There are various types of muscle contractions: a *concentric contraction* occurs simultaneously as the length of the muscle decreases (for example, the biceps during flexion of the forearm); a *static contraction* occurs while muscle length remains constant (the biceps when the forearm is flexed and held without any movement); and an *eccentric contraction* occurs as the length of the muscle increases (the biceps during the extension of the forearm). A muscle can cause movement only while its length is shortening. If the length of a muscle increases during a particular activity, then the tension generated by the muscle contraction is aimed at controlling the movement of the body segments associated with that muscle. If a muscle contracts but there is no segmental motion, then the tension in the muscle balances the effects of applied forces such as those due to gravity.

The skeletal muscles can also be named according to the functions they serve during a particular activity. For example, a muscle is called *agonist* if it causes movement through concentric contraction. An *antagonist* muscle, as a result of eccentric contraction, controls the movement.

6.3 Basic Considerations

In this chapter, we want to apply the principles of statics to investigate the forces involved in various muscle groups and joints for various postural positions of the human body and its segments. Our immediate purpose is to provide answers to questions such as: what tension must the neck extensor muscles exert on the head to support the head in a specified position? When a person bends, what would be the force exerted by the erector spinae on the fifth lumbar vertebra? How does the compression at the elbow, knee, and ankle joints vary with externally applied forces and with different segmental arrangements? How does the force on the femoral head vary with loads carried in the hand? What are the forces involved in various muscle groups and joints during different exercise conditions?

The forces involved in the human body can be grouped as internal and external. Internal forces are those associated with muscles, ligaments, and tendons. Externally applied forces include the effect of gravitational acceleration on the body or on its segments, manually and/or mechanically applied forces on the body during exercise and stretching, and forces applied to the body by prostheses and implements. In general, the unknowns in static problems involving the musculoskeletal system are the joint reaction forces and muscle tensions. Mechanical analysis of a joint requires that we know the vector characteristics of tension in the muscle including the proper locations of muscle attachments, the weights or masses of body segments, the centers of gravity of the body segments, and the anatomical axis of rotation of the joint.

6.4 Basic Assumptions and Limitations

The complete analysis of muscle forces required to sustain various postural positions is difficult because of the complex arrangement of muscles within the human body. The relative motion of body segments about a given joint is controlled by more than one muscle group. To be able to reduce a specific problem of biomechanics to one that is statically determinate and apply the equations of equilibrium, only the muscle group that is the primary source of control over the joint must be taken into consideration. Possible contributions of other muscle groups to the load-bearing mechanism of the joint must be ignored. Note however that approximations of the effect of other muscles can be made by considering their cross-sectional areas and their relative positions in relation to the joint. Also, if the phasic activity of muscles is known via some experiments such as the electromyography (EMG) measurements of muscle signals, then the tension in different muscle groups can be estimated.

To apply the principles of statics to analyze the mechanics of human joints, we shall adopt the following assumptions and limitations:

- The anatomical axes of rotation of joints are known.
- Only one muscle group is controlling the movement of a joint.
- The locations of muscle attachments are known.
- The line of action of muscle tension is known.
- Segmental weights and their centers of gravity are known.
- Frictional factors at the joints are negligible.
- Dynamic aspects of the problems will be ignored.
- Only two-dimensional problems will be considered.
- Deformability of muscles, tendons, and bones will be ignored.

These analyses require that the anthropometric data about the segment to be analyzed must be available. For this purpose, there are tables listing anthropometric information including average weights, lengths, and centers of gravity of body segments. See Chaffin and Andersson (1991) and Winter (1990) for a review of the anthropometric data availavle.

It is clear from this discussion that we shall analyze certain idealized problems of biomechanics. Based on the results obtained and experience gained, these models may be expanded by taking additional factors into consideration. However, a given problem will become more complex as more factors are considered.

In the following sections, the principles of statics will be applied to analyze forces involved at and around the major joints of the human body. First, a brief functional anatomy of each joint and related muscles will be provided, and specific biomechanical problems will be constructed. For a more complete discussion about the functional anatomy of joints, see texts such as Kirby and Roberts (1985), Nordin and Frankel (1989), and Thompson (1989). Next, an analogy will be formed between muscles, bones, and human joints, and certain mechanical elements such as cables, beams, and mechanical joints. This will enable us to construct a mechanical model of the biological system under consideration. Finally, the procedure outlined in Chapter 5.4 will be applied to analyze the mechanical model thus constructed. See Williams and Lissner (1977) and Wiktorin and Nordin (1982) for additional examples of the application of the principles of statics to biomechanics.

6.5 Mechanics of the Elbow

The elbow joint is composed of three separate articulations (Figure 6.2). The *humeroulnar joint* is a hinge (ginglymus) joint formed by the articulation between the spool-shaped trochlea of the distal humerus and the concave trochlear fossa of the proximal ulna. The structure of the humeroulnar joint is such that it allows only uniaxial rotations, confining the movements about the elbow joint to flexion (movement of the forearm towards the upper arm) and extension (movement of the forearm away from the upper arm). The *humeroradial joint* is also a hinge joint formed between the capitulum of the distal humerus and the head of the radius. The *proximal radioulnar joint* is a pivot joint formed by the head of the radius and the radial notch of the proximal ulna. This articulation allows the radius and ulna to undergo relative rotation about the longitudinal axis of one or the other bone, giving rise to pronation (the movement experienced while going from the palm-up to the palm-down) or supination (the movement experienced while going from the palm-down to the palm-up).

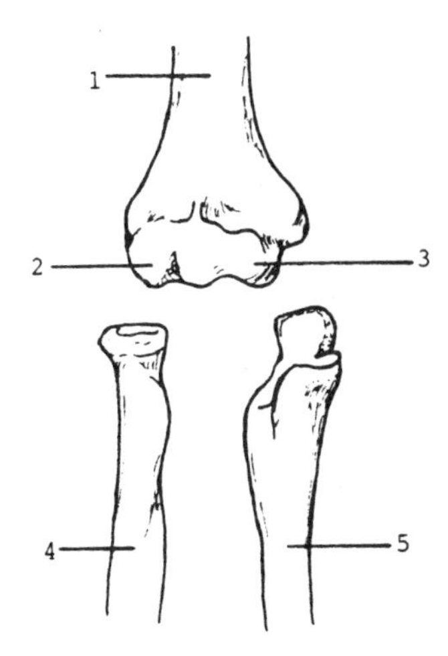

Figure 6.2 *Bones of the elbow: (1) humerus, (2) capitulum, (3) (3) trochlea, (4) radius, (5) ulna.*

The muscles coordinating and controlling the movement of the elbow joint are illustrated in Figure 6.3. The biceps brachii muscle is the most powerful flexor of the elbow joint, particularly when the elbow joint is in a supinated position. On the distal side, the biceps is attached to the tuberosity of the radius, and on the proximal side, it has attachments at the top of the coracoid process and upper lip of the glenoid fossa. Another important flexor is the brachialis muscle which has attachments at the lower half of the anterior portion of the humerus and the coronoid proc-ess of the ulna. The most important muscle controlling the extension movement of the elbow is the triceps brachii muscle. It has attachments at the lower head of the glenoid cavity of the scapula, the upper half of the posterior surface of the humerus, the lower two thirds of the posterior surface of the humerus, and the olecranon process of the ulna. Pronation and supination movements of the forearm are performed mainly by the pronator teres and supinator muscles, respectively. The pronator teres is attached to the lower part of the inner condyloid ridge of the humerus, the medial side of the ulna, and the middle third of the outer surface of the radius. The supinator muscle has attachments at the outer condyloid ridge of the humerus, the neighboring part of the ulna and the outer surface of the upper third of the radius.

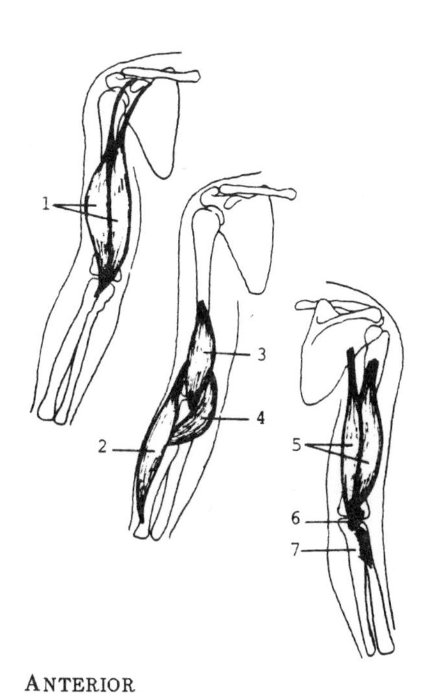

Figure 6.3 *Muscles of the elbow: (1) biceps, (2) brachioradialis, (3) brachialis, (4) pronator teres, (5) triceps brachii, (6) anconeus, (7) supinator.*

Common injuries of the elbow include fractures and dislocations. Fractures usually occur at the epicondyles of the humerus and the olecranon process of the ulna. Another elbow injury is known as "tennis elbow," which occurs as a result of repeated and forceful pronation and supination movements of the elbow.

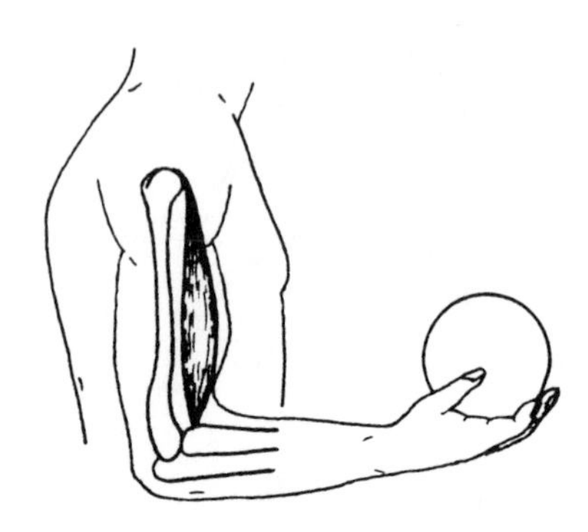

Figure 6.4 *Example 6.1.*

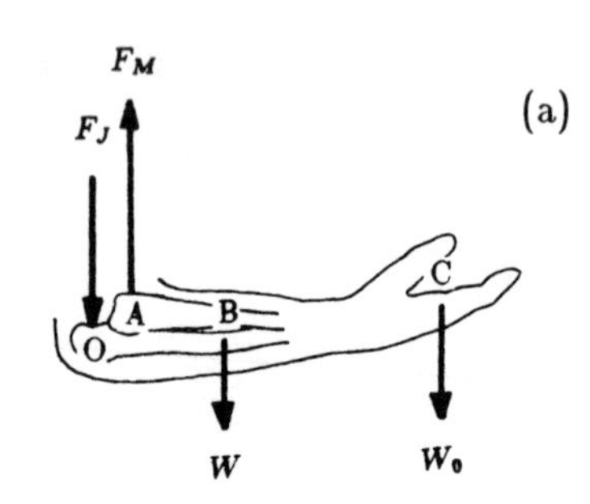

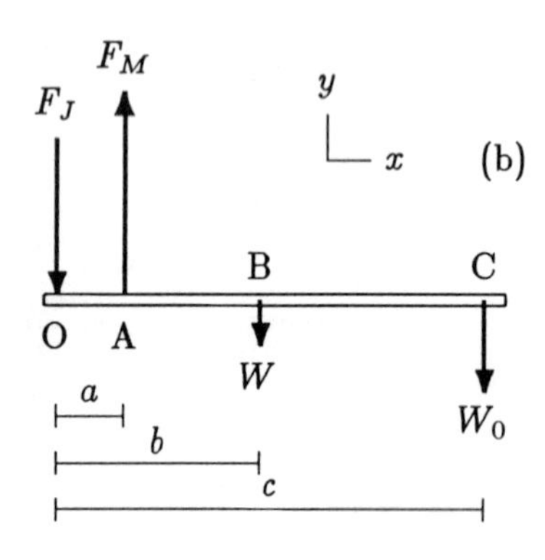

Figure 6.5 *Forces acting on the lower arm.*

Example 6.1 Consider the arm shown in Figure 6.4. The elbow is flexed to a right angle, and an object is held in the hand. The forces acting on the forearm are shown in Figure 6.5(a), and the free-body diagram of the forearm is shown on a mechanical model in Figure 6.5(b). This model assumes that the biceps is the major flexor and that the line of action of the tension (line of pull) in the biceps is vertical.

Point O designates the axis of rotation of the elbow joint, which is assumed to be fixed for practical purposes. Point A is the attachment of the biceps muscle on the radius, B is the center of gravity of the forearm, and C is a point on the forearm that lies along a vertical line passing through the center of gravity of the weight in the hand. The distances between O and A, B, and C are measured as a, b, and c, respectively. W_O is the weight of the object held in the hand and W is the total weight of the forearm. F_M is the magnitude of the force exerted by the biceps on the radius, and F_J is the magnitude of the reaction force at the elbow joint. Notice that the line of action of the muscle force is assumed to be vertical. The gravitational forces are vertical as well. Therefore, the line of action of the joint reaction force must also be vertical (a parallel force system).

The task in this example is to determine the magnitudes of the muscle tension and the joint reaction force at the elbow.

Solution: We have a parallel force system, and the unknowns are the magnitudes F_M and F_J of the muscle and joint reaction forces. Considering the rotational equilibrium of the forearm about the elbow joint will yield:

$$\sum M_O = 0: \qquad F_M = \frac{1}{a}\,(b\,W + c\,W_O) \qquad (i)$$

For the translational equilibrium of the forearm in the y direction:

$$\sum F_y = 0: \qquad F_J = F_M - W - W_O \qquad (ii)$$

For given values of geometric parameters a, b, and c, and weights W and W_O, Eqs. (i) and (ii) can be solved for the magnitudes of the muscle and joint reaction forces. For example, assume that these parameters are given as follows: a=4 cm, b=15 cm, c=35 cm, W=20 N and W_O=80 N. Then from Eqs. (i) and (ii):

$$F_M = \frac{1}{0.04}\Big[(0.15)(20) + (0.35)(80)\Big] = 775 \text{ N} \quad (+y)$$

$$F_J = 775 - 20 - 80 = 675 \text{ N} \quad (-y)$$

Remarks:

- The numerical results indicate that the force exerted by the muscle is about ten times larger than the weight of the object held in the position considered. Relative to the axis of the elbow joint, the length a of the lever arm of the muscle force is much smaller than the length c of the lever arm enjoyed by the load held in the hand. The smaller the lever arm, the greater the muscle tension required to balance the rotational effect of the load about the elbow joint. Therefore, during lifting, it is disadvantageous to have a muscle attachment close to the elbow joint. However, the closer the muscle is to the joint, the larger the range of motion of elbow flexion-extension, and the faster the distal end (hand) of the forearm can reach its goal of moving towards the upper arm or the shoulder.

- The angle (angle of pull) between the line of action of the muscle force and the long axis of the bone upon which the muscle force is exerted is critical in determining the effectiveness of the muscle force. When the forearm is flexed to a right angle, the muscle tension has only a rotational effect on the forearm about the elbow joint, because the line of action of the muscle force is at right angles with the longitudinal axis of the forearm. For other flexed positions of the forearm, the muscle force can have a translational (stabilizing or sliding) component as well as a rotational component.

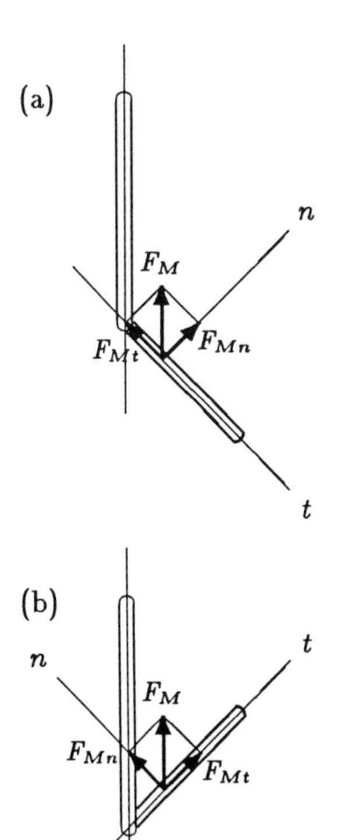

Figure 6.6 *Rotational (F_{Mn}) and stabilizing or sliding (F_{Mt}) components of the muscle force.*

Assume that the linkage system shown in Figure 6.6(a) illustrates the position of the forearm relative to the upper arm. n designates a direction perpendicular (normal) to the long axis of the forearm and t is tangent to it. Assuming that the line of action of the muscle force remains parallel to the long axis of the humerus, $\underline{F}_M$ can be decomposed into its rectangular components $\underline{F}_{Mn}$ and $\underline{F}_{Mt}$. In this case, $\underline{F}_{Mn}$ is the ***rotational*** (rotatory) component of the muscle force because its primary function is to rotate the forearm about the elbow joint. The tangential component $\underline{F}_{Mt}$ of the muscle force acts to compress the elbow joint and is called the ***stabilizing*** component of the muscle force. As the angle of pull approaches 90°, the magnitude of the rotational component of the muscle force increases while its stabilizing component decreases, and less and less energy is "wasted" to compress the elbow joint. As illustrated in Figure 6.6(b), the stabilizing role of $\underline{F}_{Mt}$ changes into a ***sliding***, or ***dislocating***, role when the angle between the long axes of the forearm and upper arm becomes less than 90°.

- The elbow is a diarthrodial joint. A ligamentous capsule encloses an articular cavity which is filled with synovial fluid. Synovial fluid is a thick, viscous material. The primary function of this fluid is to lubricate the articulating surfaces, thereby reducing the frictional forces that may develop while one articulating surface moves over the other. The synovial fluid also nourishes the articulating cartilages. Another function of the synovial fluid is to distribute the forces involved at the joint over a relatively large surface area (Figure 6.7). A common property of fluids is that they exert pressures (force per unit area) that are distributed over the surfaces they touch. The fluid pressure always acts in a direction towards and perpendicular to the surface it touches having a compressive effect on the surface. Note that in Figure 6.7, the small vectors indicating the fluid pressure have components in the horizontal and vertical directions. We determined that the joint reaction force at the elbow acts vertically downward on the ulna. This implies that the horizontal components of these vectors cancel out (i.e., half pointing to the left and half pointing to the right), but their vertical components (on the ulna, almost all of them are pointing downward) add up to form the resultant force F_J (shown with a dashed arrow in Figure 6.7c). Therefore, the joint reaction force F_J corresponds to the resultant of the distributed force system (pressure) applied through the synovial fluid.

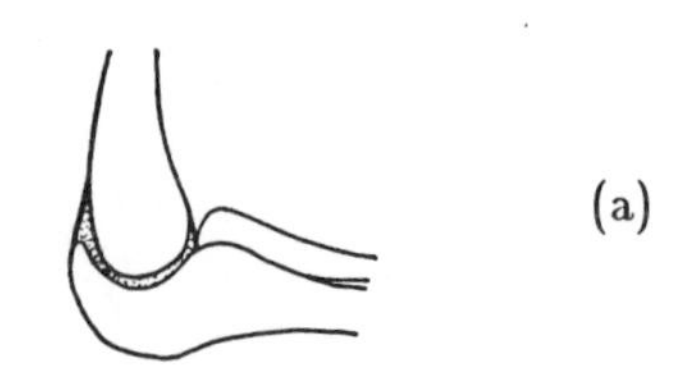

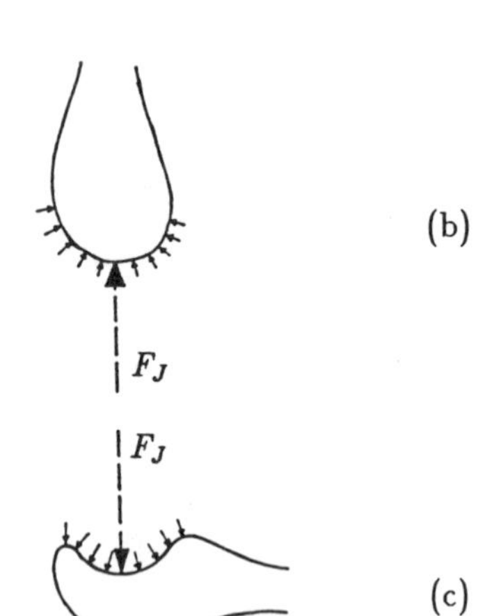

Figure 6.7 *Explaining the joint reaction force at the elbow.*

- The most critical simplification made in the preceding example is that the biceps was assumed to be the single muscle group responsible for maintaining the flexed configuration of the forearm. The reason for making such an assumption was to reduce the system under consideration to one that is statically determinate. In reality, in addition to the biceps, the brachialis and the brachioradialis are also primary elbow flexor muscles.

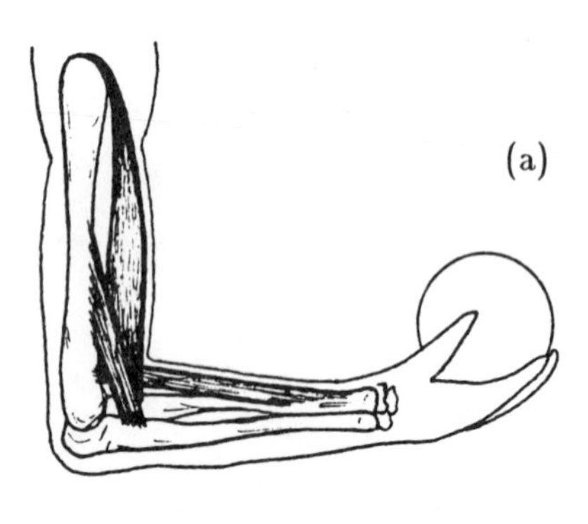

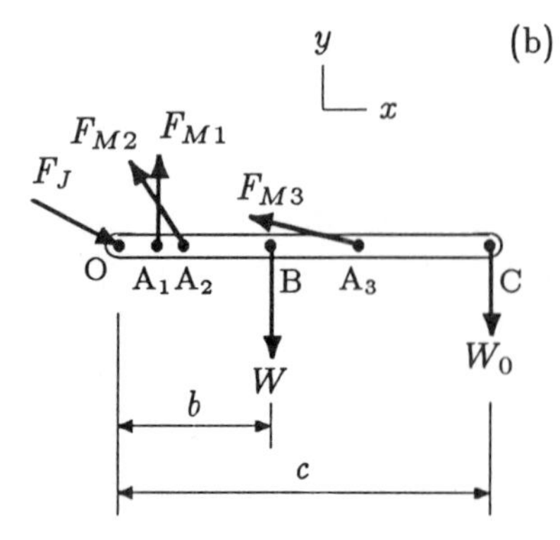

Figure 6.8 *Three-muscle system.*

Consider the flexed position of the arm shown in Figure 6.8(a). The free-body diagram of the forearm is shown in Figure 6.8(b). F_{M1}, F_{M2}, and F_{M3} are the magnitudes of the forces exerted on the forearm by the biceps, the brachialis, and the brachioradialis muscles with attachments at A_1, A_2, and A_3, respectively. Let θ_1, θ_2, and θ_3 be the angles the biceps, the brachialis and the brachioradialis muscles make with the long axis of the lower arm. As compared to the single-muscle system which consisted of two unknowns (F_M and F_J), the analysis of this three-muscle system is quite complex. First of all, this is not a simple parallel force system. Even if we assume that the locations of muscle attachments (A_1, A_2, and A_3), their angles of pull (θ_1, θ_2, and θ_3) and the lengths of their moment arms (a_1, a_2, and a_3) as measured from the elbow joint are known, there are still five unknowns in the problem (F_{M1}, F_{M2}, F_{M3}, F_J, and β). The total number of equations available from statics is three:

$$\sum M_O = 0: \quad a_1 F_{M1} + a_2 F_{M2} + a_3 F_{M3} = b\,W + c\,W_0 \qquad (iii)$$

$$\sum F_x = 0: \quad F_{Jx} = F_{M1x} + F_{M2x} + F_{M3x} \qquad (iv)$$

$$\sum F_y = 0: \quad F_{Jy} = F_{M1y} + F_{M2y} + F_{M3y} - W - W_0 \qquad (v)$$

Note that once the muscle forces are determined, Eqs. (*iv*) and

(v) will yield the components of the joint reaction force $\underline{F}_J$. As far as the muscle forces are concerned, we have only Eq. (iii) with three unknowns. In other words, we have a statically indeterminate problem. To obtain a unique solution, we need additional information relating F_{M1}, F_{M2}, and F_{M3}.

There may be several approaches to the solution of this problem. The criteria for estimating the force distribution among different muscle groups may be established by: (1) using cross-sectional areas of muscles, (2) using electromyography (EMG) measurements of muscle signals, and (3) applying certain optimization techniques. It may be assumed that each muscle exerts a force proportional to its cross-sectional area. If A_1, A_2, and A_3 are the cross-sectional areas of the biceps, the brachialis, and the brachioradialis, then this criteria may be applied by expressing muscle forces in the following manner:

$$F_{M2} = k_{21}\, F_{M1} \quad \text{with} \quad k_{21} = \frac{A_2}{A_1} \qquad (vi)$$

$$F_{M3} = k_{31}\, F_{M1} \quad \text{with} \quad k_{31} = \frac{A_3}{A_1} \qquad (vii)$$

If constants k_{21} and k_{31} are known, then Eqs. (vi) and (vii) can be substituted into Eq. (iii), which can then be solved for F_{M1}:

$$F_{M1} = \frac{b\,W + c\,W_0}{a_1 + a_2 k_{21} + a_3 k_{31}}$$

Substituting F_{M1} back into Eqs. (vi) and (vii) will then yield the magnitudes of the forces in the brachialis and the brachioradialis muscles. The values of k_{21} and k_{31} may also be estimated by using the amplitudes of muscle EMG signals.

This statically indeterminate problem may also be solved by considering some optimization techniques. If the purpose is to accomplish a certain task (static or dynamic) in the most efficient manner, then the muscles of the body must act to minimize the forces exerted, the moments about the joints (for dynamic situations), and/or the work done by the muscles. The question is, what force distribution among the various muscles facilitates the maximum efficiency? These concepts and relevant references will be presented in Section 6.11.

6.6 Mechanics of the Shoulder

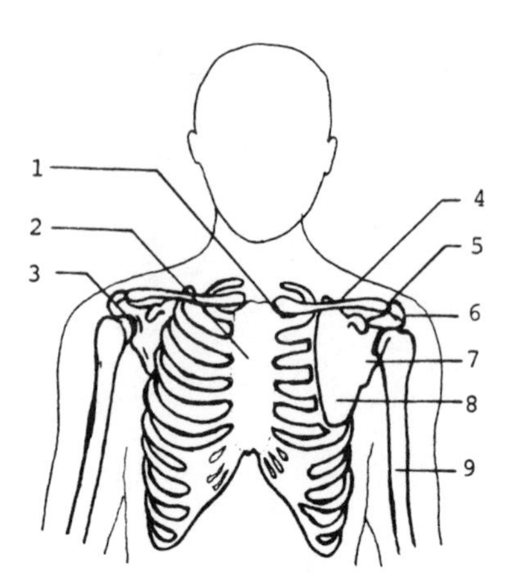

Figure 6.9 *The shoulder: (1) sternoclavicular joint, (2) sternum, (3) glenohumeral joint, (4) clavicle, (5) acromioclavicular joint, (6) acromion process, (7) glenoid fossa, (8) scapula, (9) humerus.*

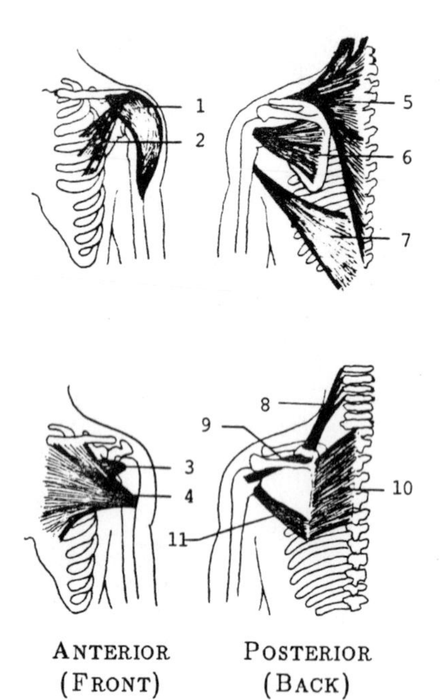

Figure 6.10 *Shoulder muscles: (1) deltoideus, (2) pectoralis minor, (3) subscapularis, (4) pectoralis major, (5) trapezius, (6) infraspinatus & teres minor, (7) latissimus dorsi, (8) levator scapulae, (9) supraspinatus, (10) rhomboideus, (11) teres major.*

The bony structure and the muscles of the shoulder complex are illustrated in Figures 6.9 and 6.10. The shoulder forms the base for all upper extremity movements. The complex structure of the shoulder can be divided into two: the shoulder joint and the shoulder girdle.

The shoulder joint, also known as the *glenohumeral articulation*, is a ball-and-socket joint between the nearly hemispherical humeral head (ball) and the shallowly concave glenoid fossa (socket) of the scapula. The shallowness of the glenoid fossa allows a significant freedom of movement of the humeral head on the articulating surface of the glenoid. The movements allowed are: flexion (movement of the humerus to the front — a forward upward movement) and extension (return from flexion), abduction (horizontal upward movement of the humerus to the side) and adduction (return from abduction), and outward rotation (movement of the humerus around its long axis to the lateral side) and inward rotation (return from outward rotation). The configuration of the articulating surfaces of the shoulder joint also makes the joint more susceptible to instability and injury, such as dislocation. The stability of the joint is provided by the glenohumeral and coracohumeral ligaments, and by the muscles crossing the joint. The major muscles of the shoulder joint are: deltoideus, supraspinatus, pectoralis major, coracobrachialis, latissimus dorsi, teres major, teres minor, infraspinatus, and subscapularis.

The bony structure of the shoulder girdle consists of the clavicle (collarbone) and the scapula (shoulder blade). The *acromioclavicular joint* is a small synovial articulation between the distal clavicle and the proximal acromion of the scapula. The stability of this joint is reinforced by the coracoclavicular ligaments. The *sternoclavicular joint* is the articulation between the manubrium of the sternum and the proximal clavicle. The stability of this joint is enhanced by the costoclavicular ligament. Both the acromioclavicular joint and the sternoclavicular joint have layers of cartilage, called *menisci*, interposed between their bony surfaces. There are four pairs of scapular movements: elevation (movement of the scapula in the frontal plane) and depression (return from elevation), upward rotation (turning the glenoid fossa upward and the lower medial border of the scapula away from the spinal column) and downward rotation (return from upward rotation), protraction (movement of the distal end of the clavicle forward) and retraction (return from protraction), and forward and backward rotation (rotation of the scapula about the shaft of the clavicle). There are six muscles that control and coordinate these movements: trapezius, levator scapulae, rhomboid, pectoralis minor, serratus anterior, and subclavius.

Example 6.2 Consider a person strengthening the shoulder muscles by means of dumbbell exercises. Figure 6.11 illustrates the position of the left arm when the arm is abducted to horizontal. The free-body diagram of the arm is shown in Figure 6.12 along with a mechanical model of the arm. Also in Figure 6.12, the forces acting on the arm are resolved into their rectangular components along the horizontal and vertical directions. O corresponds to the axis of the shoulder joint, A is where the deltoid muscle is attached to the humerus, B is the center of gravity of the entire arm, and C is the center of gravity of the dumbbell. W is the weight of the arm, W_0 is the weight of the dumbbell, F_M is the magnitude of the tension in the deltoid muscle, and F_J is the joint reaction force at the shoulder. The resultant of the deltoid muscle force makes an angle θ with the horizontal. The distances between O and A, B and C are measured as a, b, and c, respectively.

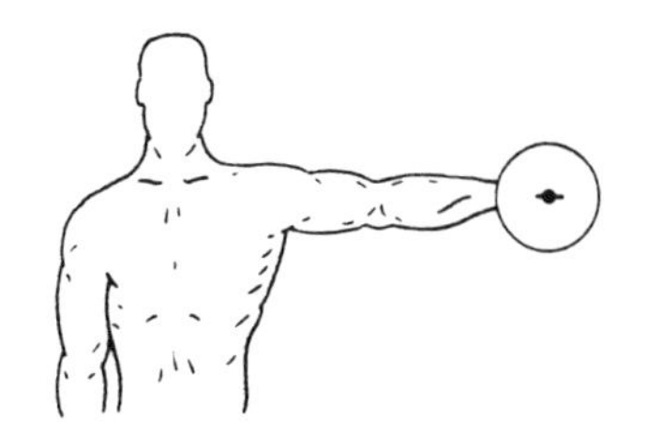

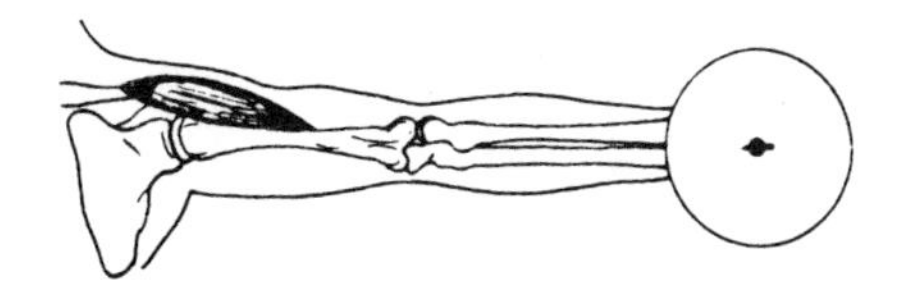

Figure 6.11 *The arm is abducted to horizontal.*

Determine the magnitude F_M of the force exerted by the deltoid muscle to hold the arm at the position shown. Also determine the magnitude and direction of the reaction force at the shoulder joint in terms of specified parameters.

Solution: With respect to the xy coordinate frame, the muscle and joint reaction forces have two components while the weights of the arm and the dumbbell act in the negative y direction. The components of the muscle force are:

$$F_{Mx} = F_M \cos\theta \quad (-x) \qquad (i)$$
$$F_{My} = F_M \sin\theta \quad (+y) \qquad (ii)$$

Components of the joint reaction force are:

$$F_{Jx} = F_J \cos\beta \quad (+x) \qquad (iii)$$
$$F_{Jy} = F_J \sin\beta \quad (-y) \qquad (iv)$$

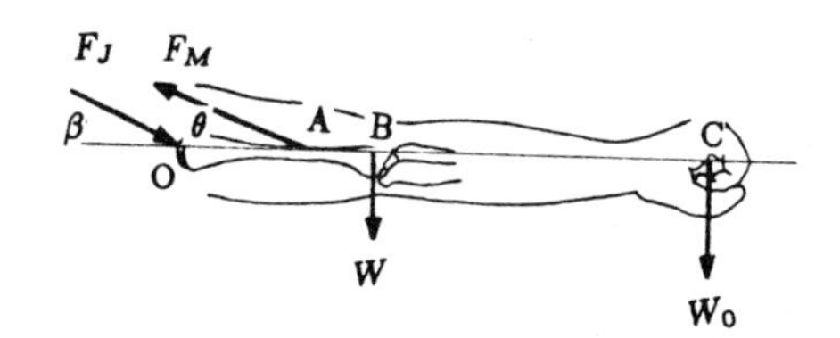

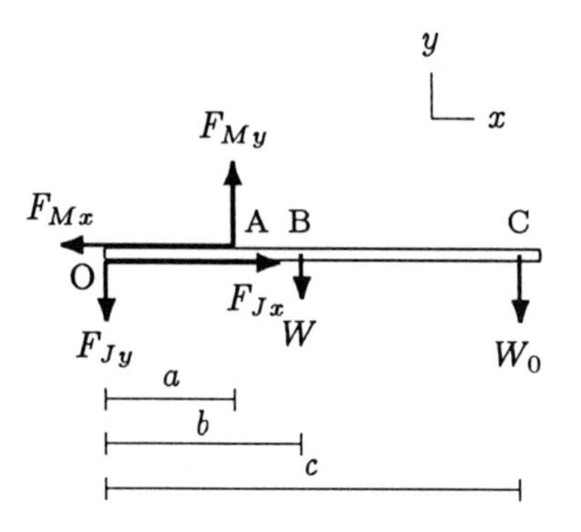

Figure 6.12 *Forces acting on the arm and a mechanical model representing the arm.*

β is the angle that the joint reaction force makes with the horizontal. The line of action and direction (in terms of θ) of the force exerted by the muscle on the arm are known. However, the magnitude F_M of the muscle force, the magnitude F_J, and the direction (β) of the joint reaction force are unknowns. We have a total of three unknowns, F_M, F_J, and β (or F_M, F_{Jx}, and F_{Jy}). To be able to solve this two-dimensional problem, we have to utilize all three equilibrium equations.

First, consider the rotational equilibrium of the arm about the shoulder joint at O. The joint reaction force produces no torque about O because its line of action passes through O. For practical purposes, we can neglect the possible contribution of the horizontal component of the muscle force to the moment generated about O by assuming that its line of action also passes through O. Note that this is not a critical or necessary assumption to solve this problem. If we knew the length of its moment arm (i.e., the vertical distance between O and A), we could easily incorporate the torque generated by F_{Mx} about O into the analysis. Under these considerations, there are only three moment producing forces about O. For the rotational equilibrium of the arm, the net moment about O must be equal to zero. Taking counterclockwise moments to be positive:

$$\sum M_O = 0: \qquad a\,F_{My} - b\,W - c\,W_0 = 0$$

$$F_{My} = \frac{1}{a}(b\,W + c\,W_0) \qquad (v)$$

For given a, b, c, W, and W_0, Eq. (v) can be used to determine the vertical component of the force exerted by the deltoid muscle. Eq. (ii) can now be used to determine the total force exerted by the muscle:

$$F_M = \frac{F_{My}}{\sin\theta} \qquad (vi)$$

Knowing F_M, Eq. (i) will yield the horizontal component of the tension in the muscle:

$$F_{Mx} = F_M\ \cos\theta \qquad (vii)$$

The components of the joint reaction force can be determined by considering the translational equilibrium of the arm in the horizontal and vertical directions:

$$\sum F_x = 0: \qquad F_{Jx} = F_{Mx} \qquad (viii)$$

$$\sum F_y = 0: \qquad F_{Jy} = F_{My} - W - W_0 \qquad (ix)$$

Knowing the rectangular components of the joint reaction force enables us to compute the magnitude of the force itself and the angle its line of action makes with the horizontal:

$$F_J = \sqrt{(F_{Jx})^2 + (F_{Jy})^2} \qquad (x)$$

$$\beta = \tan^{-1}\left(\frac{F_{Jy}}{F_{Jx}}\right) \qquad (xi)$$

Now consider that a=15 cm, b=30 cm, c=60 cm, θ=15°, W=40 N, and W_0=60 N. Then:

$$F_{My} = \frac{1}{0.15}\left[(0.30)(40) + (0.60)(60)\right] = 320 \text{ N} \quad (+y)$$

$$F_M = \frac{320}{\sin 15^\circ} = 1236 \text{ N}$$

$$F_{Mx} = (1236)(\cos 15^\circ) = 1194 \text{ N} \quad (-x)$$

$$F_{Jx} = 1194 \text{ N} \quad (+x)$$

$$F_{Jy} = 320 - 40 - 60 = 220 \text{ N} \quad (-y)$$

$$F_J = \sqrt{(1194)^2 + (220)^2} = 1214 \text{ N}$$

$$\beta = \tan^{-1}\left(\frac{220}{1194}\right) = 10^\circ$$

Remarks:

- F_{Mx} is the stabilizing component and F_{My} is the rotational component of the deltoid muscle. F_{Mx} is approximately four times larger than F_{My}. A large stabilizing component suggests that the horizontal position of the arm is not stable, and that the muscle needs to exert a high horizontal force to stabilize it.

- It is believed that the joint reaction force at the shoulder is the resultant of two components: an active contraction of the infraspinatous muscle and a force generated at the joint in terms of pressure and friction.

- The human shoulder is very susceptible to injuries. The most common injuries are dislocations of the shoulder joint and the fracture of the humerus. Since the socket of the glenohumeral joint is shallow, the head of the humerus is relatively free to rotate about the articulating surface of the glenoid fossa. This freedom of movement is achieved, however, by reduced joint stability. The humeral head may be displaced in various ways, depending on the strength or weakness of the muscular and ligamentous structure of the shoulder, and depending on the physical activity. Humeral fractures are another common type of injuries. The humerus is particularly vulnerable to injuries because of its unprotected configuration.

- Average ranges of motion of the arm about the shoulder joint are 230° during flexion-extension, and 170° in both abduction-adduction and inward-outward rotation.

6.7 Mechanics of the Spinal Column

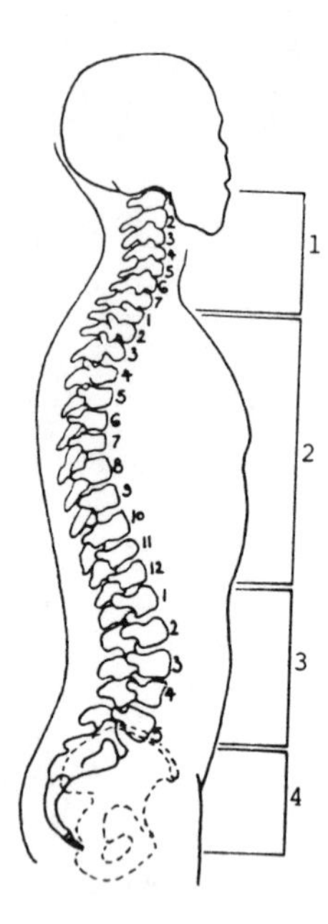

Figure 6.13 *The spinal column: (1) cervical vertabrae, (2) thoracic vertabrae, (3) lumbar vertabrae, (4) sacrum.*

The human spinal column is the most complex part of the human musculoskeletal system. The principal functions of the spinal column are to protect the spinal cord; to support the head, neck, and upper extremities; to transfer loads from the head and trunk to the pelvis; and to permit a variety of movements. The spinal column consists of the cervical (neck), thoracic (chest), lumbar (lower back), sacral, and coccygeal regions. The thoracic and lumbar sections of the spinal column make up the trunk. The sacral and coccygeal regions are united with the pelvis and can be considered parts of the pelvic girdle.

The vertebral column consists of 24 intricate and complex vertebrae (Figure 6.13). The articulations between the vertebrae are amphiarthrodial joints. A fibrocartilaginous disk is interposed between each pair of vertebrae. The primary functions of these intervertebral disks are to sustain loads transmitted from segments above, act as shock absorbers, eliminate bone-to-bone contact, and reduce the effects of impact forces by preventing direct contact between the bony structures of the vertabrae. The articulations of each vertebra with the adjacent vertabrae permit movement in three planes, and the entire spine functions like a single ball-and-socket joint. The structure of the spine allows a wide variety of movements including flexion-extension, lateral flexion, and rotation.

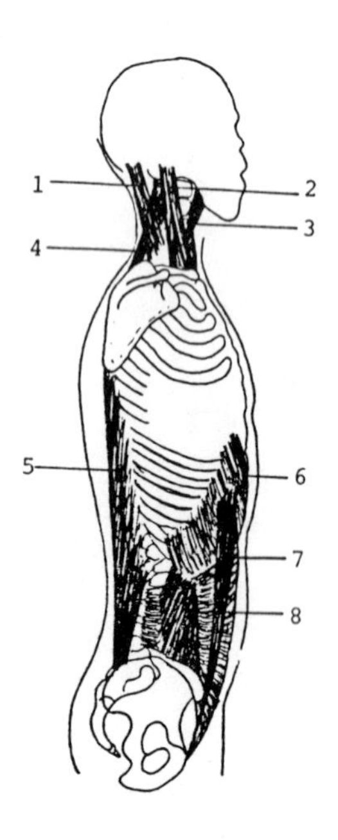

Figure 6.14 *Selected muscles of the neck and spine: (1) splenius, (2) sternocleidomastoid, (3) hyoid, (4) levator scapula, (5) erector spinae, (6) obliques, (7) rectus abdominis, (8) transversus abdominis.*

Two particularly important joints of the spinal column are those with the head. The *atlantooccipital joint* is the union between the first cervical vertebra (the atlas) and the occipital bone of the head. This is a double condyloid joint and permits movements of the head in the sagittal and frontal planes. The *atlantoaxial joint* is the union between the atlas and the odontoid process of the head. It is a pivot joint, enabling the head to rotate in the transverse plane. The muscle groups providing, controlling, and coordinating the movement of the head and the neck are the prevertebrals (anterior), hyoids (anterior), sternocleidomastoid (anterior-lateral), scaleni (lateral), levator scapulae (lateral), suboccipitals (posterior), and spleni (posterior).

The spine gains its stability from the intervertebral discs and from the surrounding ligaments and muscles. The discs and ligaments provide intrinsic stability, and the muscles supply extrinsic support. The muscles of the spine exist in pairs. The anterior portion of the spine contains the abdominal muscles: the rectus abdominis, external obliques, and internal obliques. These muscles provide the necessary force for trunk flexion and maintain the internal organs in proper position. There are three layers of posterior trunk muscles: the erector spinae, the semispinalis, and the deep posterior spinal muscle groups. The primary function of the muscles located at the posterior portion of the spine is to provide trunk extension. These muscles also support the spine against the effects of gravity. The quadratus lumborum muscle is important in lateral trunk flexion. It also stabilizes the pelvis and lumbar spine. The lateral flexion of the trunk results from the actions of the abdominal and posterior muscles. The rotational movement of the trunk is controlled by the simultaneous action of anterior and posterior muscles.

The spinal column is vulnerable to various injuries. The most severe injury involves the spinal cord, which is immersed in fluid and protected by the bony structure. Other critical injuries include fractured vertebrae and herniated intervertebral disks. Lower back pain may also result from strains in the lower regions of the spine.

Example 6.3 Consider the position of the head and the neck shown in Figure 6.15. Also shown are the forces acting on the head. The head weighs W=50 N and its center of gravity is located at C. F_M is the magnitude of the resultant force exerted by the neck extensor muscles, which is applied on the skull at A. The atlantooccipital joint is located at B. For this flexed position of the head, it is estimated that the line of action of the neck muscle force makes an angle θ=30° and the line of action of the joint reaction force makes an angle β=60° with the horizontal.

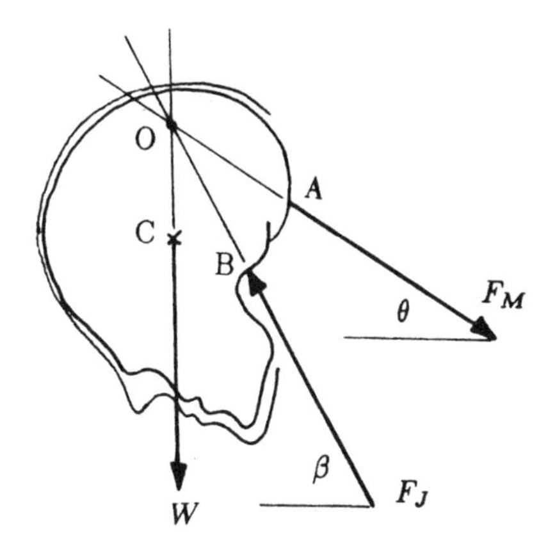

Figure 6.15 *Forces on the skull form a concurrent system.*

What tension must the neck extensor muscles exert to support the head? What is the compressive force applied on the first cervical vertebra at the atlantooccipital joint?

Solution: We have a three-force system with two unknowns: magnitudes F_M and F_J of the muscle and joint reaction forces. Since the problem has a relatively complicated geometry, it is convenient to utilize the condition that for a body to be in equilibrium the force system acting on it must be either concurrent or parallel. In this case, it is clear that the forces involved do not form a parallel force system. Therefore, the system of forces under consideration must be concurrent. Recall that a system of forces is concurrent if the lines of action of all forces have a common point of intersection.

In Figure 6.15, the lines of action of all three forces acting on the head are extended to meet at point O. In Figure 6.16, the forces $\underline{W}$, $\underline{F}_M$, and $\underline{F}_J$ acting on the skull are translated to O, which is also chosen to be the origin of the xy coordinate frame. The rectangular components of the muscle and joint reaction forces in the x and y directions are:

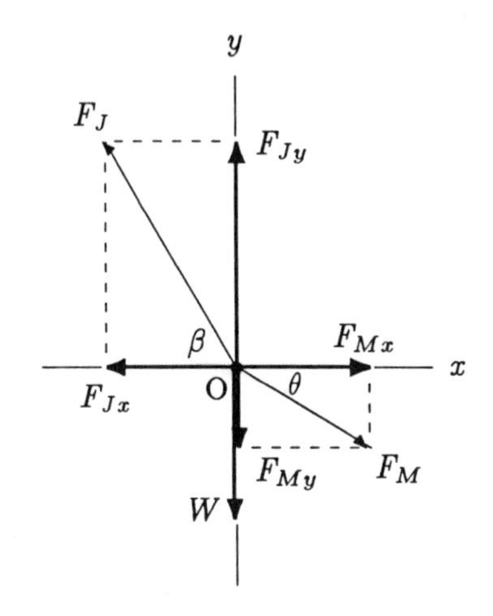

Figure 6.16 *Components of the forces acting on the head.*

$$F_{Mx} = F_M \cos\theta \qquad (i)$$

$$F_{My} = F_M \sin\theta \qquad (ii)$$

$$F_{Jx} = F_J \cos\beta \qquad (iii)$$

$$F_{Jy} = F_J \sin\beta \qquad (iv)$$

The translational equilibrium conditions in the x and y directions will yield:

$$\sum F_x = 0: \qquad F_{Jx} = F_{Mx} \qquad (v)$$

$$\sum F_y = 0: \qquad F_{Jy} = W + F_{My} \qquad (vi)$$

Substitute Eqs. (i) and (iii) into Eq. (v):

$$F_J \cos\beta = F_M \cos\theta \qquad (vii)$$

Substitute Eqs. (ii) and (iv) into Eq. (vi):

$$F_J \sin\beta = W + F_M \sin\theta \qquad (viii)$$

Divide Eq. $(viii)$ by Eq. (vii) so as to eliminate F_J:

$$\tan\beta = \frac{W + F_M \sin\theta}{F_M \cos\theta} \qquad (ix)$$

Eq. (ix) can now be solved for the unknown muscle force F_M:

$$F_M \cos\theta \tan\beta = W + F_M \sin\theta$$

$$F_M (\cos\theta \tan\beta - \sin\theta) = W$$

$$F_M = \frac{W}{\cos\theta \tan\beta - \sin\theta} \qquad (x)$$

Eq. (x) gives the tension in the muscle as a function of the weight W of the head and the angles θ and β that the lines of action of the muscle and joint reaction forces make with the horizontal. Substituting the numerical values of W, θ, and β will yield:

$$F_M = \frac{50}{(\cos 30^\circ)(\tan 60^\circ) - (\sin 30^\circ)} = 50 \text{ N}$$

From Eqs. (i) and (ii):

$$F_{Mx} = (50)(\cos 30^\circ) = 43 \text{ N} \quad (+x)$$

$$F_{My} = (50)(\sin 30^\circ) = 25 \text{ N} \quad (-y)$$

From Eqs. (v) and (vi):

$$F_{Jx} = 43 \text{ N} \quad (-x)$$

$$F_{Jy} = 50 + 25 = 75 \text{ N} \quad (+y)$$

The resultant of the joint reaction force can be computed either from Eq. (iii) or (iv). Using Eq. (iii):

$$F_J = \frac{F_{Jx}}{\cos\beta} = \frac{43}{\cos 60^\circ} = 86 \text{ N}$$

Remarks:

- The extensor muscles of the head must apply a force of 50 N to support the head in the position considered. The reaction force developed at the atlantooccipital joint is about 86 N.

- The joint reaction force can be resolved into two rectangular components, as shown in Figure 6.17. F_{Jn} is the magnitude of the normal component of $\underline{F}_J$ compressing the articulating joint surface, and F_{Jt} is the magnitude of its tangential component having a shearing effect on the joint surfaces. Forces in the muscles and ligaments of the neck operate in a manner to counterbalance this shearing effect.

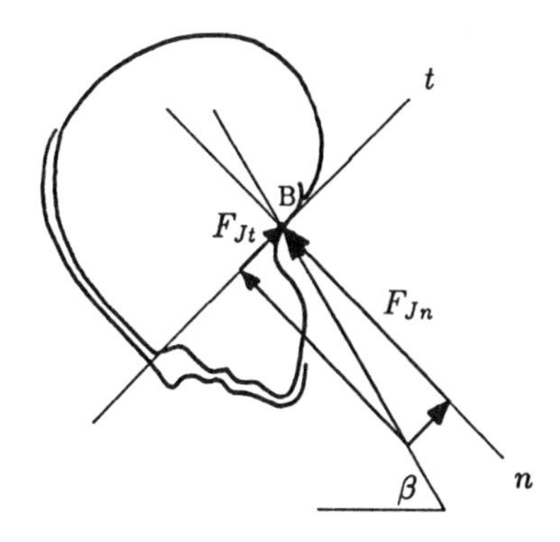

Figure 6.17 *Normal and shear components of the joint reaction force.*

Example 6.4 Consider the weight lifter illustrated in Figure 6.18, who is bent forward and lifting a weight W_0. At the position shown, the athlete's trunk is flexed by an angle θ as measured from the upright (vertical) position.

The forces acting on the lower portion of the athlete's body are shown in Figure 6.19 by considering a section passing through the fifth lumbar vertebra. A mechanical model of the athlete's lower body (the pelvis and legs) is illustrated in Figure 6.20 along with the geometric parameters of the problem under consideration. W is the total weight of the athlete, W_1 is the weight of the legs including the pelvis, $W + W_0$ is the total ground reaction force applied to the athlete through the feet (at C), F_M is the magnitude of the resultant force exerted by the erector spinae muscles supporting the trunk, and F_J is the magnitude of the compressive force generated at the union (O) of the sacrum and the fifth lumbar vertebra. The center of gravity of the legs including the pelvis is located at B. Relative to O, the lengths of the lever arms of the muscle force, lower body weight, and ground reaction force are measured as a, b, and c, respectively.

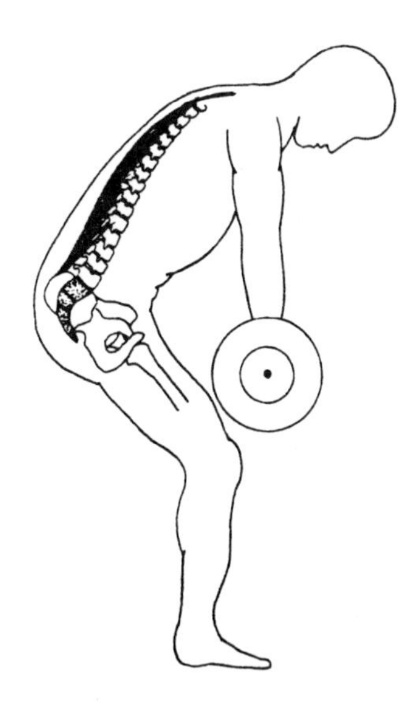

Figure 6.18 *A weight lifter.*

Assuming that the line of pull of the resultant muscle force exerted by the erector spinae muscles is parallel to the trunk (i.e., making an angle θ with the vertical), determine F_M and F_J in terms of a, b, c, θ, W_0, W_1, and W.

Solution: In this case, there are three unknowns: F_M, F_{Vx}, and F_{Vy}. The lengths of the lever arms of the muscle force, ground reaction force, and the gravitational force of the legs including the pelvis are given as measured from O. Therefore, we can apply the rotational equilibrium condition about O to determine the magnitude F_M of the resultant force exerted by the erector spinae muscles. Considering clockwise moments to be positive:

$$\sum M_O = 0: \qquad a\,F_M + b\,W_1 - c\,(W + W_0) = 0$$

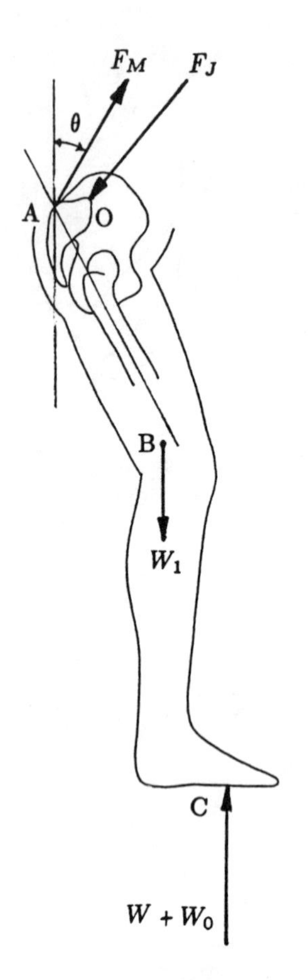

Figure 6.19 *Forces acting on the lower body of the athlete.*

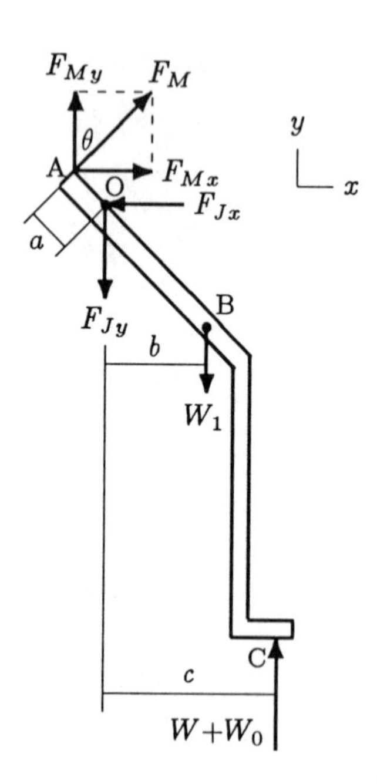

Figure 6.20 *Free-body diagram.*

Solving this equation for F_M will yield:

$$F_M = \frac{c\,(W + W_O) - b\,W_1}{a} \qquad (i)$$

For given numerical values of a, b, c, W, W_O, and W_1, Eq. (i) can be used to determine the magnitude of the resultant muscle force. Once F_M is calculated, its components in the x and y directions can be determined using:

$$F_{Mx} = F_M \sin\theta \qquad (ii)$$

$$F_{My} = F_M \cos\theta \qquad (iii)$$

The horizontal and vertical components of the reaction force developed at the sacrum can now be determined by utilizing the translational equilibrium conditions of the lower body of the athlete in the x and y directions:

$$\sum F_x = 0: \quad F_{Jx} = F_{Mx} \qquad (iv)$$

$$\sum F_y = 0: \quad F_{Jy} = F_{My} + W + W_O - W_1 \qquad (v)$$

Assume that at an instant the athlete is bent so that his trunk makes an angle θ=45° with the vertical, and that the lengths of the lever arms are measured in terms of the height h of the athlete and the weights are given in terms of the weight W of the athlete as: a=0.02h, b=0.08h, c=0.12h, W_O=W and W_1=0.4W. Using Eq. (i):

$$F_M = \frac{(0.12\,h)(W + W) - (0.08\,h)(0.4\,W)}{0.02\,h} = 10.4\,W$$

From Eqs. (ii) and (iii):

$$F_{Mx} = (10.4\,W)(\sin 45°) = 7.4\,W$$

$$F_{My} = (10.4\,W)(\sin 45°) = 7.4\,W$$

From Eqs. (iv) and (v):

$$F_{Jx} = 7.4\,W$$

$$F_{Jy} = 7.4\,W + W + W - 0.4\,W = 9.0\,W$$

Therefore, the magnitude of the resultant force on the sacrum is:

$$F_J = \sqrt{(F_{Jx})^2 + (F_{Jy})^2} = 11.7\,W$$

A remark:

- The results obtained are quite significant. While the athlete is bent forward by 45° and lifting a weight with magnitude equal to his own body weight, the erector spinae muscles exert a force more than 10 times the weight of the athlete and the force applied to the union of the sacrum and the fifth lumbar vertebra is about 12 times that of the body weight.

6.8 Mechanics of the Hip

The articulation between the head of the femur and the acetabulum of the pelvis (Figure 6.21) forms a diarthrodial joint. The stability of the hip joint is provided by its relatively rigid ball-and-socket type of configuration, its ligaments, and by the large and strong muscles crossing it. The femoral head fits well into the deep socket of the acetabulum. The ligaments of the hip joint, such as the tranverse and teres femoris ligaments, support and hold the femoral head in the acetabulum as the femoral head moves. The construction of the hip joint is such that it is very stable and has a great deal of mobility, thereby allowing a wide range of motion required for activities such as walking, sitting, and squatting. Movements of the femur about the hip joint include flexion and extension, abduction and adduction, and inward and outward rotation. In some instances, the extent of these movements is constrained by ligaments, muscles, and/or the bony structure of the hip. The articulating surfaces of the femoral head and the acetabulum are lined with hyaline cartilage. Derangements of the hip can produce altered force distributions in the joint cartilage, leading to degenerative arthritis.

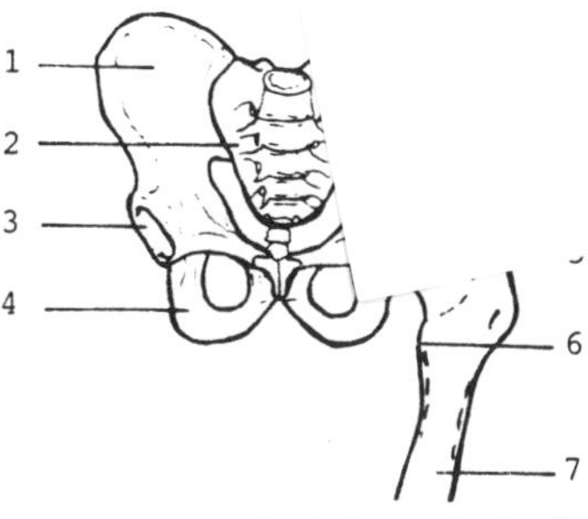

Figure 6.21 *Pelvis and the hip: (1) ilium, (2) sacrum, (3) acetabulum (4) ischium, (5) greater trochanter, (6) lesser trochanter, (7) femur.*

The pelvis consists of the ilium, ischium, and pubis bones, and the sacrum. At birth and during growth the bones of the pelvis are distinct. In adults the bones of the pelvis are fused and form synarthrodial joints which allow no movement. The pelvis is located between the spine and the two femurs. The position of the pelvis makes it relatively less stable. Movements of the pelvis occur primarily for the purpose of facilitating the movements of the spine or the femurs. There are no muscles whose primary purpose is to move the pelvis. Movements of the pelvis are caused by the muscles of the trunk and the hip.

Based on their primary actions, the muscles of the hip joint can be divided into several groups (Figure 6.22). The psoas, iliacus, rectus femoris, pectineus, and tensor fascia latae are the primary hip flexors. They are also used to carry out activities such as running or kicking. The gluteus maximus and the hamstring muscles (the biceps femoris, semitendinosus, and semimembranosus) are hip extensors. The hamstring muscles also function as knee flexors. The gluteus medius and gluteus minimus are hip abductor muscles providing for the inward rotation of the femur. The gluteus medius is also the primary muscle group stabilizing the pelvis in the frontal plane. The adductor longus, adductor brevis, adductor magnus, and gracilis muscles are the hip adductors. There are also small, deeply placed muscles (outward rotators) which provide for the outward rotation of the femur.

The hip muscles predominantly suffer contusions and strains occurring in the pelvis region.

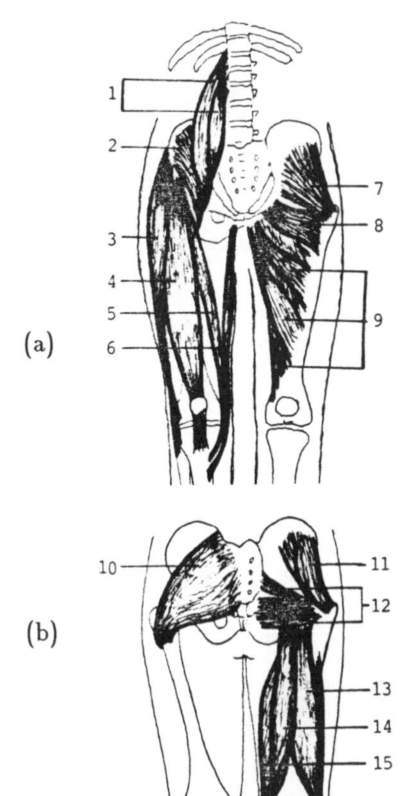

Figure 6.22 *Muscles of the hip: (a) anterior & (b) posterior views. (1) psoas, (2) iliacus, (3) tensor fascia latae, (4) rectus femoris, (5) sartorius, (6) gracilis, (7) gluteus minimis, (8) pectineus, (9) adductors (10),(11) gluteus maximus & medius (12) lateral rotators, (13) biceps femoris, (14) semitendinosus, (15) semimembranosus.*

Example 6.5 During walking and running, we momentarily put all of our body weight on one leg (the right leg in Figure 6.23). The forces acting on the leg carrying the total body weight are shown in Figure 6.24 during such a single-leg stance. F_M is the magnitude of the resultant force exerted by the hip abductor muscles, F_J is the magnitude of the joint reaction force applied by the pelvis on the femur, W_1 is the weight of the leg, W is the total weight of the body applied as a normal force by the ground on the leg. The angle between the line of action of the resultant muscle force and the horizontal is designated by θ.

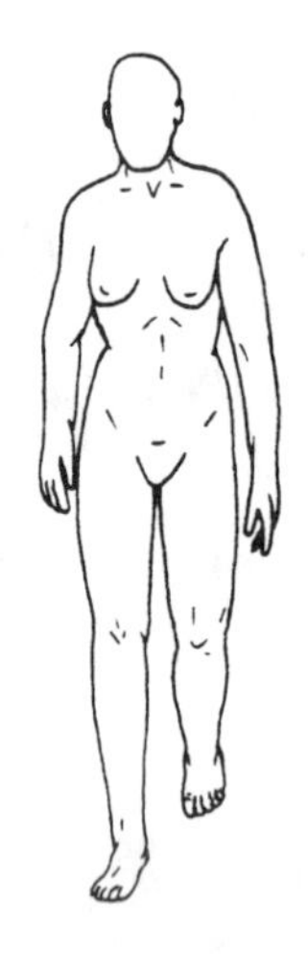

Figure 6.23 *Single-leg stance.*

A mechanical model of the leg, rectangular components of the forces acting on it, and the parameters necessary to define the geometry of the problem are shown in Figure 6.25. O is a point along the instantaneous axis of rotation of the hip joint, A is where the hip abductor muscles are attached to the femur, B is the center of gravity of the leg, and C is where the ground reaction force is applied on the foot. The distances between A and O, B, and C are specified as a, b, and c, respectively. α is the angle of inclination of the femoral neck to the horizontal, and β is the angle that the long axis of the femoral shaft makes with the horizontal. Therefore, $\alpha+\beta$ is approximately equal to the total neck-to-shaft angle of the femur.

Determine the force exerted by the hip abductor muscles and the joint reaction force at the hip to support the leg and the hip in the position shown.

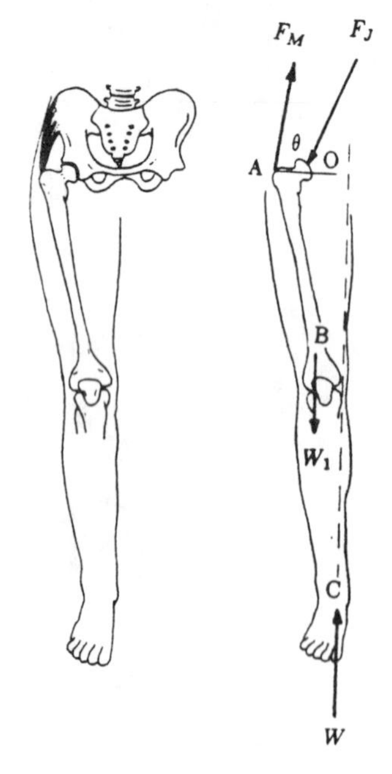

Figure 6.24 *Forces acting on the right leg carrying the entire weight of the body.*

Solution 1: Utilizing the free-body diagram of the leg.
For the solution of the problem, we can utilize the free-body diagram of the right leg supporting the entire weight of the person. In Figure 6.25(a), the muscle and joint reaction forces are shown in terms of their components in the x and y directions. The resultant muscle force has a line of action that makes an angle θ with the horizontal. Therefore:

$$F_{Mx} = F_M \cos\theta \qquad (i)$$

$$F_{My} = F_M \sin\theta \qquad (ii)$$

Since angle θ is specified (given as a measured quantity), the only unknown for the muscle force is its magnitude F_M. For the joint reaction force, neither the magnitude nor the direction is known. With respect to the axis of the hip joint located at O, a_x in Figure 6.25(b) is the moment arm of the vertical component F_{My} of the muscle force, and a_y is the moment arm of F_{Mx}. Similarly, $(b_x - a_x)$ is the moment arm for W_1 and $(c_x - a_x)$ is the moment arm for the force W applied by the ground on the leg.

From the geometry of the problem:

$$a_x = a \cos\alpha \qquad (iii)$$
$$a_y = a \sin\alpha \qquad (iv)$$
$$b_x = b \cos\beta \qquad (v)$$
$$c_x = c \cos\beta \qquad (vi)$$

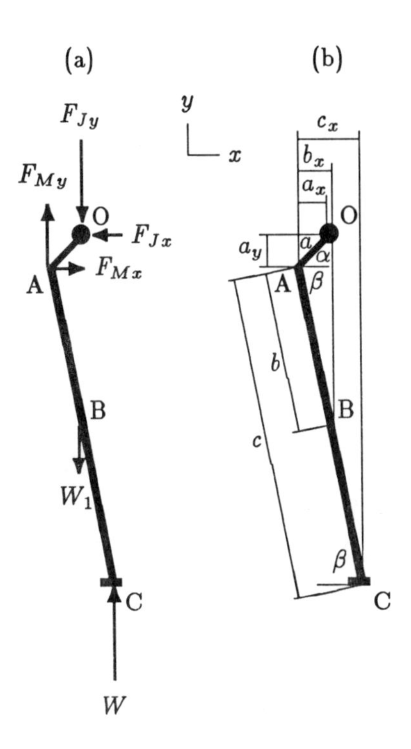

Figure 6.25 *Free-body diagram of the leg (a), and the geometric parameters (b).*

Now that the horizontal and vertical components of all forces involved, and their moment arms with respect to O are established, the condition for the rotational equilibrium of the leg about O can be utilized to determine the magnitude of the resultant muscle force applied at A. Assuming that the clockwise moments are positive:

$$\sum M_O = 0: \quad a_x F_{My} - a_y F_{Mx} - (c_x - a_x) W + (b_x - a_x) W_1 = 0$$

Substituting Eqs. (*i*) through (*vi*) into the above equation:

$$(a \cos\alpha)(F_M \sin\theta) - (a \sin\alpha)(F_M \cos\theta) - (c \cos\beta - a \cos\alpha)\, W + (b \cos\beta - a \cos\alpha)\, W_1 = 0$$

Solving this equation for the muscle force:

$$F_M = \frac{(c\,W - b\,W_1) \cos\beta - a\,(W - W_1) \cos\alpha}{a\,(\cos\alpha \sin\theta - \sin\alpha \cos\theta)} \qquad (vii)$$

Notice that the denominator of Eq. (*vii*) can be simplified as $a \sin(\theta - \alpha)$. To determine the components of the joint reaction force, we can utilize the horizontal and vertical equilibrium conditions of the leg:

$$\sum F_x = 0: \quad F_{Jx} = F_{Mx} = F_M \cos\theta \qquad (viii)$$
$$\sum F_y = 0: \quad F_{Jy} = F_{My} + W - W_1$$
$$F_{Jy} = F_M \sin\theta + W - W_1 \qquad (ix)$$

Therefore, the resultant force acting at the hip joint is:

$$F_J = \sqrt{(F_{Jx})^2 + (F_{Jy})^2} \qquad (x)$$

Assume that the geometric parameters of the problem and the weight of the leg are measured in terms of the person's height h and total weight W as follows: a=0.05h, b=0.20h, c=0.52h, α=45°, β=80°, θ=70°, and W_1=0.17W. The solution of the above equations for the muscle and joint reaction forces will yield F_M=2.6W and F_J=3.4W, the joint reaction force making an angle $\phi=\tan^{-1}(F_{Jy}/F_{Jx})$=74.8°.

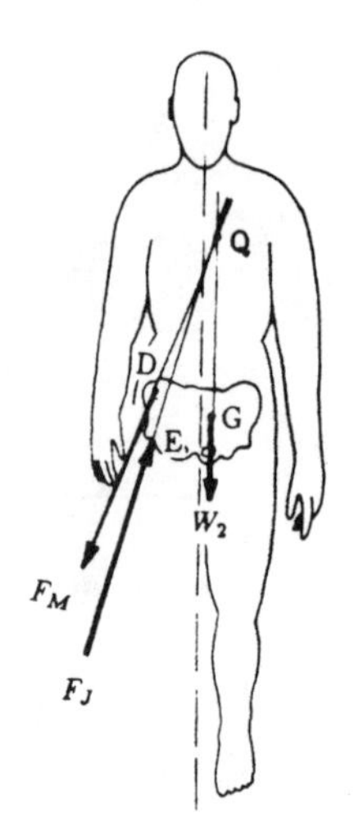

Figure 6.26 *Forces acting on the pelvis during a single-leg (right leg) stance.*

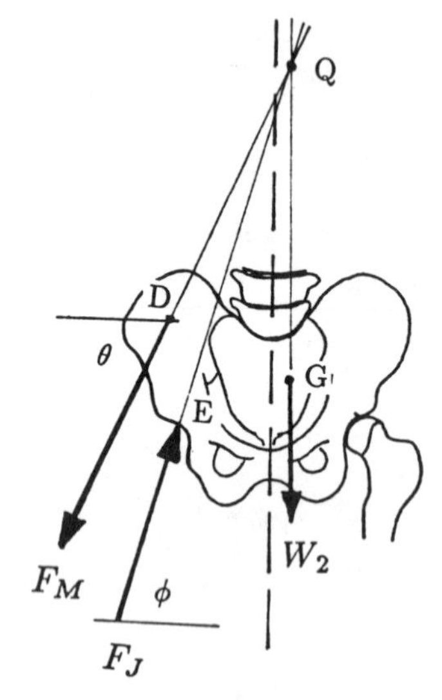

Figure 6.27 *Forces involved form a concurrent system.*

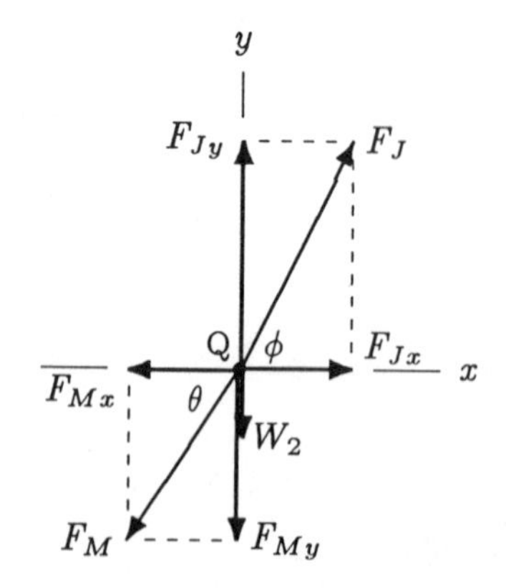

Figure 6.28 *Resolution of the forces into their components.*

Solution 2: Utilizing the free-body diagram of the upper body. Here we have an alternative approach to the solution of the same problem. In this case, instead of the free-body diagram of the right leg, the free-body diagram of the upper body (including the left leg) is utilized. The forces acting on the upper body are shown in Figures 6.26 and 6.27. F_M is the magnitude of the resultant force exerted by the hip abductor muscles applied on the pelvis at D. θ is again the angle between the line of action of the resultant muscle force and the horizontal. F_J is the magnitude of the reaction force applied by the head of the femur on the hip joint at E. W_2=W–W_1 (total body weight minus the weight of the right leg) is the weight of the upper body and the left leg acting as a concentrated force at G. Note that G is not the center of gravity of the entire body. Since the right leg is not included in the free-body, the left-hand side of the body is "heavier" than the right-hand side, and G is located to the right of the original center of gravity (a point along the vertical dashed line in Figure 6.27) of the person. The location of G can be determined utilizing the method provided in Section 5.14.

By combining the individual weights of the segments constituting the body under consideration, the problem is reduced to a three-force system. It is clear from the geometry of the problem that the forces involved do not form a parallel system. Therefore, for the equilibrium of the body, they have to form a concurrent system of forces. This implies that the lines of action of the forces must have a common point of intersection (Q in Figure 6.27), which can be obtained by extending the lines of action of $\underline{W}_2$ and $\underline{F}_M$. A line passing through points Q and E designates the line of action of the joint reaction force $\underline{F}_J$. The angle ϕ that $\underline{F}_J$ makes with the horizontal can now be measured from the geometry of the problem. Since the direction of $\underline{F}_J$ is determined through certain geometric considerations, the number of unknowns is reduced by one. As illustrated in Figure 6.28, the unknown magnitudes F_M and F_J of the muscle and joint reaction forces can now be determined simply by translating $\underline{W}_2$, $\underline{F}_M$, and $\underline{F}_J$ to point Q, and decomposing them into their components along the horizontal (x) and vertical (y) directions:

$$F_{Mx} = F_M \cos\theta$$
$$F_{My} = F_M \sin\theta$$
$$F_{Jx} = F_J \cos\phi$$
$$F_{Jy} = F_J \sin\phi$$

For the translational equilibrium in the x and y directions:

$$\sum F_x = 0: \qquad F_{Jx} = F_{Mx}$$
$$\sum F_y = 0: \qquad F_{Jy} = F_{My} + W_2$$

Simultaneous solutions of these equations will yield:

$$F_M = \frac{\cos\phi\, W_2}{\cos\theta \sin\phi - \sin\theta\cos\phi} \qquad (xi)$$

$$F_J = \frac{\cos\theta\, W_2}{\cos\theta \sin\phi - \sin\theta\cos\phi} \qquad (xii)$$

For example, if θ=70°, ϕ=74.8°, and W_2=0.83W (W is the total weight of the person), then Eqs. (xi) and (xii) will yield F_M=2.6W and F_J=3.4W.

How would the muscle and hip joint reaction forces vary if the person is carrying a load of W_0 in each hand during single-leg stance (Figure 6.29)?

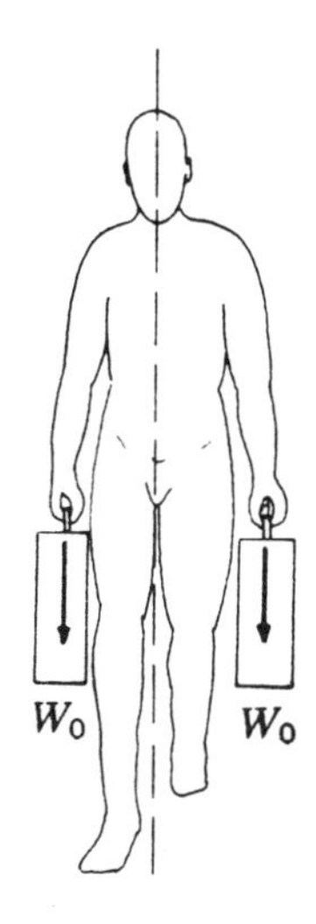

Figure 6.29 *Carrying a load in each hand.*

The free-body diagram of the upper body while the person is carrying a load of W_0 in each hand is shown in Figure 6.30. The system to be analyzed consists of the upper body of the person (including the left leg) and the loads carried in each hand. To counter-balance both the rotational and translational (downward) effects of the extra loads, the hip abductor muscles will exert additional forces, and there will be larger compressive forces generated at the hip joint.

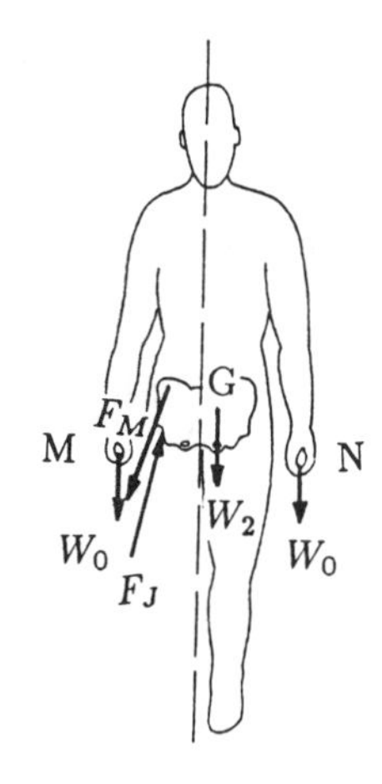

Figure 6.30 *Forces acting on the upper body.*

In this case, the number of forces is five. The gravitational pull on the upper body (W_2) and on the masses carried in the hands (W_0) form a parallel force system. If these parallel forces can be replaced by a single resultant force, then the number of forces can be reduced to three, and the problem can be solved by applying the same technique explained above (Solution 2). For this purpose, consider the force system shown in Figure 6.31. M and N correspond to the right and left hands of the person where external forces of equal magnitude (W_0) are applied. G is the center of gravity of the upper body including the left leg. The vertical dashed line shows the symmetry axis (midline) of the person in the frontal plane, and G is located to the left of this axis. Note that the distance ℓ_1 between M and G is greater than the distance ℓ_2 between N and G. If ℓ_1, ℓ_2, W_2, and W_0 are given, then a new center of gravity (G′) can be determined by applying the technique of finding the center of gravity of a system composed of a number of parts whose centers of gravity are known (see Section 5.14). By intuition, G′ is located somewhere between the symmetry axis and G. In other words, G′ is closer to the right hip joint, and therefore, the length of the moment arm of the total weight as measured from the right hip joint is shorter as compared to the case when there is no load carried in the hands. On the other hand, the magnitude of the resultant gravitational force is W_3=W_2+2W_0, which over compensates for the advantage gained by the reduction of the moment arm.

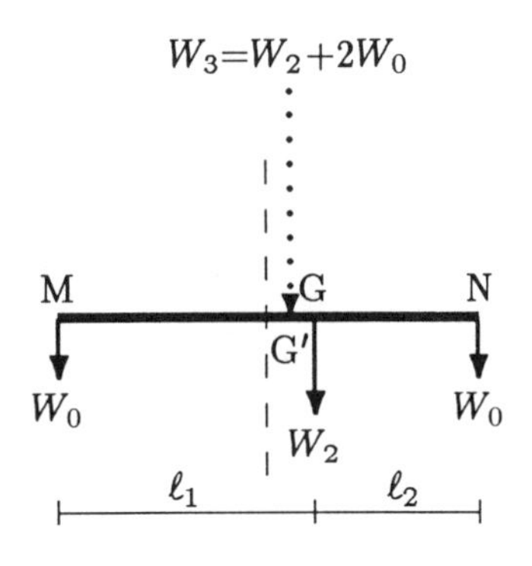

Figure 6.31 $\underline{W}_3$ *is the resultant of the three-force system.*

Once the new center of gravity of the upper body is determined, including the left leg and the loads carried in each hand, Eqs. (xi)

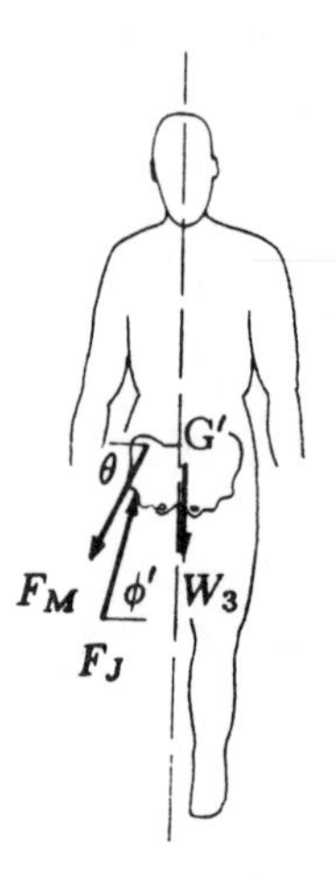

Figure 6.32 *The problem is reduced to a three-force concurrent system.*

and (*xii*) can be utilized to calculate the resultant force exerted by the hip abductor muscles and the reaction force generated at the hip joint:

$$F_M = \frac{\cos\phi'\,(W_2 + 2W_0)}{\cos\theta\sin\phi' - \sin\theta\cos\phi'}$$

$$F_J = \frac{\cos\theta\,(W_2 + 2W_0)}{\cos\theta\sin\phi' - \sin\theta\cos\phi'}$$

Here, Eqs. (*xi*) and (*xii*) are modified by replacing the weight W_2 of the upper body with the new total weight $W_3{=}W_2{+}2W_0$, and by replacing the angle ϕ that the line of action of the joint reaction force makes with the horizontal with the new angle ϕ' (Figure 6.32). ϕ' is slightly larger than ϕ because of the shift of the center of gravity from G to G′ towards the right of the person. Also, it is assumed that the angle θ between the line of action of the muscle force and the horizontal remains unchanged.

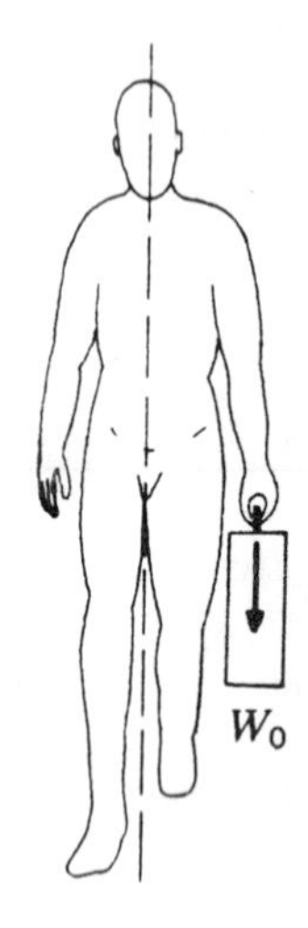

Figure 6.33 *Carrying a load in one hand.*

What happens if the person is carrying a load of W_0 in the left hand during a right-leg stance (Figure 6.33)?

Assuming that the system we are analyzing consists of the upper body, left leg, and the load in hand, the extra load W_0 carried in the left hand will shift the center of gravity of the system from G to G″ towards the left of the person. Consequently the length of the lever arm of the total gravitational force $W_4{=}W_2{+}W_0$ as measured from the right hip joint (Figure 6.34) will increase. This will require larger hip abductor muscle forces to counterbalance the clockwise rotational effect of W_4 and also increase the compressive forces at the right hip joint.

It can be observed from the geometry of the system analyzed that a shift in the center of gravity from G to G″ towards the left of the person will decrease the angle between the line of action of the joint reaction force and the horizontal from ϕ to ϕ''. For the new configuration of the free-body shown in Figure 6.34, Eqs. (*xi*) and (*xii*) can again be utilized to calculate the required hip abductor muscle force and joint reaction force produced at the right hip (opposite to the side where the load is carried):

$$F_M = \frac{\cos\phi''\,(W_2 + W_0)}{\cos\theta\sin\phi'' - \sin\theta\cos\phi''}$$

$$F_J = \frac{\cos\theta\,(W_2 + W_0)}{\cos\theta\sin\phi'' - \sin\theta\cos\phi''}$$

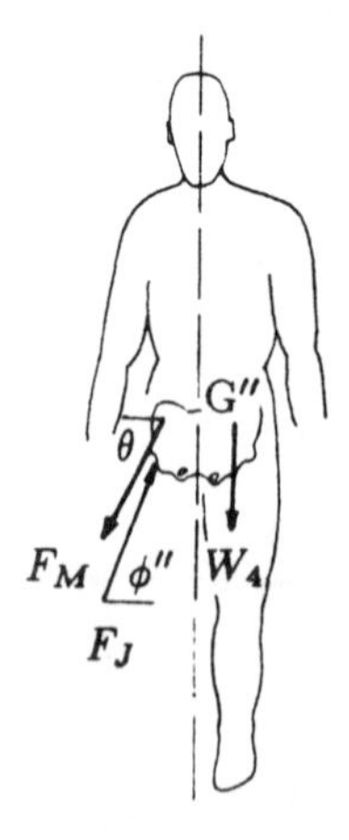

Figure 6.34 *Forces acting on the upper body.*

Remarks:

- When the body weight is supported equally on both feet, half of the supra-femoral weight falls on each hip joint. During walking and running, the entire mass of the body is momentarily supported by one joint, and we have analyzed some of these cases.

- The above analyses indicate that the supporting forces required at the hip joint are greater when a load is carried on the opposite side of the body as compared to the forces required to carry the load when it is distributed on either side. Carrying loads by using both hands and by bringing the loads closer to the midline of the body is effective in reducing required musculoskeletal forces.

- While carrying a load on one side, people tend to lean towards the other side. This brings the center of gravity of the upper body and the load being carried in the hand closer to the midline of the body, thereby reducing the length of the moment arm of the resultant gravitational force as measured from the hip joint distal to the load.

- People with weak hip abductor muscles and/or painful hip joints usually lean towards the weaker side and walk with a so-called abductor gait. Leaning the trunk sideways towards the affected hip shifts the center of gravity of the body closer to that hip joint, and consequently reduces the rotational action of the moment of the body weight about the hip joint by reducing its moment arm. This in return reduces the magnitude of the forces exerted by the hip abductor muscles required to stabilize the pelvis.

- Abductor gait can be corrected more effectively with a cane held in the hand opposite to the weak hip, as compared to the cane held in the hand on the same side as the weak hip.

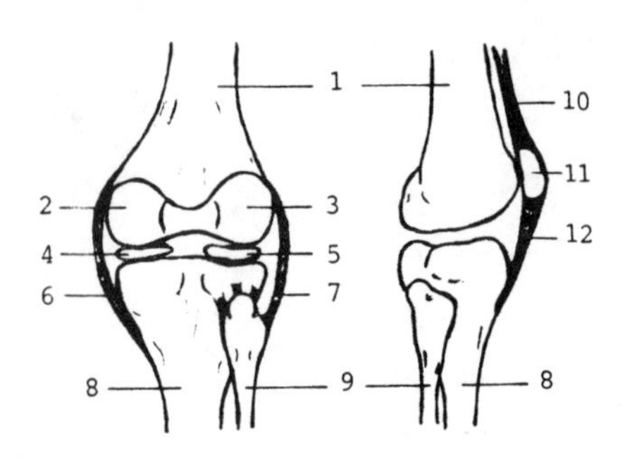

Figure 6.35 *The knee: (1) femur, (2), medial condyle, (3) lateral condyle, (4) medial meniscus, (5) lateral meniscus, (6) tibial collateral ligament, (7) fibular collateral ligament, (8) tibia, (9) fibula, (10) quadriceps tendon, (11) patella, (12) patellar ligament.*

6.9 Mechanics of the Knee

The knee is the largest joint in the body. It is a modified hinge joint. In addition to flexion and extension action of the leg in the sagittal plane, the knee joint permits some inward and outward rotation. The knee joint is designed to sustain large loads. It is an essential component of the linkage system responsible for human locomotion. The knee is extremely vulnerable to injuries.

The knee is a two-joint structure composed of the tibiofemoral joint and the patellofemoral joint (Figure 6.35). The *tibiofemoral joint* has two distinct articulations between the medial and lateral condyles of the femur and the tibia. These articulations are separated by layers of cartilage, called *menisci*. The lateral and medial menisci eliminate bone-to-bone contact between the femur and the tibia, and function as shock absorbers. The *patellofemoral joint* is the articulation between the patella and the anterior end of the femoral condyles. The patella is a "floating" bone kept in position by the quadriceps tendon and the patellar ligament. It protects the knee from impact-related injuries and improves the pulling effect of the quadriceps muscles on the tibia via the patellar tendon. The stability of the knee is provided by an intricate ligamentous structure, the menisci and the muscles crossing the joint. Most knee injuries are characterized by ligament and cartilage damage occurring on the medial side.

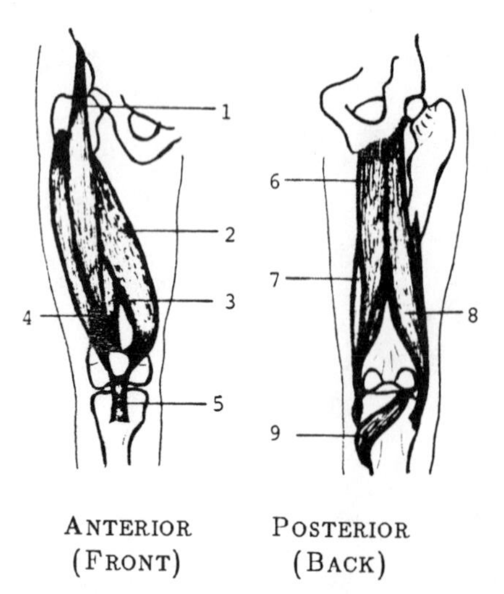

Figure 6.36 *Muscles of the knee: (1) rectus femoris, (2) vastus medialis, (3) vastus intermedius, (4) vastus lateralis, (5) patellar ligament, (6) semitendinosus, (7) semimembranosus, (8) biceps femoris, (9) gastrocnemius.*

The muscles crossing the knee protect it, provide internal forces for movement, and/or control its movement. The muscular control of the knee is produced primarily by the quadriceps muscles and the hamstring muscle group (Figure 6.36). The quadriceps muscle group is composed of the rectus femoris, vastus lateralis, vastus medialis, and vastus intermedius muscles. The rectus femoris muscle has attachments at the anterior-inferior iliac spine and the patella, and its primary actions are the flexion of the hip and the extension of the knee. The vastus lateralis, medialis, and intermedius muscles connect the femur and tibia through the patella, and they are all knee extensors. The biceps femoris, semitendinosus, and semimembranosus muscles make up the hamstring muscle group, which help control the extension of the hip, flexion of the knee, and some inward-outward rotation of the tibia. Semitendinosus and semimembranosus muscles have proximal attachments on the pelvic bone and distal attachments on the tibia. The biceps femoris has proximal attachments on the pelvic bone and the femur, and distal attachments on the tibia and fibula. There is also the popliteus muscle that has attachments on the femur and tibia. The primary function of this muscle is knee flexion. The other muscles of the knee are sartorius, gracilis, gastrocnemius, and plantaris.

Example 6.6 Consider a person wearing a weight boot, and from a sitting position, doing lower leg flexion/extension exercises to strengthen the quadriceps muscles (Figure 6.37).

Forces acting on the lower leg and a simple mechanical model of the leg are illustrated in Figure 6.38. W_1 is the weight of the lower leg, W_0 is the weight of boot, F_M is the magnitude of the tensile force exerted by the quadriceps muscle on the tibia through the patellar tendon, and F_J is the magnitude of the tibiofemoral joint reaction force applied by the femur on the tibial plateau. The tibiofemoral joint is located at O, the patellar tendon is attached to the tibia at A, the center of gravity of the lower leg is located at B, and the center of gravity of the weight boot is located at C. The distances between O and A, B, and C are measured as a, b, and c, respectively. For the position of the lower leg shown, the long axis of the tibia makes an angle β with the horizontal, and the line of action of the quadriceps muscle force makes an angle θ with the long axis of the tibia.

Assuming that points O, A, B, and C all lie along a straight line, determine F_M and F_J in terms of a, b, c, θ, β, W_1, and W_0.

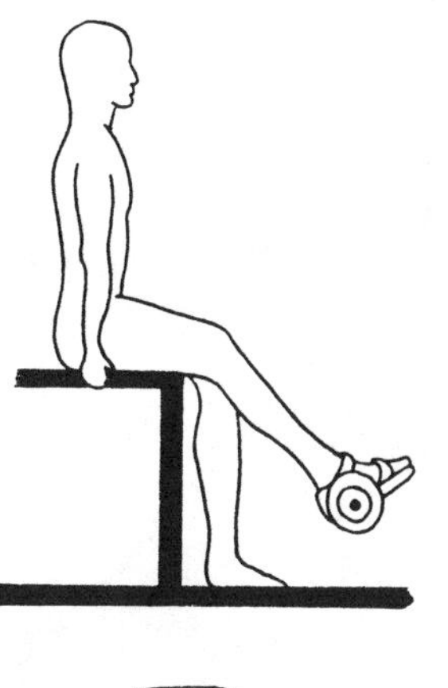

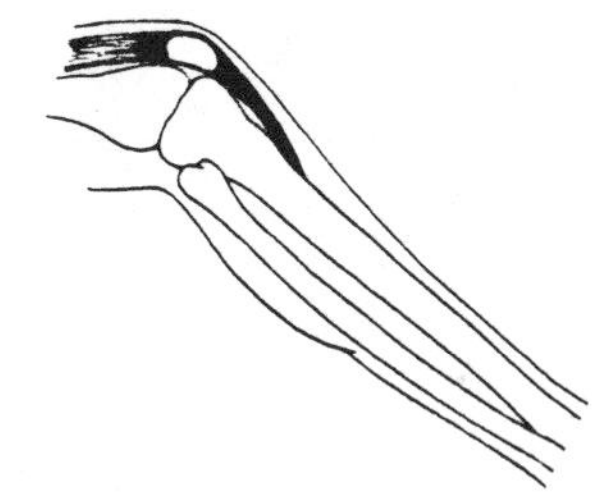

Figure 6.37 *Exercising the muscles around the knee joint.*

Solution: Horizontal (x) and vertical (y) components of the forces acting on the trunk and their lever arms as measured from the knee joint located at O are shown in Figure 6.39. The components of the muscle force are:

$$F_{Mx} = F_M \cos(\theta + \beta) \qquad (i)$$

$$F_{My} = F_M \sin(\theta + \beta) \qquad (ii)$$

There are three unknowns, namely F_M, F_{Jx}, and F_{Jy}. For the solution of this two-dimensional (plane) problem, all three equilibrium conditions must be utilized. Assuming that the counterclockwise moments are positive, consider the rotational equilibrium of the lower leg about O:

$$\sum M_O = 0: \quad (a\cos\beta)\,F_{My} - (a\sin\beta)\,F_{Mx} - (b\cos\beta)\,W_1 - (c\cos\beta)\,W_0 = 0$$

Substituting Eqs. (i) and (ii) into the above equation, and solving it for F_M will yield:

$$F_M = \frac{(b\,W_1 + c\,W_0)\cos\beta}{a\left[\cos\beta\sin(\theta+\beta) - \sin\beta\cos(\theta+\beta)\right]} \qquad (iii)$$

Note that this equation can be simplified by considering that $[\cos\beta\sin(\theta+\beta) - \sin\beta\cos(\theta+\beta)] = \sin\theta$.

Equation (iii) yields the magnitude of the force that must be exerted by the quadriceps muscles to support the leg when it is

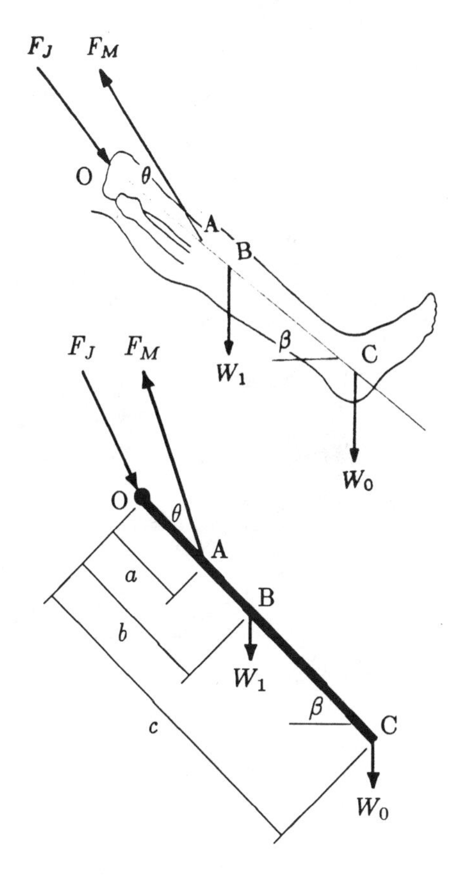

Figure 6.38 *Forces acting on the lower leg.*

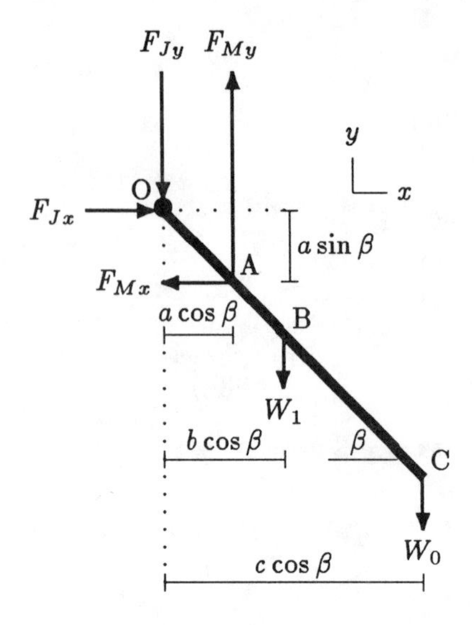

Figure 6.39 *Force components, and their lever arms.*

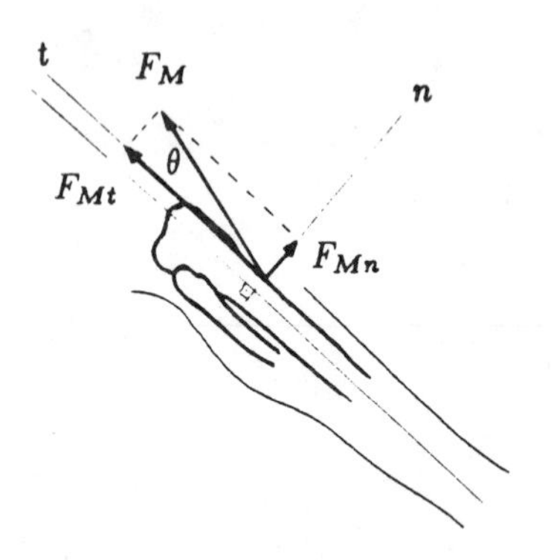

Figure 6.40 *Rotational and translatory components of $\underline{F}_M$.*

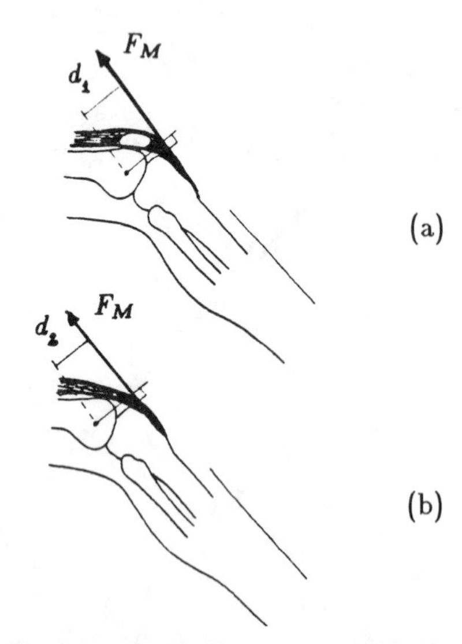

Figure 6.41 *Patella increases the length of the lever arm.*

extended forward making an angle β with the horizontal. Once F_M is determined, the components of the reaction force developed at the knee joint along the horizontal and vertical directions can also be evaluated by considering the translational equilibrium of the lower leg in the x and y directions:

$$\sum F_x = 0: \qquad F_{Jx} = F_{Mx} = F_M \cos(\theta + \beta)$$

$$\sum F_y = 0: \qquad F_{Jy} = F_{My} - W_0 - W_1$$

$$F_{Jy} = F_M \sin(\theta + \beta) - W_0 - W_1$$

The magnitude of the resultant compressive force applied on the tibial plateau at the knee joint is:

$$F_J = \sqrt{(F_{Jx})^2 + (F_{Jy})^2} \tag{iv}$$

Assume that the geometric parameters and the weights involved are given in terms of the height h and total weight W of the person such that a=0.08h, b=0.14h, c=0.28h, θ=15°, β=45°, W_1=W_0=0.06W. By substituting these parameters into Eqs. (*iii*) and (*iv*), F_M and F_J can be determined in terms of the person's weight:

$$F_M = 0.86\,W \qquad F_J = 0.97\,W$$

Remarks:

- The force $\underline{F}_M$ exerted by the quadriceps muscle on the tibia through the patellar tendon can be expressed in terms of two components normal and tangential to the long axis of the tibia (Figure 6.40). The primary function of the normal component $\underline{F}_{Mn}$ of the muscle force is to rotate the tibia about the knee joint, while its tangential component $\underline{F}_{Mt}$ tends to translate the lower leg in a direction collinear with the long axis of the tibia and applies a compressive force on the articulating surfaces of the tibiofemoral joint. Since the normal component of $\underline{F}_M$ is a sine function of angle θ, a larger angle between the patellar tendon and the long axis of the tibia indicates a larger rotational effect of the muscle exertion. This implies that for large θ, less muscle force is wasted to compress the knee joint, and a larger portion of the muscle tension is utilized to rotate the lower leg about the knee joint.

- One of the most important biomechanical functions of the patella is to provide anterior displacement of the quadriceps and patellar tendons, thus lengthening the lever arm of the knee extensor muscle forces with respect to the center of rotation of the knee by increasing angle θ (Figure 6.41a). Surgical removal of the patella brings the patellar tendon closer to the center of rotation of the knee joint (Figure 6.41b), which causes the length of the lever arm of the muscle force to decrease ($d_2 < d_1$). Losing the advantage of having a relatively long lever arm, the quadriceps muscle has to exert more force than normal to rotate the lower leg about the knee joint.

- The human knee has a two-joint structure composed of the tibiofemoral and patellofemoral joints. Notice that the quadriceps muscle goes over the patella, and the patella and the muscle form a pulley-rope arrangement. The higher the tension in the muscle, the larger the compressive force (pressure) the patella exerts on the patellofemoral joint.

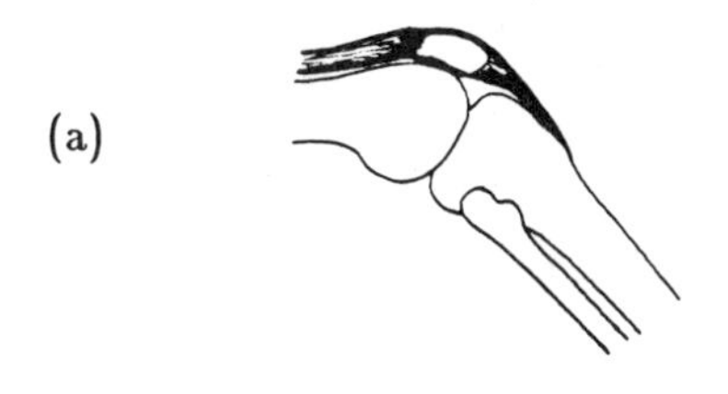

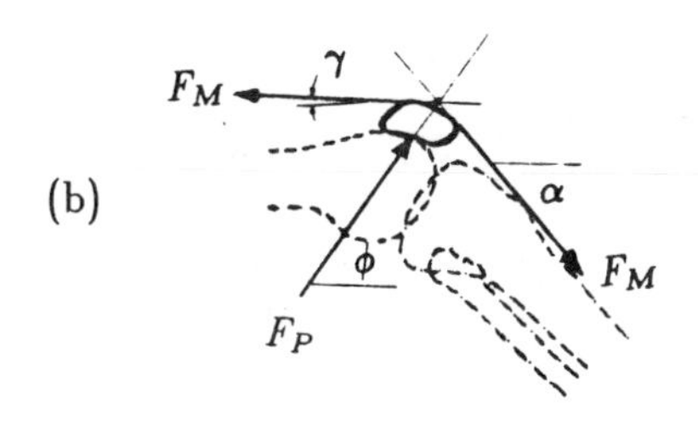

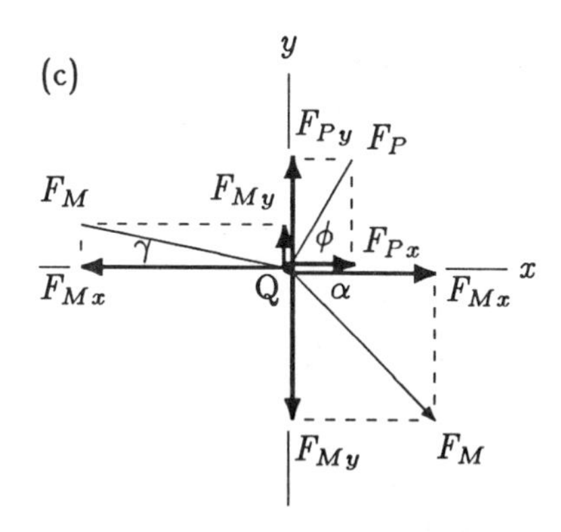

Figure 6.42 *Static analysis of the forces acting on the patella.*

We have analyzed the forces involved around the tibiofemoral joint by considering the free-body diagram of the lower leg. Having determined the tension in the patellar tendon, and assuming that the tension is uniform throughout the quadriceps, we can calculate the compressive force applied on the patellofemoral joint by considering the free-body diagram of the patella (Figure 6.42). Let F_M be the uniform magnitude of the tensile force in the patellar and quadriceps tendons, F_P be the magnitude of the force exerted on the patellofemoral joint, α be the angle between the patellar tendon and the horizontal, γ be the angle between the quadriceps tendon and the horizontal, and ϕ be the unknown angle between the line of action of the compressive reaction force at the joint (Figure 6.42b). We have a three-force system and for the equilibrium of the patella it has to be concurrent. We can first determine the common point of intersection Q by extending the lines of action of patellar and quadriceps tendon forces. A line connecting point Q and the point of application of $\underline{F}_P$ will correspond to the line of action of $\underline{F}_P$. The forces can then be translated to Q (Figure 6.42c), and the equilibrium equations can be applied. For the equilibrium of the patella in the x and y directions:

$$\sum F_x = 0: \qquad F_P \cos\phi = F_M(\cos\gamma - \cos\alpha)$$

$$\sum F_y = 0: \qquad F_P \sin\phi = F_M(\sin\alpha - \sin\gamma)$$

These equations can be solved simultaneously for angle ϕ and the magnitude F_P of the compressive force applied by the femur on the patella at the patellofemoral joint:

$$F_P = \left(\frac{\cos\gamma - \cos\alpha}{\cos\phi}\right) F_M$$

$$\phi = \tan^{-1}\left(\frac{\sin\alpha - \sin\gamma}{\cos\gamma - \cos\alpha}\right)$$

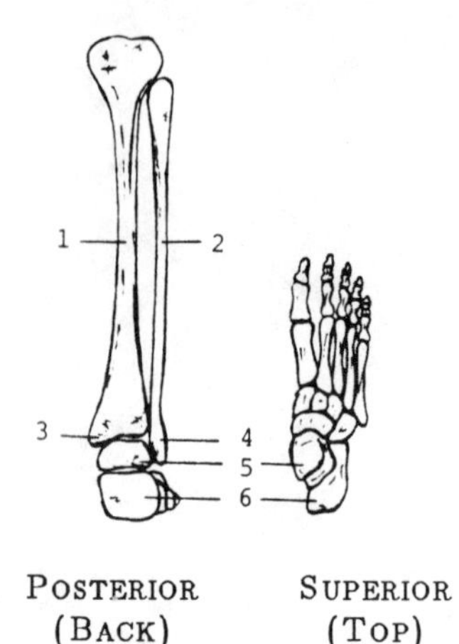

Figure 6.43 *The ankle and foot: (1) tibia, (2), fibula, (3) medial malleolus, (4) lateral malleolus, (5) talus, (6) calcaneus.*

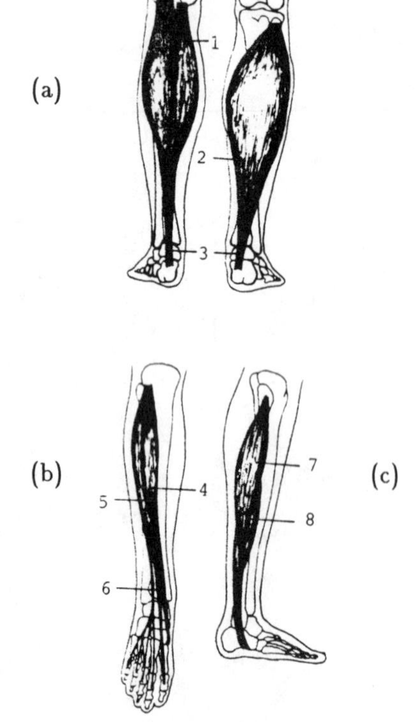

Figure 6.44 *Ankle muscles. (a) posterior, (b) anterior and (c) lateral views. (1) gastrocnemius, (2) soleus, (3) Achilles tendon, (4) tibialis anterior, (5) extensor digitorum longus, (6) extensor hallucis longus, (7) peroneus longus, (8) peroneus brevis.*

6.10 Mechanics of the Ankle

The ankle is the union of three bones: the tibia, fibula, and the talus of the foot (Figure 6.43). Like other major joints in the lower extremity, the ankle is responsible for load-bearing and kinematic functions. The anatomical configuration of the ankle joint is similar to that of the hip. The ankle joint is inherently more stable than the knee joint which requires ligamentous and muscular restraints for its stability.

The ankle joint complex consists of the tibiotalar, fibulotalar, and distal tibiofibular articulations. The ***ankle (tibiotalar) joint*** is a hinge or ginglymus-type articulation between the spool-like convex surface of the trochlea of the talus and the concave distal end of the tibia. The ***fibulotalar joint*** is the articulation between the internal malleolus of the tibia and the external malleolus of the fibula, and the ***distal tibiofibular joint*** is the articulation between the external malleolus of the fibula and the medial and lateral surfaces of the trochlea of the talus. Being a hinge joint, the ankle permits only flexion-extension (dorsiflexion-plantar flexion) movement of the foot in the sagittal plane. Other foot movements include inversion and eversion, inward and outward rotation, and pronation and supination. These movements occur about the foot joints such as the subtalar and transverse tarsal joints between the talus and calcaneus.

The ankle mortise is maintained by the shape of the three articulations, and the ligaments and muscles crossing the joint. The integrity of the ankle joint is improved by the medial (deltoid) and lateral collateral ligament systems, and the interosseous ligaments. There are numerous muscle groups crossing the ankle. The most important ankle plantar flexors are the gastrocnemius and soleus muscles (Figure 6.44). Both the gastrocnemius and soleus muscles have attachments to the posterior surface of the calcaneus via the Achilles tendon. The gastrocnemius muscle is more effective as a knee flexor if the foot is elevated, and more effective as a plantar flexor of the foot if the knee is held in extension. The plantar extensors are posterior muscles. There are also anterior (tibialis anterior, extensor digitorum longus, extensor hallucis longus, peroneus tertius) and lateral (peroneus longus, peroneus brevis) muscles whose primary function is to provide pronation and supination, and inward and outward rotation of the foot.

The ankle joint responds poorly to small changes in its anatomical configuration. Loss of kinematic and structural restraints due to severe sprains can seriously affect ankle stability and can produce malalignment of the ankle joint surfaces. The most common ankle injury, inversion sprain, occurs when the body weight is forcefully transmitted to the ankle while the foot is inverted (the sole of the foot facing inward).

Example 6.7 Consider a person standing on tiptoe on one foot (a strenuous position illustrated). The forces acting on the foot during this instant are shown in Figure 6.45. W is the person's weight applied on the foot as the ground reaction force, F_M is the magnitude of the tensile force exerted by the gastrocnemius and soleus muscles on the calcaneus through the Achilles tendon, and F_J is the magnitude of the ankle joint reaction force applied by the tibia on the dome of the talus. The Achilles tendon is attached to the calcaneus at A, the ankle joint is located at B, and the ground reaction force is applied on the foot at C. For this position of the foot, it is estimated that the line of action of the tensile force in the Achilles tendon makes an angle θ with the horizontal, and the line of action of the ankle joint reaction force makes an angle β with the horizontal.

Assuming that the relative positions of A, B, and C are known, determine expressions for the tension in the Achilles tendon and the magnitude of the reaction force at the ankle joint.

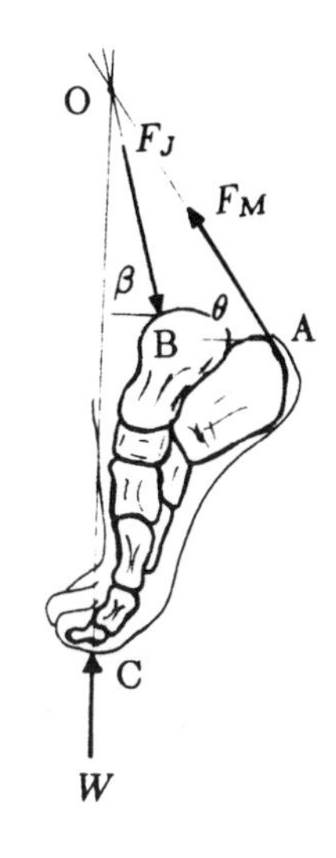

Figure 6.45 *Forces acting on the foot form a concurrent system of forces.*

Solution: We have a three-force system composed of muscle force $\underline{F}_M$, joint reaction force $\underline{F}_J$ and the ground reaction force $\underline{W}$. From the geometry of the problem, it is obvious that for the position of the foot shown, the forces acting on the foot do not form a parallel force system. Therefore, the force system must be a concurrent one. The common point of intersection (O in Figure 6.45) of these forces can be determined by extending the lines of action of $\underline{W}$ and $\underline{F}_M$. A straight line passing through both points O and B represents the line of action of the joint reaction force. Assuming that the relative positions of points A, B, and C are known (as stated in the problem), the angle (say β) of the line of action of the joint reaction force can be measured.

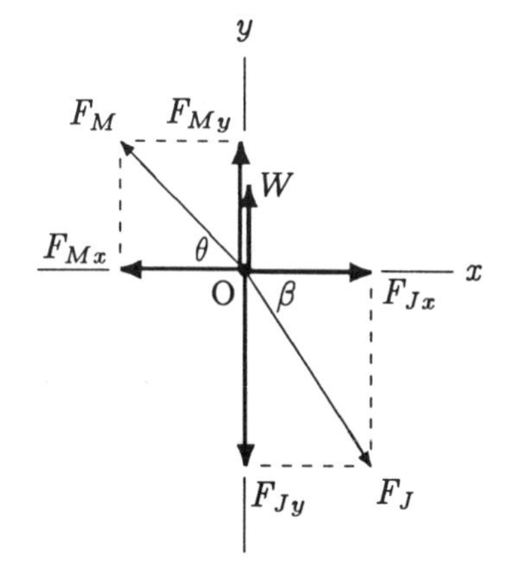

Figure 6.46 *Components of the forces acting on the foot.*

Once the line of action of the joint reaction force is determined by graphical means, the magnitudes of the joint reaction and muscle forces can be calculated by translating all three forces involved to the common point of intersection at O (Figure 6.46). The two unknowns F_M and F_J can now be determined by applying the translational equilibrium conditions in the horizontal (x) and vertical (y) directions. For this purpose, the joint reaction and muscle forces must be decomposed into their rectangular components first:

$$F_{Mx} = F_M \cos\theta$$
$$F_{My} = F_M \sin\theta$$

$$F_{Jx} = F_J \cos\beta$$
$$F_{Jy} = F_J \sin\beta$$

For the translational equilibrium of the foot in the horizontal and vertical directions:

$$\sum F_x = 0: \qquad F_{Jx} = F_{Mx}$$

$$\sum F_y = 0: \qquad F_{Jy} = F_{My} + W$$

Simultaneous solutions of these equations will yield:

$$F_M = \frac{\cos\beta\, W}{\cos\theta\sin\beta - \sin\theta\cos\beta}$$

$$F_J = \frac{\cos\theta\, W}{\cos\theta\sin\beta - \sin\theta\cos\beta}$$

For example, assume that θ=45° and β=60°. Then:

$$F_M = 1.93\, W \qquad F_J = 2.73\, W$$

6.11 Discussion

Equations of equilibrium are specific forms of equations of motion that are derived from Newton's second law of motion. In Chapter 5, the equations of equilibrium are utilized to analyze relatively simple mechanical systems. In Chapter 6, we formed some analogies between mechanical elements and parts of the human body, and analyzed muscle and joint reaction forces at different parts of the human musculoskeletal system. We made several assumptions and reduced the system under consideration to a statically determinate problem to enable an analysis within the limitations of the equations of equilibrium.

Researchers in the field of biomechanics have suggested various mathematical models to evaluate the forces involved in different muscles, bones, and joints of the human body during its various activities and static postures. Some of these models attempt to take into consideration the contributions of more than one muscle group. Different researchers utilize different formulation techniques and different criteria for the prediction of muscle forces. For example, Seireg and Arvikar (1973) developed a mathematical model for evaluating the muscle forces necessary to maintain the human body equilibrium in standing, leaning, and stooping. These researchers utilized linear optimization techniques in the prediction of muscle forces and reported that their analytical solutions showed good correlation with electromyographic muscle signals. Penrod et al. (1974) investigated the problem of distribution of forces among muscles at a joint. In order to arrive at a unique solution for the redundant (statically indeterminate) system, they suggested that the solution must be based on optimizing the total muscle effort. They applied the resulting mathematical formulation to a two tendon model. Crowninshield and Brand (1981) developed a nonlinear model for predicting the muscle activity during locomotion. These investigators utilized the criterion of maximum endurance of musculoskeletal function to develop a model based on the inversely nonlinear relationship of muscle force and contraction endurance. In a review paper, Dul et al. (1984a) compared the characteristics and performance of several linear and nonlinear criteria for load sharing between synergistic muscles reported in the literature. Based on the assumption that during a physical activity the muscular fatigue is minimized, the same group of researchers (Dul et al., 1984b) developed a model for predicting the load-sharing mechanisms between muscles.

This chapter concludes the first part of this text. The purpose of this part was to introduce the basic concepts of mechanics, along with analyses of systems in static equilibrium. The next six chapters are devoted to dynamics and motion analyses. Concepts such as position, velocity, acceleration, work, kinetic energy, potential energy, power, impulse, and momentum will be defined and utilized in the following chapters. Techniques for kinetic and kinematic analyses of systems undergoing translational and rotational motion will be provided.

6.12 References Cited

Chaffin, D.B., and Andersson, G.B.J. 1991. *Occupational Biomechanics.* 2nd ed. New York: John Wiley & Sons.

Crowninshield, R.D., and Brand, R.A. 1981. A physiologically based criterion on muscle force prediction in locomotion. *J. Biomechanics* 14:793-801.

Dul, J., Townsend, M.A., Shiavi, R., and Johnson, G.E. 1984a. Muscular synergism—I. On the criteria for load sharing between synergistic muscles. *J. Biomechanics* 17:663-673.

Dul, J., Johnson, G.E., Shiavi, R., and Townsend, M.A. 1984b. Muscular synergism—II. A minimum-fatigue criterion for load sharing between synergistic muscles. *J. Biomechanics* 17:663-673.

Kirby, R., and Roberts, J.A. 1985. *Introductory Biomechanics.* Ithaca, NY: Mouvement Publications.

Nordin, M., and Frankel, V.H. 1989. *Basic Biomechanics of the Musculoskeletal System.* 2nd ed. Philadelphia: Lea & Febiger

Penrod, D.D, Davy, D.T., and Singh, D.P. 1974. An optimization approach to tendon force analysis. *J. Biomechanics* 7:123-129.

Seireg, A., and Arvikar, R.J. 1973. A mathematical model for evaluation of force in lower extrimeties of the musculo-skeletal system. *J. Biomechanics* 6:313-326.

Thompson, C.W. 1989. *Manual of Structural Kinesiology.* 11th ed. St. Louis, MO: Times Mirror/Mosby.

Williams, M., and Lissner, H.R. 1977. *Biomechanics of Human Motion.* 2nd ed. (B. LeVeau, ed.) Philadelphia: Saunders.

Wictorin, C.H., and Nordin, M. 1982. *Introduction to Problem Solving in Biomechanics.* Philadelphia: Lea & Febiger

Winter, D.A. 1990. *Biomechanics and Motor Control of Human Behavior.* 2nd ed. New York: John Wiley & Sons.

Chapter 7

INTRODUCTION TO DYNAMICS

Dynamics is the study of bodies in motion. Dynamics is concerned with describing motion and explaining its causes. The general field of dynamics consists of two major areas: kinematics and kinetics. Each of these areas can be further divided to describe and explain linear, angular, or general motion of bodies. The fundamental concepts in dynamics are space (relative position or displacement), time, mass, and force. Other important concepts include velocity, acceleration, torque, moment, work done, energy, power, impulse, and momentum.

The broad definitions of basic terms and concepts in dynamics will be introduced in this chapter. The details of kinematic and kinetic characteristics of bodies will be covered in the following chapters.

7.1 Kinematics and Kinetics

The field of *kinematics* is concerned with the description of geometric and time-dependent aspects of motion without dealing with the forces causing the motion. Kinematic analyses are based on the relationship between position, velocity, and acceleration which are usually in the form of differential or integral equations.

The field of *kinetics* is based on kinematics, and it incorporates into the analyses the effects of forces and torques which cause the motion. Kinetic analyses utilize Newton's second law of motion which can take various mathematical forms. There are a number of different approaches to the solutions of problems in kinetics, which are based on the equations of motion, work, and energy methods, and impulse and momentum methods. Different methods may be applied to different situations, or depending on what is required to be determined. For example, the equations of motion are used for problems requiring the analysis of acceleration. Energy methods are suitable when a problem requires the analysis of forces related to changes in speed and displacement. Momentum methods are applied if the forces involved are impulsive, which is the case during impact and collision.

7.2 Linear, Angular, and General Motion

To study both kinematics and kinetics in an organized manner, it is a common practice to divide them into branches according to whether the motion is translational, rotational, or both. *Linear motion*, which is also known as *translational motion*, occurs if all parts of a body move the same distance at the same time and in the same direction. For example, if a block is pushed on a horizontal surface, the block will undergo translational motion only (Figure 7.1). Another typical example of translational motion is the vertical motion of an elevator in a shaft.

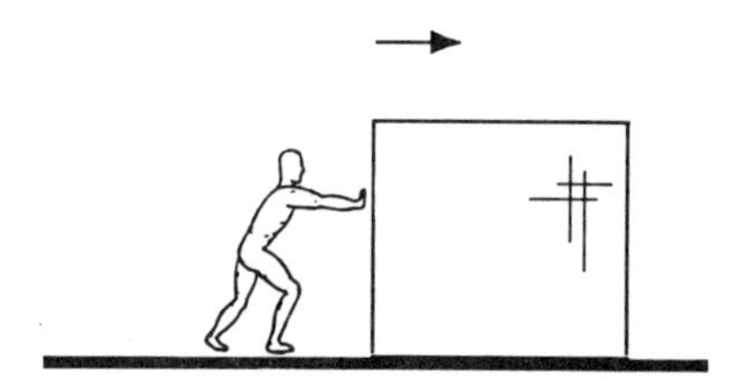

Figure 7.1 *Translation motion.*

Angular motion occurs when a body moves in a circular path such that all parts of the body move through the same angle at the same time. The angular motion is also known as *rotational motion*. The angular motion occurs about a central line known as the *axis of rotation*, which lies perpendicular to the plane of motion. For example, a gymnast doing giant circles undergoes rotational motion with the centerline of the bar acting as the axis of rotation of the motion (Figure 7.2).

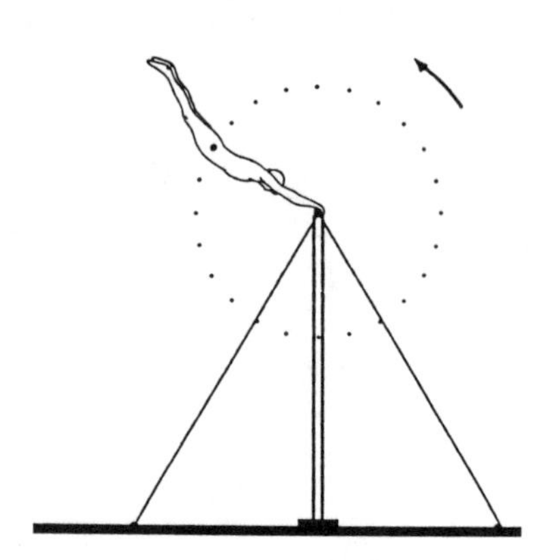

Figure 7.2 *Rotational motion.*

The third class of motion is called *general motion* which occurs if a body undergoes both translational and rotational motions. It is more complex to analyze motions composed of both translation and rotation as compared to a pure translational or rotational motion. The diver illustrated in Figure 7.3 is a typical example of a body undergoing general motion. Most of the human body segmental motions are of the general type. For example, while walking, the lower extremities both translate (move forward) and rotate. In this case, rotation occurs about the hip, knee, and ankle joints.

Figure 7.3 *General motion.*

The branch of kinematics that deals with the description of translational motion is known as *linear kinematics*; the branch that deals with rotational motion is *angular kinematics*. Similarly, the field of kinetics can be divided into *linear* and *angular kinetics*.

7.3 Particle and Rigid Body Mechanics

The general field of mechanics can also be divided into *particle mechanics* and *rigid body mechanics*. In this case, the distinction is made according to the size and shape of the object whose motion characteristics are to be analyzed. Particle mechanics undermines the shape of the body and only considers its mass at the center of gravity. It is relatively easy to implement particle mechanics. The particle concept is rather a hypothetical one, and it assumes that the object has no volume but a finite mass. In many problems, the size and shape of an object under investigation are not pertinent to the discussion of certain aspects of its motion. This is particularly true if the object is undergoing a translational motion only. For example, what is significant for a person pushing a wheelchair is the total mass of the wheelchair, not its size or shape. Therefore, the wheelchair may be treated as a particle with a mass equal to the total mass of the wheelchair, and proceed with relatively simple analyses. The size and shape of the object may become important if the object undergoes a rotational motion.

7.4 Reference Frames and Coordinate Systems

To be able to describe the motion of a body properly, a *reference frame* must be adopted and all quantities involved must be measured with respect to that reference frame. The *rectangular coordinate* system, also known as the *Cartesian coordinate* system, is composed of three mutually perpendicular directions. It is the most suitable reference frame for describing linear motions. The axes of this system are commonly labeled with x, y, and z. For two-dimensional problems the number of axes can be reduced to two by eliminating the z axis (Figure 7.4).

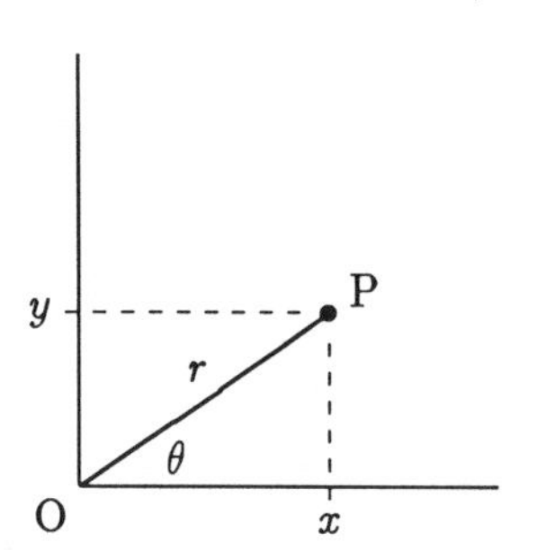

Figure 7.4 *Rectangular (x, y) and polar (r, θ) coordinates.*

Another commonly used reference frame is based on *polar coordinates*, which is better suited for analyzing angular motions. As shown in Figure 7.4, the polar coordinates of a point P are defined by parameters r and θ. r is the distance between the origin O of the coordinate frame and P, and θ is the angle line OP makes with the horizontal. The details of polar coordinates will be provided in Chapter 9.

In addition to rectangular and polar coordinates, there are other coordinate systems. These include the cylindrical coordinate system which is an extension of the polar coordinates, and the spherical coordinate system.

7.5 Distance and Displacement

In mechanics, *distance* is defined as the total length of the path followed while moving from one point to another, and *displacement* is the length of the straight line joining the two points along with some indication of direction involved. Distance is a scalar quantity (has only a magnitude) and displacement is a vector quantity (has both a magnitude and a direction).

To understand the differences between distance and displacement, consider a person who lives in an apartment building located at the corner of Third Avenue and 18th Street, and walks to work in a building located at the corner of Second Avenue and 17th Street in New York City. In Figure 7.5, A represents the corner of Third Avenue and 18th Street, B represents the corner of Second Avenue and 18th Street, and C represents the corner of Second Avenue and 17th Street. Every morning this person walks towards the east from A to B, and then towards the south from B to C. Assume that the length of the straight line between A and B is 100 meters, and between B and C is 50 meters. Therefore, the total distance the person walks every morning 150 meters. On the other hand, the door-to-door southeasterly displacement of the person is equal to the length of the straight line joining A and C, which is $\sqrt{(100)^2 + (50)^2}$=112 m.

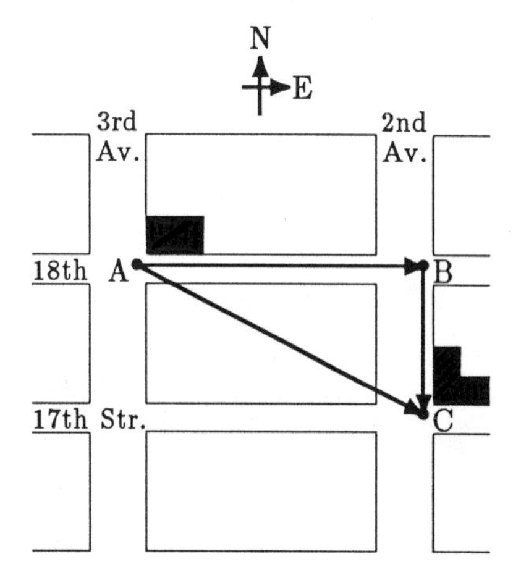

Figure 7.5 *Distance versus displacement.*

7.6 Velocity and Acceleration

Velocity is defined as the time rate of change of position. Velocity is a vector quantity having both a magnitude and direction. While the words speed and velocity are used interchangeably in ordinary language, they have distinctly different meanings in mechanics. *Speed* is a scalar quantity equal to the magnitude of the velocity vector. The rate at which the velocity of a body changes over time is known as *acceleration*. Like velocity from which it is derived, acceleration is also a vector quantity.

7.7 Force and Torque

The field of kinetics is based on the relationship between applied forces and corresponding motions. The rules of this relationship are defined by Newton's second law of motion which yields the equations of motion. For translational motions, the equations of motion can be stated in terms of rectangular parameters and applied forces. If the motion is rotational only, then the angular equivalent of the equations of motion must be used. Equations for angular motion relate angular parameters to applied torque.

7.8 Mathematics

The prerequisites for dynamic analysis are vector algebra, differential calculus, and integral calculus. The fundamentals of vector algebra have already been introduced in Chapter 2. The principles of the "derivative" and the "integral" are provided in Appendix C, along with the definitions and properties of commonly encountered functions that form the basis of calculus. Appendix C should be reviewed before proceeding to the following chapters.

Chapter 8

LINEAR KINEMATICS

8.1 Uniaxial (One-Dimensional) Motion Analysis

Uniaxial linear or *translational* motion is one in which the motion occurs only in one-direction or along a straight line. There may be many situations in which the motion of bodies occur only in one direction. For example, a car traveling on a straight highway, an elevator going up and down in a shaft, and a sprinter running a 100-meter race.

Kinematic analyses are based on the relationship between the position (displacement), velocity and acceleration, which are vector quantities. For uniaxial motion analyses, it is practical to define a coordinate axis (such as x) to coincide with direction of motion, define kinematic parameters in that direction, and carry out the analyses as if they are scalar quantities.

8.2 Position, Velocity, and Acceleration

Suppose that an object initially (at time t_0) located at O starts moving to the right along a straight horizontal path (Figure 8.1). Assume that at some time t_1 the object is observed to be at A and at a later time t_2 it is located at B. O, A, and B represent *positions* of the object at different times, and O represents the *initial position* of the object. It is a common practice to start measuring time beginning with the instant when the motion starts, in which case $t_0=0$.

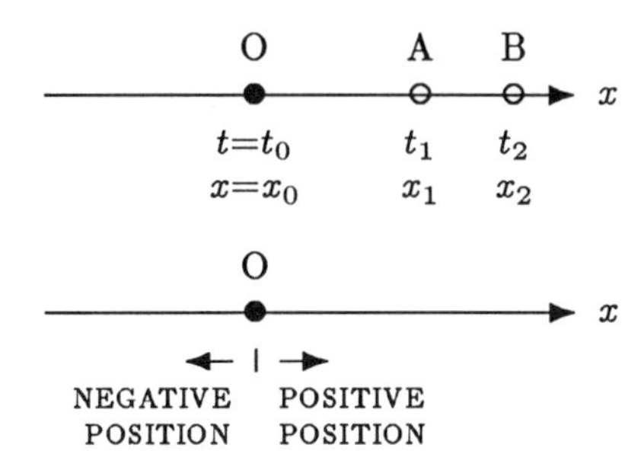

Figure 8.1 *Object is located at O, A, and B at times t_0, t_1, and t_2.*

The position of the object at different times must be measured with respect to a point in space. Let x be a coordinate axis such that its origin is located at the initial position of the object and that the positive x direction is in the direction of motion. Then the initial position of the object is $x_0=0$. If A and B are located at x_1 and x_2 distances away from O, then x_1 and x_2 define the relative positions of the object at times t_1 and t_2. Therefore, x is a measure of position of the object relative to O, and since the relative position of the object is changing with time, x is a function of time t. In the time interval between t_1 and t_2, the position of the object in Figure 8.1 changes by an amount $\Delta x=x_2-x_1$, where Δ (capital delta) implies change. This change in position is also known as the *displacement* of the object in the time interval $\Delta t=t_2-t_1$.

During a uniaxial horizontal motion, an object may be located on the right or left of the origin O of the x axis. Assuming that the positive x axis is towards the right, then the position of the object is positive if it is located on the right of O, and negative if it is on the left of O (Figure 8.1).

Velocity is defined as the time rate of change of relative position. If the position (measured in terms of x) of an object moving in the x direction is known as a function of time t, then the ***instantaneous velocity*** of the object can be determined by considering the derivative of x with respect to t:

$$v = \frac{dx}{dt} \tag{8.1}$$

If required, the ***average velocity*** of the object in any time interval can be determined by considering the ratio of change in position of the object and the time it takes to make that change:

$$\bar{v} = \frac{\Delta x}{\Delta t} = \frac{x_2 - x_1}{t_2 - t_1} \tag{8.2}$$

In Eq. (8.2), the "bar" over v implies average, and x_1 and x_2 are the relative positions of the object at times t_1 and t_2, respectively.

Velocity is a vector quantity and may take positive and negative values, indicating the direction of motion with respect to the chosen positive direction of position. The velocity will be positive if the object is moving away from the origin in the positive x direction, and it will be negative if the object is moving in the negative x direction. The absolute value of the velocity is called the *speed*, which is always a positive quantity.

The instantaneous velocity of an object may vary during a particular motion. ***Acceleration*** is defined as the time rate of change of velocity. If the velocity v of an object is known as a function of time t, then its ***instantaneous acceleration*** can be determined by considering the derivative of v with respect to t:

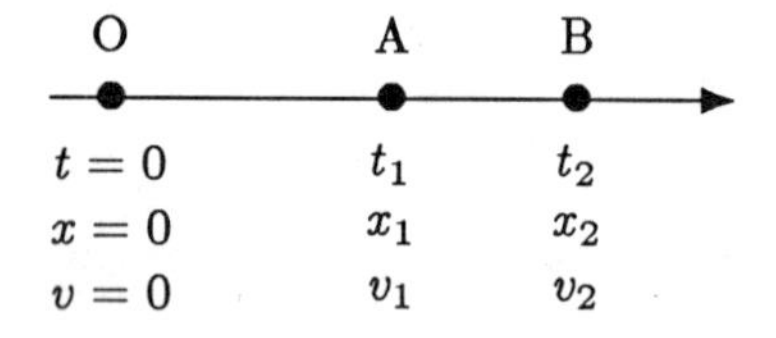

Figure 8.2 *v_1 and v_2 are the instantaneous velocities of the object at times t_1 and t_2.*

$$a = \frac{dv}{dt} \tag{8.3}$$

If required, the ***average acceleration*** $\bar{a}$ of the object in a given time interval can be determined by considering the ratio of the change in velocity of the object and the time it takes to make that change:

$$\bar{a} = \frac{\Delta v}{\Delta t} = \frac{v_2 - v_1}{t_2 - t_1} \tag{8.4}$$

In Eq. (8.4), v_1 and v_2 are the instantaneous velocities of the object at times t_1 and t_2, respectively.

Acceleration is a vector quantity and may be positive or negative. Positive acceleration does not necessarily mean that the object is increasing its velocity, and negative acceleration does not necessarily mean that the object is slowing down. At a given instant, if the velocity and acceleration have the same sign then the object is said to be "speeding up" or "accelerating" at that instant. For a uniaxial motion in the x direction, the object moves in the positive x direction with an increasing speed if both its velocity and acceleration are positive. Similarly, the object moves in the negative x direction with an increasing speed if both its velocity and acceleration are negative. At any instant, if the velocity and acceleration of the object have opposite signs then the object is "slowing down" or "decelerating" at that instant. If the acceleration is zero then the object is said to have a "constant" or "uniform" velocity.

Note that by definition, acceleration is the time rate of change of velocity, and velocity is the time rate of change of position. Since acceleration is derived from velocity, which is itself derived from position, there must be a way to relate acceleration and position directly:

$$v = \frac{dx}{dt} = \dot{x}$$

$$a = \frac{dv}{dt} = \frac{d}{dt}\left(\frac{dx}{dt}\right) = \frac{d^2x}{dt^2} = \ddot{x}$$

The "dots" over x in the above equations imply differentiation with respect to time. One dot signifies the first derivative with respect to time once, and two dots imply the second derivative.

8.3 Dimensions and Units

The change in position or displacement of a body is measured in units of length. By definition, velocity is the time rate of change of relative position, and acceleration is the time rate of change of velocity. Therefore, displacement has the dimension of length, velocity has the dimension of length divided by time, and acceleration has a dimension of velocity divided by time:

$$[\text{DISPLACEMENT}] = \text{L}$$

$$[\text{VELOCITY}] = \frac{[\text{DISPLACEMENT}]}{[\text{TIME}]} = \frac{\text{L}}{\text{T}}$$

$$[\text{ACCELERATION}] = \frac{[\text{VELOCITY}]}{[\text{TIME}]} = \frac{\text{L/T}}{\text{T}} = \frac{\text{L}}{\text{T}^2}$$

Based on these dimensions, the units of displacement, velocity and acceleration in different unit systems can be determined. Some of these units are listed in Table 8.1.

UNIT SYSTEM	DISPLACEMENT	VELOCITY	ACCELERATION
SI	meter (m)	m/s	m/s^2
c-g-s	centimeter (cm)	cm/s	cm/s^2
British	foot (ft)	ft/s	ft/s^2

Table 8.1 *Units of displacement, velocity and acceleration.*

8.4 Measured and Derived Quantities

In practice, it is possible to measure position, velocity, and acceleration over time. From any one of the three, the other two quantities can be determined by employing proper differentiation and/or integration, or by using some graphical and numerical techniques. These three possible cases will be discussed next.

Note here that there are several conditions that a function must meet before we can take its derivative and/or integral. For example, a function must be continuous over the region in which its derivative or integral is to be determined.

8.4.1 Position measured

If the position of an object undergoing uniaxial motion in the x direction is measured and corresponding times are recorded, then the position can be expressed as a function of time. Once the function representing the position of the object is established, then the velocity and acceleration of the object at different times can be calculated using:

$$v = \frac{dx}{dt} \tag{8.5}$$

$$a = \frac{dv}{dt} = \frac{d^2x}{dt^2} \tag{8.6}$$

Example 8.1 The short distance runner illustrated in Figure 8.3(a) completed a 100 meter race in 10 seconds. The time it took for the runner to reach the first 10 meters and each successive 10 meters were recorded by 10 observers using stopwatches. The data collected were then plotted to obtain a position-time graph shown in Figure 8.3(b). It is suggested that the data might be represented with the following function:

$$x = f(t) = 0.46\, t^{7/3}$$

Here, x is measured in meters and t in seconds. Determine the velocity and acceleration of the runner as functions of time, and the instantaneous velocity and acceleration of the runner 5 seconds after the start.

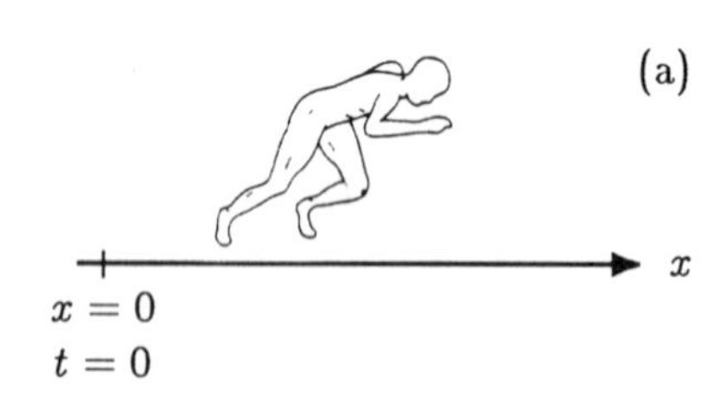

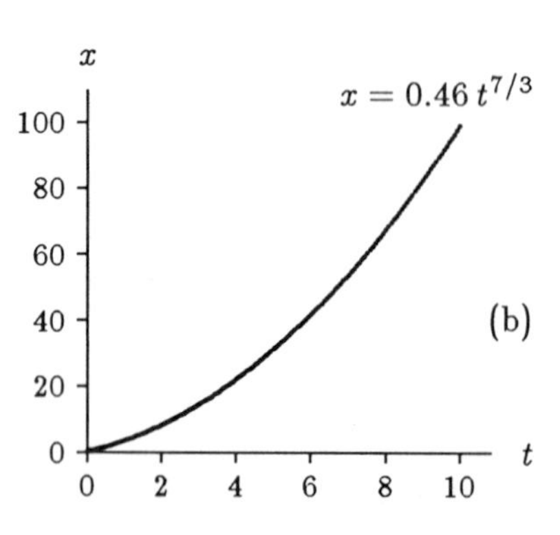

Figure 8.3 *Relative position x (m) of the runner measured over time t (s).*

Solution: Since the function representing the position of the runner is known, it can be differentiated with respect to time once to determine the velocity, and twice to determine the acceleration:

$$v = \frac{dx}{dt} = \frac{d}{dt}\left(0.46\, t^{7/3}\right) = 1.07\, t^{4/3}$$

$$a = \frac{dv}{dt} = \frac{d}{dt}\left(1.07\, t^{4/3}\right) = 1.43\, t^{1/3}$$

The graphs of these functions are shown in Figure 8.4.

To evaluate the velocity and acceleration of the runner 5 seconds after the start, substitute t=5 s in the above equations and carry out the calculations:

$$v = 9.15 \text{ m/s}$$
$$a = 2.45 \text{ m/s}^2$$

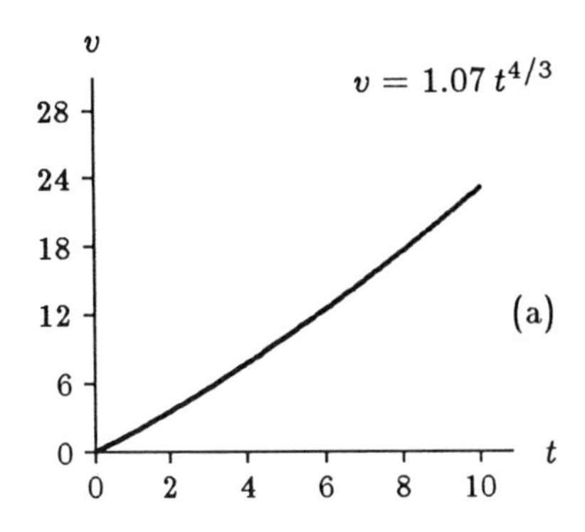

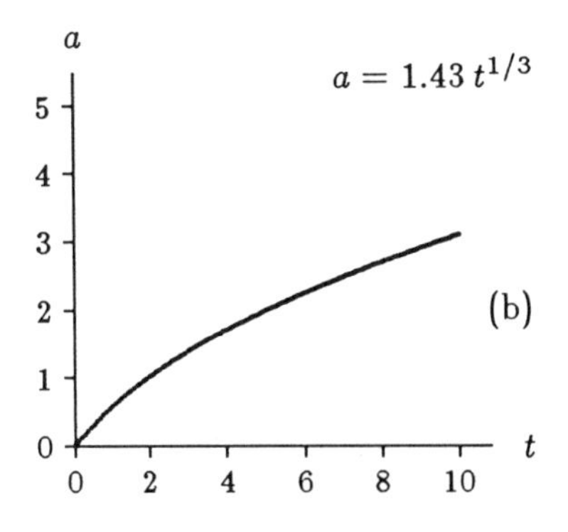

Figure 8.4 *Velocity v (m/s) and acceleration a (m/s^2) versus time t (s) curves for the runner.*

8.4.2 Velocity measured

If the velocity of an object undergoing uniaxial motion in the x direction is measured and expressed as a function of time, then the relative position and instantaneous acceleration of the object can be calculated using:

$$x = x_0 + \int_{t_0}^{t} v\, dt \tag{8.7}$$

$$a = \frac{dv}{dt} \tag{8.8}$$

The lower limit of integration, t_0, in Eq. (8.7) corresponds to the time at which the first measurements are taken, and the upper limit of integration corresponds to any time t. x_0 is the *initial position* of the object at time t=t_0. For practical purposes, t_0 can be taken to be zero. This would mean that all time measurements are made relative to the instant when the motion began.

Example 8.2 The speedometer reading of a test car driven on a straight highway is recorded for a total time interval of three minutes. The data collected are represented with a velocity versus time curve shown in Figure 8.5. Between times t_0=0 and t_1=30 s, the velocity of the car increases linearly with time from v_0=0 to v_1=72 km/h. Between times t_1=30 s and t_2=120 s, the velocity of the car remains constant at 72 km/h. Beginning at time t_2=120 s, the driver applies the brakes, decreases the velocity of the car linearly with time, and brings the car to a stop in 60 seconds.

Determine the variations of displacement and acceleration of the car with time. Calculate the total distance traveled by the car in three minutes.

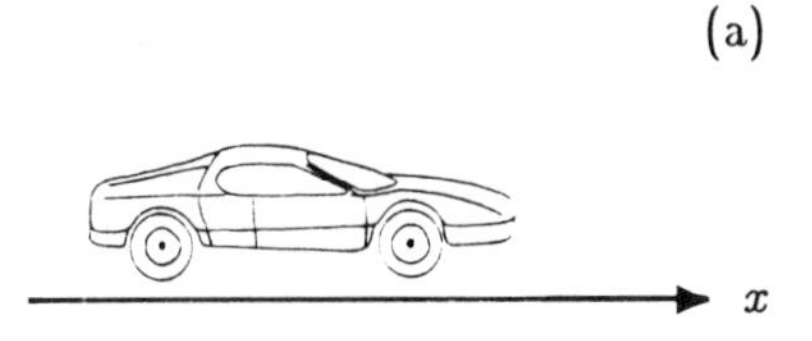

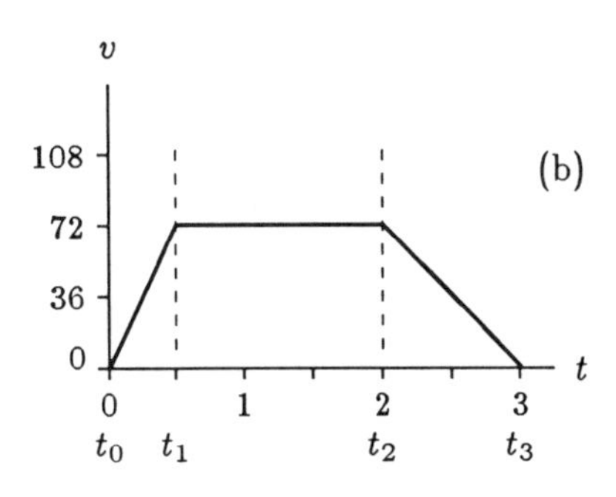

Figure 8.5 *Velocity v (km/h) versus time t (min) diagram for the car.*

Solution: The velocity versus time graph in Figure 8.5 has three distinct regions, and there is no single function that can represent the entire graph. Therefore, this problem should be analyzed in three phases.

Phase 1. Between t_0 and t_1, the velocity of the car increases linearly with time from 0 to 72 km/h in 30 s. To begin with, we must convert the velocity given in kilometers per hour into meters per second. This can be achieved by noting that 1 kilometer is 1000 meters, and that there are 3600 seconds in 1 hour:

$$72 \text{ km/h} = 72 \times \frac{1000}{3600} = 20 \text{ m/s}$$

The function representing the relationship between velocity v of the car and time t can be determined using:

$$v = v_0 + \left(\frac{v_1 - v_0}{t_1 - t_0}\right)(t - t_0)$$

Substituting v_0=0, v_1=20 m/s, t_0=0 and t_1=30 s into the above equation will yield the function v=$f(t)$ for phase 1:

$$v = 0.667\, t \qquad (i)$$

Here, velocity is in meters per second, and time in seconds.

Now, Eqs. (8.7) and (8.8) can be utilized to determine the displacement and acceleration of the car in this phase. If we measure the displacement from the starting point, then the initial position of the car is x_0=0. Therefore:

$$x = x_0 + \int_0^t v\, dt = \int_0^t (0.667\, t)\, dt = 0.667 \left[\frac{t^2}{2}\right]_0^t = 0.333\, t^2 \qquad (ii)$$

$$a = \frac{dv}{dt} = \frac{d}{dt}(0.67\, t) = 0.67 \qquad (iii)$$

From Eq. (iii), the acceleration of the car in phase 1 is constant at 0.667 m/s^2. The total distance traveled by the car at the end of phase 1 can be determined by substituting t=30 s into Eq. (ii):

$$x_1 = 0.333\, t^2 = 0.333\, (30)^2 = 300 \text{ m}$$

Phase 2. Between 1 and 2, the velocity of the car is constant at 20 m/s. Therefore, the function representing the velocity in phase 2 is:

$$v = 20 \qquad (iv)$$

The total distance traveled by the car in phase 1 was computed as x_1=300 m. Therefore, x_1=300 m designates the initial position of the car in phase 2. Also, phase 2 begins when time is t_1=30 s. Now, using Eq. (8.7):

$$x = x_1 + \int_{t_1}^t v\, dt = 300 + \int_{30}^t 20\, dt = 300 + 20\, [t]_{30}^t$$

$$= 300 + 20\,(t - 30) \qquad (v)$$

From Eq. (8.8):

$$a = \frac{dv}{dt} = 0 \qquad (vi)$$

Therefore, the acceleration of the car in phase 2 is zero. The total distance traveled by the car at the end of phase 2 can be determined by substituting t=120 s into Eq. (iv):

$$x_2 = 300 + 20\,(120 - 30) = 2100 \text{ m}$$

Phase 3. Between t_2 and t_3, the velocity of the car decreases linearly with time and to zero in 60 seconds. The function representing the relationship between velocity v of the car and time t can be determined using:

$$v = v_2 + \left(\frac{v_3 - v_2}{t_3 - t_2}\right)(t - t_2)$$

Substituting v_2=20 m/s, v_3=0, t_2=120 s, and t_3=180 into the above equation will yield function $v=f(t)$ for phase 3:

$$v = 60 - 0.333\,t \qquad (vii)$$

Phase 3 begins when time is t_2=120 s, and the initial position of the car at phase 3 is x_2=2100 m. Using Eq. (8.7):

$$\begin{aligned} x &= x_2 + \int_{t_2}^{t} v_3\,dt = 2100 + \int_{120}^{t} (60 - 0.333\,t)\,dt \\ &= 2100 + \left[60\,t - 0.167\,t^2\right]_{120}^{t} \\ &= 2100 + \left[60\,t - 0.167\,t^2\right] - \left[60\,(120) - 0.167\,(120)^2\right] \\ &= -2700 + \left(60\,t - 0.167\,t^2\right) \qquad (viii) \end{aligned}$$

Using Eq. (8.8):

$$a = \frac{dv}{dt} = \frac{d}{dt}(60 - 0.333\,t) = -0.333 \qquad (ix)$$

In phase 3, the car is decelerating at a rate of 0.333 m/s^2 in the positive x direction. The total distance traveled by the car can be determined by substituting time t_3=180 s into Eq. (ix):

$$x_3 = -2700 + \left[60\,(180) - 0.167\,(180)^2\right] = 2690 \text{ m} = 2.69 \text{ km}$$

The functions given in Eqs. (i) through (ix) are plotted in Figure 8.6 to obtain displacement, velocity, and acceleration versus time curves for the car. In phase 1, both the velocity and acceleration are positive and the car is accelerating and its speed is increasing in the positive x direction. In phase 3, the velocity is positive but the acceleration of the car is negative. This implies that the car is decelerating in the positive x direction, or accelerating in the negative x direction.

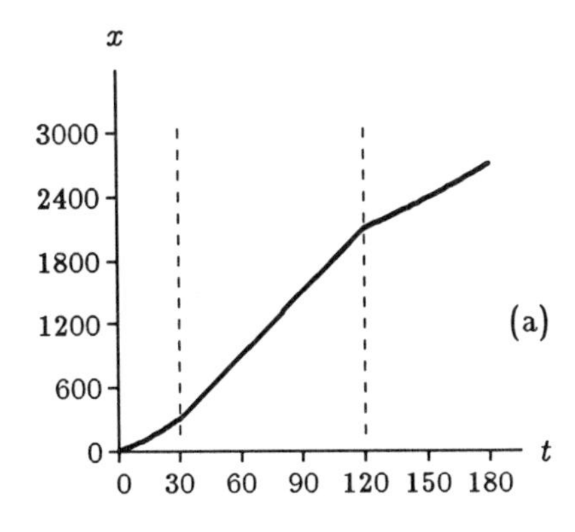

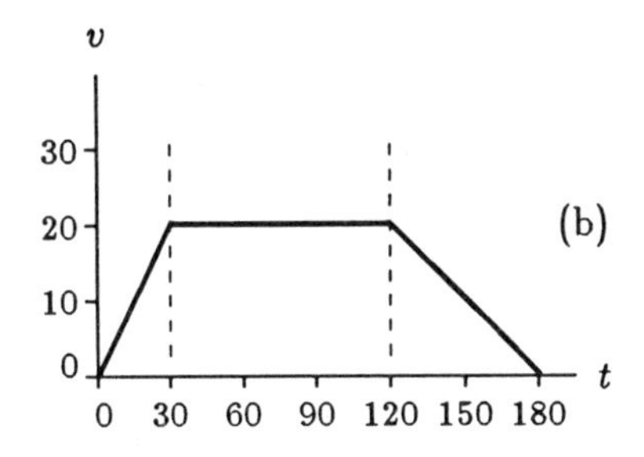

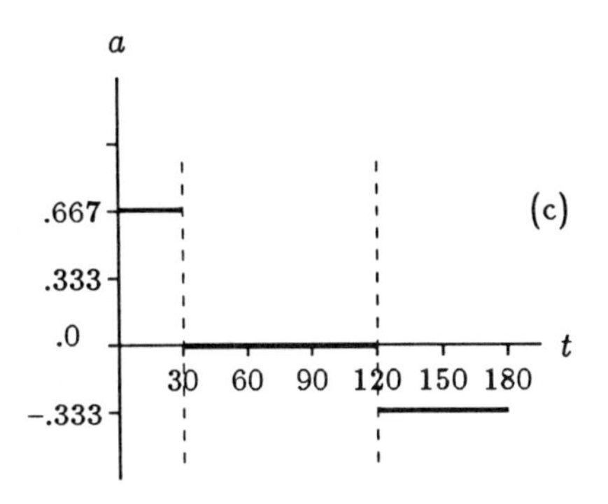

Figure 8.6 *Displacement x (m), velocity v (m/s) and acceleration a (m/s^2) versus time t (s) graphs for the car.*

8.4.3 Acceleration measured

If the acceleration of an object is measured and expressed as a function of time, then the instantaneous velocity and relative position of the object can be calculated using:

$$v = v_0 + \int_{t_0}^{t} a\, dt \tag{8.9}$$

$$x = x_0 + \int_{t_0}^{t} v\, dt \tag{8.10}$$

Here, x_0 and v_0 correspond to the initial position and initial velocity of the object at time t_0.

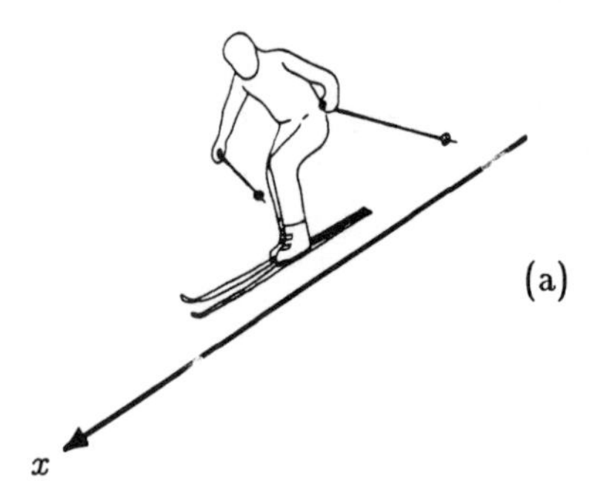

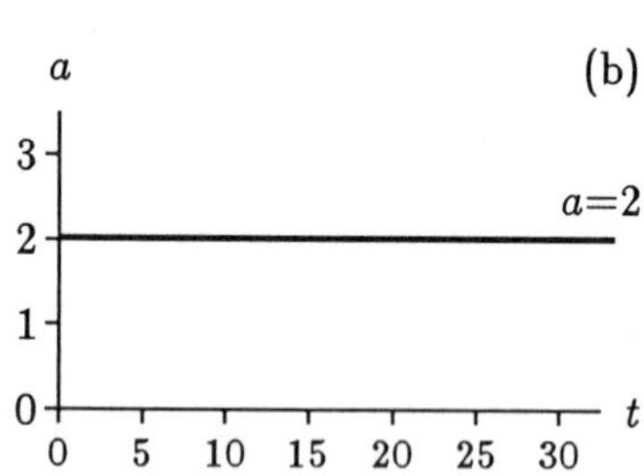

Figure 8.7 *A skier moving down the slope with a constant acceleration of a=2 m/s^2.*

Example 8.3 The skier in Figure 8.7(a) has just begun descending a straight slope. Assuming that the skier has a constant acceleration of 2 m/s^2 in the direction parallel to the slope (Figure 8.7b), determine the velocity and relative position of the skier as functions of time, and calculate the instantaneous velocity and relative position of the skier 15 seconds from the start.

Solution: Eqs. (8.9) and (8.10) can be used to analyze this problem. In this case, we can assume that all measurements are made relative to the start. In other words, we can take t_0=0, x_0=0, and v_0=0. Furthermore, the acceleration of the skier is constant and a in Eq. (8.9) can be taken outside the integral sign. Therefore:

$$v = \int_0^t a\, dt = a \int_0^t dt = a\, \big[t\big]_0^t = a\, t$$

$$x = \int_0^t v\, dt = \int_0^t (a\, t)\, dt = a \int_0^t t\, dt = a \left[\frac{t^2}{2}\right]_0^t = a\, \frac{t^2}{2}$$

Substituting a=2 m/s^2:

$$v = 2\, t$$

$$x = t^2$$

Here, for t measured in seconds, v is in meters per second and x is in meters. These functions are plotted in Figure 8.8. Substituting t=15 s into the above equations will yield:

$$v = (2)\,(15) = 30 \text{ m/s}$$

$$x = (15)^2 = 225 \text{ m}$$

Note that x is the distance traveled by the skier in the direction parallel to the sking slope.

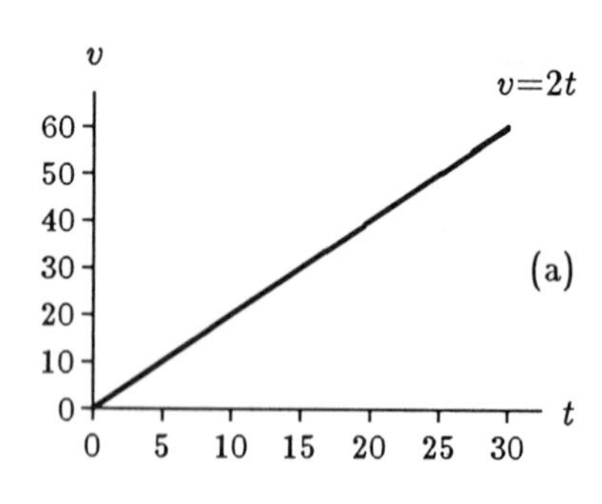

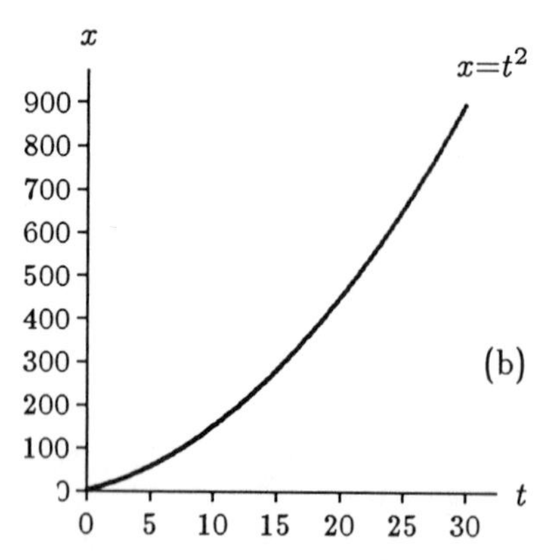

Figure 8.8 *Velocity v (m/s) and displacement x (m) as functions of time t (s).*

8.5 Uniaxial Motion with Constant Acceleration

A common type of uniaxial motion occurs when the acceleration is constant (uniform) so that the acceleration versus time graph is a straight line (Figure 8.9). Let a_0 be the constant acceleration of an object with initial velocity v_0 measured at time $t_0=0$. From Eq. (8.9):

$$v = v_0 + a_0\, t \tag{8.11}$$

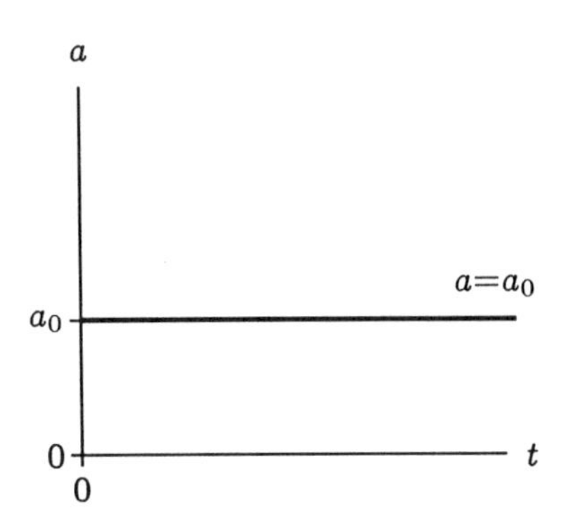

Figure 8.9 *Constant (uniform) acceleration.*

Similarly, if x_0 is the initial position of the object, then from Eq. (8.10):

$$x = x_0 + v_0\, t + \frac{1}{2}\, a_0\, t^2 \tag{8.12}$$

For a given initial position, initial velocity, and constant acceleration of an object undergoing uniaxial motion, Eqs. (8.11) and (8.12) can be used to determine the velocity and displacement of the body as functions of time.

The velocity in Eq. (8.11) is a linear function of time, and the velocity versus time graph is a straight line whose slope is constant and equal to acceleration a_0 (Figure 8.10). This is consistent with the fact that the slope of a function is equal to the derivative of that function, and that the derivative of velocity with respect to time is equal to acceleration. In Eq. (8.12), the displacement is a quadratic function of time, and as shown in Figure 8.11, the graph of this function is a parabola. At any given time, the slope of this function is equal to the velocity of the object at that instant.

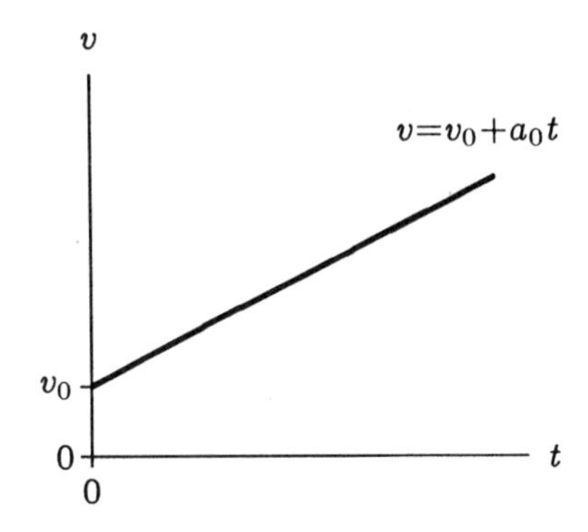

Figure 8.10 *For constant acceleration a_0, velocity v is a linear function of time t.*

For a uniaxial motion with constant acceleration, it is also possible to derive an expression between velocity, displacement, and time by solving Eq. (8.11) for a_0 and substituting it into Eq. (8.12). This will yield:

$$x = x_0 + \frac{1}{2}\,(v + v_0)\, t \tag{8.13}$$

Similarly, an expression between velocity, displacement, and acceleration can be derived by solving Eq. (8.11) for t and substituting it into Eq. (8.12):

$$v^2 = {v_0}^2 + 2\, a_0\, (x - x_0) \tag{8.14}$$

It must be noted here that Eqs. (8.11) through (8.14) are valid if the acceleration is constant. Furthermore, the acceleration in these equations must be handled properly. For one dimensional motion analyses, if the direction of acceleration is opposite to the direction of motion, then the "plus" sign in front of the terms carrying acceleration must be changed to a "minus" sign.

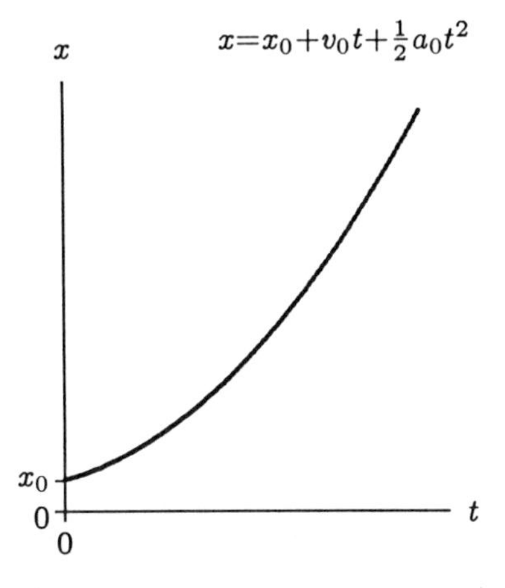

Figure 8.11 *For constant acceleration a_0, displacement x is a quadratic function of time t.*

8.6 Free Fall

One of the most common examples of uniformly accelerated motion is that of an object allowed to fall vertically to the ground. This type of motion is called *free fall.* At a given location above ground level and in the absence of air resistance, all objects fall with the same constant acceleration. This is called the *gravitational acceleration*, and is denoted with the symbol $\underline{g}$. Notice that $\underline{g}$ is a vector quantity. The direction of $\underline{g}$ is towards the center of Earth, or vertically downward. The magnitude g of the gravitational acceleration on Earth is about 9.8 m/s^2, which may vary slightly according to latitude and elevation.

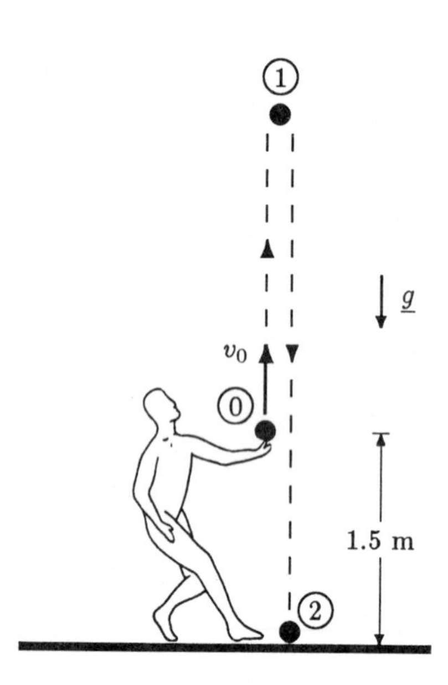

Figure 8.12 *A ball thrown upward into the air.*

Example 8.4 Consider a person throwing a ball upward into the air with an initial velocity of v_0=10 m/s (Figure 8.12). Assume that at the instant when the ball is released, the person's hand is at a height of 1.5 m above the ground.

Determine the maximum height that the ball can reach, the total time it would take for the ball to hit the ground, and the velocity of the ball just before impact.

Solution: This problem should be analyzed by identifying several stages in the motion of the ball. Let 0, 1, and 2 refer to the release, maximum height, and impact stages of the ball.

Between 0 and 1: As illustrated in Figure 8.13, if we locate the origin of the y coordinate to coincide with the ground level and positive y direction to be upward, then the elevation of the ball at the instant of release is y_0=1.5 m. The initial velocity of the ball is given as v_0=10 m/s upward. Let y_1 be the maximum elevation that the ball can reach, and t_1 be the time it would take for the ball to reach that maximum. At the maximum elevation, the velocity of the ball is zero (i.e., v_1=0). Between 0 and 1, the fact that the direction of gravitational acceleration (downward) is opposite to the positive direction of y (upward) must be taken into consideration. This can be achieved by assuming that a_0 is equal to $-g$.

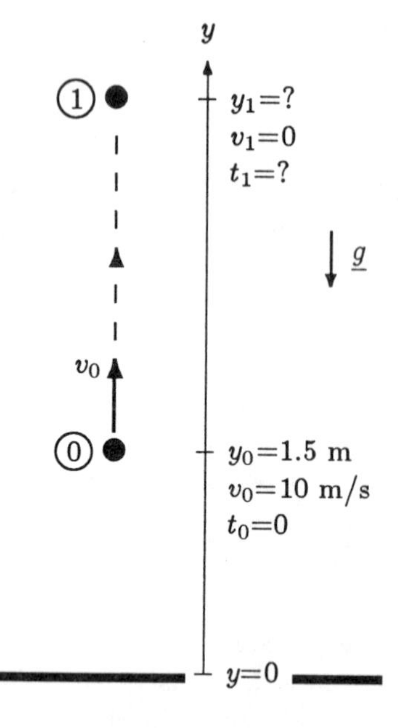

Figure 8.13 *Motion between 0 and 1.*

Now, Eqs. (8.11) through (8.14) can be used to solve the problem. In this case, x in these equations must be replaced by y. From Eq. (8.11):

$$v_1 = v_0 - g\,t_1 = 0$$

Solving this equation for t_1 will yield:

$$t_1 = \frac{v_0}{g} = \frac{10}{9.8} = 1.02 \text{ s}$$

From Eq. (8.12):

$$y_1 = y_0 + v_0\,t_1 - \frac{1}{2}\,g\,t_1^{\,2}$$

Substituting known parameters:

$$y_1 = 1.5 + (10)(1.02) - \frac{1}{2}(9.8)(1.02)^2 = 6.6 \text{ m}$$

Between 1 and 2: To analyze the motion of the ball between 1 and 2 (free fall), it is more convenient to locate the origin of the y axis to coincide with point 1 and choose the positive y direction to be downward (Figure 8.14). Under these assumptions, y_1=0 and y_2=6.6 m. Furthermore, v_1=0 and a_0=g=9.8 m/s^2 because the positive direction of y is the same as the direction of gravitational acceleration. From Eq. (8.12):

$$y_2 = y_1 + v_1\, t_2 + \frac{1}{2}\, g\, {t_2}^2$$
$$= 0 + 0\, t_2 + \frac{1}{2}\, g\, {t_2}^2$$

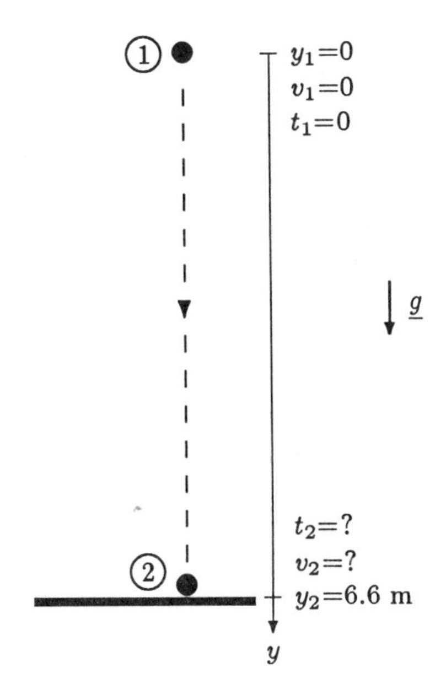

Figure 8.14 *Motion between 1 and 2 (free fall).*

Solving this equation for t_2, and substituting y_2=6.6 m and g=9.8 m/s^2 will yield:

$$t_2 = \sqrt{\frac{2y_2}{g}} = \sqrt{\frac{2\,(6.6)}{9.8}} = 1.16 \text{ s}$$

Using Eq. (8.11):

$$v_2 = v_1 + g\, t_2 = 0 + g\, t = g\, t$$

Substituting g=9.8 m/s^2 and t_2=1.16 s:

$$v_2 = (9.8)(1.16) = 11.4 \text{ m/s}$$

v_2 is the speed of the ball at the instant it hits the ground. The total time elapsed between time of release and time of impact is t_1+t_2=2.18 s.

8.7 Biaxial (Two-Dimensional) Motion Analysis

The one-dimensional linear motion characteristics of an object are completely known if the position of the object in the direction of motion is known as a function of time. The concepts introduced and the results obtained for unidirectional motion analysis can be expanded to analyze motion in two dimensions. This can be achieved by considering the vectorial properties of position, velocity, and acceleration.

8.8 Position, Velocity, and Acceleration Vectors

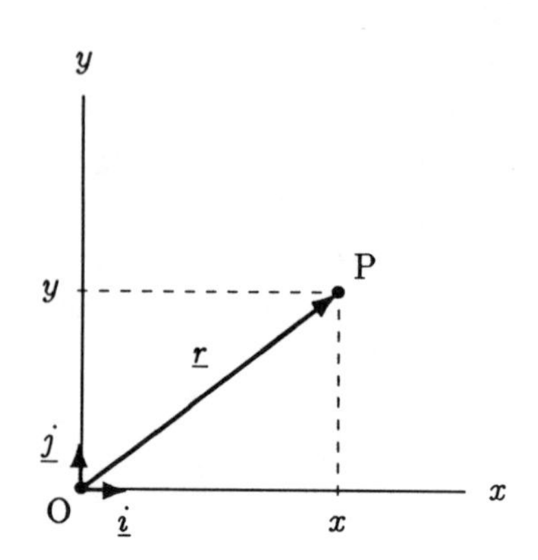

Figure 8.15 $\underline{r}$ *is the position vector of point P.*

For one-dimensional problems, the position of an object is defined by using a single coordinate. For plane problems, two coordinates must be specified to define the position of an object uniquely. As shown in Figure 8.15, let x and y represent the usual Cartesian (rectangular) coordinate directions with unit vectors $\underline{i}$ and $\underline{j}$, and the origin located at O. The ***position vector***, $\underline{r}$, of a point P in this xy-plane is a vector drawn from the origin of the coordinate frame towards that point (Figure 8.15). The position vector can be represented in terms of x and y coordinates of point P:

$$\underline{r} = x\,\underline{i} + y\,\underline{j} \tag{8.15}$$

The magnitude r of vector $\underline{r}$ is equal to the length of the line connecting O and P, and it can be calculated using:

$$r = \sqrt{x^2 + y^2} \tag{8.16}$$

If point P represents the location of a moving object at some time t, then $\underline{r}$ is the ***instantaneous position*** of that object at that time. This implies that $\underline{r}$ is changing with time, or that the x and y coordinates of the object are functions of time.

By definition, velocity is the time rate of change of position. Therefore, the ***velocity vector*** is equal to the derivative of the position vector with respect to time:

$$\underline{v} = \frac{d}{dt}(\underline{r}) = \frac{d}{dt}(x\,\underline{i} + y\,\underline{j}) = \frac{dx}{dt}\,\underline{i} + \frac{dy}{dt}\,\underline{j} \tag{8.17}$$

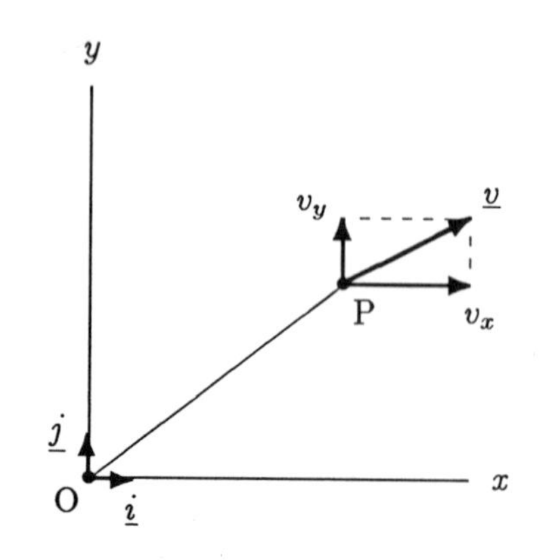

Figure 8.16 *Velocity vector.*

For two-dimensional problems, the velocity vector may have up to two components (Figure 8.16). If v_x and v_y refer to the scalar components of $\underline{v}$ in the x and y directions, respectively, then the velocity vector can also be expressed as:

$$\underline{v} = v_x\,\underline{i} + v_y\,\underline{j} \tag{8.18}$$

By comparing Eqs. (8.17) and (8.18), it can be concluded that:

$$v_x = \frac{dx}{dt} = \dot{x} \tag{8.19}$$

$$v_y = \frac{dy}{dt} = \dot{y} \tag{8.20}$$

Here, v_x and v_y are also known as the ***rectangular components*** of $\underline{v}$, and they indicate how fast the object is moving in the x and y directions, respectively. If the components v_x and v_y of the velocity vector are known, then the magnitude v of their resultant can also be determined:

$$v = \sqrt{{v_x}^2 + {v_y}^2} \tag{8.21}$$

Note that v is a positive scalar quantity also known as the *speed*. The direction of the velocity vector is always tangent to the path of the motion and pointing in the direction of motion.

By definition, acceleration is the time rate of change of velocity. Therefore, if the velocity vector of an object is known as a function of time, then its *acceleration vector*, $\underline{a}$, can also be determined by taking the derivative of $\underline{v}$ with respect to time:

$$\underline{a} = \frac{d}{dt}(\underline{v}) = \frac{d}{dt}(v_x\,\underline{i} + v_y\,\underline{j}) = \frac{dv_x}{dt}\,\underline{i} + \frac{dv_y}{dt}\,\underline{j} \tag{8.22}$$

The acceleration vector can also be expressed in terms of its components in the x and y directions (Figure 8.17):

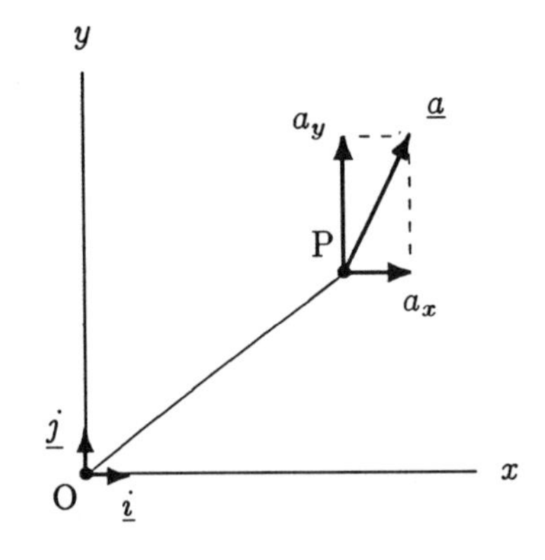

Figure 8.17 *Acceleration vector.*

$$\underline{a} = a_x\,\underline{i} + a_y\,\underline{j} \tag{8.23}$$

By comparing Eqs. (8.22) and (8.23), the rectangular components of the acceleration vector can alternatively be written as:

$$a_x = \frac{dv_x}{dt} = \frac{d^2x}{dt^2} = \ddot{x} \tag{8.24}$$

$$a_y = \frac{dv_y}{dt} = \frac{d^2y}{dt^2} = \ddot{y} \tag{8.25}$$

The magnitude a of the net (resultant) acceleration is:

$$a = \sqrt{{a_x}^2 + {a_y}^2} \tag{8.26}$$

8.9 Biaxial Motion with Constant Acceleration

Two-dimensional linear motion of an object in the xy plane can be analyzed in two stages by first considering its motion in the x and y directions separately, and then combining the results obtained using the vectorial properties of the parameters involved. The parameters defining the motion in the x direction are x, its first time derivative v_x, and its second time derivative a_x. Similarly, y, v_y, and a_y are the parameters that define the motion of the object in the y direction. If the acceleration of an object undergoing two-dimensional linear motion is constant, then a_x and a_y must also be constants. The details of uniaxial motion with constant acceleration were analyzed in Section 8.5. The results of these analyses can readily be adopted to analyze two-dimensional motions with constant acceleration.

In the x direction, Eqs. (8.11) and (8.12) can be rewritten in the following more specific forms:

$$v_x = v_{x0} + a_{x0}\,t \tag{8.27}$$

$$x = x_0 + v_{x0}\,t + \frac{1}{2}\,a_{x0}\,t^2 \tag{8.28}$$

Similarly, in the y direction:

$$v_y = v_{y0} + a_{y0}\, t \tag{8.29}$$

$$y = y_0 + v_{y0}\, t + \frac{1}{2}\, a_{y0}\, t^2 \tag{8.30}$$

Here, x_0 and y_0 are the initial coordinates of the object, v_{x0} and v_{y0} are the initial velocity components in the x and y directions, and a_{x0} and a_{y0} are the constant components of the acceleration vector. For given x_0, y_0, v_{x0}, v_{y0}, a_{x0}, and a_{y0}, Eqs. (8.27) through (8.30) can be used to calculate the relative position of the object and its velocity at any time t.

8.10 Projectile Motion

When an object is thrown into the air in any direction other than the vertical, it will move in a curved path under the influence of gravity and air resistance. The gravity of Earth will pull the object downward with a constant gravitational acceleration of about 9.8 m/s^2, and the air resistance will tend to retard its horizontal motion. This very common form of motion, called the *projectile motion*, is relatively simple to analyze once the effect of air resistance is neglected.

Figure 8.18 *A ball kicked will undergo a projectile motion.*

Projectile motion is a particular form of two-dimensional linear motion with constant acceleration. To be able to define the basic concepts involved in all projectile motions, consider a soccer ball kicked by a player into the air (Figure 8.18). The ball will first ascend, reach a peak, start descending, and finally land back on the field. The curved flight path of the ball is called the *trajectory* of motion. Let O refer to the initial position of the ball, P be the peak it reaches, and L be the location of landing (Figure 8.19). Also, let v_0 be the magnitude of the initial velocity of the ball, and θ be the angle the initial velocity vector makes with the horizontal. v_0 is called the *speed of release* or *takeoff speed*, and angle θ is called the *angle of release*. If h is the vertical distance between O and P, and ℓ is the horizontal distance between O and L, then h is called the *maximum height*, and ℓ is called the *horizontal range of motion*. The total time the ball remains in the air is called the *time of flight*.

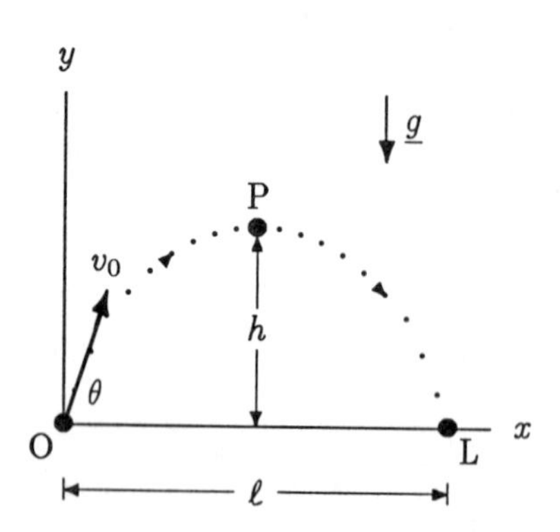

Figure 8.19 *Projectile motion.*

The equations necessary to analyze projectile motions can be derived from Eqs. (8.27) through (8.30). If the speed and angle of release of the projectile are known, then components of the velocity vector along the horizontal (x) and vertical (y) directions can be calculated:

$$v_{x0} = v_0 \cos\theta$$

$$v_{y0} = v_0 \sin\theta$$

Assuming that the air resistance on the ball is negligible, the acceleration of the ball in the x direction is zero (a_{x0}=0). The gravitational acceleration, $\underline{g}$, acts downward. Assuming that the y axis is positive upward, the gravitational acceleration acts in the negative y direction. To account for the negative direction of gravitational acceleration, the plus signs in front of the terms carrying a_{y0} in Eqs. (8.29) and (8.30) must be changed to minus, or for practical purposes, it can be assumed that $a_{y0}=-g$. Under these considerations, Eqs. (8.27) through (8.30) take the following special forms for the projectile motion:

$$x = x_0 + (v_0 \cos\theta)\, t \tag{8.31}$$

$$y = y_0 + (v_0 \sin\theta)\, t - \frac{1}{2}\, g\, t^2 \tag{8.32}$$

$$v_x = v_0 \cos\theta \tag{8.33}$$

$$v_y = v_0 \sin\theta - g\, t \tag{8.34}$$

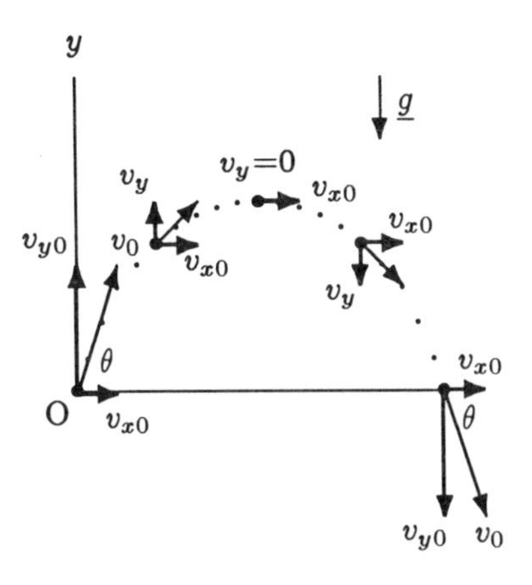

Figure 8.20 *v_{x0} remains constant while v_y changes with time. The resultant velocity vector is always tangent to the trajectory of the projectile.*

Here, if the origin of the xy coordinate frame is chosen to coincide with the initial position of the ball, then x_0=0 and y_0=0. Notice from Eq. (8.33) that the magnitude of the horizontal component of the velocity vector is not a function of time. Therefore, $v_x=v_{x0}$ remains constant throughout the projectile motion (Figure 8.20). The magnitude of the vertical component velocity vector is a linear function of time. It is positive (upward) initially, decreases in time as the object ascends, and drops to zero at the peak. After reaching the peak, the vertical component of the velocity vector changes its direction from upward to downward, while its magnitude increases until it lands on the ground. At any instant during the flight, the resultant velocity vector is tangent to the trajectory of the projectile motion.

In some cases, the objective of the projectile motion may be to increase the horizontal range of motion to a maximum. This is particularly true for a ski jumper, for example. Other situations may require a control over the height as well as the horizontal range of the projectile. Therefore, it may be useful to derive some expressions for the horizontal range and maximum height of the projectile motion. Consider Figure 8.21 which shows the trajectory of a projectile motion along with some of the parameters involved. x_0=0 and y_0=0 because the origin of the xy coordinate frame is chosen to coincide with the initial position of the object. Let t_1 be the time it takes to reach the peak, and t_2 be the total time of flight. At the peak, $y=h$, v_y=0, and from the symmetry of the motion $x=\ell/2$. At the point where the object lands, $x=\ell$ and y=0. The maximum height, h, reached by the projectile can now be determined by noting that v_y=0 at the peak. Writing Eq. (8.34) between O and P:

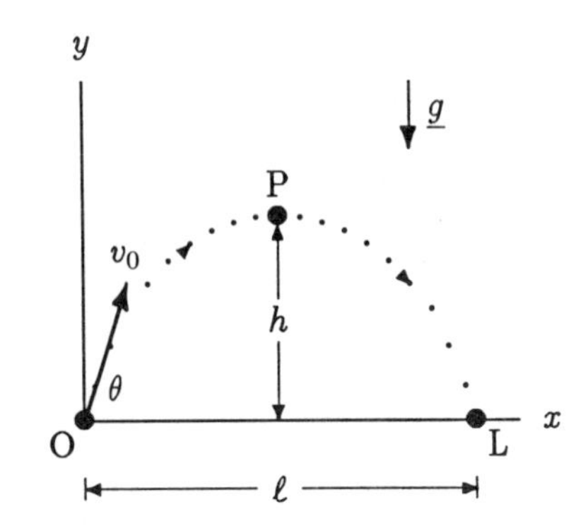

Figure 8.21 *h is the maximum height and ℓ is the horizontal range of motion of the projectile.*

$$0 = v_0 \sin\theta - g\, t_1$$

Solving this equation for t_1 will yield:

$$t_1 = \frac{v_0 \sin\theta}{g} \tag{8.35}$$

Between O and P, Eq. (8.32) will take the following form:

$$h = 0 + (v_0 \sin\theta)\, t_1 - \frac{1}{2}\, g\, t_1^{\,2}$$

Substituting Eq. (8.35) into this equation will yield:

$$h = \frac{v_0^{\,2}\, \sin^2\theta}{2g} \qquad (8.36)$$

Again from the symmetry of motion, the time lapses during the ascent must be equal to the time lapses during the descent. In other words, $t_2 = 2t_1$. This can be proven by writing Eq. (8.32) between O and L, and solving it for t_2:

$$t_2 = \frac{2\, v_0\, \sin\theta}{g} = 2\, t_1 \qquad (8.37)$$

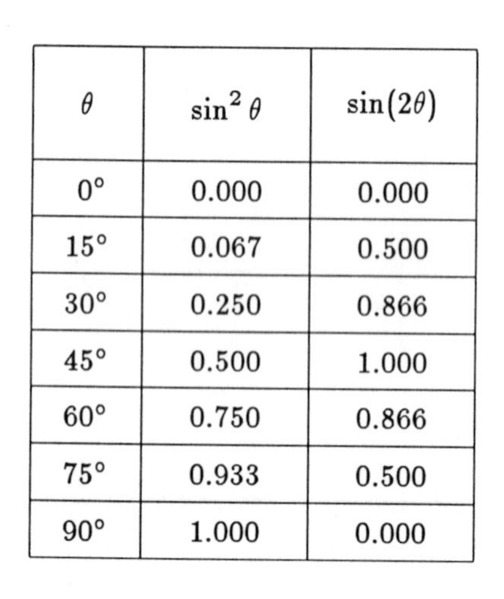

θ	$\sin^2\theta$	$\sin(2\theta)$
0°	0.000	0.000
15°	0.067	0.500
30°	0.250	0.866
45°	0.500	1.000
60°	0.750	0.866
75°	0.933	0.500
90°	1.000	0.000

Table 8.2 *Variation of $\sin^2\theta$ and $\sin(2\theta)$ in the range $0° \leq \theta \leq 90°$. The maximum height h of the projectile is a function of $\sin^2\theta$, and range ℓ depends on $\sin(2\theta)$.*

Between O and L, Eq. (8.31) can be written as:

$$\ell = 0 + (v_0\, \cos\theta)\, t_2$$

Substituting Eq. (8.37) into the above equation, and noting that $2\cos\theta\sin\theta = \sin(2\theta)$ (see Appendix C):

$$\ell = \frac{v_0^{\,2}\, \sin(2\theta)}{g} \qquad (8.38)$$

Eqs. (8.35) through (8.38) are special forms of more general equations for projectile motions as given in Eqs. (8.31) through (8.34). From Eqs. (8.36) and (8.38), it is clear that for a given v_0, the maximum height and the horizontal range of motion of the projectile are functions of the angle of release, θ. To see the variations of h and ℓ with θ, $\sin^2\theta$ and $\sin(2\theta)$ are computed for θ between 0° and 90°. The values calculated are listed in Table 8.2. In Figure 8.22, a value of 10 m/s is assigned to v_0 and corresponding trajectories are calculated for θ=30°, 45°, and 60°. The significance of these results is that, for given v_0, the range of motion ℓ is maximum when θ=45°. Therefore, if the purpose is to maximize the horizontal range of motion of the projectile, then the angle of release or takeoff should be close to 45°.

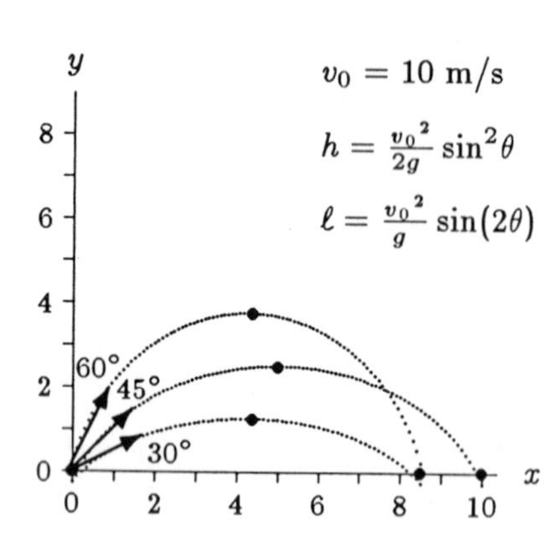

Figure 8.22 *Trajectories for v_0=10 m/s and θ=30°, 45°, 60°. (Both x and y are in meters.)*

Sometimes it is easier to measure the range of motion ℓ and the maximum height h of the projectile. In such cases, the unknowns are the takeoff speed v_0 and the angle of release θ. The relationship between these parameters can be shown to be:

$$\theta = \arctan\left(\frac{4\, h}{\ell}\right) \qquad (8.39)$$

$$v_0 = \frac{\sqrt{2\, g\, h}}{\sin\theta} \qquad (8.40)$$

8.11 Applications to Sports Mechanics

The concept of projectile motion may have many applications in athletics and sports mechanics. These applications include the motion analyses of athletes doing long jumping, high jumping, ski jumping, diving, and gymnastics, and the motion analyses of the discus, javelin, shot, baseball, basketball, football, and golf ball. The following examples illustrate some of these applications.

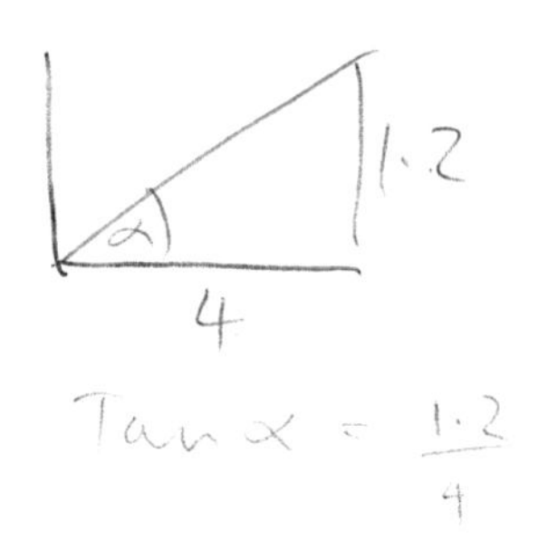

Example 8.5 Based on the assumption that the air resistance is negligible, it is suggested that the motion characteristic of a long jumper can be analyzed by assuming that the center of gravity of the athlete undergoes a projectile motion (Figure 8.23).

Consider an athlete who jumps a horizontal distance of 8 meters after reaching a maximum height of 1.2 meters (Figure 8.24). What is the takeoff speed of the athelete? How can the athlete improve his/her performance?

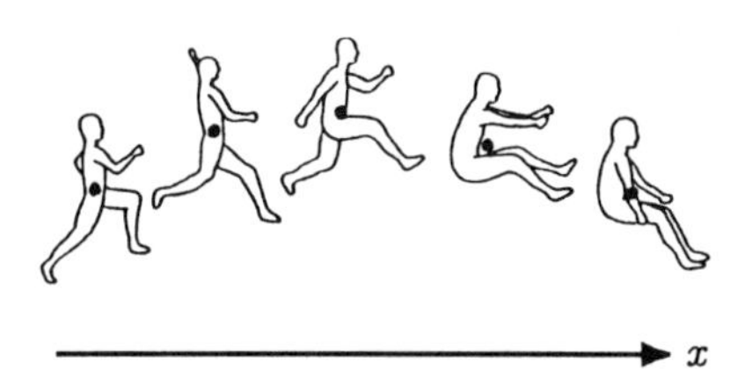

Figure 8.23 *A long-jumper.*

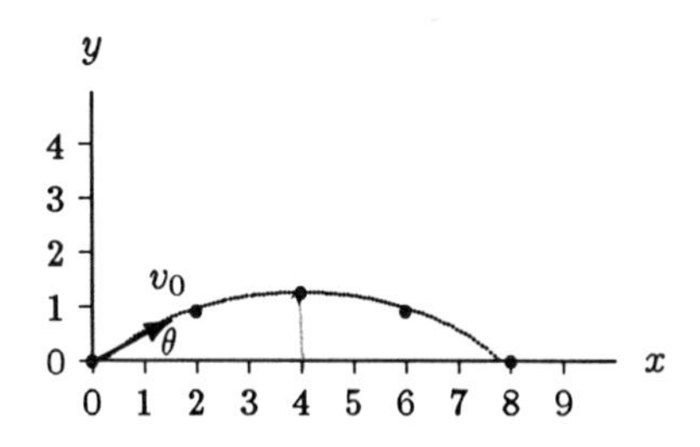

Figure 8.24 *Trajectory of the center of gravity. (Both x and y are in meters.)*

Solution: In this problem, the maximum height of the projectile is given as h=1.2 m and the horizontal range of the projectile is measured as ℓ=8 m. To determine the takeoff speed of the athlete, we can utilize Eqs. (8.39) and (8.40). Eq. (8.39) will give us the angle of takeoff and Eq. (8.40) will yield the takeoff speed of the athlete. From Eq. (8.39):

$$\tan\theta = \frac{4\,h}{\ell}$$

Substituting the numerical values:

$$\tan\theta = \frac{4\,(1.2)}{8} = 0.6$$

Taking the inverse tangent of 0.6:

$$\theta = 31^\circ$$

From Eq. (8.40):

$$v_0 = \frac{\sqrt{2\,g\,h}}{\sin\theta}$$

Substituting the numerical values:

$$v_0 = \frac{\sqrt{2\,(9.8)(1.2)}}{\sin 31^\circ} = 9.4 \text{ m/s}$$

The athlete can improve his or her performance by increasing the takeoff angle θ from 31° towards 45°.

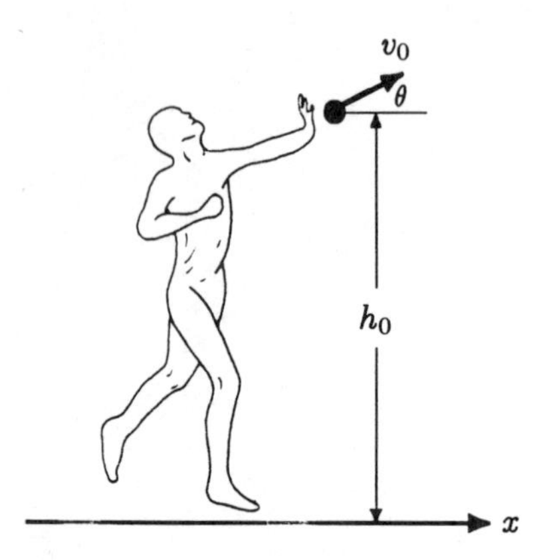

Figure 8.25 *Shot-putter.*

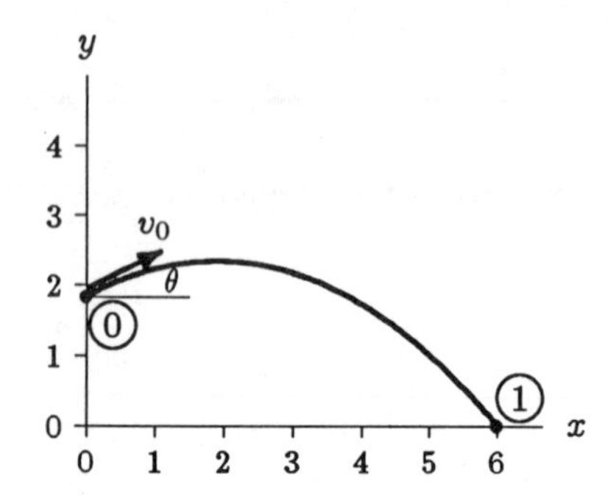

Figure 8.26 *Trajectory of the shot. (x and y are in meters.)*

Example 8.6 During a practice, a shot-putter puts the shot at a distance ℓ=6 m. At the instant the athlete releases the shot, the elevation of the shot is h_0=1.8 m as measured from ground level, and the angle of release is θ=30° (Figure 8.25).

Determine the speed at which the athlete releases the shot.

Solution: Eqs. (8.31) and (8.32) can be utilized to solve this problem. In Figure 8.26, the origin of the xy coordinate frame is located at the ground level below the point of release which is designated as 0. The shot follows a curved path and lands on the field at point 1. With respect to the coordinate frame adopted, the initial and landing coordinates of the shot are x_0=0, y_0=h_0, x_1=ℓ, and y_1=0. If t_1 refers to the total time of flight of the shot, then Eq. (8.31) can be written between 0 and 1 as:

$$x_1 = x_0 + (v_0 \cos\theta)\, t_1$$
$$\ell = 0 + (v_0 \cos\theta)\, t_1$$

Solving this equation for t_1 will yield:

$$t_1 = \frac{\ell}{v_0 \cos\theta} \qquad (i)$$

Similarly, writing Eq. (8.32) between 0 and 1:

$$y_1 = y_0 + (v_0 \sin\theta)\, t_1 - \frac{1}{2}\, g\, {t_1}^2$$
$$0 = h_0 + (v_0 \sin\theta)\, t_1 - \frac{1}{2}\, g\, {t_1}^2$$

Substituting Eq. (i):

$$0 = h_0 + (v_0 \sin\theta)\left(\frac{\ell}{v_0 \cos\theta}\right) - \frac{1}{2}\, g \left(\frac{\ell}{v_0 \cos\theta}\right)^2$$

Eliminating repeated terms and solving this equation for v_0:

$$v_0 = \frac{\ell}{\cos\theta}\sqrt{\frac{g}{2\,(h_0 + \ell\, \tan\theta)}} \qquad (ii)$$

Substituting the numerical values, and carrying out the computations:

$$v_0 = 6.7 \text{ m/s}$$

Knowing v_0, time of flight t_1 can also be calculated using Eq. (i). Furthermore, Eqs. (8.33) and (8.34) can be used to determine the landing velocity of the shot.

Example 8.7 The ski jumper shown in Figure 8.27, leaves the ramp with a horizontal velocity of v_0 and lands on a slope which makes an angle β=45° with the horizontal.

Neglecting air resistance, determine the takeoff speed, landing velocity, and the flight time of the ski jumper if the skier touches down at a distance d=50 m (measured parallel to the slope) from the ramp.

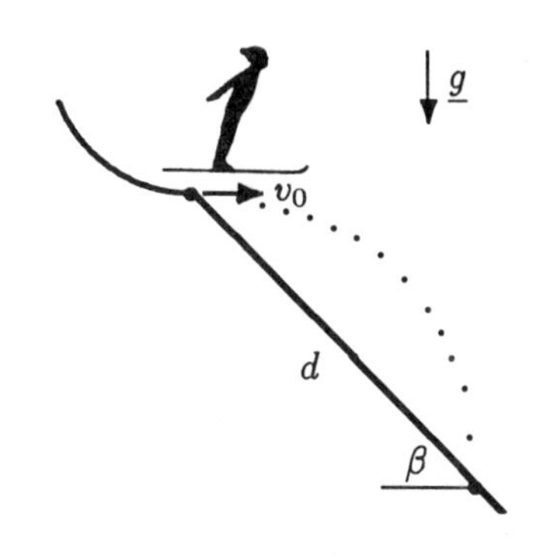

Figure 8.27 *A ski jumper.*

Solution: Eqs. (8.31) through (8.34) can be utilized for the solution of this problem. Since the ski jumper leaves the track with a horizontal velocity, the angle of takeoff is zero. Therefore, if x and y refer to the horizontal and vertical direction, then at the instant of takeoff the velocity of the ski jumper has the following components:

$$v_{x0} = v_0$$
$$v_{y0} = 0$$

As shown in Figure 8.28, let 0 and 1 refer to the points of takeoff and landing. The distance between 0 and 1 is measured along the slope, which is given as d=50 m. The slope makes an angle β=45° with the horizontal. Therefore, the properties of right-triangles can be utilized to calculate b and h, which are the horizontal and vertical distances between the points of takeoff and landing:

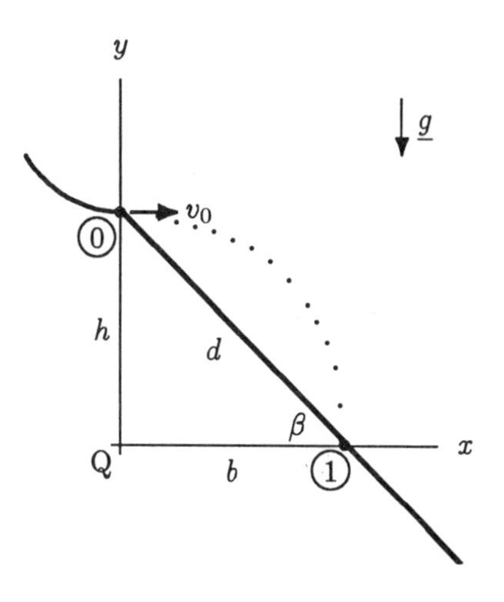

Figure 8.28 *Geometry of the problem.*

$$b = d \cos\beta = (50)(\cos 45°) = 35.4 \text{ m}$$
$$h = d \sin\beta = (50)(\sin 45°) = 35.4 \text{ m}$$

If we place the origin Q of the xy coordinate frame at a position h distance below 0 (Figure 8.28), then x_0=0, y_0=h, x_1=b, and y_1=0. Also if t_1 designates the total time the ski jumper remains in the air, then Eq. (8.31) can be written between 0 and 1 as:

$$x_1 = x_0 + (v_0 \cos\theta)\, t_1$$
$$b = 0 + (v_0 \cos 0°)\, t_1$$
$$b = v_0\, t_1$$

Solving this equation for v_0:

$$v_0 = \frac{b}{t_1} \qquad (i)$$

Similarly, writing Eq. (8.32) between 0 and 1:

$$y_1 = y_0 + (v_0 \sin\theta)\, t_1 - \frac{1}{2}\, g\, {t_1}^2$$
$$0 = h + (v_0 \sin 0°)\, t_1 - \frac{1}{2}\, g\, {t_1}^2$$
$$0 = h + 0 - \frac{1}{2}\, g\, {t_1}^2$$

Solving this equation for t_1 will yield:

$$t_1 = \sqrt{\frac{2\,h}{g}} \tag{ii}$$

Note that an alternative expression for v_0 in terms of given parameters can be obtained by substituting Eq. (*ii*) into Eq. (*i*):

$$v_0 = b\sqrt{\frac{g}{2\,h}} \tag{iii}$$

Substituting the numerical values of parameters involved into Eqs. (*ii*) and (*iii*) or (*i*):

$$t_1 = \sqrt{\frac{2\,(35.4)}{9.8}} = 2.7 \text{ s}$$

$$v_0 = 35.4\sqrt{\frac{9.8}{2\,(35.4)}} = 13.2 \text{ m/s}$$

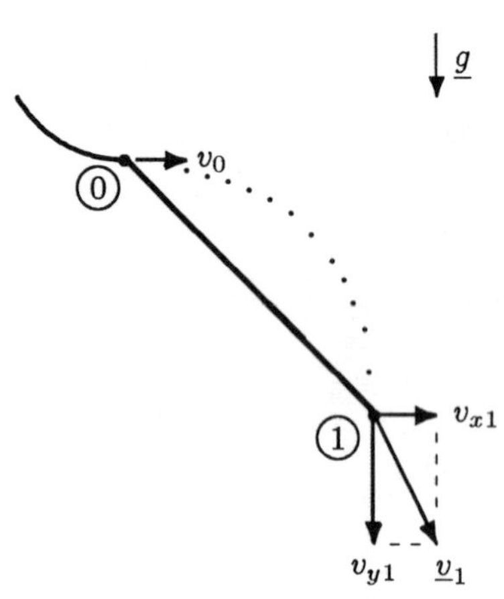

Figure 8.29 *$\underline{v}_1$ is the landing velocity of the ski jumper.*

To determine the landing velocity of the ski jumper, we can use Eqs. (8.33) and (8.34). From Eq. (8.33):

$$v_{x1} = v_{x0} = v_0 = 13.2 \text{ m/s}$$
$$v_{y1} = v_{y0} - g\,t_1 = 0 - (9.8)(2.7) = -26.5 \text{ m/s}$$

The negative sign in front of 26.5 indicates that the direction of vertical component of the landing velocity is opposite to that of the positive y direction. In other words, it is downward. The resultant landing velocity (Figure 8.29) of the skier can be expressed in vector form as follows:

$$\underline{v}_1 = 13.2\,\underline{i} - 26.5\,\underline{j}$$

Therefore, the speed of landing is:

$$v_1 = \sqrt{(13.2)^2 + (26.5)^2} = 29.6 \text{ m/s}$$

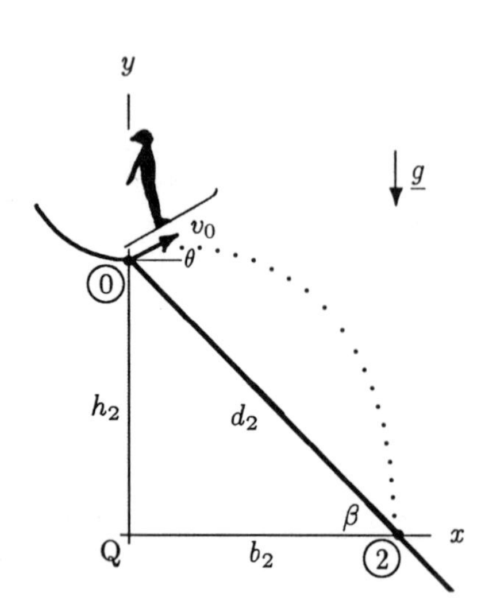

Figure 8.30 *The ski jumper leaves the ramp at an angle θ.*

Now, assume that in a second trial the ski jumper manages to maintain the takeoff speed at v_0=13.2 m/s, but leaves the ramp at an angle θ=10° with the horizontal (Figure 8.30). Has the ski jumper improved his/her performance? If so, by how much?

Because of the change in the angle of takeoff, the ski jumper will probably land at a different point on the slope. Let 2 designate the new landing point, b_2 and h_2 be the horizontal and vertical distances between 0 and 2, t_2 be the total time of flight, and place the origin Q of the xy coordinate frame at a height h_2 below 0. In this case, the takeoff speed and angle are known, and b_2 and h_2 can be calculated using Eqs. (8.31) and (8.32).

From the geometry of the problem:

$$h_2 = b_2 \tan\beta \qquad (iv)$$

Writing Eq. (8.32) between 0 and 2:

$$y_2 = y_0 + (v_0 \sin\theta)\, t_2 - \frac{1}{2}\, g\, {t_2}^2$$
$$0 = h_2 + (v_0 \sin\theta)\, t_2 - \frac{1}{2}\, g\, {t_2}^2$$

Solving this equation for h_2:

$$h_2 = -(v_0 \sin\theta)\, t_2 + \frac{1}{2}\, g\, {t_2}^2 \qquad (v)$$

Writing Eq. (8.31) between 0 and 2:

$$x_2 = x_0 + (v_0 \cos\theta)\, t_2$$
$$b_2 = 0 + (v_0 \cos\theta)\, t_2$$
$$b_2 = (v_0 \cos\theta)\, t_2 \qquad (vi)$$

Substituting Eqs. (v) and (vi) into Eq. (iv):

$$-(v_0 \sin\theta)\, t_2 + \frac{1}{2}\, g\, {t_2}^2 = (v_0 \cos\theta)\, t_2 \tan\beta$$

This equation can be simplified by eliminating t_2 which appears in all terms, and then solved for the unknown t_2:

$$t_2 = \frac{2\, v_0}{g} (\sin\theta + \cos\theta \tan\beta) \qquad (vii)$$

Substituting θ=10°, β=45°, v_0=13.2 m/s, and g=9.8 m/s^2 into Eqs. (vii), (vi), and (v), and performing necessary calculations will yield:

$$t_2 = 3.1 \text{ s}$$
$$b_2 = 40.3 \text{ m}$$
$$h_2 = 40.3 \text{ m}$$

Using h_2 and b_2, the distance d_2 traveled by the ski jumper along the slope can also be calculated:

$$d_2 = \sqrt{{b_2}^2 + {h_2}^2} = 57.0 \text{ m}$$

Comparing the results of two trials, we see that a 10° takeoff angle produces about 7 meters improvement in the performance of the ski jumper. In percentages:

$$\frac{d_2 - d}{d} \times 100 = 14.0$$

Therefore, the improvement is 14%.

Figure 8.31 *A diver.*

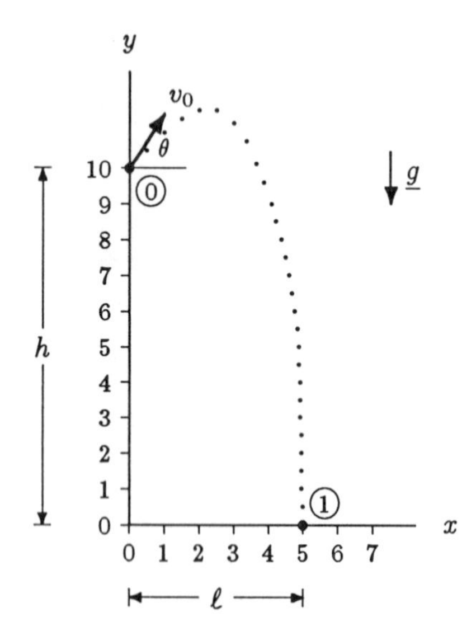

Figure 8.32 *Trajectory of the center of gravity of the diver. (Both x and y are in meters.)*

Example 8.8 The diver illustrated in Figure 8.31 undergoes both rotational and translational (i.e., general) motion. The translational motion of the diver can be analyzed by observing the trajectory of the diver's center of gravity which can be assumed to undergo a projectile motion.

Consider a case in which a diver takes off from a diving board which is located at a height h=10 m above the water level, and the diver enters the water at a horizontal distance ℓ=5 m from the end of the board. If the total time the diver spends in the air is 2.5 seconds, calculate the speed and angle of takeoff of the diver.

Solution: The trajectory of the center of gravity of the diver is shown in Figure 8.32. In this case, the angle (θ) and the velocity (v_0) of takeoff are not known, but are to be determined. Let "0" be the point of takeoff, "1" refer to the point the diver enters the water, and t_1=2.5 s be the total time the diver remains in the air. Between 0 and 1, Eqs. (8.31) can be written as:

$$x_1 = x_0 + (v_0 \cos\theta)\, t_1$$
$$\ell = 0 + (v_0 \cos\theta)\, t_1$$

Solving this equation for $v_0 \cos\theta$:

$$v_0 \cos\theta = \frac{\ell}{t_1} \qquad (i)$$

Similarly, from Eq. (8.32):

$$y_1 = y_0 + (v_0 \sin\theta)\, t_1 - \frac{1}{2} g\, {t_1}^2$$
$$0 = h + (v_0 \sin\theta)\, t_1 - \frac{1}{2} g\, {t_1}^2$$

Solving this equation for $v_0 \sin\theta$:

$$v_0 \sin\theta = \frac{1}{2} g\, t_1 - \frac{h}{t_1} \qquad (ii)$$

Dividing Eq. (ii) by Eq. (i):

$$\tan\theta = \frac{g\, {t_1}^2}{2\ell} - \frac{h}{\ell} \qquad (iii)$$

Substituting h=10 m, ℓ=5 m, t_1=2.5 s, and g=9.8 m/s^2 into Eq. (iii), performing necessary calculations, and taking the inverse tangent will yield the angle of takeoff:

$$\theta = 76.4^\circ$$

The speed of takeoff can now be determined from Eq. (i):

$$v_0 = \frac{\ell}{t_1 \cos\theta} = 8.5 \text{ m/s}$$

Chapter 9

ANGULAR KINEMATICS

9.1 Polar Coordinates

Two-dimensional angular motions of bodies are commonly described in terms of a pair of parameters, r and θ (theta), which are called the *polar coordinates.* Polar coordinates are particularly well suited for analyzing motions restricted to circular paths. As illustrated in Figure 9.1, let O and P be two points on a two-dimensional surface. The location of P with respect to O can be specified in many different ways. For example, in terms of rectangular coordinates, P is a point whose coordinates are x and y. Point P is also located at a distance r from point O making an angle θ with the horizontal. Both x and y, and r and θ specify the position of P with respect to O uniquely, and O forms the origin of both the rectangular and polar coordinate systems. x, y, r, and θ are not independent. In other words, if one pair is known, the other pair can be calculated. The equations relating these parameters can be derived by noting that they form a right-triangle with r being the hypotenuse, θ being one of the two acute angles, x and y being the lengths of the adjacent and opposite sides of the right-triangle with respect to angle θ. Therefore:

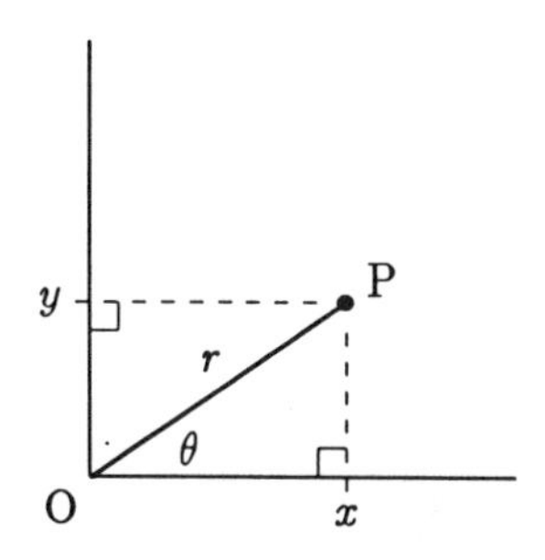

Figure 9.1 *Rectangular and polar coordinates of point P.*

$$\begin{aligned} x &= r\cos\theta \\ y &= r\sin\theta \end{aligned} \qquad (9.1)$$

Expressing r and θ in terms of x and y:

$$\begin{aligned} r &= \sqrt{x^2+y^2} \\ \theta &= \arctan\left(\frac{y}{x}\right) \end{aligned} \qquad (9.2)$$

9.2 Angular Position and Displacement

Consider an object undergoing a rotational motion in the xy plane about a fixed axis (Figure 9.2). Let O be a point in the xy plane along the axis of rotation of the object, and P be a fixed point on the object located at a distance r from O. Point P will move in a circular path of radius r and center located at O (see Section 9.6). Assume that at some time t_1, the point is located at P_1 which makes an angle θ_1 with the horizontal. At a later time t_2, the point is at P_2 which makes an angle θ_2 with the horizontal. θ_1 and θ_2 define the *angular positions* of the point at times t_1 and t_2, respectively. If θ denotes the change in angular position of the point in the time interval between t_1 and t_2, then $\theta=\theta_2-\theta_1$ is called the *angular displacement* of the point in the time interval between t_1 and t_2. In the same time interval, the point travels a distance s measured along the circular path. The equation relating the radius r of the circle, angle θ, and arc length s is:

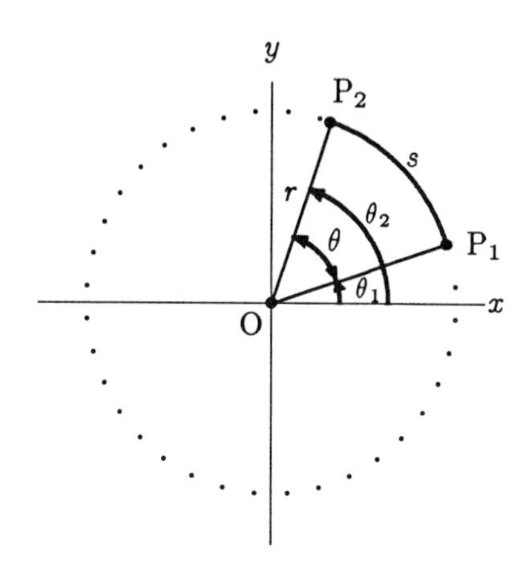

Figure 9.2 *θ is the angular displacement of the object between P_1 and P_2.*

$$s = r\,\theta \qquad \text{or} \qquad \theta = \frac{s}{r} \qquad (9.3)$$

In Eq. (9.3) angle θ must be measured in radians, rather than in degrees. As reviewed in Appendix C, radians and degrees are related in that there are 360 degrees in a complete circle which must correspond to an arc length equal to the circumference $s=2\pi r$ of the circle, with $\pi=3.14$ approximately. Therefore, $\theta=s/r=2\pi r/r=2\pi$ for a complete circle, or $360°=2\pi$. One radian is then equal to $360°/2\pi=57.3°$. The following formula can be used to convert angles given in degrees to corresponding angles in radians:

$$\theta\ (\text{radians}) = \frac{\pi}{180}\,\theta\ (\text{degrees})$$

Some selected angles and their equivalents in radians are listed in Table 9.1.

DEGREES (°)	RADIANS (rad)
30	$\pi/6=0.524$
45	$\pi/4=0.785$
60	$\pi/3=1.047$
90	$\pi/2=1.571$
180	$\pi=3.142$
270	$3\pi/2=4.712$
360	$2\pi=6.283$

Table 9.1 *Selected angles in degrees and radians.*

9.3 Angular Velocity

The time rate of change of angular position is called *angular velocity*, and it is commonly denoted by the symbol ω (omega). If the angular position of an object is known as a function of time, its angular velocity can be determined by taking its derivative with respect to time:

$$\omega = \frac{d\theta}{dt} = \dot{\theta} \tag{9.4}$$

The *average angular velocity* ($\bar{\omega}$) of an object in the time interval between t_1 and t_2 is defined by the ratio of change in angular position of the object divided by the time interval:

$$\bar{\omega} = \frac{\Delta\theta}{\Delta t} = \frac{\theta_2 - \theta_1}{t_2 - t_1} \tag{9.5}$$

In Eq. (9.5), θ_1 and θ_2 are the angular positions of the object at times t_1 and t_2, respectively.

9.4 Angular Acceleration

The angular velocity of an object may vary during motion. The variation of angular velocity over time is called *angular acceleration* which is usually denoted by the symbol α (alpha). If the angular velocity of a body is given as a function of time, then its angular acceleration can be determined by considering the derivative of the angular velocity with respect to time:

$$\alpha = \frac{d\omega}{dt} \tag{9.6}$$

The *average angular acceleration*, $\bar{\alpha}$, is the ratio of change in angular velocity and the time interval required to make that change. If ω_1 and ω_2 are the instantaneous angular velocities of a body measured at times t_1 and t_2, respectively, then the average angular acceleration of the body in the time interval between t_1 and t_2 is:

$$\bar{\alpha} = \frac{\Delta\omega}{\Delta t} = \frac{\omega_2 - \omega_1}{t_2 - t_1} \tag{9.7}$$

Note that using the definition of angular velocity in Eq. (9.4), angular acceleration can alternatively be expressed in the following forms:

$$\alpha = \frac{d\omega}{dt} = \frac{d}{dt}\left(\frac{d\theta}{dt}\right) = \frac{d^2\theta}{dt^2} = \ddot{\theta} \tag{9.8}$$

Angular displacement, velocity, and acceleration are vector quantities. Therefore, their directions must be stated as well as their magnitudes. For plane problems, the motion is either in the clockwise or counterclockwise direction. In other words, the direction of motion is either "into" or "out of" the plane of motion. While analyzing plane problems, first assume a positive direction (either clockwise or counterclockwise) preferably in the direction of motion. ω is positive if the motion is in the chosen direction, negative otherwise. The direction of angular acceleration is the same as angular velocity. α is positive when ω is increasing in time, and it is negative when ω is decreasing in time.

9.5 Dimensions and Units

The angular displacement is measured in radians which is a unit of angle. From Eq. (9.3), angular displacement of an object undergoing circular motion is equal to the ratio of the arc length and radius, and both arc length and radius have the dimension of length. Therefore, the dimension of angular displacement is 1, or it is a *dimensionless* quantity. By definition, angular velocity is the time rate of change of angular position, and angular acceleration is the time rate of change of angular velocity. Therefore, angular velocity has the dimension of 1 divided by time, and angular acceleration has the dimension of angular velocity divided by time:

$$[\text{ANG. DISPLACEMENT}] = \frac{\text{L}}{\text{L}} = 1$$

$$[\text{ANG. VELOCITY}] = \frac{[\text{ANG. POSITION}]}{[\text{TIME}]} = \frac{1}{\text{T}}$$

$$[\text{ANG. ACCELERATION}] = \frac{[\text{ANG. VELOCITY}]}{[\text{TIME}]} = \frac{1/\text{T}}{\text{T}} = \frac{1}{\text{T}^2}$$

Note that angular quantities θ, ω, and α differ dimensionally from linear quantities x, v, and a by a length factor, and that the units of angular quantities in different unit systems are the same (Table 9.2).

ANGULAR DISPLACEMENT	ANGULAR VELOCITY	ANGULAR ACCELERATION
radian (rad)	rad/s (s^{-1})	rad/s^2 (s^{-2})

Table 9.2 *Units of angular quantities.*

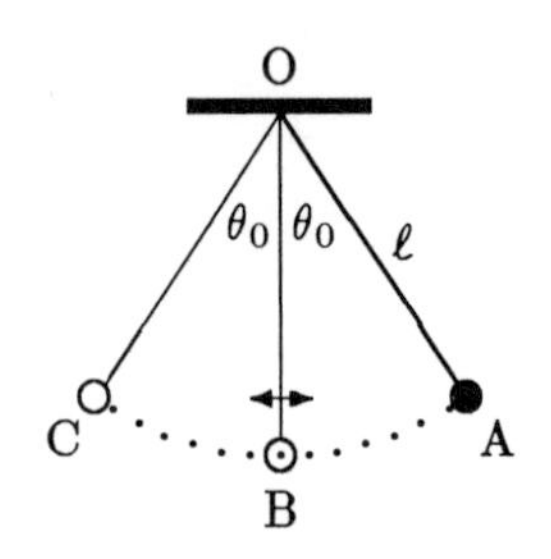

Figure 9.3 *Pendulum oscillating about its neutral position (B).*

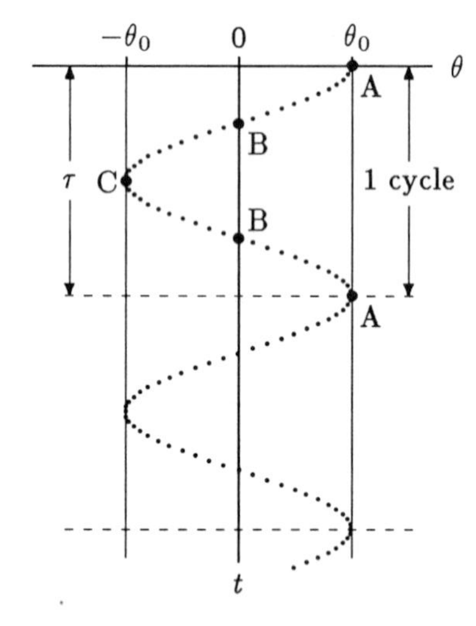

Figure 9.4 *Simple harmonic motion.*

Example 9.1 *Pendulum*

Figure 9.3 illustrates a pendulum which consists of a string of length ℓ fixed to the ceiling at one end (at O) and a mass m suspended at the other end. If the mass is pulled to the side (to A) so that the string makes an angle θ_0 with the vertical and is then released, the mass will oscillate or swing back and forth about its neutral or equilibrium position (B) in a circular arc path of radius ℓ. Due to internal friction and air resistance, the oscillations will die out over time and eventually the pendulum will come to a stop in its neutral position.

Analysis of the motion characteristics of this relatively simple system can give us considerable insight into the nature of other more complex dynamic systems.

Case A. Simple harmonic motion

We shall first ignore the air resistance and frictional effects, and assume that once the pendulum is excited, it will oscillate forever. This implies that in each swing (between A and C) the pendulum will cover a total angle equal to twice that of angle θ_0. If θ represents the angle the pendulum makes with the vertical at any time t, then θ is essentially a measure of the instantaneous angular position of the pendulum. θ can be expressed as a sine or cosine function of time. Assuming that θ is positive between A and B, zero at B, and negative between B and C we can write:

$$\theta = \theta_0 \cos(\phi\, t) \qquad (i)$$

In this case, θ_0 and ϕ (phi) in Eq. (i) are some constants. Part of the graph of this function is shown in Figure 9.4. The motion described in Figure 9.4 is known as the *simple harmonic motion.* At time t=0, the mass is located at A which makes an angle θ_0 with the vertical. Time t=0 corresponds to the instant when the mass is first released. The mass swings, passes through point B where θ=0, and reaches point C where θ=$-\theta_0$. At point C, the mass momentarily stops and then reverses its direction of motion from clockwise to counterclockwise. It passes through point B again and returns to point A, thus completing its one full *cycle* in a time interval of τ (tau). These series of events are repeated over and over in τ time intervals.

The total angle covered by the pendulum between A and C is called the *range of motion* and is equal to $2\theta_0$. τ is called the *period* measured in seconds and θ_0 is the *amplitude* of the harmonic oscillations measured in radians. ϕ is called the *angular frequency* measured in radians per second (rad/s). The period and angular frequency are related through:

$$\phi = \frac{2\pi}{\tau} \qquad \text{(with } \pi = 3.1416\text{)}$$

In this case, both the amplitude and period of the harmonic motion are constants. Therefore, the angular frequency and range of motion of harmonic oscillations are constants as well.

For oscillatory motions, the reciprocal of the period is called the *frequency*, f, which represents the total number of cycles occurring in a second:

$$f = \frac{1}{\tau} = \frac{\phi}{2\pi}$$

Frequency, f, is expressed in cycles per second (cps) with units 1/s or s^{-1}. In the SI system of units, f is denoted by Hertz (Hz).

Now that we have defined most of the important parameters involved, we can also determine the angular velocity and angular acceleration of the pendulum. Utilizing Eqs. (9.4) and (9.6):

$$\omega = \frac{d\theta}{dt} = -\theta_0 \, \phi \, \sin(\phi \, t) \qquad (ii)$$

$$\alpha = \frac{d\omega}{dt} = -\theta_0 \, \phi^2 \, \cos(\phi \, t) \qquad (iii)$$

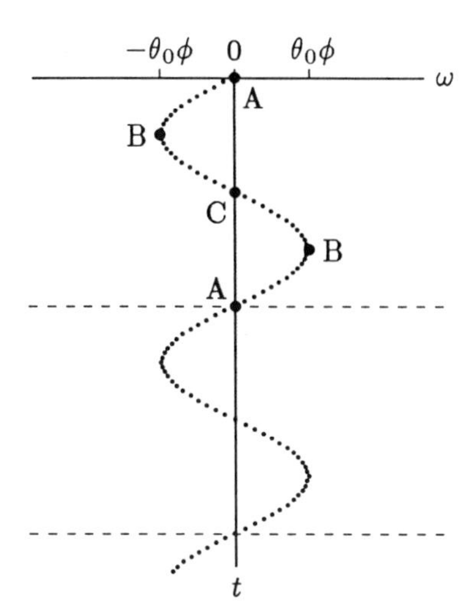

Figure 9.5 *Angular velocity of the pendulum as a function of time.*

These functions are plotted in Figures 9.5 and 9.6. Notice that the amplitude of the angular velocity of the pendulum is $\theta_0\phi$, and the amplitude of its angular acceleration is $\theta_0\phi^2$. At A, the angular velocity is zero. Between A and B, the mass accelerates and the magnitude of its angular velocity increases in the clockwise direction. The angular velocity reaches a peak value of $\theta_0\phi$ at B. The angular velocity is negative and the angular acceleration is positive between B and C. Therefore, the mass decaleretes (i.e., its angular velocity decreases in the clockwise direction) between B and C. The angular velocity reduces to zero at C. In the meantime, the magnitude of the angular acceleration reaches its peak value of $\theta_0\phi^2$. Between C and B, the mass accelerates in the counterclockwise direction, the magnitude of its angular velocity returns to a peak at B, slows down between B and C, and momentarily comes to rest at A. This series of events is repeated over time, which can be observed in Figures 9.5 and 9.6.

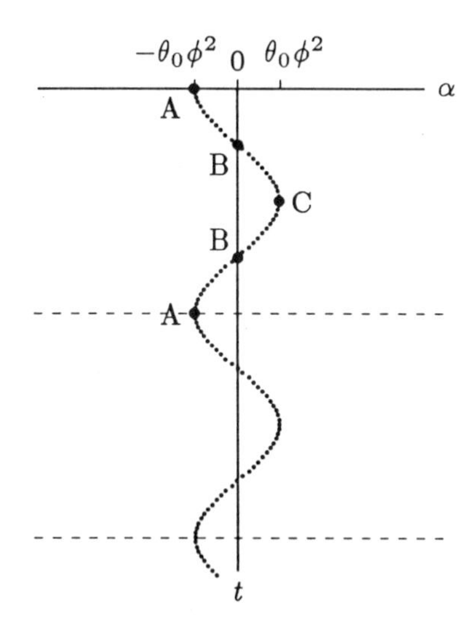

Figure 9.6 *Angular acceleration of the pendulum as a function of time.*

Case B. Damped oscillations

Consider that the mass is again pulled to the side (to A) so that the pendulum makes an angle θ_0 with the vertical and is released (Figure 9.7). In this case, we shall take into consideration the air resistance which will cause the pendulum to come to a stop. To simplify the problem, we shall assume that the θ versus time graph of the pendulum is as shown in Figure 9.8.

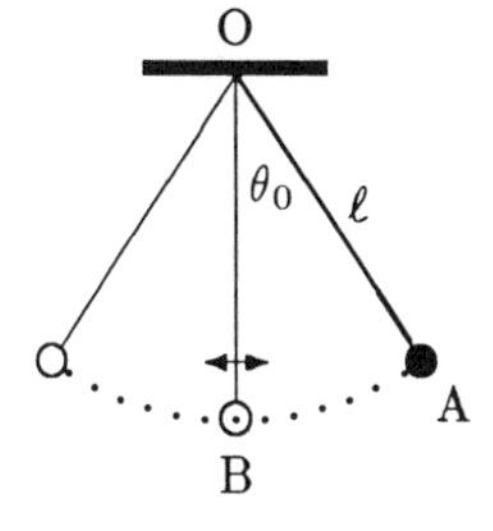

Figure 9.7 *Because of internal friction and air resistance, the oscillations will die out in time.*

As observed in Figure 9.8, the pendulum completes four full cycles in t_f seconds before coming to a stop. The period of each cycle is equal, but the amplitude of the harmonic oscillations decreases linearly with time to zero at time t_f. That is, we have a harmonic motion with constant period but varying amplitude.

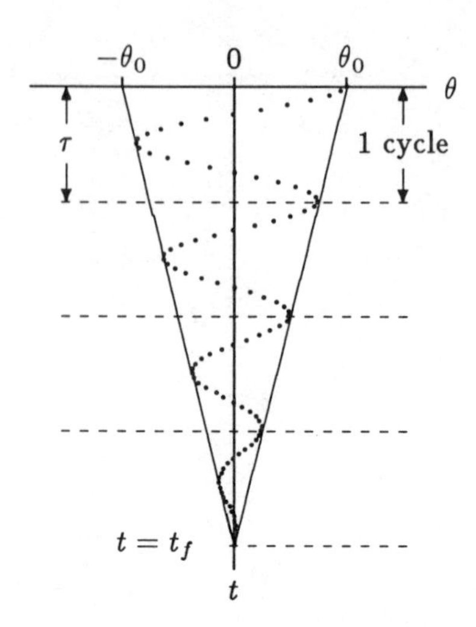

Figure 9.8 *Damped oscillations.*

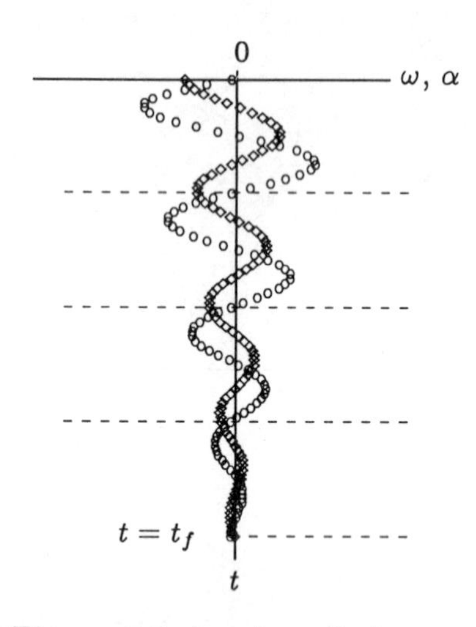

Figure 9.9 *Angular velocity ω ($\circ$) and angular acceleration α ($\diamond$) as functions of time t.*

For measured θ_0, τ, and t_f, the θ versus time graph shown in Figure 9.8 can be represented by the following function:

$$\theta = \theta_0 \left(1 - \frac{t}{t_f}\right) \cos(\phi\, t) \qquad (iv)$$

In Eq. (iv), ϕ is again the angular frequency of harmonic oscillations and is equal to $2\pi/\tau$. What is different in this case is that the harmonic oscillations of the pendulum are enveloped between two converging straight lines which can be represented by the functions $\theta{=}\theta_0(1-t/t_f)$ and $\theta{=}{-}\theta_0(1-t/t_f)$, and that the oscillations of the pendulum are "damped-out" by air resistance. Knowing the angular position as a function of time enables us to determine the angular velocity and acceleration of the pendulum. Using Eqs. (9.4) and (9.6), and applying the product and chain rules of differentiation (see Appendix C):

$$\omega = \frac{d\theta}{dt} = -\frac{\theta_0}{t_f}\cos(\phi\, t) - \theta_0\, \phi \left(1 - \frac{t}{t_f}\right) \sin(\phi\, t) \qquad (v)$$

$$\alpha = \frac{d\omega}{dt} = \frac{2\,\theta_0\, \phi}{t_f} \sin(\phi\, t) - \theta_0\, \phi^2 \left(1 - \frac{t}{t_f}\right) \cos(\phi\, t) \qquad (vi)$$

The functions given in Eqs. (v) and (vi) are relatively complex. Their graphs are shown in Figure 9.9, which are obtained simply by assigning values to t, determining corresponding expressions for ω and α from Eqs. (v) and (vi), and plotting them. Notice that amplitudes of ω and α decrease in time and reduce to zero at time t_f.

Example 9.2 *Shoulder abduction*

Figure 9.10 shows a person doing shoulder abduction in the frontal plane. O represents the axis of rotation of the shoulder joint in the frontal plane, line OA represents the position of the arm when it is stretched out parallel to the ground (horizontal), line OB represents the position of the arm when the hand is at its highest elevation, and line OC represents the position of the arm when the hand is closest to the body. In other words, for this activity, OB and OC are the arm's limits of range of motion. Assume that the angle between OA and OB is equal to the angle between OA and OC, which are represented by angle θ_0. The motion of the arm is symmetric with respect to line OA. Also assume that the time it takes for the arm to cover the angles between OA and OB, OB and OA, OA and OC, and OC and OA are approximately equal.

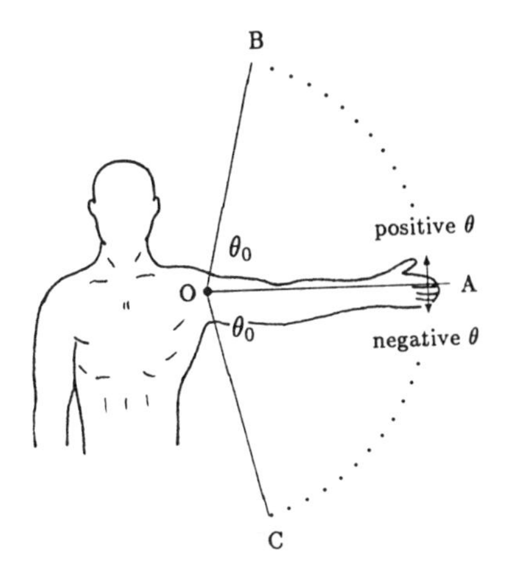

Figure 9.10 *Shoulder abduction.*

Derive expressions for the angular displacement, velocity, and acceleration of the arm. Take the period of angular motion of the arm to be 3 seconds and the angle θ_0 to be 80°.

Solution: Notice the similarities between the arm in this example and the pendulum analyzed in Example 9.1. In this case, angle θ_0 represents the amplitude of the angular displacement of the arm while undergoing a harmonic motion about line OA. The range of motion (ROM) of the arm is equal to twice that of angle θ_0. The period of the angular motion is given as τ=3 s, and the angular frequency of harmonic oscillations of the arm about line OA (the horizontal) can be calculated as $\phi=2\pi/\tau=2.09$ rad/s. If we let θ represent the angular displacement of the arm measured relative to the position defined by line OA, then θ can be written as a sine function of time in the following form:

$$\theta = \theta_0 \sin(\phi t) \qquad (i)$$

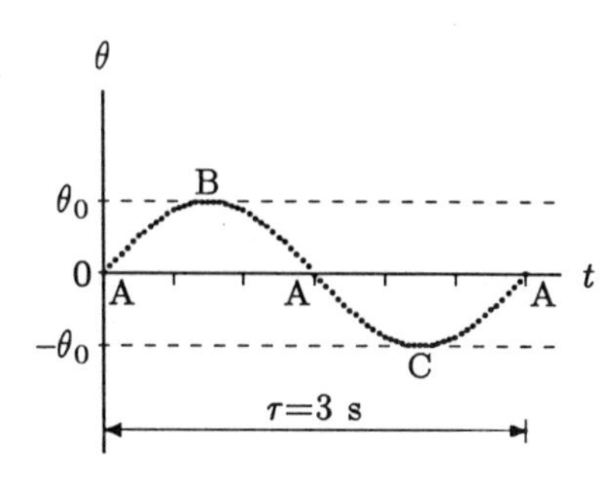

Figure 9.11 *Graph of function* $\theta = \theta_0 \sin(\phi t)$ *with* $\phi = 2\pi/\tau$.

The angular displacement of the arm as given in Eq. (i) is plotted as a function of time in Figure 9.11. Notice that θ is zero when the arm is at position A. θ assumes positive values between A and B, and θ is negative while the arm is covering the angular distance between A and C. θ reaches its peak at B and C, and θ_0 is the amplitude of angular displacement of the arm. Since all of these are consistent with the information provided in the statement of the problem, Eq. (i) (whose graph is shown in Figure 9.11) does represent the angular displacement of the arm.

To derive expressions for the angular velocity and acceleration of the arm, we have to consider time derivatives of the function given in Eq. (i). Time rate of change of angular displacement is defined as angular velocity:

$$\omega = \frac{d\theta}{dt} = \theta_0 \, \phi \, \cos(\phi \, t) \qquad (ii)$$

Time rate of change of angular velocity is angular acceleration:

$$\alpha = \frac{d\omega}{dt} = -\theta_0\, \phi^2\, \sin(\phi\, t) \qquad (iii)$$

Equations (*ii*) and (*iii*) can alternatively be written as:

$$\omega = \omega_0\, \cos(\phi\, t) \qquad (iv)$$

$$\alpha = -\,\alpha_0\, \sin(\phi\, t) \qquad (v)$$

Here, ω_0 is the amplitude of the angular velocity and α_0 is the amplitude of the angular acceleration of the arm, such that:

$$\omega_0 = \theta_0\, \phi = \theta_0 \frac{2\pi}{\tau}$$

$$\alpha_0 = \theta_0\, \phi^2 = \theta_0 \frac{4\pi^2}{\tau^2}$$

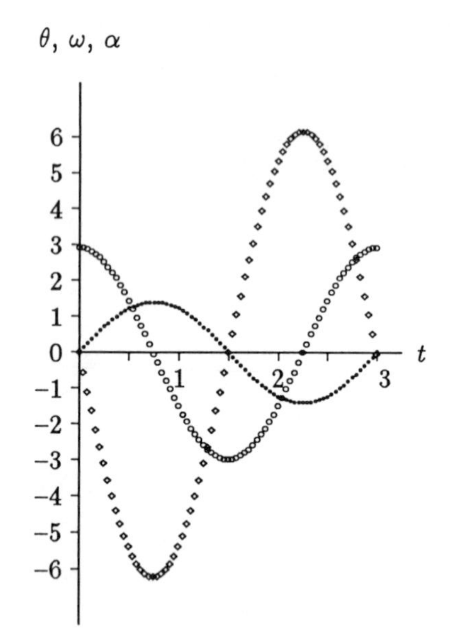

Figure 9.12 *Variations of angular displacement* (·), velocity (∘) and *acceleration* (⋄) *with time t.* (θ *in rad*, ω *in rad/s*, α *in rad/s²*.)

Notice that the amplitude of the angular velocity is a linear function of the angular frequency, and the amplitude of angular acceleration is a quadratic function of angular frequency. Angular frequency, on the other hand, is inversely proportional with the period of harmonic oscillations. Therefore, decreasing period means increasing frequency, and increasing velocity and acceleration amplitudes.

We can use the numerical values of θ_0=80°=1.40 rad and ϕ=2.09 rad/s to calculate ω_0 and α_0 as 2.93 rad/s and 6.12 rad/s^2, respectively. Eqs. (*i*), (*iv*), and (*v*) can be rewritten as:

$$\theta = 1.40\, \sin(2.09\, t) \qquad (vi)$$

$$\omega = 2.93\, \cos(2.09\, t) \qquad (vii)$$

$$\alpha = -\,6.12\, \sin(2.09\, t) \qquad (viii)$$

By assigning values to time t, computing corresponding angular displacement, velocity, and acceleration using Eqs. (*vi*) through (*viii*), and by plotting them, we can obtain θ versus t, ω versus t, and α versus t graphs. A set of sample graphs are shown in Figure 9.12 for a single cycle.

Example 9.3 *Flexion-extension test*

Figure 9.13 illustrates a computer-controlled dynamometer which can be used to make angular displacement, angular velocity, and torque measurements. During a repetitive flexion-extension test in the sagittal plane, a subject is placed in the dynamometer, positioned in the machine so that the subject's fifth lumbar vertebra (L5/S1) is aligned with the flexion-extension axis (denoted as O) of the machine, tied to the equipment firmly, and asked to perform trunk flexion and extension as long as possible, exerting as much effort as possible. The angular position of the subject's trunk relative to the upright position is measured and recorded. The data collected is then plotted to obtain an angular displacement, θ, versus time, t, graph. The curves obtained for this particular subject in different cycles are observed to be qualitatively and quantitatively similar except for the first and the last few cycles. A couple of sample cycles are provided in Figure 9.14, in which the angular displacement of the trunk measured in degrees is plotted as a function of time measured in seconds.

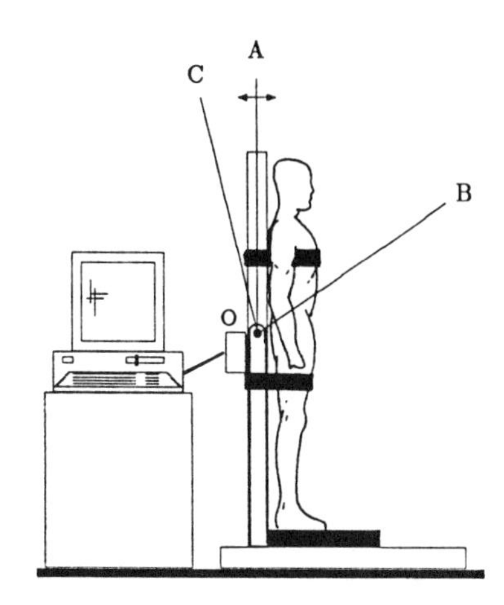

Figure 9.13 *Computer-controlled dynamometer.*

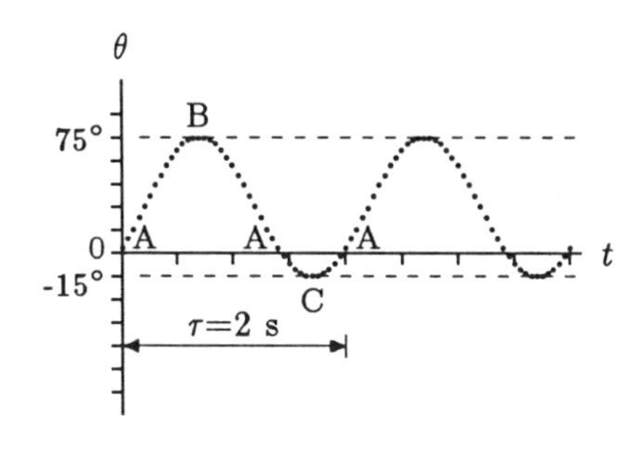

Figure 9.14 *Angular displacement versus time graph. (θ is in degrees and t is in seconds.)*

The angular position measurements are made relative to the upright position in which the angular displacement of the trunk is zero. The subject flexes between A and B, and reaches a peak flexion at B. The extension phase is identified with the motion of the trunk from B towards A. The angular displacement of the trunk is positive between A and B. Between A and C the trunk undergoes hyperextension, and reaches a peak extension at C. In this range, the angular displacement of the trunk assumes negative values.

The purpose of this example is to illustrate the means of analyzing experimentally collected data. The specific task is to find a function that can express the angular displacement of the subject's trunk as a function of time, from which we can derive expressions for the angular velocity and acceleration of the trunk.

Solution: The problem may be easier to visualize if we form an analogy between the upper body and a mechanical system called the ***inverted pendulum***, shown in Figure 9.15. An inverted pendulum consists of a concentrated mass, m, attached to a very light rod of length, ℓ, which is hinged to the ground through an axis about which it is allowed to rotate. In this case, the concentrated mass represents the total mass of the upper body. The hinge corresponds to the disk between the fifth lumbar vertebra and the sacrum, about which the upper body rotation occurs in the sagittal plane. Length ℓ is the distance between the fifth lumbar vertebra and the center of gravity of the upper body.

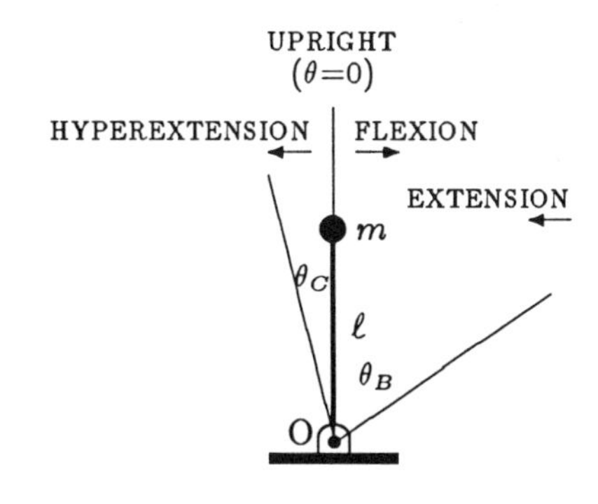

Figure 9.15 *Inverted pendulum.*

It is clear from Figure 9.14 that θ is a harmonic (sine or cosine) function of time. From Figure 9.14, it is possible to read the peak angles the trunk makes with the upright position during flexion and extension phases, and the period of harmonic motions. However, it is not easy to determine exactly how θ varies with time. To obtain a function relating θ and t, we must work through several steps.

Let τ be the period of harmonic oscillations, and θ_B and θ_C the peak angular displacements of the trunk in the flexion and extension phases, respectively. From Figure 9.14 or using the experimentally obtained raw data, τ=2 s, θ_B=75°, and θ_C=−15°. Knowing the period, the angular frequency of the harmonic oscillations can be determined:

$$\phi = \frac{2\pi}{\tau} = \frac{2\pi}{2} = \pi \ \text{rad/s}$$

Using θ_B and θ_C, we can also calculate the range of motion of the trunk. By definition, range of motion is the total angle covered by the rotating object. Therefore:

$$\text{ROM} = \theta_B - \theta_C = 75° + 15° = 90°$$

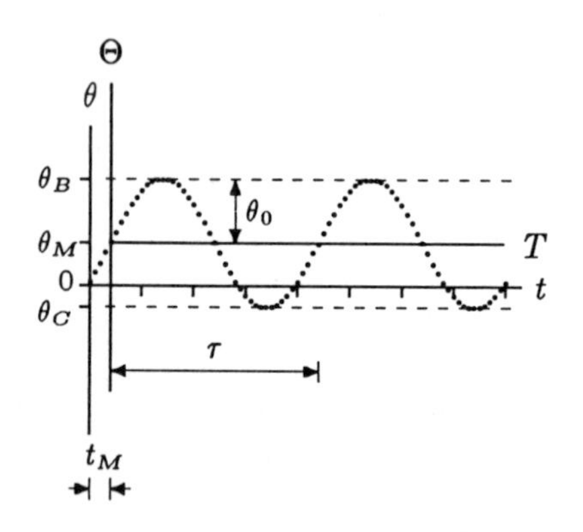

Figure 9.16 *Θ versus T coordinate frame is obtained by translating the origin of θ versus t coordinate frame.*

Figure 9.14 is redrawn in Figure 9.16 in which two sets of coordinates are used. In addition to θ and t, we have a second set of coordinates Θ (capital theta) and T which is obtained by translating the origin of the θ versus t coordinate system to a point whose coordinates are identified by t=t_M and θ=θ_M. Here, θ_M designates the mean angular displacement which can be calculated as:

$$\theta_M = \frac{\theta_B + \theta_C}{2} = \frac{75° - 15°}{2} = 30° \quad \left(\frac{\pi}{6} \ \text{rad}\right)$$

Time t_M corresponds to the time when θ=θ_M. t_M can be determined from the experimentally collected data. In this case, t_M=0.232 s.

We define a second set of coordinates so that, with respect to Θ and T, the function representing the angular displacement versus time curve is simply a sine function:

$$\Theta = \theta_0 \sin(\phi T) \qquad (i)$$

In Eq. (i), θ_0 is the amplitude of the harmonic oscillations and is equal to one half of the range of motion:

$$\theta_0 = \frac{\text{ROM}}{2} = \frac{90°}{2} = 45° \quad \left(\frac{\pi}{4} \ \text{rad}\right)$$

We now have a function representing the experimentally obtained curve in terms of Θ and T. If we can relate Θ to θ and T to t, then we can derive a function in terms of θ and t. This can be achieved by employing ***coordinate transformation.*** Notice that $\Theta=0$ when $\theta=\theta_M$. Therefore:

$$\Theta = \theta - \theta_M \qquad (ii)$$

Also notice that $T=0$ when $t=t_M$. Hence:

$$T = t - t_M \qquad (iii)$$

Substituting Eqs. (ii) and (iii) into Eq. (i) will yield:

$$\theta = \theta_M + \theta_0 \sin[\phi(t - t_M)] \qquad (iv)$$

In Eq. (iv), the angular displacement of the trunk is defined as a function of time, representing the experimentally obtained curve shown in Figure 9.14. We can also obtain expressions for the angular velocity and acceleration of the trunk by considering the time derivatives of θ in Eq. (iv):

$$\omega = \frac{d\theta}{dt} = \theta_0\, \phi \cos[\phi(t - t_M)] \qquad (v)$$

$$\alpha = \frac{d\omega}{dt} = -\theta_0\, \phi^2 \sin[\phi(t - t_M)] \qquad (vi)$$

The numerical values of θ_M, θ_0, t_M, and ϕ can be substituted into the above equations to obtain:

$$\theta = \frac{\pi}{6} + \frac{\pi}{4} \sin[\pi(t - 0.232)] \qquad (vii)$$

$$\omega = \frac{\pi^2}{4} \cos[\pi(t - 0.232)] \qquad (viii)$$

$$\alpha = -\frac{\pi^3}{4} \sin[\pi(t - 0.232)] \qquad (ix)$$

These functions are plotted in Figure 9.17 to obtain angular displacement, velocity, and acceleration versus time graphs for the trunk.

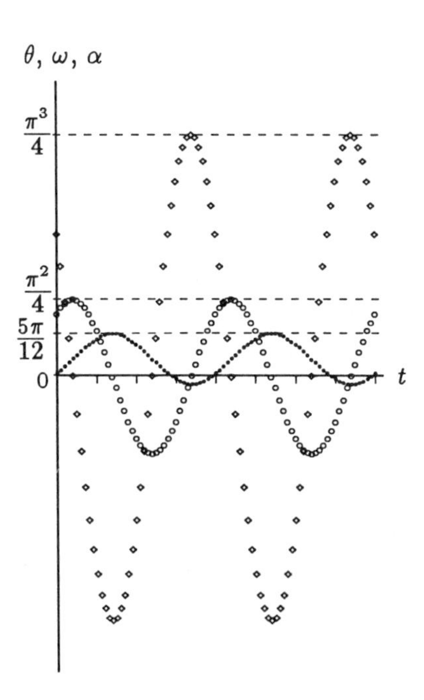

Figure 9.17 *Variations of angular displacement* (·), *velocity* (∘), *and acceleration* (◇) *with time.*

Note that the validity of Eq. (vii) can be checked by assigning values to t and computing corresponding θ values using Eq. (vii). For example, $\theta=0$ when $t=0$ and $t=\tau=2$ s, and $\theta=\pi/6=0.52$ rad or 30° when $t=t_M=0.232$ s which are consistent with the initial data presented in Figure 9.14.

Also note that angular velocity is a cosine function of time. The amplitude of the ω versus t curve shown in Figure 9.17 is equal to the coefficient $\pi^2/4=2.47$ rad/s in front of the cosine function in Eq. $(viii)$. Similarly, the amplitude of the angular acceleration is $\pi^3/4=7.75$ rad/s^2.

9.6 Rotational Motion About a Fixed Axis: Circular Motion

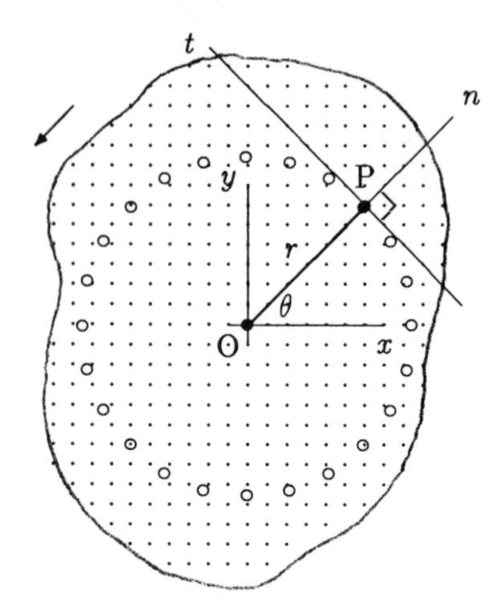

Figure 9.18 *n and t designate normal (radial) and tangential directions at a point P undergoing a circular motion.*

Consider the arbitrarily shaped object in Figure 9.18. Assume that the object is undergoing a rotational motion in the xy plane about a fixed axis perpendicular to the xy plane. Let O and P be two points in the xy plane, such that O is along the axis of rotation of the object and P is a fixed point on the rotating object located at a distance r from O. Note that point O can be inside or outside the rotating object. Due to the rotation of the object, point P will experience a circular motion with r being the radius of its circular path.

To describe circular motions, it is usually convenient to define velocity and acceleration vectors with respect to two mutually perpendicular directions normal (designated by n) and tangential (designated by t) to the circular path. By definition, the velocity vector $\underline{v}$ is always tangent to the path of motion. Therefore, for a circular motion, the velocity vector can have only one component tangent to the circular path (Figure 9.19). $\underline{v}$ is called the *tangential,* or *linear,* velocity. The magnitude v of this velocity can be determined by considering the time rate of change of relative position of the object along the circular path:

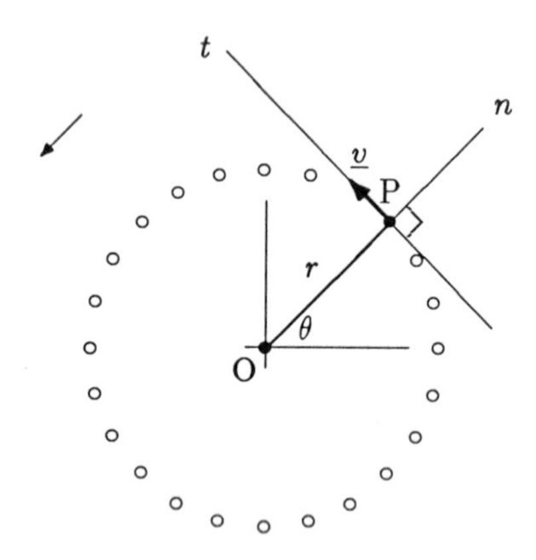

Figure 9.19 $\underline{v}$ *is always tangent to the circular path of motion.*

$$v = \frac{ds}{dt} \tag{9.9}$$

For a circular motion, the acceleration vector can have both tangential and normal components (Figure 9.20). The *tangential acceleration,* $\underline{a}_t$, is related to the change in magnitude of the velocity vector and has a magnitude:

$$a_t = \frac{dv}{dt} \tag{9.10}$$

The *normal acceleration,* $\underline{a}_n$, is related to the change in direction of the velocity vector and has a magnitude:

$$a_n = \frac{v^2}{r} \tag{9.11}$$

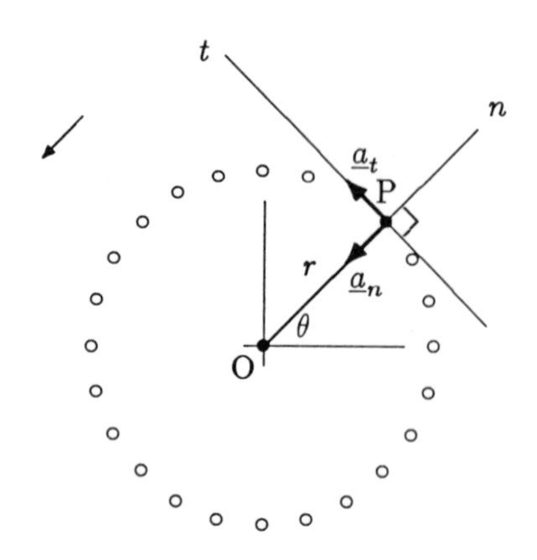

Figure 9.20 $\underline{a}_t$ *and* $\underline{a}_n$ *are tangential and normal components of the acceleration vector.*

The direction of $\underline{a}_t$ is the same as the direction of $\underline{v}$ if v is increasing, or opposite to that of $\underline{v}$ if v is decreasing over time. The normal component of the acceleration vector is also known as *radial* or *centripetal* ("center-seeking"), and it is always directed towards the center of rotation of the body. (The derivations of the acceleration components are provided in Section 9.12.)

If the tangential and normal acceleration components are known, then the net (resultant or total) acceleration of a point on a body rotating about a fixed axis can also be determined (Figure 9.21). If $\underline{t}$ and $\underline{n}$ designate unit vectors in the tangential and normal directions, respectively, then the resultant acceleration vector can be written as follows:

$$\underline{a} = \underline{a}_t + \underline{a}_n = a_t\,\underline{t} + a_n\,\underline{n} \qquad (9.12)$$

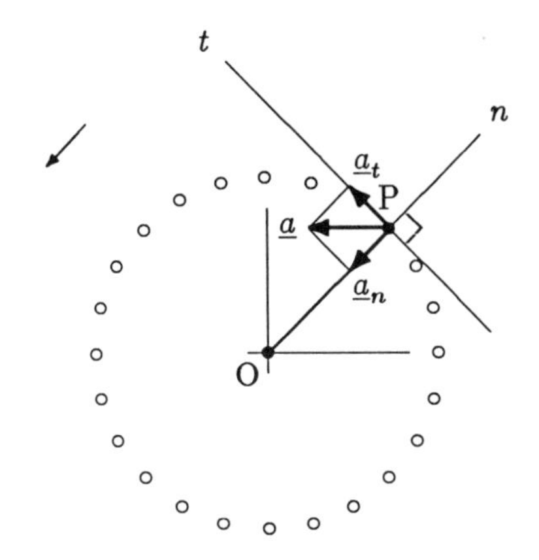

Figure 9.21 *$\underline{a}$ is the resultant linear acceleration vector.*

The magnitude of the resultant acceleration vector is:

$$a = \sqrt{{a_t}^2 + {a_n}^2} \qquad (9.13)$$

On the other hand, the velocity vector can be expressed as:

$$\underline{v} = v\,\underline{t} \qquad (9.14)$$

Note that v and a are linear quantities. v has the dimension of length divided by time, and both a_t and a_n have the dimension of length divided by time squared, and therefore they have the units of linear quantities listed in Table 8.1.

Also note that it is customary to take the positive normal direction (the direction of $\underline{n}$) to be outward (i.e., from the center of rotation towards the rim), and the positive tangential direction (the direction of $\underline{t}$) to be counterclockwise.

9.7 Relationships Between Linear and Angular Quantities

Recall from Eq. (9.3) that $s=r\theta$. For a circular motion, radius r is constant and Eq. (9.9) can be evaluated as follows:

$$v = \frac{d}{dt}(r\,\theta) = r\,\frac{d\theta}{dt}$$

By definition, time rate of change of angular displacement is angular velocity. Therefore:

$$v = r\,\omega \qquad (9.15)$$

Eq. (9.15) states that the magnitude of the tangential (linear) velocity of a point in a body that is undergoing a rotational motion about a fixed point is equal to the distance of that point from the center of rotation multiplied by the angular velocity of the body. Notice that at a given instant, every point on the body has the same angular velocity but may have different linear velocities. The magnitude of the linear velocity increases with increasing radial distance, or as one moves outward from the center of rotation toward the rim.

Using the relationship given in Eq. (9.15), Eq. (9.10) can be evaluated for a motion in a circular path as follows:

$$a_t = \frac{d}{dt}(r\,w) = r\,\frac{dw}{dt}$$

By definition, time rate of change of angular velocity is angular acceleration. Therefore:

$$a_t = r\,\alpha \tag{9.16}$$

Similarly, substituting Eq. (9.15) into Eq. (9.11) will yield:

$$a_n = r\,w^2 \tag{9.17}$$

Eqs. (9.15), (9.16), and (9.17) relate linear quantities v, a_t, and a_n to angular quantities r, ω, and α. Eq. (9.16) states that the tangential component of linear acceleration of a point on a body rotating about a fixed axis is equal to the distance of that point from the axis of rotation times the angular acceleration of the body. From Eq. (9.17), the magnitude of the normal component of linear acceleration is a linear function of the radial distance and a quadratic function of the angular velocity.

9.8 Uniform Circular Motion

If the linear velocity of a point in a body undergoing a rotational motion about a fixed axis has a constant magnitude, then the motion is called ***uniform circular motion***. Notice that for a circular motion, whether v is constant or not, $\underline{v}$ is always tangent to the circular path. The direction of $\underline{v}$ changes continuously over time, and therefore the velocity vector itself is not constant. This indicates the presence of a centripetal acceleration that changes the direction of the linear velocity. On the other hand, the magnitude of the tangential component of the linear acceleration vector is determined by taking the derivative of v with respect to time. In this case the magnitude of the linear velocity vector is constant, and the tangential acceleration is zero. Therefore, for a uniform circular motion:

$$\begin{aligned} v &= \text{constant} \\ a_t &= 0 \\ a_n &= \frac{v^2}{r} = r\,\omega^2 \end{aligned} \tag{9.18}$$

For a uniform circular motion both r and v are constants. Hence, from Eqs. (9.15) and (9.7):

$$\begin{aligned} \omega &= \frac{v}{r} \quad \text{(constant)} \\ \alpha &= 0 \end{aligned} \tag{9.19}$$

The following example will demonstrate an application of some of the concepts introduced in the preceding sections.

Example 9.4 Figure 9.22 illustrates a gymnast doing giant circles. Assume that the center of gravity of the gymnast is located at a distance r=0.8 m from the bar, undergoing a uniform circular motion with v=4 m/s.

Determine the angular velocity and acceleration of the gymnast, and the tangential and normal components of the linear acceleration of the center of gravity of the gymnast.

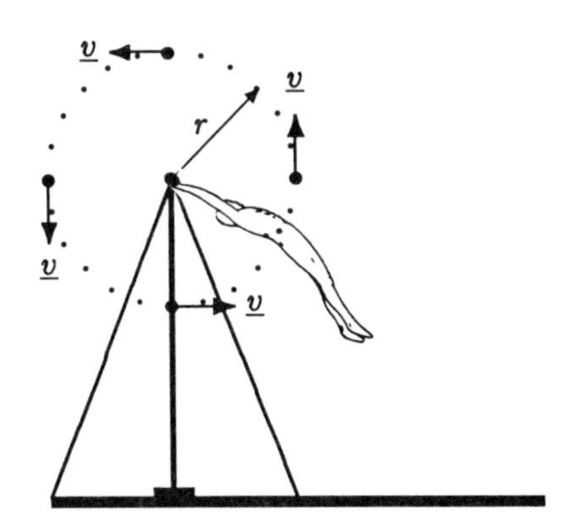

Figure 9.22 *A gymnast doing giant circles.*

Solution: The center of gravity of the gymnast undergoes a uniform circular motion with the bar forming the fixed axis of rotation, r being the radius of the circular path, and v being the constant speed of the center of gravity. This problem can be analyzed by utilizing the conditions stated by Eqs. (9.18) and (9.19):

$$v = 4 \text{ m/s (given)}$$

$$\omega = \frac{v}{r} = \frac{4}{0.8} = 5 \text{ rad/s (constant)}$$

$$\alpha = \frac{d\omega}{dt} = 0$$

$$a_n = \frac{v^2}{r} = \frac{(4)^2}{0.8} = 20 \text{ m/s}^2$$

$$a_t = r\,\alpha = 0$$

Consider that after completing several cycles around the bar, the gymnast releases the bar at an instant when the center of gravity is directly beneath the bar (Figure 9.23). At the instant of release, the center of gravity of the gymnast is located at a height 2.5 meters above the floor level. Assuming that the center of gravity of the gymnast is located 1 meter above his/her feet, that the center of gravity of the gymnast undergoes a projectile motion, and that the gymnast lands on the floor in standing position, determine how far from the bar the gymnast will land on the floor.

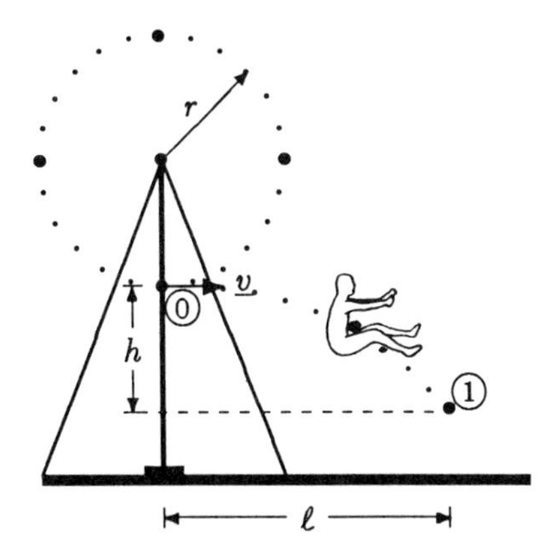

Figure 9.23 *Between 0 and 1, center of gravity of the gymnast undergoes a projectile motion.*

In this case, we can utilize the concepts introduced in Chapter 8.10. Figure 9.24 illustrates the trajectory of the center of gravity of the gymnast, where 0 and 1 designate positions of the center of gravity at the instants of release and landing. The total time, t_1, elapsed between 0 and 1 (time of flight) can be determined by writing Eq. (8.32) between 0 and 1:

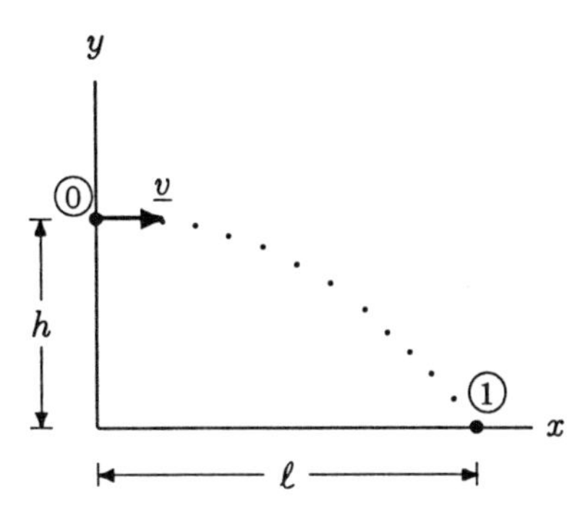

Figure 9.24 *Trajectory of the center of gravity of the gymnast.*

$$y_1 = y_0 + v_{y0}\,t_1 - \frac{1}{2}\,g\,{t_1}^2$$

Note that the velocity of the center of gravity of the gymnast has no component in the y direction. If we measure heights relative to the position of the center of gravity of the gymnast at the instant of landing, then $y_0=h=2.5-1.0=1.5$ m and $y_1=0$. Therefore:

$$0 = h + 0 - \frac{1}{2} g\, t_1^{\,2}$$

Solving this equation for t_1:

$$t_1 = \sqrt{\frac{2\,h}{g}}$$

Substituting numerical values of $h=1.5$ m and $g=9.8$ m/s^2 and carrying out the computations will yield:

$$t_1 = \sqrt{\frac{2\,(1.5)}{9.8}} = 0.55 \text{ s}$$

We can now determine the horizontal range of motion, ℓ, of the gymnast's center of gravity. Writing Eq. (8.31) between 0 and 1:

$$x_1 = x_0 + v_{x0}\, t_1$$

At the instant of release, $v_{x0}=v$ since the velocity of the center of gravity of the gymnast is acting in the x direction. $x_0=0$ and $x_1=\ell$ because the distances in the x direction are measured from the position of the center of gravity of the gymnast at the instant of release. Therefore:

$$\ell = 0 + v\, t_1$$

Substituting $v=4$ m/s and $t_1=0.55$ s into the above equation will yield:

$$\ell = (4)(0.55) = 2.2 \text{ m}$$

9.9 Rotational Motion with Constant Acceleration

In Section 8.5, a set of equations (Eqs. 8.11 through 8.14) were derived which can be used to analyze the characteristics of bodies undergoing one-dimensional linear motion with constant acceleration. A set of kinematic equations can also be derived for rotational motion about a fixed axis with constant angular acceleration, α_0. Derivations of these equations are similar to those employed in Section 8.5. Here, they are summarized as follows:

$$\omega = \omega_0 + \alpha_0\, t \qquad (9.20)$$

$$\theta = \theta_0 + \omega_0\, t + \frac{1}{2}\, \alpha_0\, t^2 \qquad (9.21)$$

$$\theta = \theta_0 + \frac{1}{2}\,(\omega + \omega_0)\, t \qquad (9.22)$$

$$\omega^2 = \omega_0^{\,2} + 2\,\alpha_0\,(\theta - \theta_0) \qquad (9.23)$$

It should be reiterated that Eqs. (9.20) through (9.23) are valid for rotational motion about a fixed axis with constant angular acceleration, and that they are similar to their linear counterparts given in Eqs. (8.11) through (8.14).

9.10 Relative Motion

A motion observed in different frames of reference may be different. For example, the motion of a train observed by a stationary person would be different than the motion of the same train observed by a passenger in a moving car. The motion of a ball thrown up into the air by a person riding in a moving vehicle would have a vertical path as observed by the person riding in the same vehicle (Figure 9.25a), but a curved path for a second, stationary person watching the ball (Figure 9.25b). The general approach in analyzing such physical situations requires defining the motion of the moving body with respect to a convenient moving coordinate frame, defining the motion of this frame with respect to a fixed coordinate frame, and combining the two.

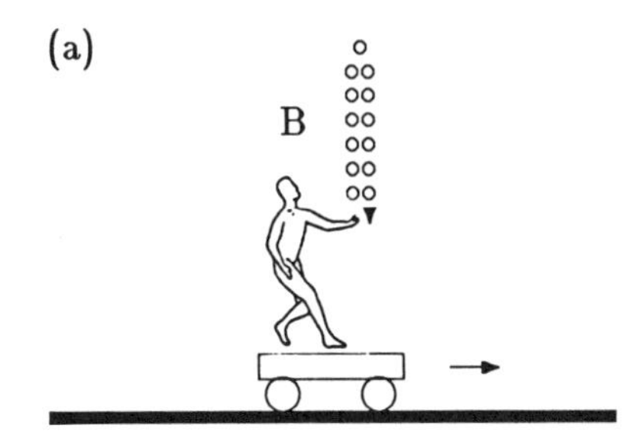

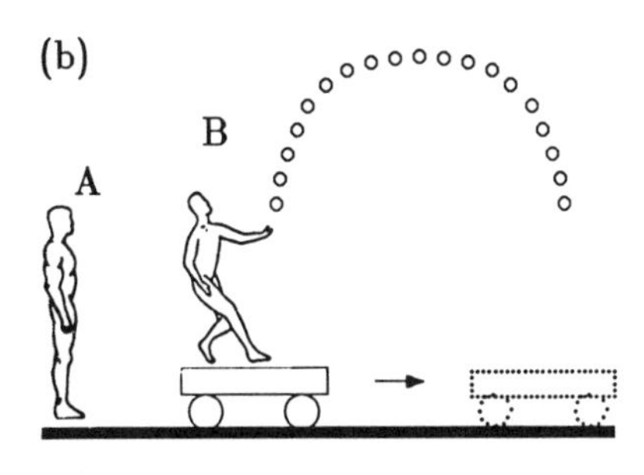

Figure 9.25 *A motion observed by different observers in different reference frames may be different.*

Assume that the motion of a point P in a moving body is to be analyzed. Let XYZ and xyz refer to two coordinate frames with origins at A and B, respectively (Figure 9.26). Assume that the XYZ frame is fixed (stationary) and the xyz frame is moving, such that the respective coordinate directions (e.g., x and X) remain parallel throughout the motion. This implies that the xyz coordinate frame is undergoing a translational motion only, and that the same set of unit vectors $\underline{i}$, $\underline{j}$, and $\underline{k}$ can be used in both reference frames. The motion of the moving xyz frame can be identified by specifying the motion of its origin B. If $\underline{r}_B$ denotes the position vector of B with respect to the fixed coordinates X, Y, and Z, then the velocity and acceleration vectors of B with respect to the XYZ coordinate frame are:

$$\underline{v}_B = \frac{d}{dt}(\underline{r}_B) = \dot{\underline{r}}_B$$

$$\underline{a}_B = \frac{d}{dt}(\underline{v}_B) = \ddot{\underline{r}}_B$$

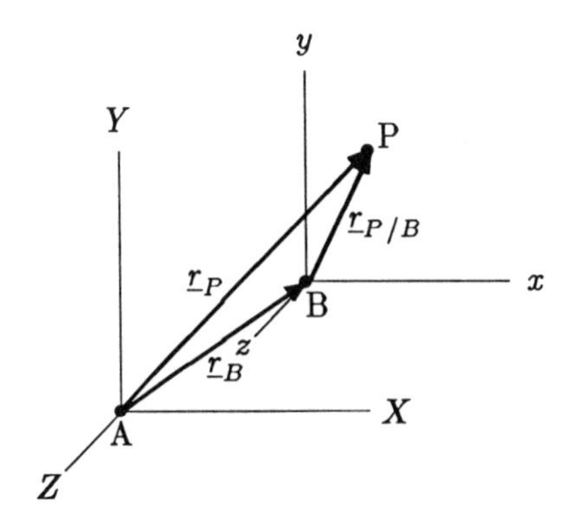

Figure 9.26 *XYZ is a fixed and xyz is a moving coordinate frame.*

Similarly, the motion of point P with respect to the moving coordinate frame xyz can be defined by the position vector $\underline{r}_{P/B}$ of point P relative to the origin B of the xyz frame. The first and second time derivatives of $\underline{r}_{P/B}$ will yield the velocity and acceleration vectors of point P relative to the xyz frame:

$$\underline{v}_{P/B} = \frac{d}{dt}(\underline{r}_{P/B}) = \dot{\underline{r}}_{P/B}$$

$$\underline{a}_{P/B} = \frac{d}{dt}(\underline{v}_{P/B}) = \ddot{\underline{r}}_{P/B}$$

Finally, the position vector $\underline{r}_P$, velocity vector $\underline{v}_P$, and acceleration vector $\underline{a}_P$ of point P with respect to the fixed coordinate frame XYZ can be obtained by superposition:

$$\underline{r}_P = \underline{r}_B + \underline{r}_{P/B} \quad (9.24)$$

$$\underline{v}_P = \underline{v}_B + \underline{v}_{P/B} \quad (9.25)$$

$$\underline{a}_P = \underline{a}_B + \underline{a}_{P/B} \quad (9.26)$$

The motion of point B (which happens to be the origin of the moving coordinate frame xyz) with respect to the fixed XYZ coordinate frame is called the ***absolute motion*** of B and is denoted by the subscript B. Similarly, the motion of point P with respect to the XYZ frame is called the absolute motion of P. The motion of point P with respect to the moving coordinate frame is called the ***relative motion*** of P and is denoted by the subscript P/B. Note here that the position vector $\underline{r}_{P/B}$ refers to a vector drawn from point P to point B. Also note that the position vector of point P relative to the XYZ frame could also be expressed as $\underline{r}_{P/A}$. However, by convention, $\underline{r}_P$ implies that the position vector is defined relative to the fixed coordinate frame.

Example 9.5 Consider the motion described in Figure 9.25. A person (B) riding on a vehicle that is moving towards the right by a constant velocity of 2 meters per second throws a ball straight up into the air with an initial velocity of 10 meters per second.

Describe the motion of the ball as observed by a stationary person (A) in the time interval between when the ball is first released and when it reaches its maximum elevation.

Solution: This is a two-dimensional problem and can be analyzed in three steps. First, let x and y represent a coordinate frame moving with the vehicle. With respect to the xy frame, the ball thrown up into the air will undergo one-dimensional linear motion (translation) in the y direction (Figure 9.27). Because of the constant downward gravitational acceleration, the ball will decelerate in the positive y direction, reach its maximum elevation, change its direction of motion, and begin to descent. With respect to the xy coordinate frame moving with the vehicle, or as observed by person B moving with the vehicle, the velocity of the ball in the y direction between the instant of release and when the ball reaches its peak elevation can be determined from (see Chapter 8):

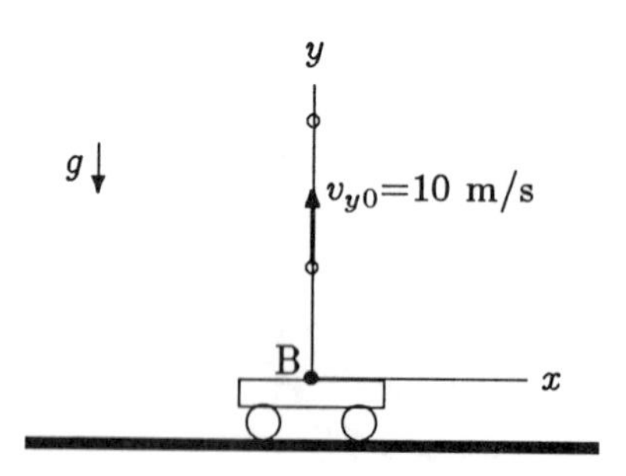

Figure 9.27 *Relative to the xy frame, the ball is undergoing a translational motion in the y direction.*

$$v_y = v_{y0} - g\,t$$

Here, v_{y0}=10 m/s is the initial velocity of the ball, $g \approx 10$ m/s^2 is the magnitude of gravitational acceleration. This equation is valid in the time interval between t=0 (the instant of release) and $t = v_{y0}/g = 10/10 = 1$ s (the time it takes for the ball to reach its maximum elevation where v_y=0). As observed by person B, the ball has no motion in the x direction. Therefore, the velocity, $\underline{v}_{P/B}$, of the ball relative to person B can be expressed as:

$$\underline{v}_{P/B} = v_y\,\underline{j}$$

Next, let X and Y represent a coordinate frame fixed to the ground. With respect to the XY frame, or with respect to the stationary person A, the vehicle is moving in the positive X direction with a constant velocity of v_{x0}=2 m/s (Figure 9.28). Therefore:

$$\underline{v}_B = v_{x0}\,\underline{i}$$

Finally, to determine the velocity of the ball relative to person A, we have to add velocity vectors $\underline{v}_B$ and $\underline{v}_{P/B}$ together:

$$\underline{v}_P = \underline{v}_B + \underline{v}_{P/B} = v_{x0}\,\underline{i} + v_y\,\underline{j}$$

Or, by substituting the known parameters:

$$\underline{v}_P = 2\,\underline{i} + (10 - 10\,t)\,\underline{j}$$

For example, half a second after the ball is released, the ball has a velocity:

$$\underline{v}_P = 2\,\underline{i} + 5\,\underline{j}$$

That is, according to person A or relative to the XY coordinate frame, the ball is moving to the right with a speed of 2 m/s and upward with a speed of 5 m/s (Figure 9.29). At this instant, the magnitude of the net velocity of the ball is $v_P=\sqrt{(2)^2+(5)^2}$=5.4 m/s.

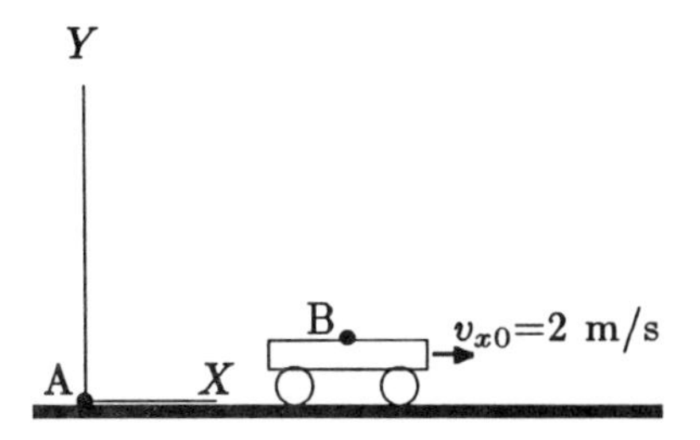

Figure 9.28 *Relative to the XY frame, the vehicle is undergoing a translational motion in the X direction with constant velocity.*

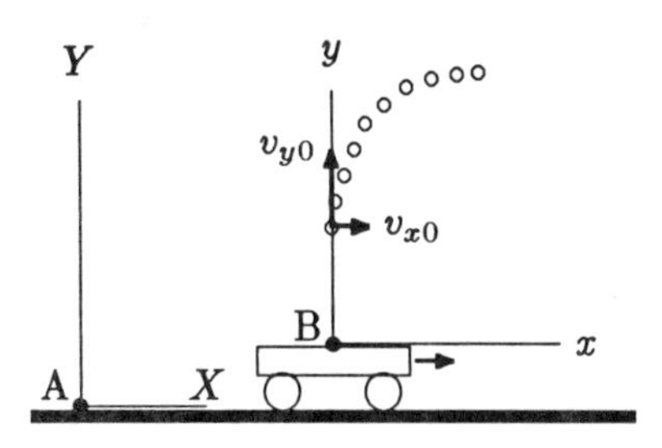

Figure 9.29 *Relative to the XY frame, the ball moves both in the X and Y directions.*

9.11 Linkage Systems

A *linkage system* is composed of several parts connected to each other and/or to the ground by means of hinges or joints, such that each part constituting the system can undergo motion relative to the other segments. An example of such a system is called the *double pendulum*. As shown in Figure 9.30, a double pendulum consists of two bars hinged together and to the ground. Linkage systems are also known as *multi-link systems*.

A particularly important concept associated with linkage systems is the number of independent coordinates necessary to describe the motion characteristics of the parts constituting the system. The number of independent coordinates required defines the *degrees of freedom* of the system. For example, the two-dimensional motion characteristics of the simple pendulum shown in Figure 9.31 can be fully described by coordinate θ that locates the position of the pendulum uniquely. Therefore, a simple pendulum has one degree of freedom. On the other hand, coordinates θ_1 and θ_2 are necessary to analyze the coplanar motion of bar BC of the double pendulum shown in Figure 9.30, and therefore, a double pendulum has two degrees of freedom.

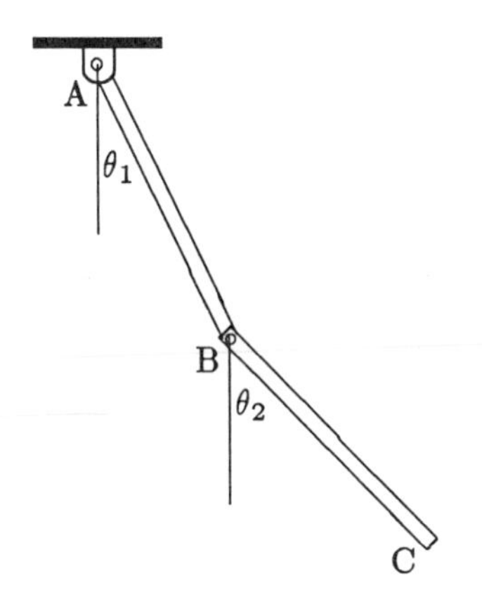

Figure 9.30 *Double pendulum.*

If the angular velocity and acceleration of individual parts are known, then the principles of relative motion can be applied to analyze the motion characteristics of each part constituting the multi-link system. The following example will illustrate the procedure of analyzing the motion of a double pendulum. However, the procedure to be introduced can be generalized to analyze any multi-link system.

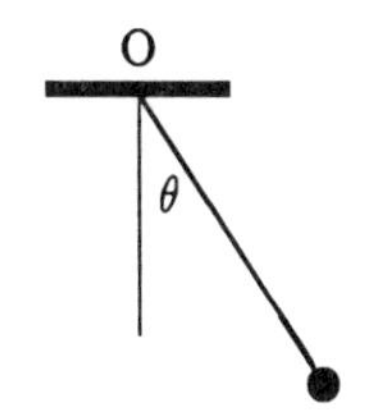

Figure 9.31 *Pendulum.*

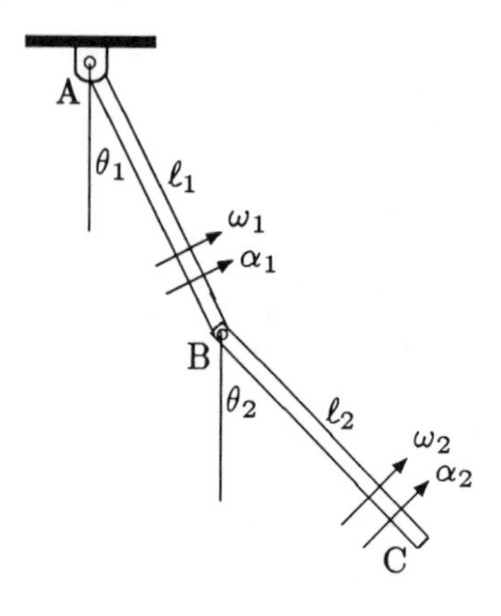

Figure 9.32 *Double pendulum.*

Example 9.6 *Double pendulum.*
Assume that arms AB and BC of the double pendulum shown in Figure 9.32 are undergoing coplanar motion. Let ℓ_1=0.3 m and ℓ_2=0.3 m be the lengths of arms AB and BC, and θ_1 and θ_2 be the angles arms AB and BC make with the vertical. The angular velocity and acceleration of arm AB are measured as ω_1=2 rad/s (counterclockwise) and α_1=0 relative to point A. The angular velocity and acceleration of arm BC is measured as ω_2=4 rad/s (counterclockwise) and α_2=0 relative to point B.

Determine the linear velocity and acceleration of point B on arm AB and point C on arm BC at an instant when θ_1=30° and θ_2=45°.

Solution: Let X and Y refer to a set of rectangular coordinates with origin located at A, and x and y be a second set of rectangular coordinates with origin at B. The XY coordinate frame is stationary, while the xy frame can move as point B moves. Since the angular velocity and acceleration of arm AB are given relative to point A, the motion characteristics of any point on arm AB can be determined with respect to the XY coordinate frame. Similarly, the motion of any point on arm BC can easily be analyzed relative to the xy coordinate frame.

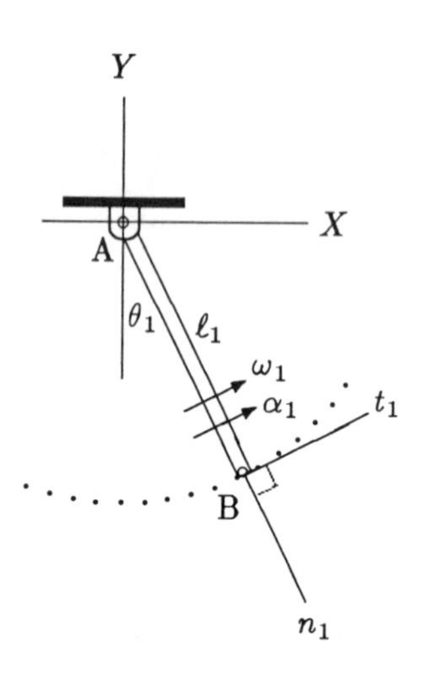

Figure 9.33 *Circular motion of B as observed from point A.*

Motion of point B as observed from point A:

Every point on arm AB undergoes a rotational motion about a fixed axis passing through point A with constant angular velocity of ω_1=2 rad/s. Every point on arm AB experiences a uniform circular motion in the counterclockwise direction. As illustrated in Figure 9.33, point B moves in a circular path of radius ℓ_1. Magnitudes of linear velocity in the tangential direction and linear acceleration in the normal direction of point B can be determined using:

$$v_B = \ell_1 \, \omega_1$$
$$a_B = \ell_1 \, {\omega_1}^2$$

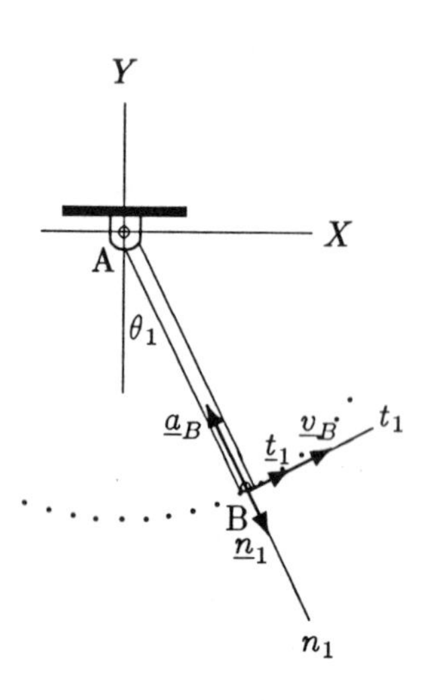

Figure 9.34 *Tangential velocity and normal acceleration of B.*

The magnitude of the tangential component of the acceleration vector is zero since ω_1 is constant or since α_1=0. Therefore, v_B and a_B are essentially the magnitudes of the resultant linear velocity and acceleration vectors. To express these quantities in vector forms, let n_1 and t_1 represent the normal and tangential directions to the circular path of point B when arm AB makes an angle θ_1 with the horizontal (Figure 9.34). Also let $\underline{n}_1$ and $\underline{t}_1$ be unit vectors in the positive n_1 and t_1 directions, such that the positive $\underline{n}_1$ direction is outward (i.e., from A towards B) and positive $\underline{t}_1$ direction is pointing in the direction of motion (i.e., counterclockwise). The normal (centripetal) acceleration is always directed towards the center of motion, and is acting in the negative $\underline{n}_1$ direction:

$$\underline{v}_B = v_B \, \underline{t}_1 = \ell_1 \, \omega_1 \, \underline{t}_1$$
$$\underline{a}_B = -\, a_B \, \underline{n}_1 = -\ell_1 \, \omega_1^{\,2} \, \underline{n}_1$$

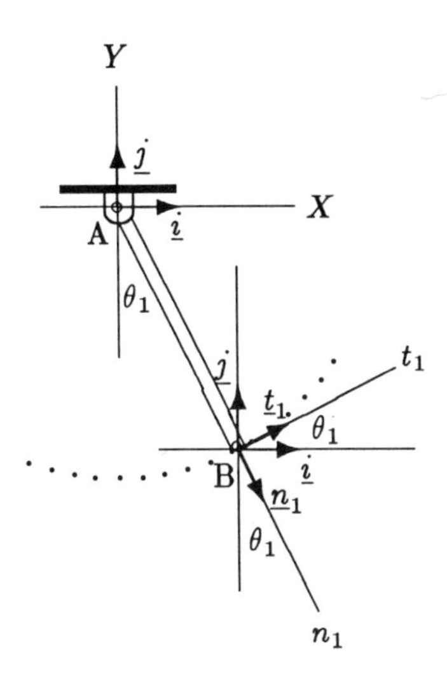

Figure 9.35 *Expressing unit vectors $\underline{n}_1$ and $\underline{t}_1$ in terms of Cartesian unit vectors $\underline{i}$ and $\underline{j}$.*

Notice that directions defined by unit vectors $\underline{n}_1$ and $\underline{t}_1$ change continuously as point B moves along its circular path. That is, $\underline{n}_1$ and $\underline{t}_1$ define a set of *local coordinate* directions which vary in time. By employing proper coordinate transformations, we can express these unit vectors in terms of Cartesian unit vectors $\underline{i}$ and $\underline{j}$. Cartesian coordinate directions, which are *global* as opposed to local, are not influenced by the motion of point B. The coordinate transformation can be done by expressing unit vectors $\underline{n}_1$ and $\underline{t}_1$ in terms of Cartesian unit vectors $\underline{i}$ and $\underline{j}$. It can be observed from the geometry of the problem that (Figure 9.35):

$$\underline{n}_1 = \sin\theta_1 \, \underline{i} - \cos\theta_1 \, \underline{j}$$
$$\underline{t}_1 = \cos\theta_1 \, \underline{i} + \sin\theta_1 \, \underline{j}$$

Therefore, the velocity and acceleration vectors of point B with respect to the XY coordinate frame and in terms of Cartesian unit vectors are:

$$\underline{v}_B = \ell_1 \, \omega_1 \, (\cos\theta_1 \, \underline{i} + \sin\theta_1 \, \underline{j})$$
$$\underline{a}_B = -\, \ell_1 \, \omega_1^{\,2} \, (\sin\theta_1 \, \underline{i} - \cos\theta_1 \, \underline{j})$$

If we substitute the numerical values of ℓ_1=0.3 m, θ_1=30°, and ω_1=2 rad/s, and carry out the necessary calculations we obtain:

$$\underline{v}_B = 0.52 \, \underline{i} + 0.30 \, \underline{j} \qquad (i)$$
$$\underline{a}_B = -\, 0.60 \, \underline{i} + 1.04 \, \underline{j} \qquad (ii)$$

<u>*Motion of point C as observed from point B:*</u>

The motion of point C as observed from point B is similar to the motion of point B as observed from point A. Point C rotates with a constant angular velocity of ω_2 in a circular path of radius ℓ_2 about point B (Figure 9.36). Therefore, the derivation of velocity and acceleration vectors for point C relative to the xy coordinate frame follows the same procedure outlined for the derivation of velocity and acceleration vectors for point B relative to the XY coordinate frame. Magnitudes of the tangential velocity and normal acceleration vectors of point C relative to B are:

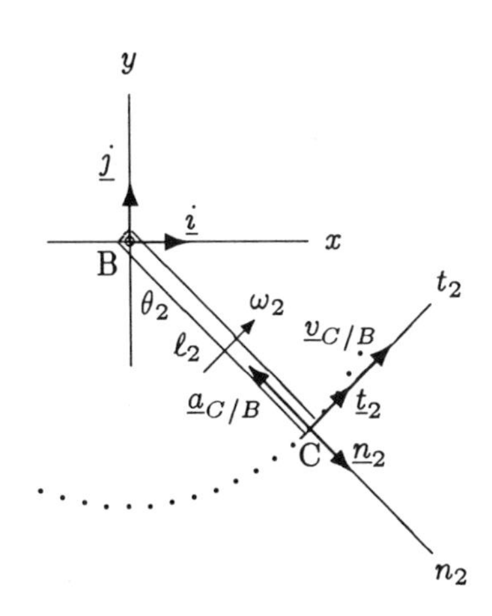

Figure 9.36 *Circular motion of C as observed from point B.*

$$v_{C/B} = \ell_2 \, \omega_2$$
$$a_{C/B} = \ell_2 \, \omega_2^{\,2}$$

If $\underline{n}_2$ and $\underline{t}_2$ are unit vectors in the normal and tangential directions to the circular path of C when arm BC makes an angle θ_2 with the vertical, then:

$$\underline{v}_{C/B} = v_{C/B}\,\underline{t}_2 = \ell_2\,\omega_2\,\underline{t}_2$$
$$\underline{a}_{C/B} = -\,a_{C/B}\,\underline{n}_2 = -\ell_2\,\omega_2^{\,2}\,\underline{n}_2$$

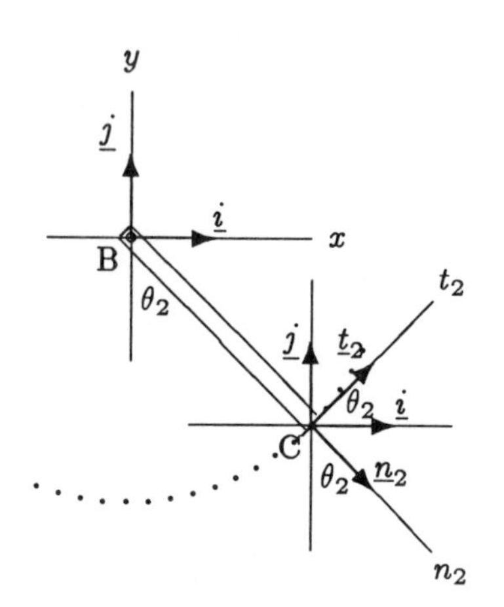

Figure 9.37 *Expressing unit vectors $\underline{n}_2$ and $\underline{t}_2$ in terms of Cartesian unit vectors $\underline{i}$ and $\underline{j}$.*

From Figure 9.37, unit vectors $\underline{n}_2$ and $\underline{t}_2$ can be expressed in terms of Cartesian unit vectors $\underline{i}$ and $\underline{j}$ as:

$$\underline{n}_2 = \sin\theta_2\,\underline{i} - \cos\theta_2\,\underline{j}$$
$$\underline{t}_2 = \cos\theta_2\,\underline{i} + \sin\theta_2\,\underline{j}$$

Therefore, the velocity and acceleration vectors of point C relative to the xy coordinate frame can be written as:

$$\underline{v}_{C/B} = \ell_2\,\omega_2\,(\cos\theta_2\,\underline{i} + \sin\theta_2\,\underline{j})$$
$$\underline{a}_{C/B} = -\,\ell_2\,\omega_2^{\,2}\,(\sin\theta_2\,\underline{i} - \cos\theta_2\,\underline{j})$$

Substituting the numerical values of ℓ_2=0.3 m, θ_2=45°, and ω_2=4 rad/s, and carrying out the necessary calculations we obtain:

$$\underline{v}_{C/B} = 0.85\,\underline{i} + 0.85\,\underline{j} \qquad (iii)$$
$$\underline{a}_{C/B} = -\,3.39\,\underline{i} + 3.39\,\underline{j} \qquad (iv)$$

<u>Motion of point C as observed from point A:</u>

We determined the velocity and acceleration of point C relative to B, and velocity and acceleration of point B with respect to A. Now, we can apply the principles of relative motion to determine the velocity and acceleration of point C as observed from point A or with respect to the XY coordinate frame:

$$\underline{v}_C = \underline{v}_B + \underline{v}_{C/B} \qquad (v)$$
$$\underline{a}_C = \underline{a}_B + \underline{a}_{C/B} \qquad (vi)$$

Since we have already expressed $\underline{v}_B$, $\underline{v}_{C/B}$, $\underline{a}_B$, and $\underline{a}_{C/B}$ in terms of Cartesian unit vectors, we can simply substitute Eqs. (*i*) and (*iii*) into Eq. (*v*), and Eqs. (*ii*) and (*iv*) into Eq. (*vi*):

$$\underline{v}_C = (0.52\,\underline{i} + 0.30\,\underline{j}) + (0.85\,\underline{i} + 0.85\,\underline{j})$$
$$\underline{a}_C = (-0.60\,\underline{i} + 1.04\,\underline{j}) + (-3.39\,\underline{i} + 3.39\,\underline{j})$$

Collecting horizontal and vertical components together:

$$\underline{v}_C = 1.37\,\underline{i} + 1.15\,\underline{j}$$
$$\underline{a}_C = -\,3.99\,\underline{i} + 4.43\,\underline{j}$$

Magnitudes of the velocity and acceleration vectors are:

$$v_C = \sqrt{(1.37)^2 + (1.15)^2} = 1.79 \text{ m/s}$$
$$a_C = \sqrt{(3.99)^2 + (4.43)^2} = 5.96 \text{ m/s}^2$$

9.12* Derivation of Linear Acceleration

Consider a particle moving in a circular path of radius r. Let P be the location of the particle at time t (Figure 9.38). At this instant the line joining the center of motion at O and point P makes an angle θ with the horizontal. Let $\underline{n}$ and $\underline{t}$ be the unit vectors in the normal and tangential directions to the circular path at P. Angle θ and unit vectors $\underline{n}$ and $\underline{t}$ are not constant, but functions of time. As the particle moves along the circular path angle θ changes as well as the directions normal and tangent to the circular path.

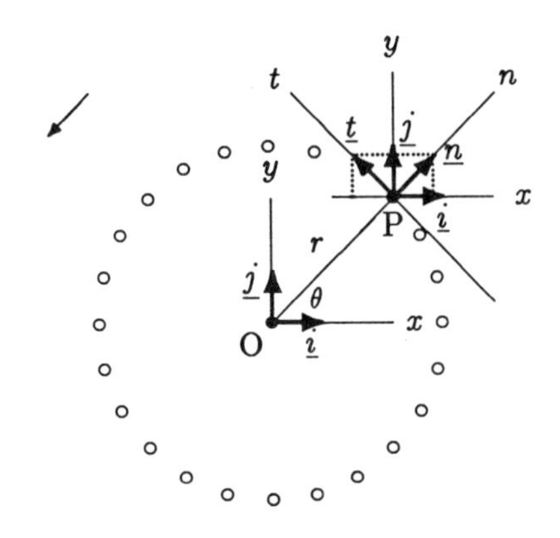

Figure 9.38 *Unit vectors in the normal and tangential directions can be expressed in terms of Cartesian unit vectors* $\underline{i}$ *and* $\underline{j}$.

Unit vectors $\underline{n}$ and $\underline{t}$ can be expressed in terms of Cartesian unit vectors $\underline{i}$ and $\underline{j}$ which are constant vectors. From the geometry of the problem:

$$\underline{n} = \cos\theta\, \underline{i} + \sin\theta\, \underline{j} \tag{9.27}$$

$$\underline{t} = -\sin\theta\, \underline{i} + \cos\theta\, \underline{j} \tag{9.28}$$

Any object undergoing a rotational motion has a velocity $\underline{v}$ always acting in a direction tangent to the path of motion:

$$\underline{v} = v\, \underline{t} \tag{9.29}$$

In general, the magnitude v of this linear velocity can be a function of time. The direction $\underline{t}$ of $\underline{v}$ changes with time as well.

By definition, acceleration is the time rate of change of velocity. By taking the derivative of Eq. (9.29) with respect to time and utilizing the product rule of differentiation, the acceleration vector can be written as:

$$\underline{a} = \frac{d\underline{v}}{dt} = \frac{d}{dt}(v\, \underline{t}) = \frac{dv}{dt}\, \underline{t} + v\, \frac{d\underline{t}}{dt} \tag{9.30}$$

The acceleration as given in Eq. (9.30) is not very useful unless the derivative of the unit tangent vector is evaluated. Using Eq. (9.28), the time rate of change of $\underline{t}$ can be expressed as:

$$\frac{d\underline{t}}{dt} = \frac{d}{dt}(-\sin\theta\, \underline{i} + \cos\theta\, \underline{j}) \tag{9.31}$$

Since $\underline{i}$ and $\underline{j}$ are constant vectors, only the derivatives of $\sin\theta$ and $\cos\theta$ should be considered. Applying the chain rule of differentiation and noting that the time rate of change of angular position is angular velocity (i.e., $\omega = d\theta/dt$):

$$\frac{d}{dt}(\sin\theta) = \cos\theta\, \frac{d\theta}{dt} = \omega\, \cos\theta$$

$$\frac{d}{dt}(\cos\theta) = -\sin\theta\, \frac{d\theta}{dt} = -\omega\, \sin\theta$$

Substituting these into Eq. (9.31) will yield:

$$\frac{d\underline{t}}{dt} = -\omega\,(\cos\theta\,\underline{i} + \sin\theta\,\underline{j}) \qquad (9.32)$$

Note that the terms in the parentheses represent the unit normal vector $\underline{n}$ as given in Eq. (9.27). Furthermore, for a circular motion, the angular and linear velocities are related through the radius of the circular path as $\omega = v/r$. Therefore:

$$\frac{d\underline{t}}{dt} = -\omega\,\underline{n} = -\frac{v}{r}\,\underline{n} \qquad (9.33)$$

Substituting Eq. (9.33) into Eq. (9.30):

$$\underline{a} = \frac{dv}{dt}\,\underline{t} - \frac{v^2}{r}\,\underline{n} \qquad (9.34)$$

In Eq. (9.34), the term carrying the unit tangent vector is the tangential acceleration $\underline{a}_t$ and the term having the unit normal vector is the normal (radial) component $\underline{a}_n$ of the acceleration vector. The acceleration vector can be expressed in the following general form:

$$\underline{a} = \underline{a}_t + \underline{a}_n = a_t\,\underline{t} + a_n\,\underline{n} \qquad (9.35)$$

By comparing Eqs. (9.34) and (9.35), the magnitudes of the acceleration vectors in the tangential and normal directions are:

$$a_t = \frac{dv}{dt} \qquad (9.36)$$

$$a_n = \frac{v^2}{r} \qquad (9.37)$$

In Eq. (9.34), the minus sign in front of the term representing the acceleration component in the normal direction implies that $\underline{a}_n$ is acting in the negative $\underline{n}$ direction. Since the direction of $\underline{n}$ is chosen to be positive outward, $\underline{a}_n$ is directed toward the center of rotation. This is why $\underline{a}_n$ is also known as the centripetal acceleration.

Chapter 10

KINETICS: EQUATIONS OF MOTION

10.1 An Overview

As studied in the previous chapters, kinematic analyses are concerned with the description of time dependent aspects of motion in terms of displacement, velocity, and acceleration without dealing with the factors causing the motion. The field of ***kinetics***, on the other hand, is based on kinematics and incorporates into the analysis the effects of forces and moments that cause the motion.

Based on the type of motion involved, the field of kinetics can be divided into linear (translational) and angular (rotational) kinetics. Translation is caused by the net force applied on an object, whereas rotation is the consequence of a net torque acting on the object. An object will translate and rotate simultaneously (i.e., undergo a general motion) if there is both a net force and a net moment acting on it. In addition to classifying a motion as translational, rotational, or general, the field of kinetics can be further distinguished as the ***kinetics of particles*** and the ***kinetics of rigid bodies***. The distinction between a particle and a rigid body is particularly important if the object is undergoing a rotational motion. If the object is sufficiently small or if the purpose is to gain some general information about the motion of an object, then the geometric properties of the object can be ignored and the object can be treated as a particle located at its center of gravity (or center of mass) with a mass equal to the total mass of the object. Particle kinetics is easier to implement than rigid body kinetics which introduces the geometric properties of the bodies into the analyses.

Kinetic analyses utilize Newton's second law of motion which can be formulated in various ways. One way of representing Newton's second law of motion is in terms of equations of motion, which are particularly suitable for solving problems requiring the analysis of acceleration. In some cases, the solution of equations of motion may be difficult. To handle such situations, alternative methods are developed based on the concepts of work, energy, impulse, and momentum, which will be discussed in Chapters 11 and 12.

10.2 Newton's Second Law of Motion

A body accelerates (i.e., its velocity changes) if there is a net or a resultant force acting on it. Newton's second law of motion states that the magnitude of the acceleration of a particle is directly proportional to the magnitude of the resultant force, and inversely proportional to its mass. The direction of acceleration is the same as the direction of the resultant force. Consider a particle of mass m acted on by a force $\underline{F}$, and let $\underline{a}$ be the resulting acceleration of the particle (Figure 10.1). Newton's second law is expressed as:

$$\underline{F} = m\,\underline{a} \tag{10.1}$$

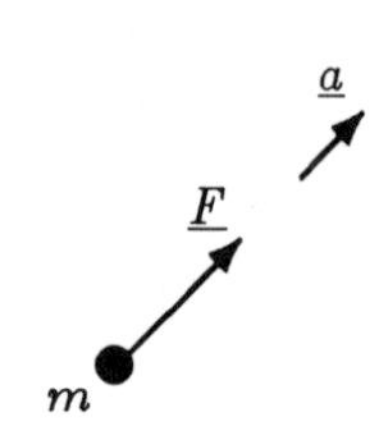

Figure 10.1 *A body accelerates in the direction of applied force.*

If there is more than one force acting on the particle (Figure 10.2), then $\underline{F}$ in Eq. (10.1) must be replaced by the net or the resultant of all forces acting on it. The resultant of a system of forces can be found by considering the vector sum of all forces. Therefore:

$$\sum \underline{F} = m\underline{a} \tag{10.2}$$

Eq. (10.2) is known as the *equation of motion*, and states that the acceleration is caused by the resultant force and that the particle has only one acceleration vector no matter how many forces act upon it.

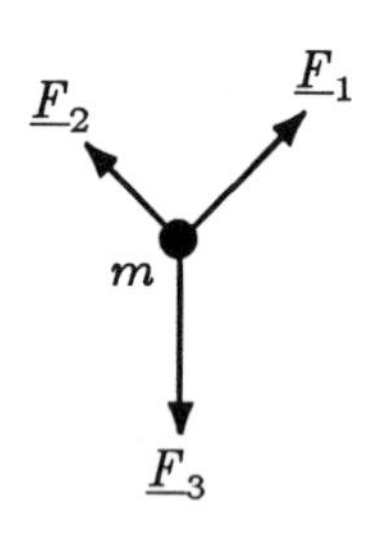

Figure 10.2 *A body has only one acceleration vector no matter how many forces act upon it.*

10.3 Kinetics of Translational Motion

Notice that Newton's second law of motion is stated for a particle. In many problems, the size and shape of an object are irrelevant in the discussion of certain aspects of its motion. This is particularly true if the object is undergoing translational motion only. For example, what is significant for a person pushing a block is the total mass of the block, not its size or shape. It can be assumed that the entire mass m of the block is concentrated at the center of mass of the block, which can then be treated as a particle of mass m. This is the underlying consideration in applying the equations of motion to analyze the translational motion of an object which has a finite size and shape.

10.4 Equations of Motion: Rectangular Components

Since force and acceleration are vector quantities, they can be expressed in terms of their components in reference to a chosen coordinate frame. For translational motion analyses it is best to use rectangular (Cartesian) coordinates. Therefore, it makes sense to express the force and acceleration vectors in terms of their components along the rectangular coordinate directions (Figure 10.3), and write equations of motion with respect to these coordinate directions.

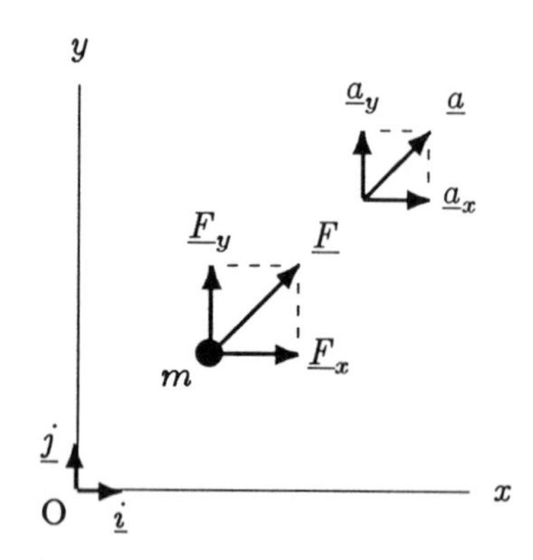

Figure 10.3 *Rectangular components of the force and acceleration vectors.*

The force and acceleration vectors can be expressed in terms of their components along the x and y directions:

$$\underline{F} = F_x\,\underline{i} + F_y\,\underline{j} \tag{10.3}$$

$$\underline{a} = a_x\,\underline{i} + a_y\,\underline{j} \tag{10.4}$$

Substituting Eqs. (10.3) and (10.4) into Eq. (10.1):

$$F_x\,\underline{i} + F_y\,\underline{j} = m\,a_x\,\underline{i} + m\,a_y\,\underline{j} \tag{10.5}$$

This vector equation will yield two scalar equations valid along the x and y directions:

$$F_x = m\,a_x \tag{10.6}$$

$$F_y = m\,a_y \tag{10.7}$$

If there is more than one force acting on the object, then F_x must be replaced by the sum of all forces acting in the x direction and F_y must be replaced by the sum of y components of all forces:

$$\sum F_x = m\,a_x \tag{10.8}$$

$$\sum F_y = m\,a_y \tag{10.9}$$

Eqs. (10.8) and (10.9) state that the sum of all forces acting in one direction is equal to the mass times the acceleration of the body in that direction. Notice that for three-dimensional motion analysis, there is an additional equation, $\sum F_z = m\,a_z$, which governs the motion in the z direction.

10.5 Procedure for Problem Solving in Kinetics

The procedure for analyzing kinetic characteristics of objects undergoing translational motion can be outlined as follows:

- Draw a simple, neat diagram of the system to be analyzed.
- Isolate the bodies of interest from their surroundings and draw their free-body diagrams by showing all external forces acting on them. Indicate the correct directions for the known forces. If the direction of a force vector is not known, assume a positive direction for it. If that force appears to have a negative value in the solution, it would mean that the assumed direction for the force vector (or the acceleration vector) was incorrect.
- Designate the direction of velocities or accelerations of each object on the sidelines (not as parts of the free-body diagrams). It is particularly important to be consistent with the assumed direction of the motion.
- Choose a convenient coordinate system. For plane translational motion, rectangular coordinates x and y are usually the most convenient.
- Apply the equations of motion. For two-dimensional motion analysis there are two governing equations, and therefore the number of unknowns to be determined cannot be more than two. In kinetics, the unknowns are either forces or accelerations.
- Include the correct directions of forces and accelerations in the solution, along with their units.
- The kinematic relations between position, velocity, and acceleration can also be utilized if the information about the velocity and/or position of the object analyzed is given or required.

10.6 Special Cases of Translational Motion

A force can be applied or measured in various ways. For example, an applied force may be constant or it may vary over time. The magnitude of a force vector can be measured as a function of the relative position of the object upon which it is applied, or it may be measured as a function of velocity. These various possible cases will be analyzed next.

Note that the simplest form of translational motion is the one along a straight line (one-dimensional or uniaxial), for which it is adequate to consider a single coordinate, x, collinear with the direction of motion, and to use the equation of motion in the x direction. To illustrate the methods of handling various cases of translational motion in a concise manner, it will be assumed that the motion is along a straight line under the action of only one applied force. However, using the vectorial properties of the parameters involved, these methods can be easily expanded to analyze two- and three-dimensional translational motions under the action of more than one externally applied force.

10.6.1 Force is constant

If a force applied on an object has a constant magnitude and direction, the object will move with a constant acceleration in the direction of the applied force. Assume that a force of magnitude F_x is applied in the x direction on an object of mass m. The magnitude of constant acceleration of the object in the x direction can be calculated using Eq. (10.6):

$$a_x = \frac{F_x}{m} = \text{constant}$$

Once the acceleration of the object is determined, the kinematic equations can be utilized to calculate the velocity and displacement of the object as well:

$$v_x = v_{x0} + \int_0^t a_x \, dt = v_{x0} + \frac{F_x}{m} \, t \tag{10.10}$$

$$x = x_0 + \int_0^t v_x \, dt = x_0 + v_{x0} \, t + \frac{1}{2} \frac{F_x}{m} \, t^2 \tag{10.11}$$

Here, v_{x0} and x_0 are the initial velocity and displacement of the object at time t=0. Note that these results can also be used for a situation in which there is a second force with constant magnitude F_y acting on the object in the y direction (this can be the component of a single force in the y direction), simply by replacing x with y throughout the equations.

10.6.2 Force is a function of time

The magnitude of a force applied on an object may vary over time, $F_x = F_x(t)$. The resulting acceleration of the object is then a function of time as well:

$$a_x(t) = \frac{F_x(t)}{m}$$

The velocity and displacement of the object can again be determined using:

$$v_x = v_{x0} + \int_0^t a_x(t)\, dt \qquad (10.12)$$

$$x = x_0 + \int_0^t v_x(t)\, dt \qquad (10.13)$$

The function $F_x(t)$ must be provided so that the integral in Eq. (10.12) can be evaluated.

10.6.3* Force is a function of velocity

Force can also be measured as a function of velocity, $F_x = F_x(v_x)$. Analyses of such cases require manipulation of the equation of motion, with the use of kinematic relations and rules of differentiation. Since acceleration is the derivative of velocity with respect to time, Eq. (10.6) can also be written as:

$$m\,\frac{dv_x}{dt} = F_x(v_x)$$

Multiplying both sides by dt and dividing them by F_x (applying separation of variables):

$$m\,\frac{dv_x}{F_x(v_x)} = dt$$

Notice that the left-hand side of this equation is a function of v_x only, whereas the right-hand side is a function of time only. Therefore, the left-hand side can be integrated with respect to v_x, and the right-hand side can be integrated with respect to t:

$$m \int_{v_{x0}}^{v_x} \frac{dv_x}{F_x} = \int_0^t dt$$

Or:

$$m \int_{v_{x0}}^{v_x} \frac{dv_x}{F_x} = t \qquad (10.14)$$

The solution of Eq. (10.14) will yield a function relating v_x and t. To be able to evaluate the integral in this equation, the exact form of $F_x(v_x)$ must be known. Once $v_x(t)$ is determined, displacement and acceleration of the object can also be determined using the kinematic equations.

10.6.4* Force is a function of displacement

Sometimes it is more convenient to express force as a function of displacement, $F_x = F_x(x)$. To start deriving some equations relating acceleration, velocity, displacement, mass, and applied force, the equation of motion expressed in terms of velocity can be used:

$$m \frac{dv_x}{dt} = F_x(x)$$

Employing the chain rule of differentiation (see Appendix C.2.6), the time derivative of velocity can be expressed as:

$$\frac{dv_x}{dt} = \frac{dv_x}{dx}\frac{dx}{dt} = \frac{dv_x}{dx} v_x$$

Therefore, the equation of motion in the x direction can be rewritten as:

$$m\, v_x \frac{dv_x}{dx} = F_x(x)$$

Multiplying both sides by dx (separating the variables):

$$m\, v_x\, dv_x = F_x(x)\, dx$$

The left-hand side of this equation is a function of v_x only and can be integrated with respect to v_x, and the right-hand side is a function of x only and can be integrated with respect to x:

$$m \int_{v_{x0}}^{v_x} v_x\, dv_x = \int_{x_0}^{x} F_x(x)\, dx$$

Evaluating the integral on the left-hand side:

$$\frac{m}{2}\left({v_x}^2 - {v_{x0}}^2\right) = \int_{x_0}^{x} F_x(x)\, dx$$

Rearranging the order of terms:

$${v_x}^2 = {v_{x0}}^2 + \frac{2}{m}\int_{x_0}^{x} F_x(x)\, dx \qquad (10.15)$$

F_x must be provided as a function of x, so that the integral in Eq. (10.15) can be evaluated. For given $F_x(x)$, Eq. (10.15) will yield $v_x(x)$. Once v_x is known the acceleration of the object can be determined using:

$$a_x = v_x \frac{dv_x}{dx} \qquad (10.16)$$

Example 10.1 As illustrated in Figure 10.4, consider a block of mass m=50 kg which is being pulled on a rough, horizontal surface by a person using a rope. Assume that the person is applying a constant force of T=150 N, the rope makes an angle θ=30° with the horizontal, and the coefficient of kinetic friction between the block and the horizontal surface is μ=0.2.

Figure 10.4 *A block is being pulled over a horizontal surface.*

Determine the acceleration of the block if the bottom surface of the block remains in full contact with the floor throughout the motion.

Solution: Consider the free-body diagram of the block shown in Figure 10.5. The positive x direction is chosen to represent the direction of motion of the block. $\underline{T}$ is the force exerted by the person which is transmitted to the block through the rope. The rope makes an angle θ=30° with the horizontal. Therefore, $\underline{T}$ has components both in the x and y directions:

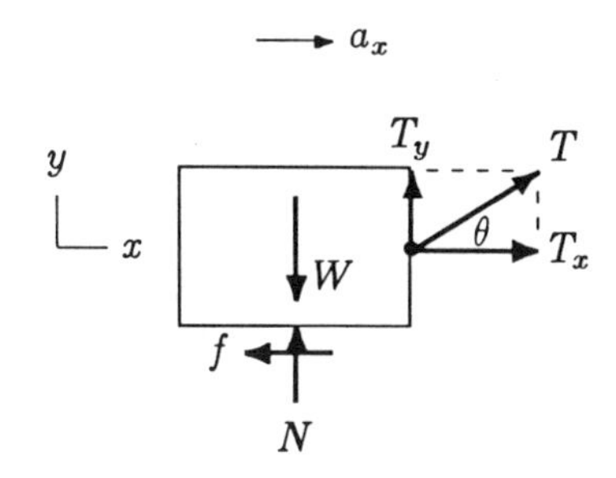

Figure 10.5 *Free-body diagram of the block.*

$$T_x = T \cos\theta \qquad (\rightarrow)$$
$$T_y = T \sin\theta \qquad (\uparrow)$$

The weight, $\underline{W}$, of the block is due to the gravitational effect of Earth on the block, and can be expressed as:

$$W = m\,g \qquad (\downarrow)$$

$\underline{f}$ is the frictional force acting in the direction opposite to the direction of motion. $\underline{N}$ is the reaction force applied by the floor on the block. The magnitude of the frictional force is proportional to the magnitude of the normal force, and they are related through the coefficient of friction between the surfaces in contact:

$$f = \mu\,N \qquad (\leftarrow) \qquad (i)$$

Equations of motion in the x and y directions can now be applied to determine an expression for the acceleration of the block. The block has no motion in the y direction (it always remains in contact with the floor). Therefore, the acceleration of the block in the y direction is zero (a_y=0), and the equation of motion in the y direction then becomes:

$$\sum F_y = 0: \qquad N + T_y - W = 0$$

Solving this equation of equilibrium for force N:

$$N = W - T_y = m\,g - T\,\sin\theta \qquad (ii)$$

Substituting Eq. (*ii*) into Eq. (*i*):

$$f = \mu\,(m\,g - T\,\sin\theta) \qquad (iii)$$

Now, the equation of motion in the x direction can be considered:

$$\sum F_x = m\,a_x : \qquad T_x - f = m\,a_x$$

Solving this equation for a_x:

$$a_x = \frac{1}{m}(T_x - f) = \frac{1}{m}\left[T\ \cos\theta - \mu(m\,g - T\ \sin\theta)\right]$$

Substituting the numerical values m=50 kg, T=150 N, θ=30°, μ=0.2, and g=9.8 m/s, and carrying out the calculations will yield:

$$a_x = 0.94\ \text{m/s}^2 \quad (\rightarrow)$$

Some remarks:

- A force applied by a person on a block has components $\underline{T}_x$ and $\underline{T}_y$ acting in the horizontal and vertical directions, respectively. Since the intended direction of motion is horizontal, $\underline{T}_x$ is the driving force. The frictional force, $\underline{f}$, is acting in the direction opposite to the direction of motion, and its magnitude is dependent on the magnitude of the normal force, $\underline{N}$, applied by the ground on the bottom surface of the block.

- The direction at which the force $\underline{T}$ is applied is very important for the efficiency of the task to be accomplished. To demonstrate the physical significance of this, consider three cases illustrated in Figure 10.6. Assume that the only difference between these cases is the direction of the applied force $\underline{T}$. To compare these cases, the normal reaction force, the frictional force, the component of the applied force in the direction of motion, and the resultant force in the direction of motion must be determined for each case.

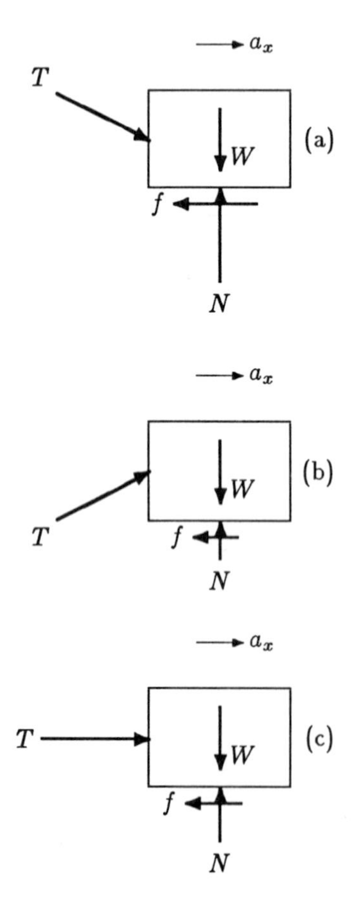

Figure 10.6 *Three ways of applying force $\underline{T}$.*

Among these three cases, the one shown in Figure 10.6(c) is the most efficient and the one shown in Figure 10.6(a) is the least efficient. The reasons are two-fold. First of all, for the case shown in Figure 10.6(a), only the horizontal component of the total force $\underline{T}$ is applied in the direction of motion. In other words, only a part of T is utilized to push the block in the intended direction of motion. This is true also for the case shown in Figure 10.6(b). Furthermore, for the case shown in Figure 10.6(a), the magnitude of the vertical component of $\underline{T}$ contributes to (i.e., increases) the magnitude of the normal force $\underline{N}$, which in return increases the magnitude of the frictional force $\underline{f}$. Since $\underline{f}$ is always acting to retard the motion, increased f does not help the person accomplish the task.

Example 10.2 Figure 10.7 illustrates a person pushing a block of mass m on a surface which makes an angle θ with the horizontal. The coefficient of kinetic friction between the block and the inclined surface is μ.

If the person is applying a force of magnitude F parallel to the incline, determine an expression for the acceleration of the block in the direction of motion.

Figure 10.7 *A person pushing a block over an inclined surface.*

Solution: The free-body diagram of the block is shown in Figure 10.8. The x coordinate is parallel to the incline representing the direction of motion. Note that the weight of the block has components both in the x and y directions. From the geometry of the problem:

$$W_x = W \sin\theta = m\,g\,\sin\theta \qquad (-x)$$
$$W_y = W \cos\theta = m\,g\,\cos\theta \qquad (-y)$$

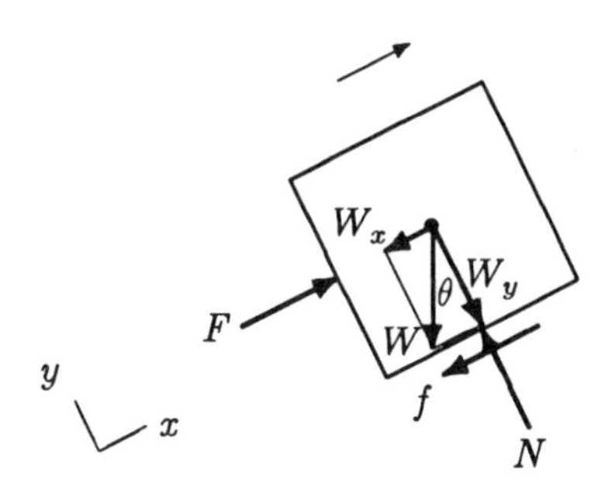

Figure 10.8 *Free-body diagram of the block.*

$\underline{N}$ is the reaction force applied by the incline on the block in the positive y direction. Since there is no motion in the y direction, the equation of motion in the y direction can be replaced by the equation of equilibrium:

$$\sum F_y = 0: \qquad N - W_y = 0$$

Solving this equation for N will yield:

$$N = W_y = m\,g\,\cos\theta \qquad (+y)$$

The frictional force, $\underline{f}$, is acting in the negative x direction. The magnitude of $\underline{f}$ is:

$$f = \mu\,N = \mu\,m\,g\,\cos\theta \qquad (-x)$$

There are a total of three forces acting on the block in the x direction: $\underline{F}$, $\underline{f}$, and $\underline{W}_x$. Now, the equation of motion in the x direction can be applied to determine an expression for the acceleration of the block:

$$\sum F_x = m\,a_x: \qquad F - f - W_x = m\,a_x$$

Solving this equation for a_x will yield:

$$a_x = \frac{1}{m}\left(F - f - W_x\right) = \frac{F}{m} - g\left(\mu\,\cos\theta + \sin\theta\right)$$

Note that the block can be pushed up the incline if and only if F is larger than the sum of f and W_x. Also note that a_x decreases with increasing θ. In other words, the larger the slope, the slower the motion of the block, provided that the magnitude of the applied force remains the same.

10.7 Equations of Motion: Normal and Tangential Components

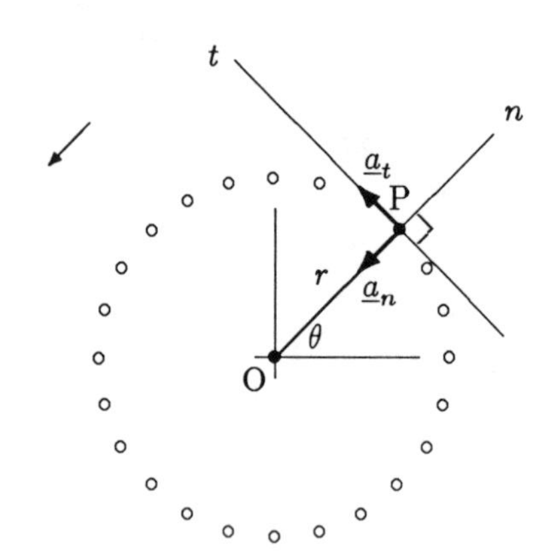

Figure 10.9 *Rotational motion of a particle about a fixed axis.*

To analyze the rotational motion of a particle about a fixed axis, the force and acceleration vectors can be expressed in terms of their components normal and tangential to the circular path of motion. If n and t designate the normal (radial) and tangential directions at any point along the circular path of motion (Figure 10.9), then the equations of motion can be expressed as:

$$\sum F_n = m\, a_n \tag{10.17}$$

$$\sum F_t = m\, a_t \tag{10.18}$$

Here, $\sum F_n$ is the net force acting in the normal direction, $\sum F_t$ is the net force acting in the tangential direction, a_n is the magnitude of the centripetal acceleration (always directed towards the center of rotation), and a_t is the magnitude of tangential acceleration. While applying Eq. (10.17), the forces acting towards the center of rotation (centripetal forces) must be taken to be positive, and the forces directed outward (centrifugal forces) must be negative.

For rotational motion about a fixed axis, the motion characteristics are completely known if the linear velocity (with magnitude v and direction tangent to the circular path) and the radius r of the circular path are known. Since $a_n = v^2/r$ and $a_t = dv/dt$, Eqs. (10.17) and (10.18) can alternatively be written as:

$$\sum F_n = m\, \frac{v^2}{r} \tag{10.19}$$

$$\sum F_t = m\, \frac{dv}{dt} \tag{10.20}$$

Recall that $v = r\omega$ and $dv/dt = r\alpha$. Therefore, if the angular velocity (ω) and acceleration (α) of the motion are known, then the kinetic characteristics of the problem can be analyzed using:

$$\sum F_n = m\, r\, w^2 \tag{10.21}$$

$$\sum F_t = m\, r\, \alpha \tag{10.22}$$

Note that for rotational motion there is always a normal component of the acceleration vector and consequently a force acting in the normal direction, but the tangential components of the force and acceleration vectors may or may not exist.

Example 10.3 Figure 10.10 illustrates a 60 kilogram gymnast swinging on a high bar. Assume that the center of gravity (center of mass) of the gymnast is located at a distance r=1 m from the high bar, and is undergoing a uniform circular motion with constant speed v=5 m/s.

Determine the forces applied on the gymnast's arms at positions 1 through 5 shown in Figure 10.10. Angle θ=45° for positions 2 and 4.

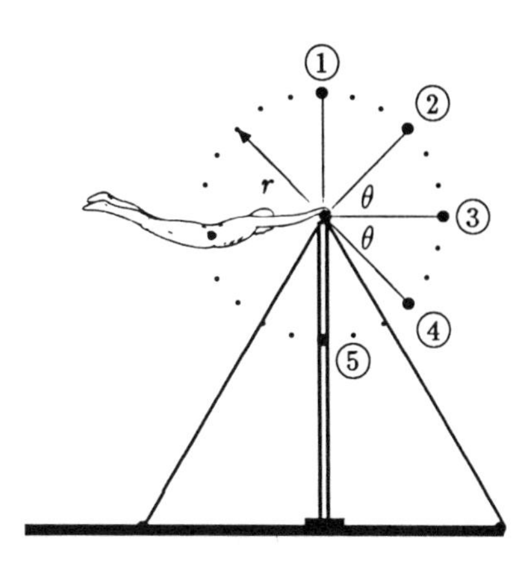

Figure 10.10 *A gymnast on the high bar.*

Solution: The uniform circular motion of the gymnast may be simplified by modeling the gymnast as a particle attached to a string such that the mass of the particle is equal to the mass of the gymnast and the length of the string is equal to the distance between the high bar and the center of gravity of the gymnast.

Free-body diagrams of the gymnast at position 1 through 5 are shown in Figure 10.11. The forces acting on the gymnast are those applied on the arms by the high bar and by gravity. At any instant, the force applied by the high bar is a centripetal force trying to pull the gymnast towards the center of rotation to balance the centrifugal effect due to the rotation of the gymnast.

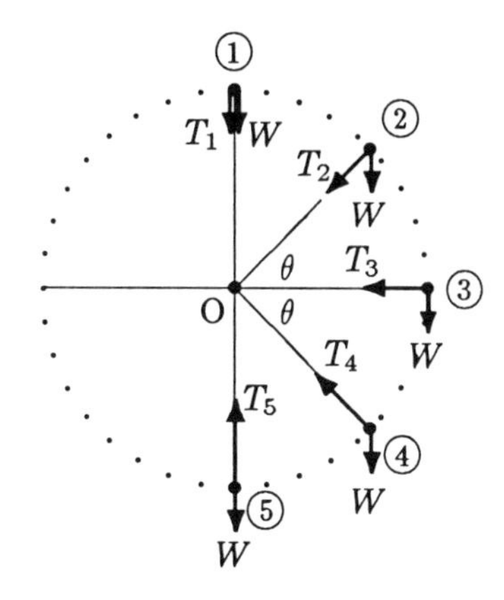

Figure 10.11 *Forces acting on the particle (gymnast's center of gravity).*

For any given position to determine the force applied on the arms of the gymnast (the tensile force in the string), it is sufficient to analyze the problem in the normal (radial) direction. Knowing the radius r and constant linear velocity of the circular motion, we can determine the magnitude of the normal component of the acceleration vector:

$$a_n = \frac{v^2}{r} = \frac{(5)^2}{1} = 25 \text{ m/s}^2$$

The equation of motion in the normal direction is:

$$\sum F_n = m\,a_n$$

If we let T_1, T_2, T_3, T_4, and T_5 be the forces applied by the high bar on the arms of the gymnast while moving along the circular path, then the equation of motion in the normal direction will yield the following solutions.

Position 1: Both $\underline{T}_1$ and the weight $\underline{W}$ are towards the center of rotation.

$$T_1 + W = m\,a_n$$

$$T_1 = m\,a_n - W = m\,(a_n - g) \qquad (i)$$

Position 2: The weight of the gymnast has a normal component $W \sin\theta$ acting towards the center of rotation.

$$T_2 + W\ \sin\theta = m\,a_n$$

$$T_2 = m\,a_n - W\ \sin\theta = m\,(a_n - g\ \sin\theta) \qquad (ii)$$

Position 3: The weight of the gymnast has no component in the normal direction.

$$T_3 = m\,a_n \qquad (iii)$$

Position 4: The weight of the gymnast has a normal component $W \sin\theta$ acting in a direction away from the center of rotation.

$$T_4 - W\ \sin\theta = m\,a_n$$

$$T_4 = m\,a_n + W\ \sin\theta = m\,(a_n + g\ \sin\theta) \qquad (iv)$$

Position 5: The entire weight of the gymnast is acting in a direction away from the center of rotation.

$$T_5 - W = m\,a_n$$

$$T_5 = m\,a_n + W = m\,(a_n + g) \qquad (v)$$

Substituting the numerical values of m=60 kg, a_n=25 m/s^2, g=9.8 m/s^2, and θ=45° into Eqs. (i) through (v), and carrying out the calculations will yield:

$$T_1 = 912\ \text{N}$$
$$T_2 = 1084\ \text{N}$$
$$T_3 = 1500\ \text{N}$$
$$T_4 = 1916\ \text{N}$$
$$T_5 = 2088\ \text{N}$$

Notice that T_1 through T_5 are the total force exerted by the high bar on two arms. Also note that the forces exerted on the arms are minimum at position 1. The force applied on the arms increases as the gymnast moves towards position 5. Between 2 and 5, in addition to overcoming the centrifugal effects of rotation, the gymnast must carry his/her body weight.

Example 10.4 Figure 10.12 illustrates the circular end region of a ski jump track. The radius of curvature of the track at this region is r=50 m.

Consider a 60 kilogram ski jumper who is decelerating at a rate of 1.5 m/s^2 due to air resistance. If the friction on the track is negligible, and if the ski jumper reaches the end of the track with a speed of v=20 m/s, determine the forces applied on the skier by the air resistance and the track.

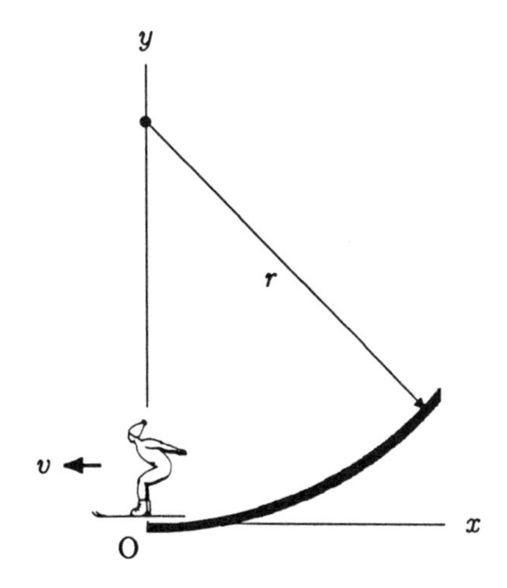

Figure 10.12 *Circular end region of a ski-jump track.*

Solution: At the circular end region of the track, the ski jumper undergoes a motion in a circular path with radius r=50 m. The speed of the ski jumper at the instant of takeoff is v=20 m/s, and therefore, the magnitude of the skier's acceleration in the normal direction is:

$$a_n = \frac{v^2}{r} = \frac{(20)^2}{50} = 8 \text{ m/s}^2$$

At the instant of takeoff, the direction normal to the circular path coincides with the vertical (y) and the tangent to the circular path is horizontal (x). It can be assumed that the ski jumper is accelerating at a rate of 8 m/s^2 in the positive y direction (towards the center of rotation) just before takeoff. Therefore:

$$a_y = a_n = 8 \text{ m/s}^2$$

At the circular end region, the ski jumper is losing speed at a rate of 1.5 m/s^2 due to the air resistance. It is reasonable to assume that at the instant of takeoff, the air resistance applies a force on the ski jumper in the horizontal direction. Therefore, the ski jumper is decelerating at a rate of 1.5 m/s^2 towards the left (in the negative x direction) or accelerating at the same rate towards the right (in the positive x direction):

$$a_x = 1.5 \text{ m/s}^2$$

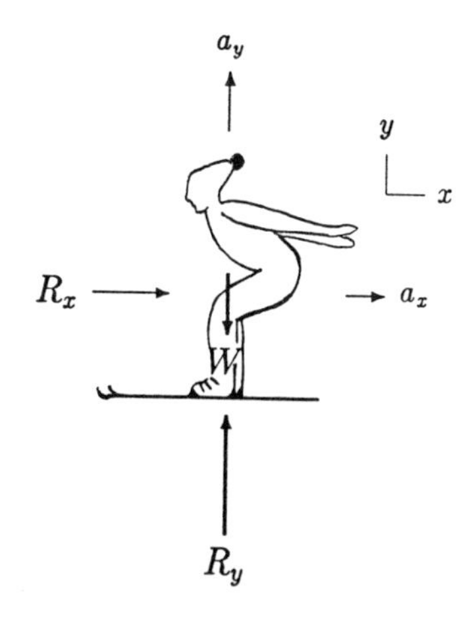

Figure 10.13 *Free-body diagram of the ski jumper before takeoff.*

The free-body diagram of the ski jumper just before takeoff is shown in Figure 10.13. The forces acting on the skier are due to gravity (W), the reaction force $\underline{R}_y$ applied by the track in the vertical direction, and the air resistance $\underline{R}_x$ applied in the horizontal direction. To solve the problem for R_x, we can utilize the equation of motion in the x direction:

$$\sum F_x = m\,a_x : \qquad R_x = m\,a_x = (60)(1.5) = 90 \text{ N}$$

Application of the equation of motion in the y direction will yield the reaction force applied by the track on the ski jumper:

$$\sum F_y = m\,a_y : \qquad R_y - W = m\,a_y$$

$$R_y = m\,(a_y + g) = (60)(8 + 9.8) = 1068 \text{ N}$$

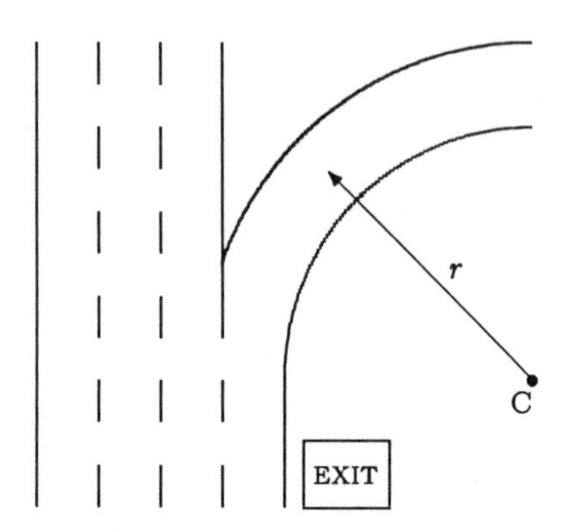

Figure 10.14 *A curved exit.*

Example 10.5 Figure 10.14 illustrates a curved exit ramp on a highway. Exit ramps are usually banked to reduce skidding due to the centrifugal effects of curved motion (Figure 10.15). Without a bank (on a level road), the centrifugal effects would have to be balanced entirely by the friction between the tires and the road.

Assume that the radius of curvature of a particular exit ramp is r, and the road is banked at an angle θ with the horizontal. If r=40 m and θ=15°, what is the optimal speed that should be posted at the entrance of the ramp?

Solution: For a given angle θ, there is an optimal speed v a car should travel along a banked, curved road. At this optimal speed, the effects of frictional forces on the tires of the car are minimal. Since the purpose of this example is to determine this optimal speed, we will assume that there is no frictional force exerted on the tires of the car.

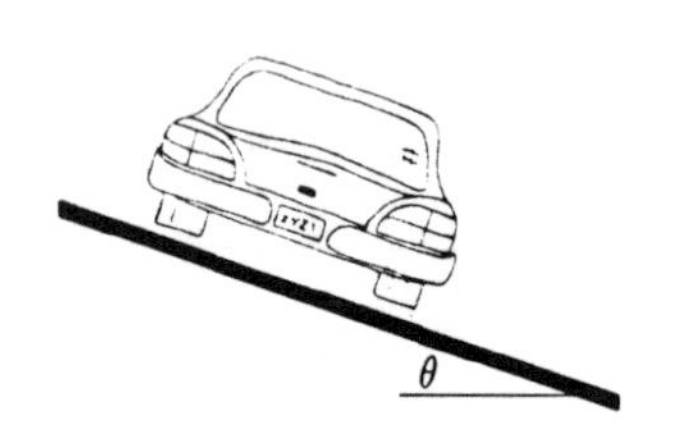

Figure 10.15 *The rear view of the car on the road banked at an angle θ.*

The forces acting on the car are shown in Figure 10.16. The x and y directions are chosen such that the positive x direction points towards the center of the curve, and the positive y direction is upward. W=mg is the weight of the car, and N is the magnitude of the normal force applied by the road on the car. Because of the bank of the road, this force has components N_x=$N\sin\theta$ and N_y=$N\cos\theta$ in the x and y directions, respectively. The car is in equilibrium in the y direction. Therefore:

$$\sum F_y = 0: \qquad N\cos\theta = W \qquad (i)$$

The radius of curvature of the road is r. If the speed of the car is v, then the car has a centripetal acceleration of a_x=v^2/r in the positive x direction. Applying the equation of motion in the x (radial) direction:

$$\sum F_x = m\,a_x: \qquad N\sin\theta = m\,\frac{v^2}{r} \qquad (ii)$$

Dividing Eq. (*ii*) by Eq. (*i*), and solving the resulting equation for v will yield:

$$v = \sqrt{r\,g\,\tan\theta}$$

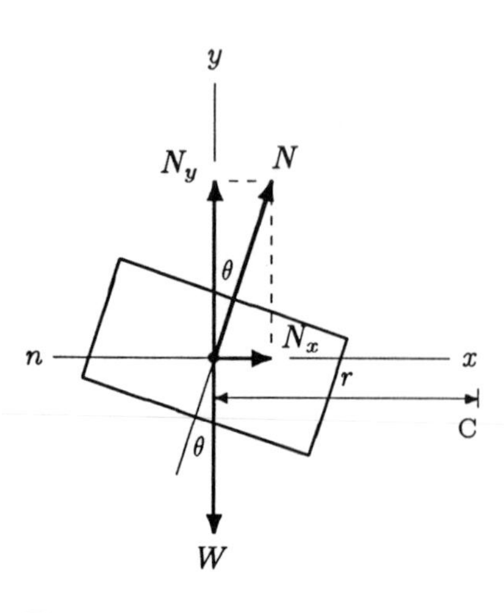

Figure 10.16 *Forces acting on the car.*

For example, if the radius of curvature of an exit ramp is r=40 m and the road is banked at an angle θ=15°, then the speed posting at a sign before the exit should be v=10 m/s or 36 km/h. A car traveling on such a road with a speed higher than 36 km/h will tend to slide up the incline (especially if the road is icy). Similarly, a car traveling slower than 36 km/h will tend to slide down the incline.

10.8 Kinetics of Angular Motion

Torque is the quantitative measure of the ability of a force to rotate an object. The mathematical definition of torque is the same as that of moment, studied in detail in Chapter 4. Consider the bolt and wrench arrangement illustrated in Figure 10.17. Force $\underline{F}$ applied on the wrench rotates the wrench, which advances the bolt into the wall by rotating it in the clockwise direction. The magnitude of torque or moment $\underline{M}$ due to force $\underline{F}$ about O can be determined using:

$$M = r\ F\ \sin\phi \tag{10.23}$$

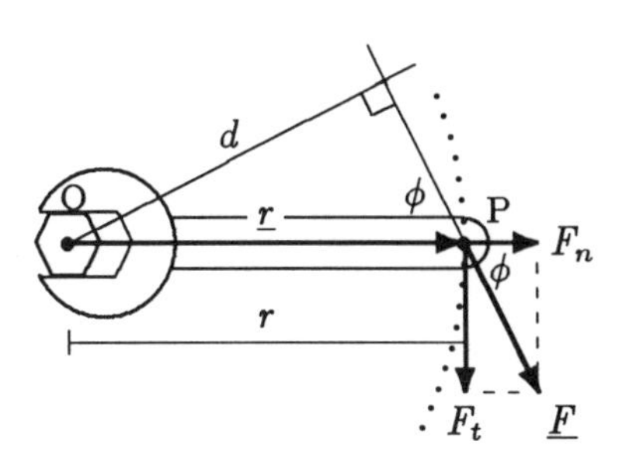

Figure 10.17 *Force $\underline{F}$ applied on the wrench produces a clockwise torque about the centerline of the bolt.*

The line of action of $\underline{M}$ is perpendicular to the plane of rotation and its direction can be determined by using the right-hand rule (in this case, clockwise).

An object would rotate about an axis if the rotational motion of the object is not constrained and if there is a net torque acting on the object about that axis. The angular acceleration of an object undergoing a rotational motion is directly proportional to the resultant torque acting on it. To derive the relationship between torque and angular acceleration, consider a particle of mass m undergoing a rotational motion about a fixed axis. Let O be a point on this axis, r be the radius of the circular path of motion, and $\underline{F}_t$ be the tangential force causing the rotational motion (Figure 10.18). The equation of motion in the tangential direction can be written as:

$$F_t = m\ a_t \tag{10.24}$$

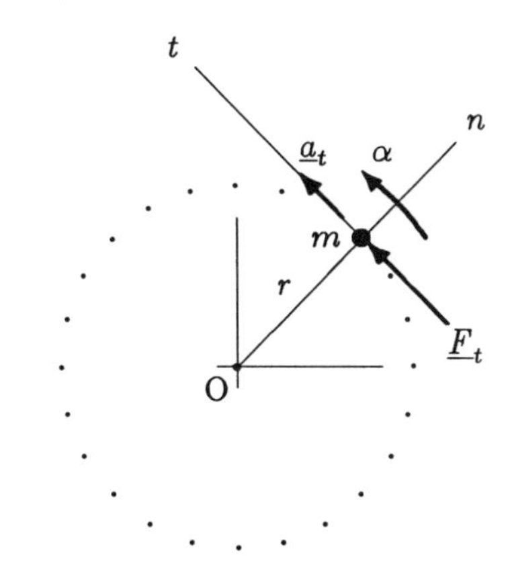

Figure 10.18 *$F_t = m a_t$ and $M_o = I_o \alpha$.*

In Eq. (10.24), a_t is the magnitude of the tangential acceleration of the particle. If the angular acceleration, α, of the particle is known, then $a_t = r\alpha$. Replacing a_t by $r\alpha$ and multiplying both sides of Eq. (10.24) by r will yield $rF_t = rmr\alpha$, or:

$$r\ F_t = (m\,r^2)\ \alpha \tag{10.25}$$

Note that the left-hand side of Eq. (10.25) is the magnitude M_o of the torque about O due to force $\underline{F}_t$. The quantity mr^2 on the right-hand side is known as the ***mass moment of inertia*** of the particle about O. Denoting the mass moment of inertia with I_o, Eq. (10.25) can also be written as:

$$M_o = I_o\ \alpha \tag{10.26}$$

If there is more than one torque-generating force applied to the particle, then M_o in Eq. (10.26) represents the net torque acting on the particle about O. The general form of Eq. (10.26) can be obtained by representing the torque and angular acceleration as vector quantities:

$$\underline{M} = I\ \underline{\alpha} \tag{10.27}$$

This is the rotational analogue of Newton's second law of motion, and states that the torque and angular acceleration are proportional, and the mass moment of inertia is the constant of proportionality.

10.9 Mass Moment of Inertia

In general, the term inertia implies resistance to change. When a rotation-causing force is applied to a pivoted body, its tendency to resist angular acceleration depends on its mass moment of inertia. The larger the mass moment of inertia of a body, the more difficult it is to accelerate it in rotation. For a particle of mass m, the mass moment of inertia about an axis is defined as the mass times the square of the shortest distance (r) between the particle and the axis about which the mass moment of inertia is to be determined:

$$I = m\,r^2 \tag{10.28}$$

A rigid body that is not a particle can be assumed to consist of many particles, the sum of masses of which is equal to the total mass of the body itself. The mass moment of inertia of the entire body can be determined by considering the integral of the mass of each particle of the body multiplied by the square of its distance from the axis of rotation.

Notice that the mass moment of inertia of a rigid body is proportional to its mass, which is a function of its density and volume. Therefore, the mass moment of inertia of a body depends upon its material and geometric properties as well as the location and orientation of the axis about which it is to be determined. The ability of a body to resist changes in its angular velocity (i.e., experience angular acceleration) is dependent not only upon the mass of the body, but also upon the distribution of the mass. The greater the concentration of mass at the periphery, the greater the mass moment of inertia and the more difficult it is to change the angular velocity. For a rigid body with a simple, symmetrical geometry and homogeneous composition, the mass moment of inertia about an axis coinciding with an axis of symmetry, called a *centroidal axis*, can be calculated relatively easily. In Table 10.1, moments of inertia for some geometric shapes are provided about their centroidal axes.

The mass moment of inertia is a scalar quantity. It has a dimension $[\mathrm{M}][\mathrm{L}^2]$, and is measured in terms of kg-m^2 in SI.

10.10 Parallel-Axis Theorem

If the moment of inertia of a body about a centroidal axis is known, then the moment of inertia of the same body about any other axis parallel to that centroidal axis can be determined using the *parallel-axis theorem*. This theorem can be stated as:

$$I = I_c + m\,{r_c}^2 \tag{10.29}$$

In Eq. (10.29), m is the total mass of the body, I_c is the mass moment of inertia of the body about one of its centroidal axes, I is the required mass moment of inertia about an axis parallel to the centroidal axis, and r_c is the perpendicular distance between the two axes. For example, consider the solid cylinder shown in Figure 10.19. From Table 10.1, the mass moment of inertia of the cylinder about AA is $I_{AA}=\frac{1}{2}mr^2$. The mass moment of inertia of the same cylinder about DD, which is parallel to AA and located at a distance r_c from AA, is:

$$I_{DD} = I_{AA} + mr_c{}^2 = \frac{1}{2}mr^2 + mr_c{}^2$$

Note that in the case of human body segments, each segment or limb rotates about the joints at either end of the moving segment rather than about its mass center or centroidal axes. Furthermore, mass moment of inertia measurements can only be made about a joint center. If needed, the parallel-axis theorem can be utilized to determine the mass moment of inertia of a segment about its mass center.

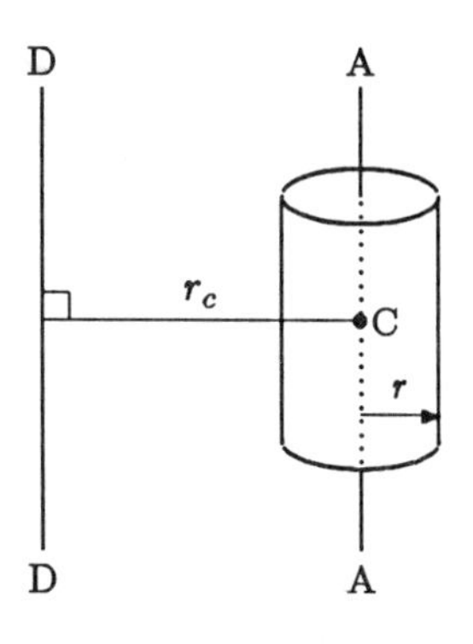

Figure 10.19 *According to the parallel-axis theorem,* $I_{DD}=I_{AA}+mr_c{}^2$.

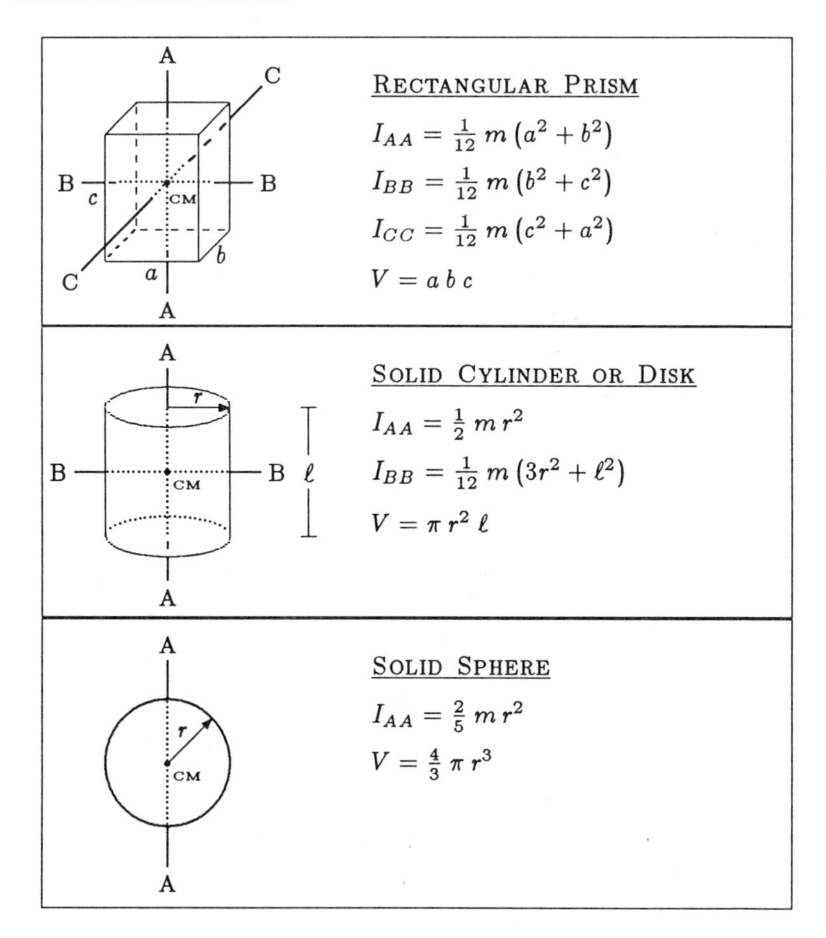

Body	Moments of inertia and volume
RECTANGULAR PRISM	$I_{AA} = \frac{1}{12}\, m\,(a^2+b^2)$ $I_{BB} = \frac{1}{12}\, m\,(b^2+c^2)$ $I_{CC} = \frac{1}{12}\, m\,(c^2+a^2)$ $V = a\,b\,c$
SOLID CYLINDER OR DISK	$I_{AA} = \frac{1}{2}\, m\,r^2$ $I_{BB} = \frac{1}{12}\, m\,(3r^2+\ell^2)$ $V = \pi\, r^2\, \ell$
SOLID SPHERE	$I_{AA} = \frac{2}{5}\, m\,r^2$ $V = \frac{4}{3}\,\pi\, r^3$

Table 10.1 *Moments of inertia of homogeneous rigid bodies with different geometries about their centroidal axes. Also provided are their volumes.*

10.11 Radius of Gyration

Consider a rigid body of mass m. Let I be the mass moment of inertia of the rigid body about a given axis AA. Also consider a point mass m located at a distance ρ (rho) from the same axis such that its mass moment of inertia ($m\rho^2$) about AA is equal to the mass moment of inertia (I) of the rigid body about AA. That is:

$$\rho = \sqrt{I/m} \tag{10.30}$$

ρ is called the ***radius of gyration***, and for rotational motion analysis, the rigid body can be treated as a point mass equal to the total mass of the body located at a distance ρ from the axis of rotation.

10.12 Segmental Motion Analysis

The information provided in the previous sections can be utilized to develop mathematical models for analyzing the motion characteristics of human body segments. Here, the general procedure for developing a dynamic model of a body segment will be outlined, which is then applied to analyze the rotational motion of the lower leg about the knee joint.

The first step of a dynamic model analysis involves defining the forces acting on the body segment. These may include the gravitational (weight), external, inertial, muscle, and joint reaction forces. The weight of the body segment can be assumed to act at its center of gravity, and therefore, the center of gravity of the segment must be known. The magnitude, point of application, and direction of any external force present must be specified. Inertial forces are those present due to the dynamics of the problem under consideration. One way of incorporating inertial effects into the model is through the use of the radius of gyration. Muscle and joint reaction forces are the unknowns to be determined as a result of these analyses. It is important to draw the free-body diagram of the segment to be analyzed, and to identify all the known and unknown forces acting on it.

The next step of dynamic analysis is the identification of measurable quantities. In general, both the angular displacement of the moving body segment and the net torque generated about its axis of rotation can be measured as functions of time over the range of segmental motion. The angular displacement measurement techniques include ganiometric, dynamometric, and photogrammetric methods. The angular displacement data can be used to calculate the angular velocity and angular acceleration of the moving segment. If the angular displacement (θ) versus time (t) curve can be represented in terms of a function, $\theta=\theta(t)$, then its angular velocity ($\dot{\theta}$ or ω) and acceleration ($\ddot{\theta}$ or α) can be determined by considering the derivatives of θ with respect to t:

$$\omega = \frac{d\theta}{dt} \qquad \alpha = \frac{d\omega}{dt} = \frac{d^2\theta}{dt^2}$$

If it is not possible to find a function representing the relationship between the angular displacement and time, then numerical differentiation techniques can also be employed. Once α is determined, the relationship between torque, mass moment of inertia, and angular acceleration can be used to calculate the net torque produced about the joint center, provided that the mass moment of inertia (or the radius of gyration) of the segment about the joint center is known.

For a two-dimensional (planar) motion analysis of a segment about its joint center, if the net torque (M) produced about the joint axis is measured as a function of time and if the mass moment of inertia (I) of the moving segment about the same axis is known, then the angular acceleration of the segment can be determined from:

$$\alpha = \frac{M}{I}$$

If needed, the angular velocity and displacement of the moving segment can also be determined by considering the integral of α with respect to time t.

It is clear from this discussion that anthropometric information about the moving segment must be available. For this purpose, anthropometric data tables listing average segmental weights, lengths and radii of gyration can be utilized. For a review of dynamic measurement techniques and the anthropometric data available, see Chaffin and Andersson (1991) or Winter (1990).

Another important consideration is that the instantaneous center of rotation of the moving segment must also be known. Note however that the instantaneous center of rotation about a given joint may vary. For example, by taking a series of X-rays throughout the range of motion and using a graphical method, Frankel, Burstein, and Brooks (1971) showed that the instantaneous center of rotation of the tibiofemoral joint for flexion and extension is not unique but forms a semicircle.

The final step of dynamic model analyses involves the computation of muscle and joint reaction forces. Note that the net torque measured or calculated includes the effects of all forces acting on the moving segment. The torque generated by the muscles crossing the joint can be determined by subtracting the effects of external and gravitational forces from the net torque measured or calculated about the joint center. If the forces generated by individual muscles are required, then additional factors must be considered. For example, the locations of muscle attachments and the lines of action (lines of pull) of the muscle forces must be known. The distribution of forces among different muscles must also be specified. If the segmental motion is achived primarily by a single muscle group, then the rotational component of the muscle force can be determined by applying Newton's second law of rotational motion ($\sum M = I\alpha$). Since the line of action of the muscle force is assumed to be known, the magnitude of the muscle force can also be determined.

Example 10.6 *Knee extension.*
A person is shown in Figure 10.20 seated on a table, with the back placed against a back rest, the knees at the edge and the lower legs hanging vertically downward. The subject was strapped to the back rest, and the right thigh was strapped firmly to the table. A well-padded sawhorse was placed in front of the subject to prevent hyper-extension at the knee joint. An electrogoniometer was attached to the subject's right leg. The arms of the goniometer were aligned with the estimated long axes of the thigh and shank, and the axis of rotation of the goniometer was aligned with the estimated axis of the knee joint. The subject was then asked to extend the lower leg as rapidly as possible. The signals received from the electrogoniometer's potentiometer were stored in a computer, and were used to calculate the angular displacement (θ) of the lower leg as measured from its initial vertical position. Using a finite difference (numerical differentiation) technique, the angular velocity (ω) and angular acceleration (α) of the lower leg were also computed.

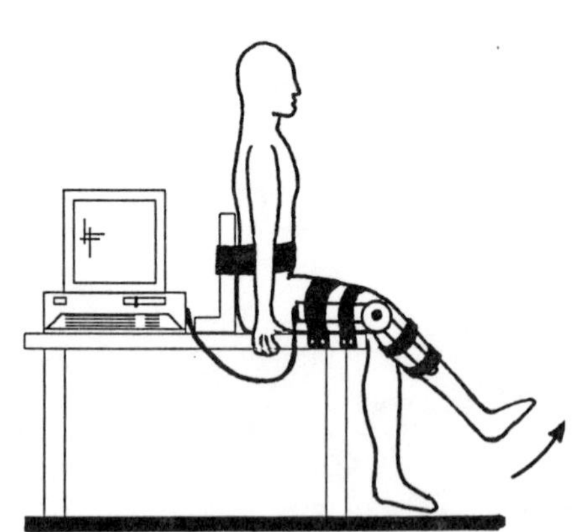

Figure 10.20 *Knee extension.*

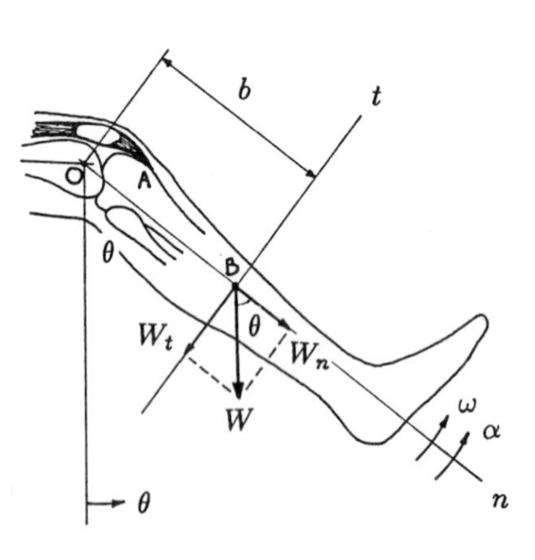

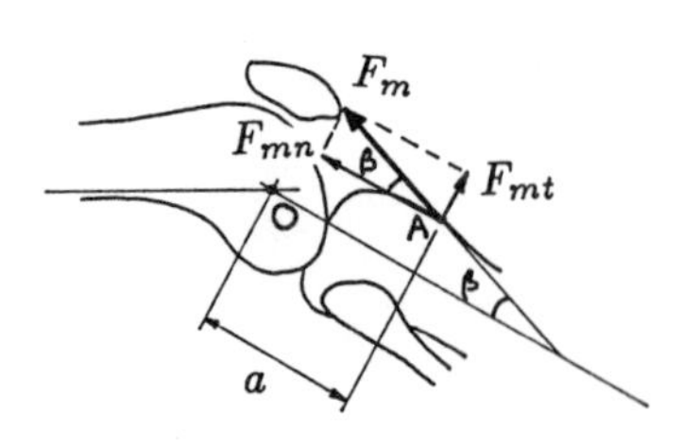

Figure 10.21 *Some of the forces acting on the lower leg.*

Some of the forces acting on the lower leg are shown in Figure 10.21, along with the geometric parameters of the model under consideration. Notice that the model is based on the assumption that the quadriceps muscle is the primary mover. The patellar tendon is attached to the tibia at A, which is located at a distance a from the instantaneous center of rotation (O) of the lower leg about the knee joint. The total weight of the lower leg is W and its center of gravity is located at B, a distance b from O. The line of pull of the patellar tendon force makes an angle β with the long axis of the tibia. The intended direction of motion is counterclockwise.

Assuming that W=50 N, a=5 cm, b=22 cm, β=24°, and the mass moment of inertia of the lower leg about the knee joint is I_o=0.25 kg-m^2, determine (a) the magnitude of the net torque produced by the knee extensor muscles, (b) the rotational component of the tension in the patellar tendon, (c) the tension in the patellar tendon, and (d) the reaction force at the knee joint at an instant when θ=60°, ω=5 rad/s, and α=200 rad/s^2.

Solution:
(a) From Newton's second law of motion, the resultant torque M_o generated about the knee joint is:

$$M_o = I_o\,\alpha = (0.25)(200) = 50 \text{ N-m}$$

Note that M_o is the magnitude of the net torque about the knee joint and it includes the rotational effect of all forces acting on the lower leg. Considering the rotational effects of these forces about the knee joint:

$$M_o = M_m - b\,W\,\sin\theta$$

Here, M_m is the net torque generated by the knee extensor muscles. Solving this equation for M_m will yield:

$$M_m = M_o + b\,W\sin\theta = 50 + (0.22)(50)(\sin 60°) = 59.5 \text{ N-m}$$

(b) Assuming that the quadriceps is the primary muscle group, then M_m must be due to the rotational component (F_{mt}) of the force exerted by the patellar tendon on the tibia. The distance between the joint center at O and the point A where the tendon is attached to the tibia is a=0.05 cm. Therefore:

$$M_m = a\,F_{mt}$$

Solving this for F_{mt} will yield:

$$F_{mt} = \frac{M_m}{a} = \frac{59.5}{0.05} = 1190 \text{ N}$$

(c) The angle between the line of pull of the patellar tendon and the long axis of the tibia is β=24°. Since F_{mt}=$F_m\sin\beta$, the magnitude F_m of the force applied by the tendon is:

$$F_m = \frac{F_{mt}}{\sin\beta} = \frac{1190}{\sin 24°} = 2926 \text{ N}$$

(d) One way of taking into account the inertial effects of a moving body, in this case a rotating body segment, is by means of the d'Alembert's principle. If the mass (m), distance (r) between the mass center and the axis of rotation, angular velocity (ω), and angular acceleration (α) of the rotating body are known, then the magnitudes of inertial forces F_{in} and F_{it} which are normal and tangent to the path of motion can be calculated using Eqs. (10.21) and (10.22) provided in Section 10.7:

$$F_{in} = m\,a_n = m\,r\,\omega^2$$
$$F_{it} = m\,a_t = m\,r\,\alpha$$

In this case, we have $m=W/g=50/9.8=5.1$ kg, $r=b=0.22$ cm, $\omega=12$ rad/s, and $\alpha=200$ rad/s^2. a_n and a_t are the scalar components of the acceleration vector of the lower leg in directions normal and tangential to its path of motion. $\underline{a}_n$ is always towards the center of rotation, and since the motion is counterclockwise, $\underline{a}_t$ is counter-clockwise. Therefore, the inertial forces $\underline{F}_{in}$ and $\underline{F}_{it}$ are such that $\underline{F}_{in}$ is centripetal (towards the center of rotation) and $\underline{F}_{it}$ is trying to rotate the leg in the counterclockwise direction. As illustrated in Figure 10.22, the d'Lambert's principle can be applied by assuming that $\underline{F}_{in}$ is a centrifugal force (rather than centripetal) trying to pull the leg outward, $\underline{F}_{it}$ is trying to rotate the lower leg in the clockwise direction (rather than in the counterclockwise direction), and the system is in static equilibrium. Now, considering the translational equilibrium of the lower leg in the directions normal and tangential to the path of its motion:

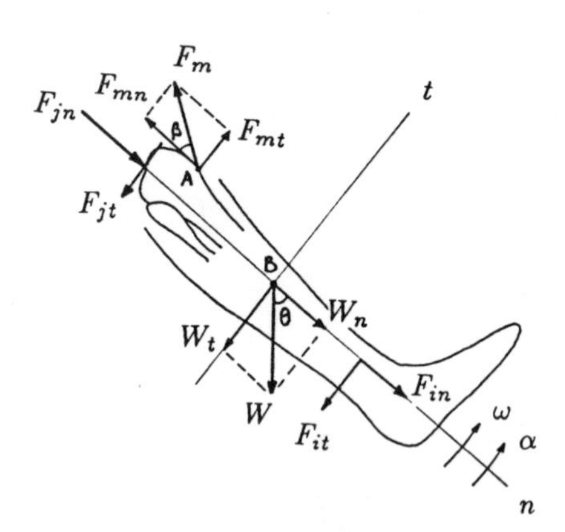

Figure 10.22 *Free-body diagram of the lower leg.*

$$\sum F_n = 0: \qquad F_{jn} - F_{mn} + F_{in} + W_n = 0$$
$$\sum F_t = 0: \qquad F_{jt} - F_{mt} + F_{it} + W_t = 0$$

Here, F_{jn} and F_{jt} are the scalar components of the joint reaction force at the knee, and W_n and W_t are the components of the weight of the lower leg in directions normal and tangent to its path of motion. Solving these equations for the components of joint reaction force will yield:

$$F_{jn} = F_{mn} - F_{in} - W_n$$
$$F_{jt} = F_{mt} - F_{it} - W_t$$

Substituting the parameters whose numerical values are given or are already calculated:

$$F_{jn} = F_m \cos\beta - m\,b\,\omega^2 - W\cos\theta$$
$$F_{jt} = F_m \sin\beta - m\,b\,\alpha - W\sin\theta$$

Substituting the numerical values and carrying out calculations:

$$F_{jn} = (2926)(\cos 24^\circ) - (5.1)(0.22)(5)^2 - (50)(\cos 60^\circ) = 2620 \text{ N}$$
$$F_{jt} = (2926)(\sin 24^\circ) - (5.1)(0.22)(200) - (50)(\sin 60^\circ) = 922 \text{ N}$$

Therefore, the magnitude of the resultant force applied by the femur on the tibia is $F_j=\sqrt{(F_{jn})^2 + (F_{jt})^2}=2777$ N.

Note that the angular motion of the lower leg about the knee joint, and the forces and torques produced by the muscles crossing the knee joint during knee flexions and extensions have been investigated by a number of researchers utilizing different experimental techniques. For example, Van Eijden, De Boer, and Verburg (1983) used a dynamometer for the measurement of the extension torques of the lower leg during static and dynamic contractions of the quadriceps muscle. Using an electrogoniometry method similar to the one described here, Bober, Putnam, and Woodworth (1987) investigated the angular velocities of the lower leg extensions about the knee joint for different ranges of motion. Using X-ray techniques and a force table, Smidt (1973) investigated the changes in the center of rotation of the knee joint and the torques generated by the knee extensor muscles during eccentric, isometric and concentric knee extensions.

Also reported in the literature are the motion analyses of forearm movements. For example, Amis, Dowson, and Wright (1980) used a high-speed cine camera to record the movement of the forearm about the elbow joint with the upper arm in a fixed position. The data collected were used to determine the angular displacement, velocity, and acceleration of the forearm. Based on a mathematical model similar that adapted here, Amis and his coworkers analyzed the forces exerted by the elbow muscle and forces applied to the articulations of the elbow during elbow flexions and extensions.

10.13 References Cited

Amis, A.A., Dowson, D., and Wright, V. 1980. Analysis of elbow forces due to high-speed forearm movements. *J. Biomechanics* 13:825-831.

Bober, T., Putnam, C.A., and Woodworth, G.G. 1987. Factors influencing the angular velocity of a human limb segment. *J. Biomechanics* 20:511-521.

Chaffin, D.B., and Andersson, G.B.J. 1991. *Occupational Biomechanics.* 2nd ed. New York: John Wiley & Sons.

Frankel, V.H., Burstein, A.H, and Brooks, C.B. 1971. Biomechanics of internal derangement of the knee. Pathomechanics as determined by analysis of the instant centers of motion. *J. Bone and Joint Surgery* 53A:945-967.

Smidt, G.L. 1973. Biomechanical analysis of knee flexion and extension. *J. Biomechanics* 6:79-92.

Van Eijden, T.M.G.J., De Boer, W., and Verburg, J. 1983. A dynamometer for the measurement of the extension torque of the lower leg during static and dynamic contractions of the quadriceps femoris muscle. *J. Biomechanics* 16:1019-1023.

Winter, D.A. 1990. *Biomechanics and Motor Control of Human Behavior.* 2nd ed. New York: John Wiley & Sons.

Chapter 11

WORK AND ENERGY METHODS

11.1 Introduction

The fundamental method of analyzing the kinetic characteristics of bodies is based on the equations of motion which are mathematical representations of Newton's second law of motion. Using the equations of motion, one can determine accelerations. Using the kinematic relationships, the velocities and displacements can be calculated as well. In some cases, particularly when the forces involved are not constant, the solution of equations of motion may be difficult. To handle such situations, alternative methods are developed which are based on the concepts of work and energy. These methods are also derived from Newton's laws, and they provide supplementary approaches for problem solving in dynamics. The work and energy methods can be applied to analyze the forces, velocities, and displacements involved in relatively complex systems without resorting to the equations of motion.

11.2 Work Done by a Constant Force

By definition, *work* is the product of force and corresponding displacement. Work is a scalar quantity. There is no line of action, direction, or point of application associated with work. To explore the definition of work, consider the block in Figure 11.1. Assume that a constant, horizontal force $\underline{F}$ is applied on the block so as to move it from position 1 to position 2 at a distance s. The work done, $\mathcal{W}$, by force $\underline{F}$ on the block to move the block from position 1 to 2 is equal to the magnitude of the force vector times the amount of displacement:

$$\mathcal{W} = F\, s \tag{11.1}$$

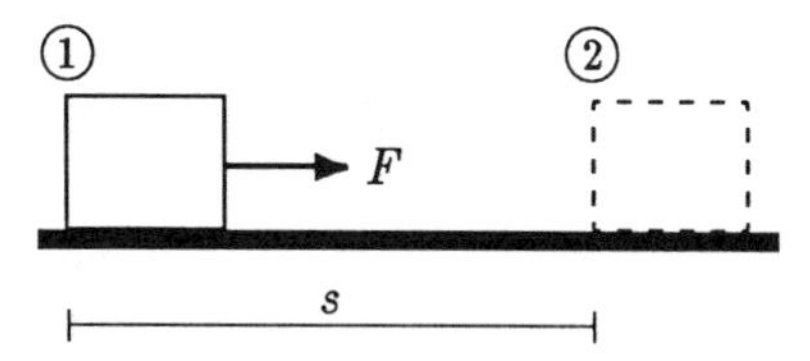

Figure 11.1 *A constant force applied on a block displaces it from position 1 to position 2.*

Consider the same block which is pulled from position 1 to 2 by another constant force $\underline{F}$ which makes an angle θ with the horizontal (Figure 11.2). The work done by $\underline{F}$ on the block is equal to the magnitude of the force component in the direction of displacement times the magnitude of the displacement itself. Since the component of $\underline{F}$ along the horizontal direction is $F_x = F\cos\theta$, the work done by $\underline{F}$ to move the block from position 1 to 2 is:

$$\mathcal{W} = F_x\, s = F\, s\, \cos\theta \tag{11.2}$$

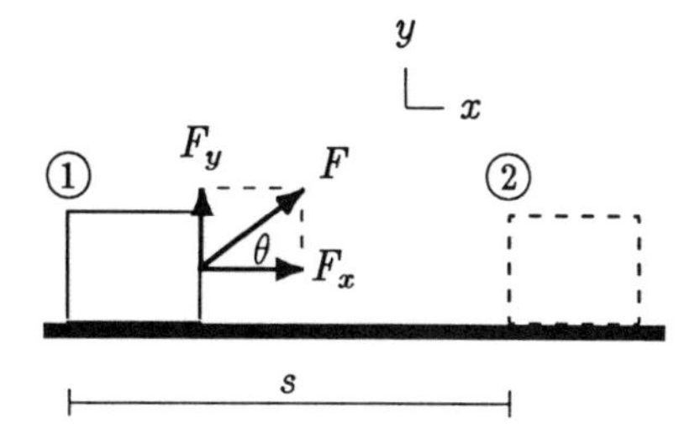

Figure 11.2 *A constant force which makes an angle θ with the horizontal is applied on the block.*

Note that Eqs. (11.1) and (11.2) are consistent with each other since $\cos\theta = 1$ when $\theta = 0°$.

For a force to do work, the body on which the force is applied must undergo a displacement and the force vector must have a non-zero component in the direction of displacement. For example, the vertical component, $F_y = F\sin\theta$, of the force vector in Figure 11.2 does no work because the block is not displaced in the vertical direction.

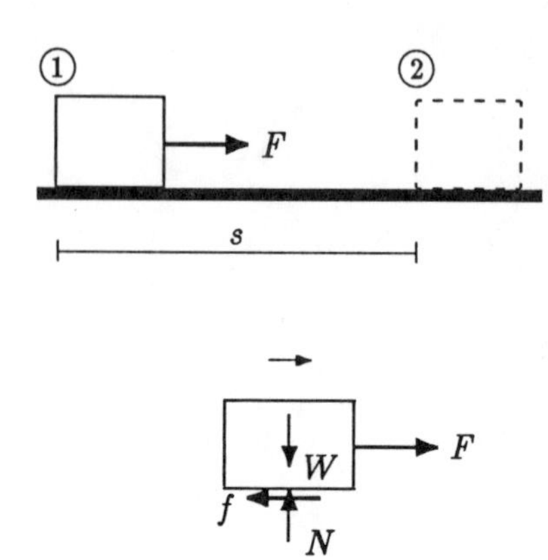

Figure 11.3 *Frictional forces do negative work.*

Work can have a positive or a negative sign. The work done by a force is positive if the force has the same direction (sense) as the displacement. If the applied force and displacement have opposite directions, then the work done by that force is negative. A typical example of negative work is the one done by a frictional force. As illustrated in Figure 11.3, assume that a block is pulled by a force $\underline{F}$ towards the right to displace the block by a distance s. The negative work done, $\mathcal{W}_f$, by the frictional force $\underline{f}$ on the block while the block was displaced by a distance s is:

$$\mathcal{W}_f = -f\ s \tag{11.3}$$

If there is more than one external force acting on a body in motion, then there is one work done for each force. The net work done is the algebraic sum of work done by individual forces. For example, the net work done, $\mathcal{W}_{net}$, for the situation shown in Figure 11.3 is:

$$\mathcal{W}_{net} = F\ s - f\ s$$

11.3 Dimension and Units of Work

By definition, work done is force times displacement. Therefore, work has the dimension of force times the dimension of length.

$$[\text{WORK}]=[\text{FORCE}]\ [\text{DISPLACEMENT}] = \text{M}\frac{\text{L}^2}{\text{T}^2}$$

The units of work (and energy) in different systems of units are provided in Table 11.1.

SYSTEM	UNITS OF WORK AND ENERGY	SPECIAL NAME
SI	Newton-meter (N-m)	Joule (J)
c-g-s	dyne-centimeter (dyn-cm)	erg
British	pound-foot (lb-ft)	

Table 11.1 *Units of work and energy.*

Example 11.1 A 20 kilogram block is pushed up a rough inclined surface by a constant force of 150 N which is applied parallel to the incline (Figure 11.4). The incline makes an angle θ=30° with the horizontal and the coefficient of friction between the incline and the block is μ=0.2.

If the block is displaced by ℓ=10 m, determine the work done on the block by force $\underline{F}$, by the force of friction, and by the force of gravity. What is the net work done on the block?

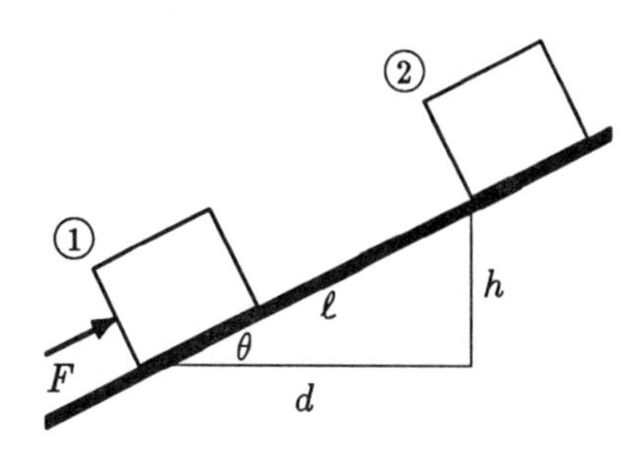

Figure 11.4 *A block is pushed up an inclined surface.*

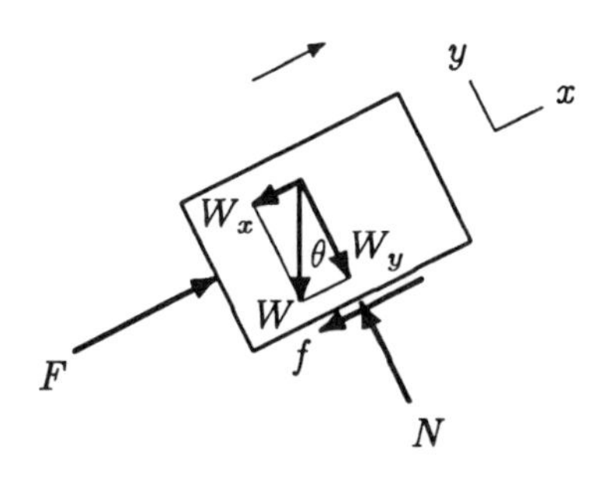

Figure 11.5 *Free-body diagram of the block.*

Solution: Consider the free-body diagram of the block shown in Figure 11.5. Force $\underline{F}$ is applied in the same direction as the displacement of the block. Therefore, the work done by $\underline{F}$ to displace the block by a distance of ℓ is:

$$\mathcal{W}_F = F\,\ell \qquad (i)$$

Since there is no motion in the y direction (the direction perpendicular to the incline), the equation of equilibrium in the y direction can be utilized to determine the magnitude of the normal reaction force applied by the incline on the block. This will yield $N{=}W_y{=}mg\cos\theta$. The magnitude of the frictional force is $f{=}\mu N{=}\mu mg\cos\theta$. The frictional force is acting in a direction parallel to the incline but opposite to that of the displacement of the block. Therefore, the work done by $\underline{f}$ on the block as the block is displaced by a distance ℓ is:

$$\mathcal{W}_f = -f\,\ell = -\mu\, m\, g\, \ell \cos\theta \qquad (ii)$$

The force of gravity (weight) always acts downward. In this case, it has a component parallel to the incline in the negative x direction with magnitude $W_x{=}mg\sin\theta$. The work done by W_x on the block as it moves from position 1 to position 2 is:

$$\mathcal{W}_g = -m\, g\, \ell \sin\theta \qquad (iii)$$

Note here that $\ell\sin\theta$ is equal to h which is the vertical displacement of the block between 1 and 2. In other words, the work done by the gravity is equal to the weight $W{=}mg$ times the vertical displacement $h{=}\ell\sin\theta$ of the block, which is consistent with the definition of work.

Knowing the work done by individual forces acting on the block, we can determine the net work $\mathcal{W}_{net}$ done on the block:

$$\mathcal{W}_{net} = \mathcal{W}_F + \mathcal{W}_f + \mathcal{W}_g = F\,\ell - m\,g\,\ell\,(\sin\theta + \mu\,\cos\theta) \qquad (iv)$$

Substituting the numerical values of the parameters involved into Eqs. (i) through (iv):

$$\begin{aligned}
\mathcal{W}_F &= (150)(10) = 1500 \text{ J} \\
\mathcal{W}_f &= -\,(0.2)(20)(9.8)(10)(\cos 30^\circ) = -340 \text{ J} \\
\mathcal{W}_g &= -\,(20)(9.8)(10)(\sin 30^\circ) = -980 \text{ J} \\
\mathcal{W}_{net} &= 1500 - 340 - 980 = 180 \text{ J}
\end{aligned}$$

11.4 Work Done by a Varying Force

Equation (11.1) can only be used to calculate the work done by a constant force. If an applied force is a function of displacement, then the work done can be calculated by considering the integral of the force over the distance it is applied.

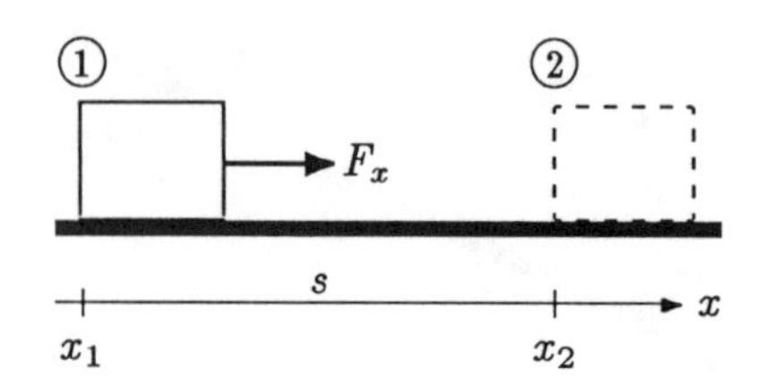

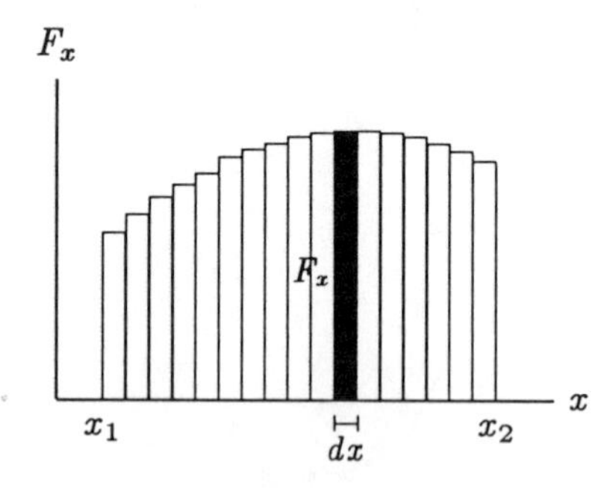

Figure 11.6 *Work is equal to the area under the force versus displacement curve.*

As illustrated in Figure 11.6, consider a block pulled along the x direction by a force F_x which varies with the displacement of the block in the x direction (i.e., $F_x = F_x(x)$). Assume that the block which was originally located at position 1 moves to position 2, at a distance s. Let x_1 and x_2 represent the original and final positions of the block. If the variation of F_x with respect to x is known, then the work done by F_x to move the block from position 1 to 2 can be determined using:

$$\mathcal{W} = \int_{x_1}^{x_2} F_x \, dx \tag{11.4}$$

Notice that $F_x dx$ in Eq. (11.4) is equal to the shaded area shown in Figure 11.6, and the definite integral gives the total area bounded by the force versus position curve and the x axis between $x = x_1$ and $x = x_2$. Also note that if the force F_x is constant, the integration in Eq. (11.4) will simply yield $\mathcal{W} = F_x(x_2 - x_1) = F_x s$.

Example 11.2 Consider that a force with magnitude F_x which varies with displacement along the x direction is applied on a body. Assume that the variation of the force is as shown in Figure 11.7.

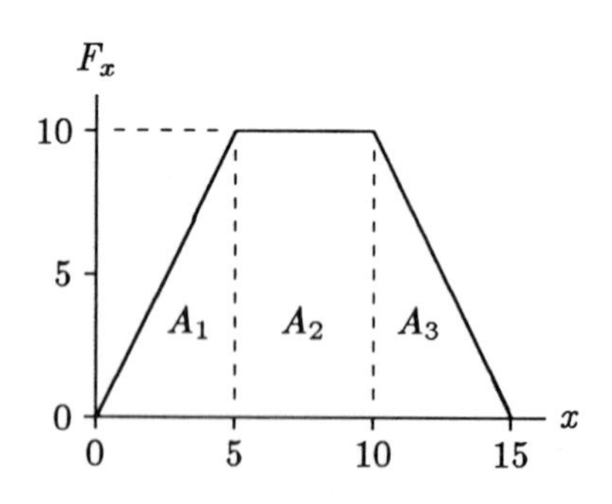

Figure 11.7 *Force measured in Newtons versus displacement in meters.*

Determine the work done by the force on the body as the body moves from $x=0$ to $x=15$ m.

Solution: The total work done by force F_x is equal to the total area under the force versus displacement curve in Figure 11.7. This area is equal to the sum of area A_1 of the initial triangular section, area A_2 of the intermediate rectangular section, and area A_3 of the final triangular section. Therefore:

$$\mathcal{W} = A_1 + A_2 + A_3$$

$$\mathcal{W} = \frac{1}{2}(5)(10) + (5)(10) + \frac{1}{2}(5)(10)$$

$$\mathcal{W} = 100 \text{ J}$$

Example 11.3 A force with varying magnitude F_x is applied on a body and the displacement of the body is recorded in terms of x. The applied force is then plotted as a function of displacement, and the curve shown in Figure 11.8 is obtained. It is observed that between x=0 and x=9 m the force is proportional to the square root of displacement:

$$F_x = c\sqrt{x}$$

The constant of proportionality is estimated to be c=6.

Determine the work done by the applied force to move the body from x=0 to x=4 m, and from x=0 to x=9 m.

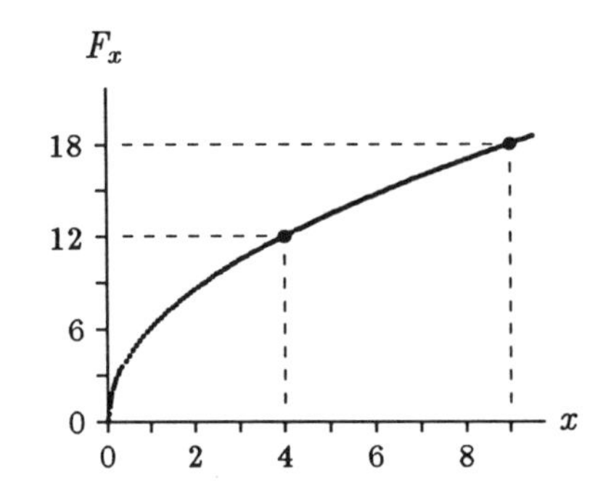

Figure 11.8 *Force measured in Newtons versus displacement in meters.*

Solution: Let x_1 and x_2 correspond to positions of the body along its path of motion. The work done by the force to move the body from x_1 to x_2 is equal to the area under the force versus displacement curve between x_1 and x_2. This area can be calculated by considering the definite integral of the function F_x=$c\sqrt{x}$ with respect to x between x_1 and x_2:

$$\mathcal{W} = \int_{x_1}^{x_2} F_x \, dx = \int_{x_1}^{x_2} c\sqrt{x} \, dx$$

Here, c is a constant and can be taken outside the integral sign, and the integral of $\sqrt{x}$=$x^{1/2}$ is equal to $(2/3)x^{3/2}$. Therefore:

$$\mathcal{W} = \int_{x_1}^{x_2} c\, x^{\frac{1}{2}} \, dx$$

$$\mathcal{W} = \frac{2}{3} c \left[x^{\frac{3}{2}} \right]_{x_1}^{x_2}$$

$$\mathcal{W} = \frac{2}{3} c \left[(x_2)^{\frac{3}{2}} - (x_1)^{\frac{3}{2}} \right]$$

Substituting c=6 and x_1=0:

$$\mathcal{W} = 4\,(x_2)^{\frac{3}{2}}$$

Therefore, the work done to move the block from x=0 to x=4 m is $\mathcal{W}$=32 J, and that done from x=0 to x=9 m is $\mathcal{W}$=108 J.

11.5* Work as a Scalar Product of Force and Displacement Vectors

For some applications it may be convenient to utilize the definition of work as the dot (scalar) product of the force and displacement vectors. As introduced in Section 2.12, the dot product of any two vectors is a scalar quantity equal to the product of magnitudes of the two vectors multiplied by the cosine of the smaller angle between the two. In the case of work done by a constant force $\underline{F}$ on a body whose displacement vector is given by $\underline{s}$:

$$\mathcal{W} = \underline{F} \cdot \underline{s} \tag{11.5}$$

If θ is the smaller angle between vectors $\underline{F}$ and $\underline{s}$, then:

$$\mathcal{W} = \underline{F} \cdot \underline{s} = F\, s\, \cos\theta \tag{11.6}$$

Note that for plane problems, vectors $\underline{F}$ and $\underline{s}$ can be expressed in terms of their Cartesian components as follows:

$$\underline{F} = F_x\, \underline{i} + F_y\, \underline{j} \tag{11.7}$$

$$\underline{s} = x\, \underline{i} + y\, \underline{j} \tag{11.8}$$

The dot product of unit vectors are such that $\underline{i} \cdot \underline{i} = \underline{j} \cdot \underline{j} = 1$ and $\underline{i} \cdot \underline{j} = \underline{j} \cdot \underline{i} = 0$. Substituting Eqs. (11.7) and (11.8) into Eq. (11.5) and carrying out the dot products of unit vectors will yield:

$$\mathcal{W} = F_x\, x + F_y\, y \tag{11.9}$$

Equation (11.9) is significant in that it represents the total work done by the components of the force vector in the x and y directions. The work done in the x direction is equal to the magnitude of the force component in the x direction times the amount of displacement (x) in the same direction. The work done in the y direction is equal to the magnitude of the force component in the y direction times the amount of displacement (y) in that direction.

Note that $\mathcal{W} = F_x\, x + F_y\, y + F_z\, z$ for three-dimensional problems.

Example 11.4 An object moves in the xy plane under the effect of a constant force given in vector form as $\underline{F} = (10\underline{i} + 25\underline{j})$ N. The displacement of the object is given by the vector $\underline{s} = (2\underline{i} + 4\underline{j})$ m. Determine the total work done on the object by force $\underline{F}$.

Solution: By definition, work done is equal to the dot product of the force and displacement vectors. Therefore:

$$\mathcal{W} = (10\underline{i} + 25\underline{j}) \cdot (2\underline{i} + 4\underline{j}) = (10)(2) + (25)(4) = 120 \text{ J}$$

11.6 Energy

The term *energy* is used to describe the capacity of a system to do work on another system. Energy can take various forms such as mechanical, thermal, chemical, and nuclear. The field of mechanics is primarily concerned with the mechanical form of energy. Mechanical energy can be categorized as potential energy and kinetic energy. Energy is a scalar quantity and has the same dimension and units as work.

11.7 Potential Energy

The *potential energy* of a system is associated with its position or elevation. It is the energy stored in the system that can be converted into kinetic energy. Kinetic energy is associated with motion. Common types of potential energy are the gravitational potential energy and the elastic potential energy.

11.7.1 Gravitational potential energy

The concept of potential energy comes from the perception that an object located at a height can do useful work if it is allowed to descend. The potential of an object to do work due to the relative height of its center of gravity is defined as *gravitational potential energy*. Consider the object with weight W shown in Figure 11.9. The object is at position 1 which is located at a height h measured relative to position 2. The gravitational potential energy, $\mathcal{E}_P$, of the object at position 1 relative to position 2 is defined as:

$$\mathcal{E}_P = W\,h = m\,g\,h \tag{11.10}$$

Notice that Wh is essentially the work that the force of gravity would do on the object to move it from position 1 to position 2, h distance apart.

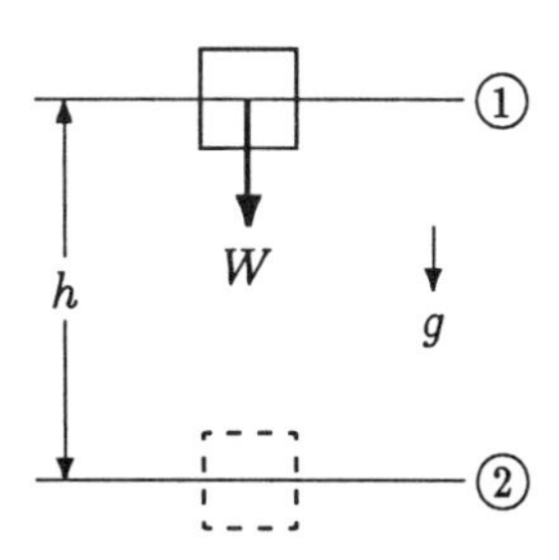

Figure 11.9 *Gravitational potential energy.*

11.7.2 Elastic potential energy

Elastic potential energy is associated with the energy storage ability of materials. Different materials have different material properties such as elastic, viscous, viscoelastic, and plastic. These concepts will be discussed in detail in later chapters. One way of analyzing the elastic behavior of materials is by means of the *theory of springs* proposed by Robert Hooke in 1642.

There are various types of springs. Consider the spring shown in Figure 11.10 which is loaded with progressively larger forces. An applied load will extend or increase the length of the spring, the amount of extension being proportional to the magnitude of the applied load. If the applied forces are not excessive, the deformations will be linearly proportional to the applied forces. Therefore, the applied force versus deformation graph of the spring can

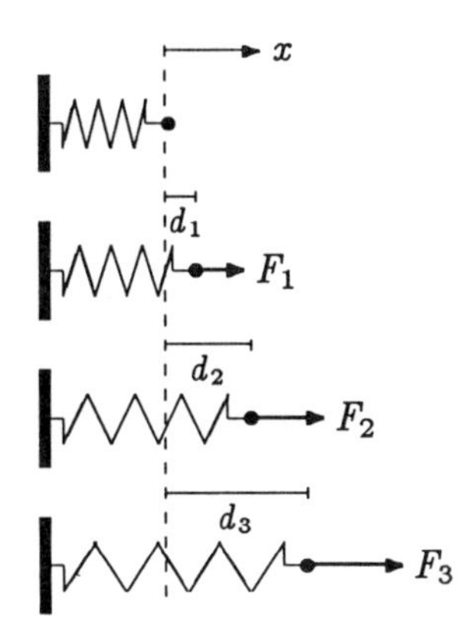

Figure 11.10 *Elastic potential energy.*

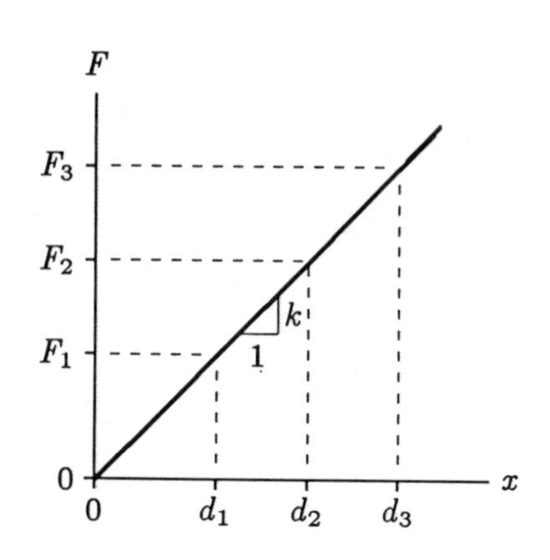

Figure 11.11 *Force-displacement relation for a linear spring.*

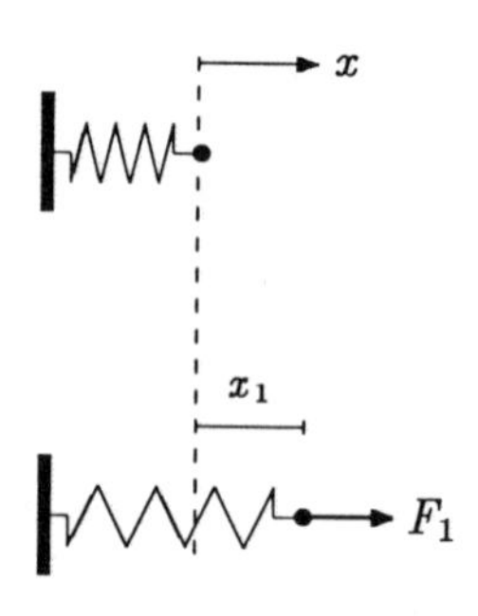

Figure 11.12 *Force F_1 extends the spring by an amount x_1.*

be represented by a straight line as illustrated in Figure 11.11, or by the function:

$$F = k\,x \tag{11.11}$$

Here, k is the constant of proportionality between the applied force, F, and the corresponding deformation, x, and it is the slope of the line representing the relationship between F and x. k is called the ***spring constant*** or the ***stiffness*** of the spring. Different springs with different material properties, wire diameter, coil diameter, and length have different spring constants.

Consider the spring shown in Figure 11.12. The force F_1 applied on the spring causes the spring to extend by an amount x_1 measured from its unstretched configuration. The work $\mathcal{W}$ done by F_1 can be determined as:

$$\mathcal{W} = \int_0^{x_1} F_1\,dx = \int_0^{x_1} k\,x\,dx = \frac{1}{2}\,k\,x_1^2$$

Note that the work done by F_1 on the spring to extend the spring by an amount x_1 is equal to the area under the F_1 versus x line between x=0 and x=x_1. The above result can be generalized for any amount of deformation x as:

$$\mathcal{W} = \frac{1}{2}\,k\,x^2 \tag{11.12}$$

Once deformed, the spring has an ability to do useful work if the external force applied on the spring is removed. In other words, the work done on the spring by deforming the spring is stored in the spring as a potential energy, and once the applied force is removed, the spring will use this energy to resume its original undeformed configuration. Therefore, the right hand side of Eq. (11.12) is also the ***elastic potential energy*** stored in the spring which is stretched or compressed by an amount x:

$$\mathcal{E}_P = \frac{1}{2}\,k\,x^2 \tag{11.13}$$

11.7.3 Total potential energy

An object attached to a spring can be subject to the effects of both gravity and the spring. Denoting the gravitational potential energy by $\mathcal{E}_{P_g}$ and the elastic potential energy by $\mathcal{E}_{P_e}$, the total potential energy $\mathcal{E}_P$ of an object subject to the action of both gravity and a spring is:

$$\mathcal{E}_P = \mathcal{E}_{P_g} + \mathcal{E}_{P_e} \tag{11.14}$$

Example 11.5 A spring is hung vertically and a one kilogram mass is attached to it (Figure 11.13). The spring-mass system came to rest after its length increased by an amount d=5 cm.

Assuming that the force-deformation relationship for the spring is linear, determine (a) the spring constant, (b) the work done by the weight of the body, (c) the work done by the spring, (d) the net work done, and (e) the potential energy stored in the spring.

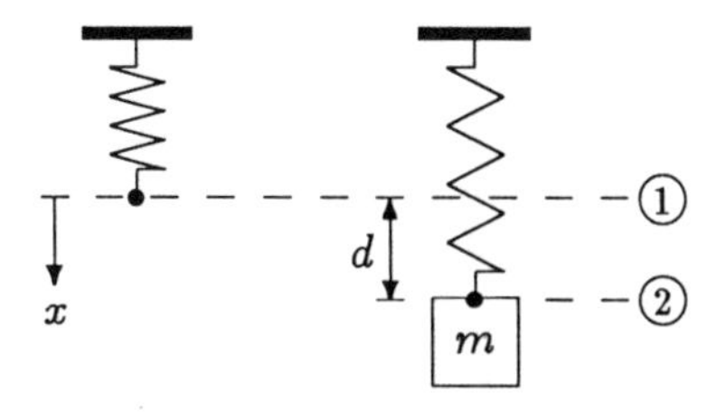

Figure 11.13 *A mass attached to a spring deforms the spring.*

Solution:
(a) At position 2 the system is in static equilibrium. The free-body diagram of the block is shown in Figure 11.14. For the equilibrium of the block in the vertical direction, the weight W of the block must be equal to the force F_s exerted by the spring on the block. Therefore:

$$F_s = W = m\,g = (1)(9.8) = 9.8 \text{ N}$$

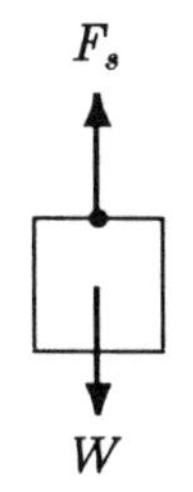

Figure 11.14 *Free-body diagram of the block in equilibrium.*

Since action must be equal to reaction, F_s is also the magnitude of the force applied by the block on the spring. The spring is stretched by an amount d. Solving F_s=kd for the spring constant k will yield:

$$k = \frac{F_s}{d} = \frac{9.8}{0.05} = 196 \text{ N/m}$$

(b) The work done by the weight of the body to bring the spring-mass system from position 1 to 2 can be determined as:

$$\mathcal{W}_W = W\,d = m\,g\,d = (1)(9.8)(0.05) = 0.490 \text{ J}$$

(c) The force F_s acting on the block is acting in a direction opposite to that of the displacement. Therefore, the work done by the spring on the block is negative:

$$\mathcal{W}_s = -\frac{1}{2}\,k\,d^2 = -\frac{1}{2}(196)(0.05)^2 = -0.245 \text{ J}$$

(d) The net work $\mathcal{W}_{net}$ done on the block is:

$$\mathcal{W}_{net} = \mathcal{W}_W + \mathcal{W}_s = 0.490 - 0.245 = 0.245 \text{ J}$$

(e) The potential energy $\mathcal{E}_{Ps}$ stored in the spring at position 2:

$$\mathcal{E}_{Ps} = \mathcal{W}_s = \frac{1}{2}\,k\,d^2 = 0.245 \text{ J}$$

Note that the work done by the spring is stored as potential energy. It is the release of this energy that would bring the spring to its undeformed position 1 if the weight is detached and the spring is released.

11.8 Kinetic Energy

Kinetic energy is associated with motion. Every moving object has a kinetic energy. The kinetic energy, $\mathcal{E}_K$, of an object with mass m moving with a speed v is equal to the product of one half of the mass and the square of the speed of the object:

$$\mathcal{E}_K = \frac{1}{2}\, m\, v^2 \tag{11.15}$$

There is a relationship between the kinetic energy and the work done. The net work done, $\mathcal{W}_{12}$, by a constant force on an object to displace the object from position 1 to position 2 is equal to the change in kinetic energy, $\Delta\mathcal{E}_K$, of the object between positions 1 and 2. This is known as the *work-energy theorem* and can be expressed as:

$$\mathcal{W}_{12} = \Delta\mathcal{E}_K = \mathcal{E}_{K2} - \mathcal{E}_{K1} \tag{11.16}$$

11.9 Conservation of Energy

In general, forces may be divided into two categories as *conservative* and *nonconservative*. A force is conservative if the work done by that force to move an object between two points is independent of the path taken. Typical examples of conservative forces are the gravitational force and the restoring force in a spring. The frictional force, on the other hand, is a nonconservative force.

The work done on a system by a conservative force is converted into kinetic and/or potential energies. The sum of kinetic and potential energies is constant at any position of a system which is moving under the action of conservative forces. This is known as the *principle of conservation of mechanical energy*, and between any two positions 1 and 2 it can be stated as:

$$\mathcal{E}_{K1} + \mathcal{E}_{P1} = \mathcal{E}_{K2} + \mathcal{E}_{P2} \tag{11.17}$$

11.10 Applications of Energy Methods

The work-energy theorem stated by Eq. (11.16) and the principle of conservation of energy stated by Eq. (11.17) provide alternative methods of problem solving in dynamics. The work-energy theorem can be used to analyze problems involving nonconservative forces. On the other hand, the principle of conservation of energy is useful when the forces involved are conservative. As compared to the applications of the equations of motion, these methods are easier to apply and are particularly useful when the information provided or to be determined is in terms of velocities rather than accelerations. The following examples will illustrate some of the applications of these methods.

Example 11.6 As illustrated in Figure 11.15, a ball of mass m is dropped from a height h above the ground (free fall). Determine the speed of the ball as a function of y measured from the ground level.

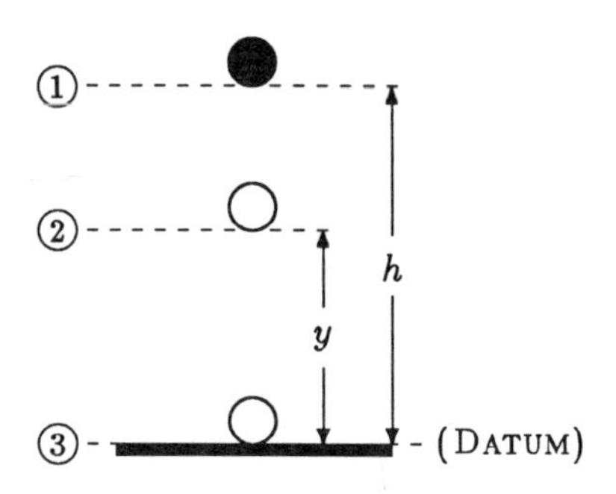

Figure 11.15 *Free fall.*

Solution: If the air resistance is neglected, the only force acting on the ball is the force of gravity which is a conservative force. The ball has zero speed and consequently zero kinetic energy at the instant of release (position 1). At this instant, the ball is at a height h above the ground and has a potential energy. At any other position 2 below position 1, which is at a height y above the ground, the ball has both kinetic and potential energies. Applying the principle of conservation of energy between 1 and 2:

$$\mathcal{E}_{K1} + \mathcal{E}_{P1} = \mathcal{E}_{K2} + \mathcal{E}_{P2}$$

$$0 + m\,g\,h = \frac{1}{2}\,m\,v^2 + m\,g\,y$$

Eliminating m and solving this equation for v will yield:

$$v = \sqrt{2\,g\,(h-y)}$$

This equation is valid for any y between y=0 and y=h. At position 1, $y=h$ and $v_1=0$. At position 3, $y=0$ and $v_3=\sqrt{2gh}$.

Note that as the ball descends the kinetic energy of the ball increases while its potential energy decreases.

Example 11.7 Consider the pendulum with mass m and length ℓ shown in Figure 11.16. Assume that the pendulum is released from position 1 which makes an angle θ with the vertical. Determine the speed v_2 of the mass when it is at position 2.

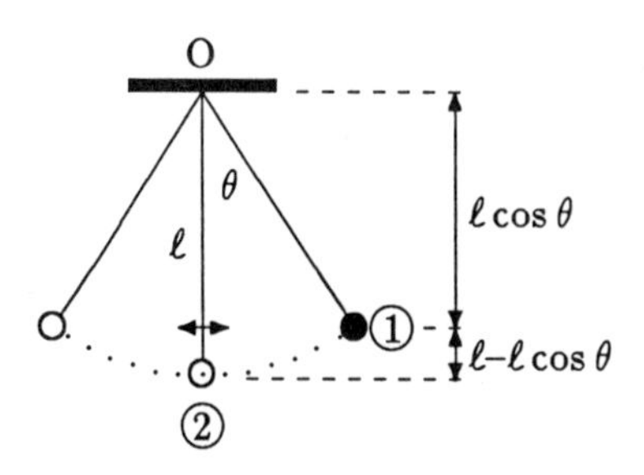

Figure 11.16 *The pendulum.*

Solution: The mass has zero velocity and no kinetic energy at the instant of release (position 1). If we choose position 2 to be the datum from which we measure heights, then the mass has a height $h_1=\ell(1-\cos\theta)$ at the instant of release, and zero height when it is at position 2. Applying the principle of conservation of energy between 1 and 2:

$$\mathcal{E}_{K1} + \mathcal{E}_{P1} = \mathcal{E}_{K2} + \mathcal{E}_{P2}$$

$$0 + m\,g\,h_1 = \frac{1}{2}\,m\,{v_2}^2 + 0$$

Solving this equation for v_2 will yield:

$$v_2 = \sqrt{2\,g\,\ell\,(1-\cos\theta)}$$

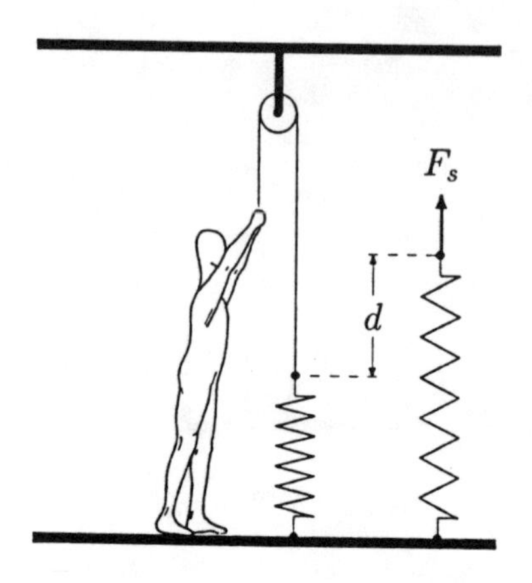

Figure 11.17 *A person stretches the spring by pulling the cable.*

Example 11.8 The simple device shown in Figure 11.17 can be used to measure force output of the arm muscles. The device consists of a linear spring fixed to the ground at one end and connected to a cable at the other end. The cable is wrapped around a pulley and has a handle.

Consider a person who stretches the spring by an amount of d=30 cm by pulling the handle down. Determine the force exerted by the person, and the potential energy stored in the spring if the spring constant is k=600 N/m.

Solution: The person pulls down the handle and stretches the spring by an amount d=0.3 m. Therefore, the force applied on the spring is:

$$F_s = k\, d = (600)(0.3) = 180 \text{ N}$$

This force is applied by the person on the spring through the cable. The potential energy stored in the spring is:

$$\mathcal{E}_P = \frac{1}{2}\, k\, d^2 = \frac{1}{2}\,(600)(0.3)^2 = 27 \text{ J}$$

Notice that the work done by the person is completely converted into elastic potential energy because the frictional effects (for example at the pulley) are not taken into consideration.

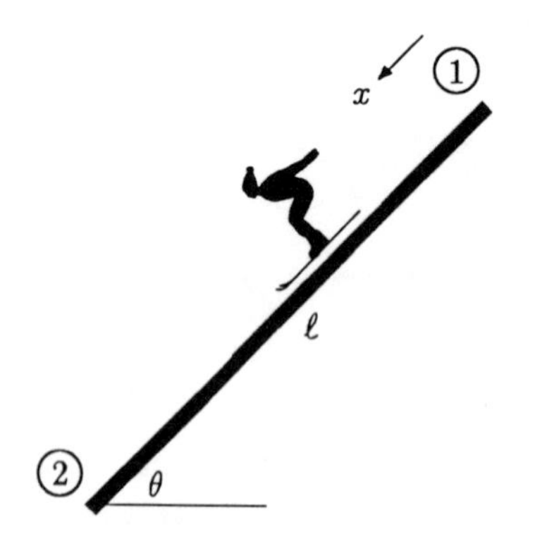

Figure 11.18 *A skier on a ski-jumping track.*

Example 11.9 As illustrated in Figure 11.18, consider a ski-jumper moving down a straight track to acquire sufficient speed to accomplish the ski-jumping task. The length of the track is ℓ=25 m, and the track makes an angle θ=45° with the horizontal.

If the skier starts at the top of the track with zero initial speed, determine the takeoff speed of the skier at the bottom of the track using (a) the work-energy theorem, (b) the conservation of energy principle, and (c) the equation of motion along with the kinematic relationships. Assume that the effects of friction and air resistance are negligible.

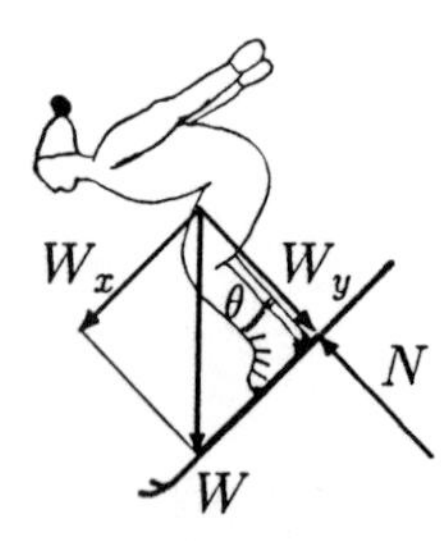

Figure 11.19 *Free-body diagram of the ski jumper.*

Solution (a): Work-energy method

The free-body diagram of the ski jumper is shown in Figure 11.19. The only force acting on the ski jumper is that of gravity. The motion of the skier is due solely to the component of this force in the x direction (the direction parallel to the track).

Labeling the top and the bottom of the track as points 1 and 2, the work done by $W_x = W \sin\theta = mg \sin\theta$ as the skier moves from 1 to 2 which are ℓ distance apart is:

$$\mathcal{W}_{12} = W_x\, \ell = m\, g\, \ell\, \sin\theta \qquad (i)$$

According to the work-energy theorem, $\mathcal{W}_{12}$ must be equal to the change in kinetic energy of the skier between 1 and 2. Therefore:

$$\mathcal{W}_{12} = \mathcal{E}_{K2} - \mathcal{E}_{K1} = \frac{1}{2}\, m\, v_2^{\,2} - \frac{1}{2}\, m\, v_1^{\,2} \qquad (ii)$$

The second term on the right hand side of Eq. (ii) is zero because the initial speed of the skier is $v_1{=}0$. Substituting Eq. (i) into Eq. (ii), eliminating the repeated parameter m (the mass of the skier), and solving Eq. (ii) for the takeoff speed, v_2, of the skier will yield:

$$v_2 = \sqrt{2\, g\, \ell\, \sin\theta} \qquad (iii)$$

Solution (b): Conservation of energy method

Since the effects of nonconservative forces due to friction and air resistance are negligible, this problem can also be analyzed by utilizing the principle of conservation of energy:

$$\mathcal{E}_{K1} + \mathcal{E}_{P1} = \mathcal{E}_{K2} + \mathcal{E}_{P2}$$

$$\frac{1}{2}\, m\, v_1^{\,2} + m\, g\, h_1 = \frac{1}{2}\, m\, v_2^{\,2} + m\, g\, h_2 \qquad (iv)$$

In Eq. (iv), the first term on the left-hand side is zero since $v_1{=}0$. If we measure heights relative to the bottom of the track (or selecting 2 to be the datum as shown in Figure 11.20), then height $h_2{=}0$ and the height of the top of the track is $h_1{=}\ell \sin\theta$. Therefore, the second term on the right-hand side of Eq. (iv) is zero as well. Substituting $h_1{=}\ell \sin\theta$ into Eq. (iv), eliminating the repeated parameter m, and solving Eq. (iv) for v_2 will again yield Eq. (iii).

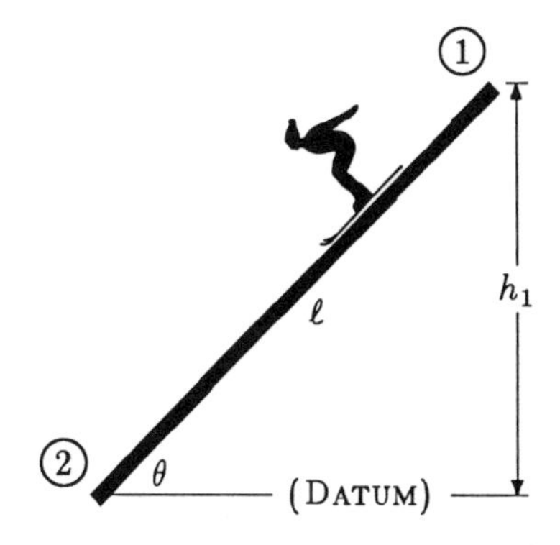

Figure 11.20 *A skier on a ski jumping track.*

Solution (c): Using the equation of motion

The equation of motion in the direction of motion (x) is:

$$\sum F_x = m\, a_x : \qquad W_x = m\, a_x \qquad (v)$$

Substituting $W_x{=}mg \sin\theta$ into Eq. (v), eliminating m, and solving Eq. (v) for the acceleration of the skier in the x direction will yield:

$$a_x = g\, \sin\theta \qquad (vi)$$

Since the acceleration of the skier is due to gravity only, what we have is a one-dimensional motion with constant acceleration. By definition, acceleration is the time rate of change of velocity, or velocity is the integral of acceleration with respect to time. Since a_x is constant and the initial velocity of the skier is zero, we can write:

$$v_x = a_x\, t \qquad (vii)$$

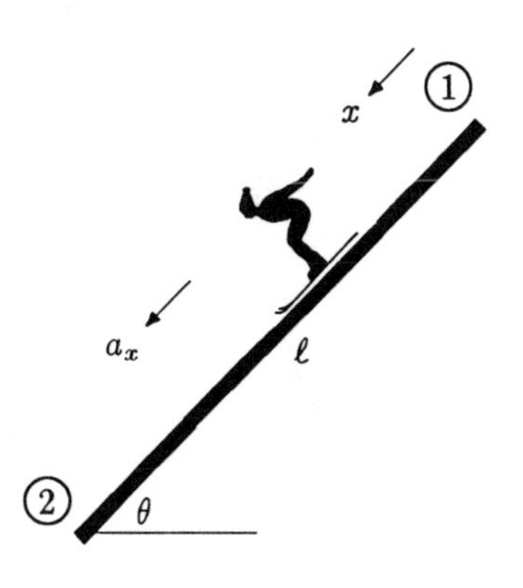

Figure 11.21 *x is the direction of motion.*

The kinematic relationship between the velocity and displacement is such that displacement is equal to the integral of velocity. If we measure the displacement relative to the position of the skier at the start, then the initial displacement is zero. Therefore, the equation relating displacement, acceleration, and time reduces to:

$$x = \frac{1}{2}\, a_x\, t^2 \qquad (viii)$$

Eq. (vii) can be solved for time $t{=}v_x/a_x$, which can then be substituted into Eq. $(viii)$ so as to eliminate t. This will yield:

$$x = \frac{1}{2}\frac{{v_x}^2}{a_x}$$

Solving this equation for v_x will give:

$$v_x = \sqrt{2\, x\, a_x} \qquad (ix)$$

This is a general solution relating the acceleration, velocity, and displacement of the skier when the skier is anywhere along the track. $x{=}\ell$ and $v_x{=}v_2$ when the skier reaches the bottom of the track, and the acceleration of the skier is always $a_x{=}g\sin\theta$. Substituting these into Eq. (ix) will again yield Eq. (iii).

Finally, if we substitute the numerical values of $g{=}9.8$ m/s^2, $\ell{=}25$ m, and $\theta{=}45°$ into Eq. (ii) and carry out the calculations, we get:

$$v_2 = 18.6 \text{ m/s}$$

Remarks:

- It is clear that for problems involving displacement, speed, and force, applications of the methods based on the work-energy theorem and the conservation of energy principle are more straightforward as compared to the application of equations of motion. In general, one should try work-energy or conservation of energy methods first before resorting to the equations of motion.

- Since the effects of nonconservative forces due to friction and air resistance are neglected, the solution of the problem is independent of the shape of the track or how the skier covers the distance between the top and bottom of the track. The most important parameter in this problem affecting the takeoff speed of the skier is the total vertical distance between 1 and 2. This implies that the problem could be simplified by noting that the skier undergoes a "free fall" between 1 and 2 which are $h_1{=}\ell\sin\theta$ distance apart. This is illustrated in Figure 11.22. Applying the principle of conservation of energy between 1 and 2 will again yield Eq. (iii).

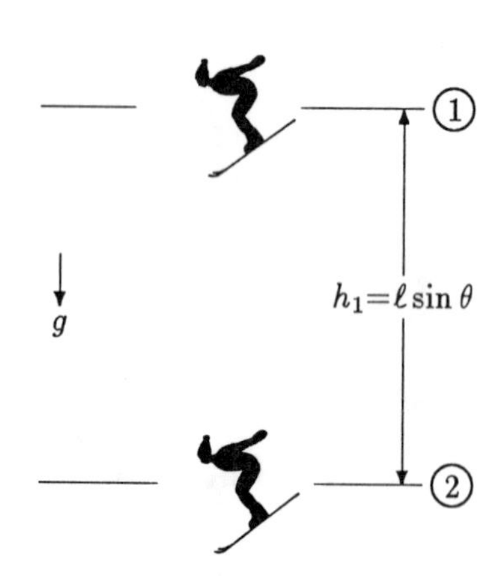

Figure 11.22 *The solution of the problem is independent of the path of motion.*

Example 11.10 As illustrated in Figure 11.23, consider a 70 kilogram gymnast doing giant circles. The motion of the gymnast may be analyzed by assuming that the total mass of the gymnast is concentrated at the center of gravity of the gymnast, and by modeling the gymnast as a particle undergoing a rotational motion about a fixed axis.

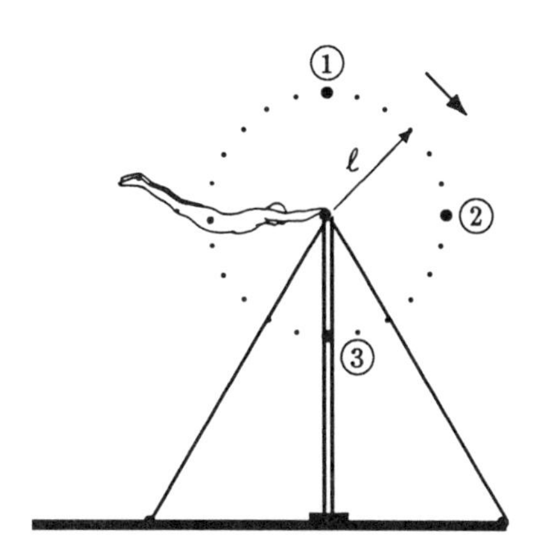

Figure 11.23 *A gymnast doing giant circles.*

Assuming that the center of gravity of the gymnast is located at a distance ℓ=1 m from the high bar and the speed of the gymnast is negligibly small at position 1, determine the speeds of the gymnast's center of gravity at positions 2 and 3 shown in Figure 11.23. Calculate the forces exerted on the arms of the gymnast at positions 1, 2, and 3.

Solution: We can utilize the conservation of energy method to determine the speeds of the gymnast's center of gravity. Figure 11.24 shows the heights of the gymnast's center of gravity at positions 1, 2 and 3 measured relative to position 3 (i.e., position 3 is chosen to be the datum). At position 1, $h_1=2\ell$ and $v_1=0$. $h_2=\ell$ is the height of the gymnast's center of gravity when it is at position 2, and $h_3=0$ since position 3 is the datum.

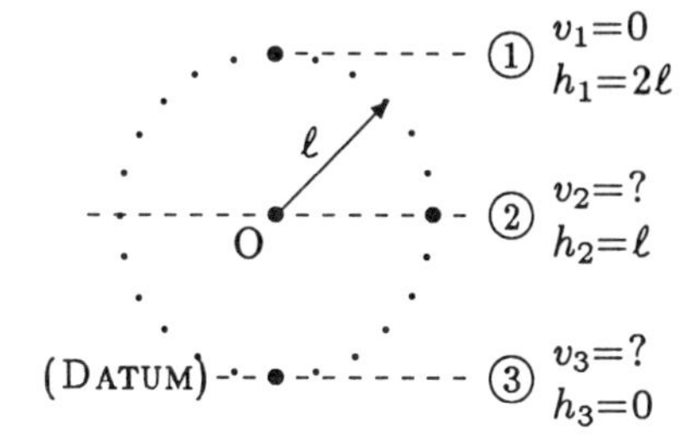

Figure 11.24 *Relative heights of the gymnast's center of gravity.*

Applying the conservation of energy principle between 1 and 2:

$$\mathcal{E}_{K1} + \mathcal{E}_{P1} = \mathcal{E}_{K2} + \mathcal{E}_{P2}$$

$$\frac{1}{2}\, m\, {v_1}^2 + m\, g\, h_1 = \frac{1}{2}\, m\, {v_2}^2 + m\, g\, h_2$$

Substituting $v_1=0$, $h_1=2\ell$, and $h_2=\ell$ into the above equation, eliminating the repeated parameter m (mass of the gymnast), and solving this equation for v_2 will yield:

$$v_2 = \sqrt{2\, g\, \ell} \qquad (i)$$

Similarly, applying the conservation of energy principle between 1 and 3:

$$\frac{1}{2}\, m\, {v_1}^2 + m\, g\, h_1 = \frac{1}{2}\, m\, {v_3}^2 + m\, g\, h_3$$

Substituting $v_1=0$, $h_1=2\ell$, and $h_3=0$ into the above equation, eliminating m, and solving it for v_3 will yield:

$$v_3 = \sqrt{4\, g\, \ell} \qquad (ii)$$

By substituting the numerical values of ℓ=1 m and g=9.8 m/s^2 into Eqs. (i) and (ii), and carrying out the calculations:

$$v_2 = 4.43 \text{ m/s}$$
$$v_3 = 6.26 \text{ m/s}$$

To determine the forces applied on the arms of the gymnast, consider the free-body diagrams of the gymnast's center of gravity (Figure 11.25). At position 1, the gymnast is momentarily in static equilibrium. Therefore, the vector sum of all forces acting in the vertical direction must be equal to zero. Since the weight with magnitude $W = mg$ is always acting downward, the reaction force $\underline{T}_1$ exerted by the high bar on the arms of the gymnast must be applied upward:

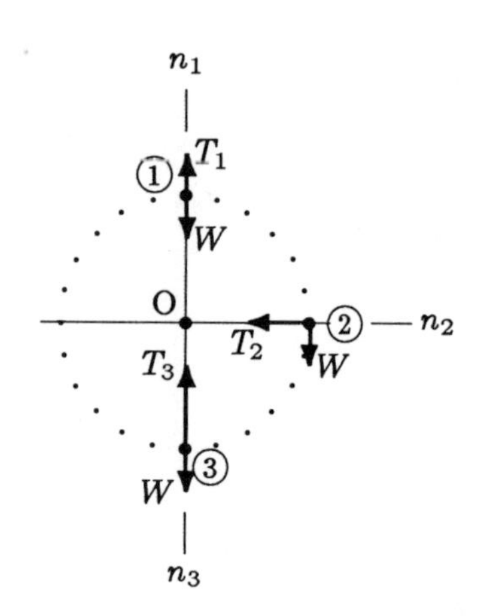

Figure 11.25 *Forces acting on the gymnast's center of gravity.*

$$T_1 = W = m\,g = (70)(9.8) = 686 \text{ N} \quad (\uparrow)$$

The gymnast's center of gravity undergoes a circular motion with radius ℓ. An object undergoing a circular motion always has an acceleration in the radial direction, called the centripetal acceleration. The magnitude of the centripetal acceleration can be determined from $a_n = v^2/\ell$ where v is the magnitude of the linear velocity of the object and ℓ is the radius of the circular path. Therefore, the centripetal acceleration of the center of gravity of the gymnast at positions 2 and 3 are:

$$a_{n2} = \frac{{v_2}^2}{\ell} = \frac{(4.43)^2}{1} = 19.60 \text{ m/s}^2$$

$$a_{n3} = \frac{{v_3}^2}{\ell} = \frac{(6.26)^2}{1} = 39.20 \text{ m/s}^2$$

When the gymnast is at position 2, the only force acting in the normal (radial) direction is the force $\underline{T}_2$ exerted by the high bar on the arms of the gymnast. The equation of motion in the radial direction is:

$$\sum F_n = m\,a_{n2}: \qquad T_2 = m\,a_{n2} \qquad (iii)$$

When the gymnast is at position 3, both the weight W of the gymnast and the force $\underline{T}_3$ exerted by the high bar on the arms are applied in the normal direction. Therefore, the equation of motion in the normal direction is:

$$\sum F_n = m\,a_{n3}: \qquad T_3 - W = m\,a_{n3}$$

$$T_3 = m\,(g + a_{n3}) \qquad (iv)$$

Substituting the numerical values of m=70 kg, a_{n2}=19.6 m/s^2, a_{n3}=39.2 m/s^2, and g=9.8 m/s^2 into Eqs. (*iii*) and (*iv*):

$$T_2 = (70)(19.60) = 1372 \text{ N} \quad (\leftarrow)$$

$$T_3 = (70)(9.8) + (70)(39.20) = 3430 \text{ N} \quad (\uparrow)$$

Notice that the forces T_1, T_2, and T_3 are shared by both arms.

11.11 Power

Power, $\mathcal{P}$, is defined as the time rate of work done:

$$\mathcal{P} = \frac{d\mathcal{W}}{dt} \tag{11.18}$$

The work done by a constant force on an object can be determined by considering the dot product of the force and displacement vectors ($\mathcal{W}=\underline{F} \cdot \underline{s}$). Therefore:

$$\mathcal{P} = \frac{d}{dt}(\underline{F} \cdot \underline{s})$$

If the force vector $\underline{F}$ is constant, then:

$$\mathcal{P} = \underline{F} \cdot \frac{d\underline{s}}{dt} = \underline{F} \cdot \underline{v} \tag{11.19}$$

In Eq. (11.19), $\underline{v}$ is the velocity vector of the object. If the applied force is collinear with the velocity, then $\mathcal{P}=Fv$. Power is a scalar quantity, and has the dimension of force times velocity ($\mathrm{ML}^2/\mathrm{T}^3$). The units of power are given in Table 11.2.

SYSTEM	UNITS OF POWER	SPECIAL NAME
SI	N-m/s=J/s	Watt (W)
c-g-s	dyn-cm/s=erg/s	
British	lb-ft/s	horsepower (hp)

Table 11.2 *Units of power.* (1 hp=550 lb-ft/s=746 W)

Example 11.11 Consider a cyclist who can produce a power of 0.25 hp. At a θ=20° ramp (Figure 11.26), the cyclist moves with a constant speed of v=10 km/h.

Neglecting frictional effects, determine the force the cyclist is producing while traveling up the ramp if the total weight of the cyclist and the bicycle is W=800 N.

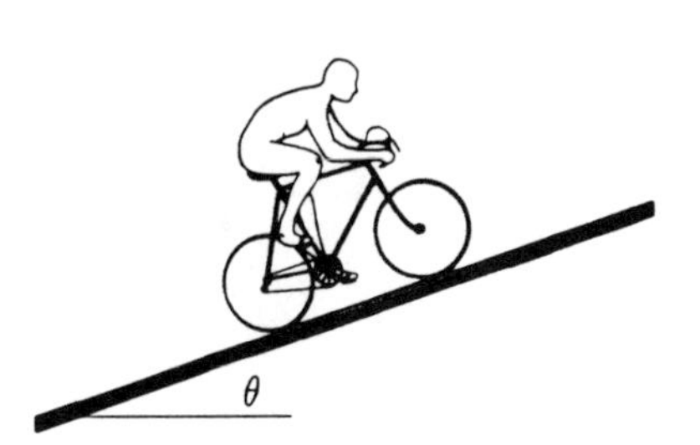

Figure 11.26 *A cyclist.*

Solution: The forces acting on the cyclist in the direction of motion (parallel to the incline) are the component $W \sin\theta$ of the weight of the rider and the bicycle, and the force F that the cyclist is supplying. The net force in the direction of motion is $F_{net}=F - W\sin\theta$, which is converted into positive power:

$$F_{net} = \frac{\mathcal{P}}{v} = \frac{(0.25 \text{ hp})}{(10 \text{ km/h})} = \frac{(0.25 \times 746 \text{ W})}{(10 \times 1000/3600 \text{ m/s})} = 67.1 \text{ N}$$

The force F generated by the cyclist is:

$$F = F_{net} + W \sin\theta = 67.4 + (800)(\sin 20°) = 341 \text{ N}$$

11.12 Rotational Kinetic Energy

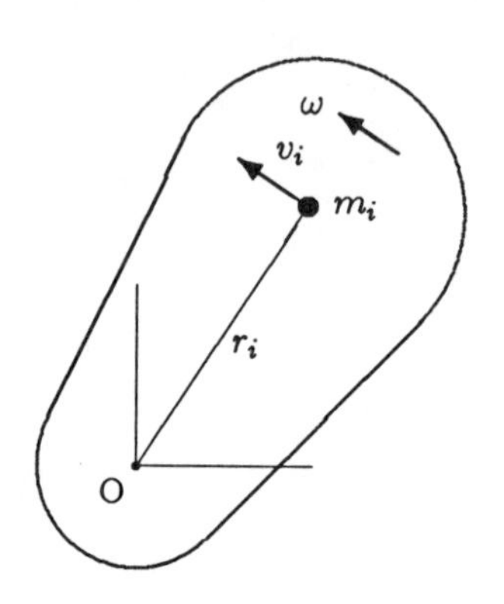

Figure 11.27 *Rotational motion of a body about a fixed axis.*

Assume that the rigid body shown in Figure 11.27 is composed of many small particles and that the body rotates about a fixed axis with an angular velocity ω. If m_i and v_i are the mass and the speed of the ith particle in the body, respectively, then the kinetic energy of the particle is:

$$\mathcal{E}_{Ki} = \frac{1}{2}\, m_i\, v_i^{\,2}$$

At any instant, every particle in the body has the same angular velocity ω, but the linear velocity of each particle depends on its distance measured from the axis of rotation. If r_i is the perpendicular distance between the ith particle and the axis of rotation (i.e., radius of the circular motion path of the ith particle), then $v_i = r_i \omega$ and its kinetic energy is $\mathcal{E}_{Ki} = \frac{1}{2} m_i r_i^{\,2} \omega^2$. Each particle in the body has a kinetic energy, and the total kinetic energy, $\mathcal{E}_K$, of the rotating body is the sum of the kinetic energies of individual particles in the body. That is, $\mathcal{E}_K = \sum \mathcal{E}_{Ki} = \frac{1}{2}(\sum m_i r_i^{\,2})\omega^2$. The quantity in parentheses is the mass moment of inertia, I, of the body. Therefore:

$$\mathcal{E}_K = \frac{1}{2}\, I\, \omega^2 \qquad (11.20)$$

Equation (11.20) defines the rotational kinetic energy of a body in terms of the mass moment of inertia and angular velocity of the body, and it is analogous to the kinetic energy $\mathcal{E}_K = \frac{1}{2} m v^2$ associated with linear motion.

11.13 Angular Work and Power

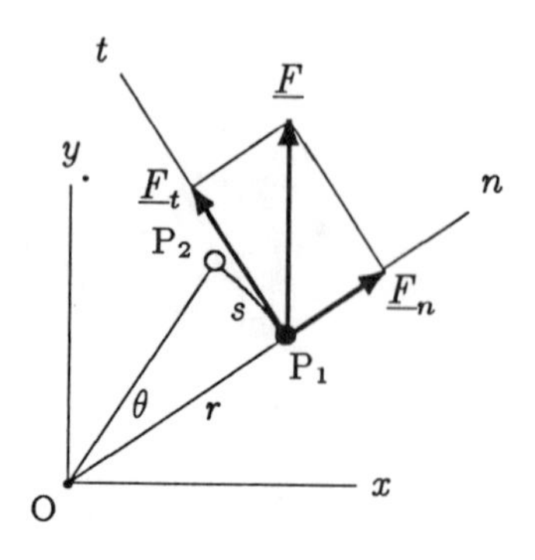

Figure 11.28 *A particle located at P_1 is displaced by an angle θ or arc length s to position P_2.*

By definition, the work done by a force is equal to the force times the corresponding displacement. The *angular work done* by a force applied on a rotating body is related to the angular displacement of the body. Consider a body rotating about a fixed axis at O due to an applied force $\underline{F}$. As illustrated in Figure 11.28, let P_1 and P_2 represent the positions of a point in the body at times t_1 and t_2, respectively. In the time interval between t_1 and t_2, the body rotates through an arc length s or angle θ. The work done by $\underline{F}$ on the body is equal to the magnitude of the component of the force vector in the direction of motion (tangential component, F_t), times the displacement s:

$$\mathcal{W} = F_t\, s$$

The arc length is related to the angular displacement through the radius of the circular path of motion as $s = r\theta$. Therefore:

$$\mathcal{W} = F_t\, r\, \theta$$

By definition, $F_t r$ is the magnitude M of the torque generated by force $\underline{F}$ about O. Hence:

$$\mathcal{W} = M\,\theta \tag{11.21}$$

In other words, the work done by a rotation-producing force is equal to the torque generated by the force times the angular displacement of the body. Notice that the normal (radial) component of the force vector does no work on a body undergoing rotational motion because there is no motion in the normal direction.

It must be pointed out here that the relationship between angular work done, torque and angular displacement given in Eq. (11.21) is valid when the torque is constant. The work done by a torque, which is a function of angular displacement, on a body to rotate the body from position 1 to 2 is:

$$\mathcal{W} = \int_{\theta_1}^{\theta_2} M\, d\theta \tag{11.22}$$

Here, θ_1 and θ_2 are the angular displacements of the body at positions 1 and 2, respectively. Equation (11.22) can also be written in terms of the change in angular velocity by noting that $M{=}I\alpha$ and $\alpha{=}d\omega/dt$. Using the chain rule of differentiation:

$$M = I\,\alpha = I\,\frac{d\omega}{dt} = I\,\frac{d\omega}{d\theta}\,\frac{d\theta}{dt} = I\,\omega\,\frac{d\omega}{d\theta}$$

Substituting this into Eq. (11.22):

$$\mathcal{W} = \int_{\omega_1}^{\omega_2} I\,\omega\, d\omega = \frac{1}{2}\,I\,{\omega_2}^2 - \frac{1}{2}\,I\,{\omega_1}^2 \tag{11.23}$$

In Eq. (11.23), ω_1 and ω_2 are the angular velocities of the body at positions 1 and 2, respectively. Equation (11.23), known as the *work-energy theorem in rotational motion*, states that the net angular work done on a rigid body in rotating the body about a fixed axis is equal to the change in the body's rotational kinetic energy.

The rate at which work is done is known as power. The *angular power* describes the rate at which angular work is done. For a constant torque:

$$\mathcal{P} = \frac{d\mathcal{W}}{dt} = M\,\frac{d\theta}{dt} = M\,\omega \tag{11.24}$$

That is, the angular power is equal to the product of the applied torque and the angular velocity of the body.

Example 11.12 Consider the knee extension problem analyzed in Example 10.6. As illustrated in Figure 11.29, the person is seated on a table. The upper body is strapped to a back rest and the right thigh is strapped firmly on the table with the lower leg hanging vertically downward. The person is then asked to extend the right lower leg. The angular displacement of the lower leg during knee extension is determined via a goniometer attached to the leg. After a series of computations, it is determined that the lower leg was extended from θ=0° to 90° in a time period of 0.5 seconds with an average angular velocity of 3 rad/s by producing an average extensor muscle torque of 90 N-m.

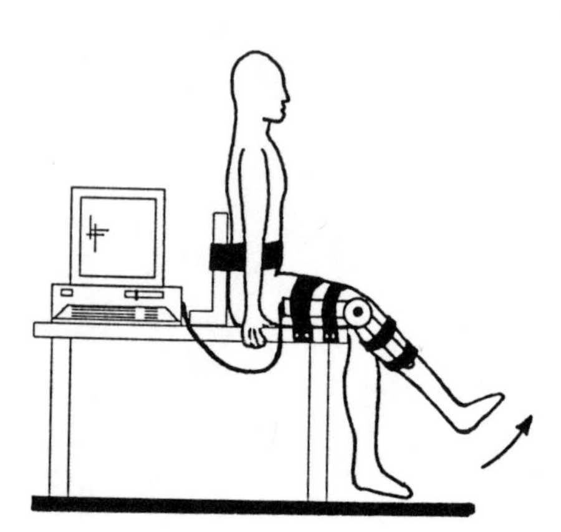

Figure 11.29 *Knee extension.*

Assuming that the mass moment of inertia of the lower leg about the center of rotation of the knee joint is 92 kg-m^2, calculate the average angular kinetic energy produced, angular work done, and angular power generated by the knee extensor muscles to extend the lower leg from θ=0° to 90°.

Solution: The range of motion of the lower leg is $\Delta\theta$=90°, which is covered in a time period of Δt=0.5 s (Figure 11.30). The mass moment of inertia of the lower leg about the knee joint is given as I_o=92 kg-m^2. The average angular velocity of the lower leg is calculated to be $\overline{\omega}$=3 rad/s and the average torque produced by the knee extensors is $\overline{M}$=90 N-m. Therefore, the average angular kinetic energy produced by the knee extensor muscles is:

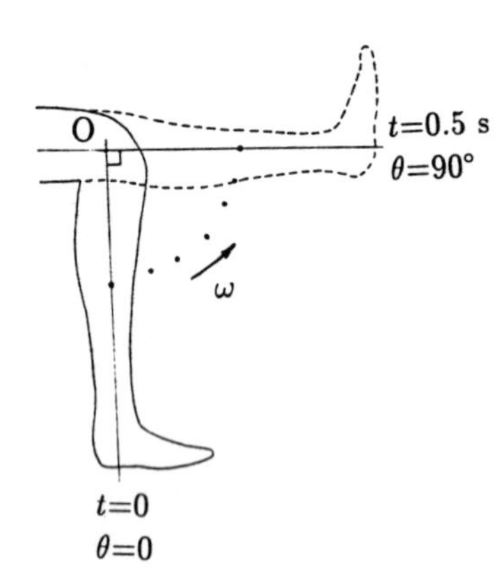

Figure 11.30 *Movement of the lower leg.*

$$\overline{\mathcal{E}_K} = \frac{1}{2}\, I_o\, \overline{\omega}^2 = \frac{1}{2}\,(92)(3)^2 = 414 \text{ J}$$

The average work done by the muscles to extend the lower leg at an angle of 90° or 90×π/180=1.57 rad is:

$$\overline{\mathcal{W}} = \overline{M}\,\Delta\theta = (92)(1.57) = 144 \text{ J}$$

The average power generated by the extensors is:

$$\overline{\mathcal{P}} = \overline{M}\,\overline{\omega} = (92)(3)^2 = 828 \text{ W}$$

Chapter 12

MOMENTUM METHODS

12.1 Introduction

In Chapter 10, Newton's second law of motion is presented in the form of "equations of motion." In Chapter 11, the concepts of work and energy are introduced. Based on the same law, "work-energy" and "conservation of energy" methods are devised to facilitate the solutions of specific problems in kinetics. In this chapter, the concepts of linear momentum and impulse will be defined. Newton's second law of motion will be reformulated to introduce other methods for kinetic analyses based on the "impulse-momentum theorem" and the principle of "conservation of linear momentum." These methods will then be applied to analyze the impact and collision of bodies.

12.2 Linear Momentum and Impulse

Consider an object with mass m acted upon by an external force $\underline{F}$. Let $\underline{a}$ be the acceleration of the object under the action of the applied force. The relationship between $\underline{F}$, m, and $\underline{a}$ is described by the equation of motion in the following vector form:

$$\underline{F} = m\,\underline{a} \tag{12.1}$$

By definition, acceleration is the time rate of change of velocity. If the mass of the object is constant, then Eq. (12.1) can be written as:

$$\underline{F} = \frac{d}{dt}(m\,\underline{v}) \tag{12.2}$$

The vector $m\underline{v}$ is called the *linear momentum* (or simply the momentum) of the object, and is denoted by $\underline{p}$:

$$\underline{p} = m\,\underline{v} \tag{12.3}$$

Momentum is a vector quantity. The line of action and direction of the momentum vector is the same as the velocity vector of the object. The magnitude of the momentum is equal to the product of the mass and the speed of the object.

The equation of motion can now be expressed in terms of momentum by substituting Eq. (12.3) into Eq. (12.2):

$$\underline{F} = \frac{d\underline{p}}{dt} \tag{12.4}$$

If there is more than one force acting on the object, then $\underline{F}$ in Eq. (12.4) must be replaced by the vector sum of all forces (the resultant force) acting on the object. Equation (12.4) states that the time rate of change of momentum of an object is equal to the resultant force on the object.

The concept of momentum is particularly useful for analyzing the effects of forces applied in very short time intervals. Such forces

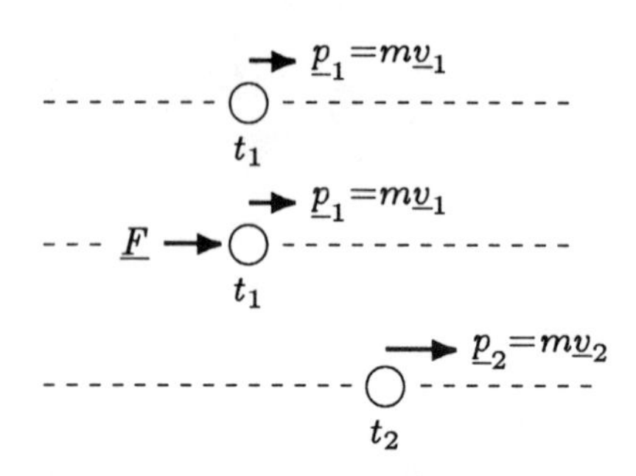

Figure 12.1 *Impulsive force $\underline{F}$ changes the momentum of the object from $\underline{p}_1$ to $\underline{p}_2$.*

are called ***impulsive forces***, and the motions associated with them are called ***impulsive motions***. Consider an object moving with velocity $\underline{v}_1$ at time t_1 (Figure 12.1). Assume that a force $\underline{F}$ is applied on the object in the time interval between t_1 and t_2, and the velocity of the object is changed to $\underline{v}_2$ at time t_2. Multiplying Eq. (12.4) by dt and integrating it between t_1 and t_2 will yield:

$$\int_{t_1}^{t_2} \underline{F}\, dt = \underline{p}_2 - \underline{p}_1 = m\, \underline{v}_2 - m\, \underline{v}_1 \tag{12.5}$$

The integral on the left-hand side of Eq. (12.5) represents the *linear impulse* of force $\underline{F}$ on the object in the time interval $\Delta t = t_2 - t_1$. The right-hand side of Eq. (12.5) is equal to the change in momentum $\Delta \underline{p} = \underline{p}_2 - \underline{p}_1$ of the object in the time interval Δt. Therefore:

$$\int_{t_1}^{t_2} \underline{F}\, dt = \Delta \underline{p} \tag{12.6}$$

This equation is the mathematical representation of the *impulse-momentum theorem*.

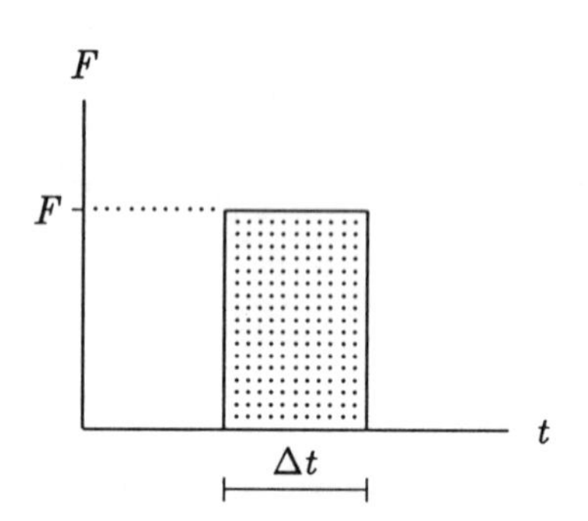

Figure 12.2 *Impulse is the area under the force versus time curve.*

Impulse is a vector quantity. It has the same line of action and direction as the impulsive force. In general, the magnitude and direction of the impulsive force may vary in the time interval Δt. If the force is known as a function of time, the impulse can be determined by integrating $\underline{F}$ with respect to time which will essentially yield the area under the force versus time curve. If the impulsive force is constant (Figure 12.2), then the impulse is simply:

$$\underline{F}\, \Delta t = \Delta \underline{p} \qquad (\underline{F} \text{ constant}) \tag{12.7}$$

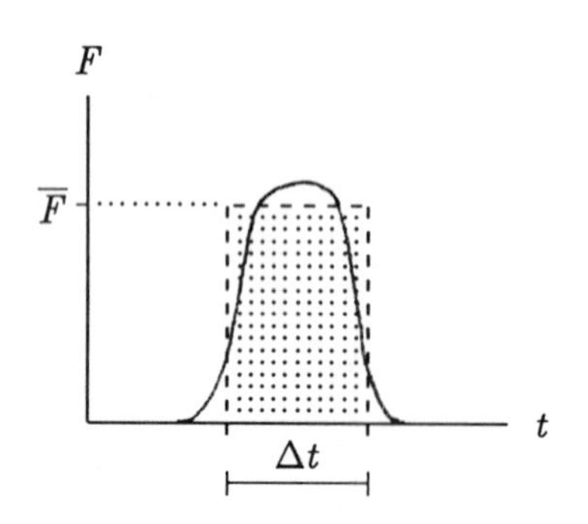

Figure 12.3 *Impulse of a varying force can be determined by considering an average force and an average time.*

Equation (12.7) can also be used to determine the impulse of a varying force by approximating the force with an average force value (Figure 12.3).

In the equations derived so far, all of the parameters involved (except for the mass m) are vector quantities. These parameters can be represented in terms of their rectangular components along the Cartesian coordinate directions x, y, and z. This approach will yield three independent scalar equations valid along the x, y, and z directions. For example, Eq. (12.5) will yield:

$$\begin{aligned} \int_{t_1}^{t_2} F_x\, dt &= \Delta p_x = p_{x2} - p_{x1} = m\, v_{x2} - m\, v_{x1} \\ \int_{t_1}^{t_2} F_y\, dt &= \Delta p_y = p_{y2} - p_{y1} = m\, v_{y2} - m\, v_{y1} \\ \int_{t_1}^{t_2} F_z\, dt &= \Delta p_z = p_{z2} - p_{z1} = m\, v_{z2} - m\, v_{z1} \end{aligned} \tag{12.8}$$

For one-dimensional problems, any one of the above equations is sufficient to analyze the problem. For plane problems, two equations may be needed. While using Eqs. (12.8), it should be kept in mind that force, velocity, momentum, and impulse are vector quantities. For example, consider an object moving in the positive x direction. Assume that between times t_1 and t_2, an impulsive force $\underline{F}_x$ is applied on the object in the direction of motion (Figure 12.4a). The momentum of the object will become:

$$p_{x2} = p_{x1} + \int_{t_1}^{t_2} F_x \, dt$$

That is, the final momentum is equal to the initial momentum plus the impulse. If the force is applied in the direction opposite to that of the motion of the object (Figure 12.4b), then the final momentum of the object is equal to the initial momentum minus the impulse:

$$p_{x2} = p_{x1} - \int_{t_1}^{t_2} F_x \, dt$$

If the motion is in one direction and the impulsive force is applied in a different direction (Figure 12.4c), then it is easier to treat impulse and momentum as vector quantities.

Impulse and momentum have the same dimension (ML/T) and units. Their units in different systems are listed in Table 12.1.

(a)

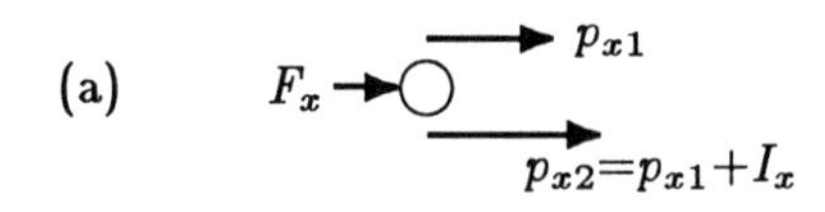

(b)

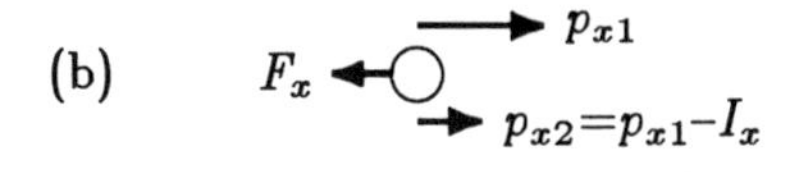

(c)

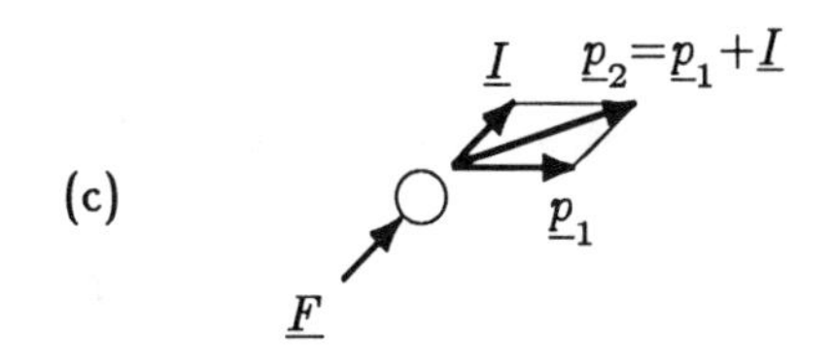

Figure 12.4 *Scalar and vector additions of impulse and momentum.*

SYSTEM	UNITS OF IMPULSE AND MOMENTUM
SI	kg-m/s=N-s
c-g-s	g-cm/s=dyn-s
British	slug-ft/s=lb-s

Table 12.1 *Units of impulse and momentum.*

12.3 Applications of the Impulse-Momentum Method

The concepts of impulse and momentum have numerous applications. These concepts may facilitate the analysis of situations where impact and impulsive forces play a significant role. The impulse-momentum method can be utilized to determine the forces involved in various athletic, sports, and daily activities. This method enables us to investigate the forces exerted by the foot on a football, or by the foot or the forehead on a soccer ball, which is in fact the same force exerted by the ball on the foot or the forehead. Forces involved while walking and running and the forces applied on various joints of the human body which are transmitted through the feet during landing after a long jump or after a routine in gymnastics, may also be analyzed with the same technique. Other applications of the method include the analyses of forces exerted by a club on a golf ball, a bat on a baseball, and a racket on a tennis ball.

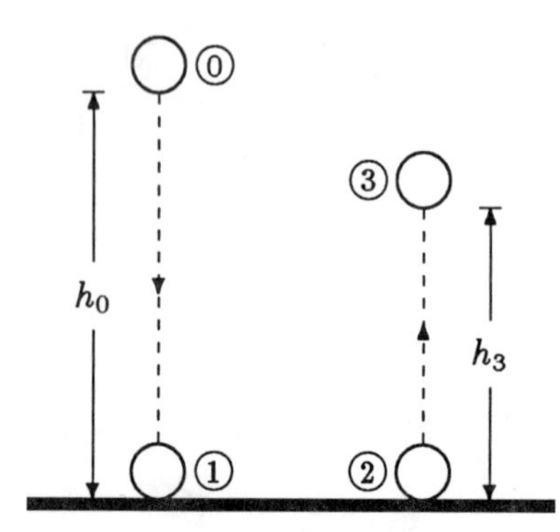

Figure 12.5 *The ball hits the floor and bounces.*

Example 12.1 As illustrated in Figure 12.5, consider a ball of mass m=0.25 kg dropped from a height h_0=1 m. The ball hits the floor, bounces, and reaches a height h_3=0.75 m.

Determine the momentum of the ball immediately before and after it collides with the floor. Assuming that the duration of collision (duration of contact) is Δt=0.01 s, determine an average force exerted by the floor on the ball during collision.

Solution: Let 0, 1, 2, and 3 correspond to each stage of the action: release, impact, rebound, and reaching the height 0.75 meter for the ball. The speed v_1 of the ball at the instant of impact can be determined by utilizing the conservation of energy principle between 0 and 1. Since v_0=0 and h_1=0:

$$m\,g\,h_0 = \frac{1}{2}\,m\,{v_1}^2$$

Solving this equation for v_1 will yield:

$$v_1 = \sqrt{2\,g\,h_0} = \sqrt{2(9.8)(1)} = 4.43 \text{ m/s}$$

Similarly, the speed v_2 of the ball immediately after collision can be determined by considering the motion of the ball between 2 and 3. Since h_2=0 and v_3=0:

$$\frac{1}{2}\,m\,{v_2}^2 = m\,g\,h_3$$

Solving this equation for v_2 will yield:

$$v_2 = \sqrt{2\,g\,h_3} = \sqrt{2(9.8)(0.75)} = 3.83 \text{ m/s}$$

Momenta of the ball immediately before and after collision are:

$$p_1 = m\,v_1 = (0.25)(4.43) = 1.11 \text{ kg-m/s} \quad (\downarrow)$$
$$p_2 = m\,v_2 = (0.25)(3.83) = 0.96 \text{ kg-m/s} \quad (\uparrow)$$

In vector form:

$$\underline{p}_1 = -\,1.11\,\underline{j} \ \ (\text{kg-m/s})$$
$$\underline{p}_2 = 0.96\,\underline{j} \ \ (\text{kg-m/s})$$

The change in momentum of the ball during the course of collision is:

$$\Delta\underline{p} = \underline{p}_2 - \underline{p}_1 = (0.96\,\underline{j}) - (-1.11\,\underline{j}) = 2.07\,\underline{j} \ \ (\text{kg-m/s})$$

The force exerted by the floor on the ball during the course of collision can be determined by assuming that the impulsive force is approximately constant during the collision. Using the impulse-momentum equation given in Eq. (12.7):

$$\underline{F}\,\Delta t = \Delta \underline{p}$$

Solving this equation for the unknown force $\underline{F}$:

$$\underline{F} = \frac{\Delta \underline{p}}{\Delta t} = \frac{2.07\,\underline{j}}{0.01} = 207\,\underline{j} \;\; \text{(N)}$$

Remarks:

- The magnitude of the average impulsive force calculated is much greater than the force of gravity on the ball which is about 2.5 N. This result suggests that for problems involving collisions, the impulsive force between the colliding bodies (in this case the ball and the floor) is the most dominant, and the effects of all other external forces which may be present can be ignored.

- The average force calculated depends heavily on the duration of contact Δt.

- The fact that v_2=3.83 m/s is less than v_1=4.43 m/s suggests that some of the kinetic energy of the ball just before collision is lost as heat during the course of collision. The amount of energy lost can be calculated by considering the difference between the kinetic energies of the ball before and after collision.

Example 12.2 Consider a soccer player kicking a stationary ball (Figure 12.6). If the effect of air resistance is negligible, the ball will undergo a projectile motion (Figure 12.7).

Figure 12.6 *The player applies an impulsive force on the ball.*

Assuming that the mass of the ball is m=0.5 kg, the horizontal range of motion of the ball is ℓ=40 m, the maximum height the ball reaches in the air is h=4 m, and the time at which the foot of the soccer player remains in contact with the ball is Δt=0.1 s, determine the momentum of the ball at the instant of takeoff and the impulsive force applied by the player on the ball.

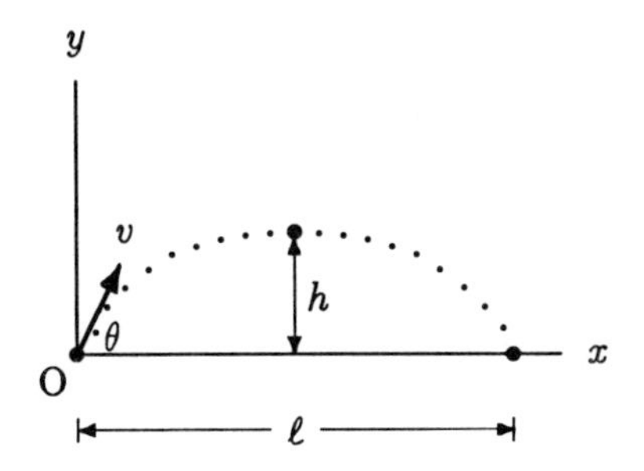

Figure 12.7 *The soccer ball undergoes a projectile motion.*

Solution: To determine the momentum of the ball at the instant of takeoff, we need to calculate the takeoff speed of the ball and its angle of takeoff first. For this purpose, we can utilize the formulas given in Section 8.10. The angle of takeoff is provided in Eq. (8.39) in terms of the maximum height and horizontal range of motion of the projectile motion in the following form:

$$\theta = \arctan\left(\frac{4\,h}{\ell}\right)$$

Substituting the numerical values of h=4 m and ℓ=40 m into the above equation, and carrying out the calculations will yield:

$$\theta = \arctan\left[\frac{4(4)}{40}\right] = 21.8^\circ$$

Using Eq. (8.40), the speed of takeoff can also be determined:

$$v = \frac{\sqrt{2\,g\,h}}{\sin\theta} = \frac{\sqrt{2(9.8)(4)}}{\sin 21.8^\circ} = 23.8 \text{ m/s}$$

The velocity of takeoff $\underline{v}$ has components both in the x and y directions with magnitudes:

$$v_x = v\,\cos\theta = (23.8)(\cos 21.8^\circ) = 22.1 \text{ m/s} \quad (\rightarrow)$$
$$v_y = v\,\sin\theta = (23.8)(\sin 21.8^\circ) = 8.8 \text{ m/s} \quad (\uparrow)$$

Therefore, the takeoff velocity of the ball in vector form is:

$$\underline{v} = 22.1\,\underline{i} + 8.8\,\underline{j} \quad (\text{m/s})$$

The momentum of the soccer ball at the instant of takeoff (immediately after impact) can be determined as:

$$\underline{p} = m\,\underline{v} = 0.5(22.1\,\underline{i} + 8.8\,\underline{j}) = 11.0\,\underline{i} + 4.4\,\underline{j} \quad (\text{kg-m/s})$$

The ball is stationary and its momentum is zero immediately before the soccer player kicks the ball. Therefore, the change in momentum of the ball during the course of contact with the player's foot is equal to its takeoff momentum. The impulse-momentum equation (Eq. 12.7) can now be utilized to determine an average impulsive force exerted by the soccer player on the ball:

$$\underline{F} = \frac{\Delta \underline{p}}{\Delta t} = \frac{11.0\,\underline{i} + 4.4\,\underline{j}}{0.1} = 110\,\underline{i} + 44\,\underline{j} \quad (\text{N})$$

Therefore, the net or the resultant force applied on the ball by the player is:

$$F = \sqrt{(110)^2 + (44)^2} = 118.5 \text{ N}$$

Note that since action and reaction must have equal magnitudes, F is also the magnitude of the force applied by the ball on the player's foot.

Example 12.3 A *force platform*, as illustrated in Figure 12.8, is a flat, rectangular, force-sensitive device which electronically records forces exerted against its upper surface. This device can be used to measure the impulsive forces involved during walking, running, jumping, and other activities.

Figure 12.8 *A force platform.*

Consider the force versus time recording shown in Figure 12.9 for an athlete making vertical jumps on a force platform. The force scale is normalized with the weight of the athlete so that the force reading is zero when the athlete is stationary (standing still or crouching). The reason for normalizing the force measurement is to be able to disregard the effect of gravitational acceleration on the impulse calculations. In this case, a positive force means a force exerted on the platform due to factors other than the weight of the person. The force versus time graph has three distinct regions. An initial "takeoff push" during which the athlete exerts a positive force on the platform, an "airborne" region during which the athlete is not in contact with the platform, and a "landing" period in which the athlete again exerts impulsive forces on the platform. These regions are approximated with rectangular areas A_1 and A_2, and a triangular area A_3. Boundaries of these regions are shown with dashed lines in Figure 12.9. The approximate force applied and the duration of takeoff push are about F_1=600 N and Δt_1=0.3 s. The force reading is about F_2=−800 N while the athlete is airborne, and the athlete remains in the air for about Δt_2=0.4 s. This suggests that the weight of the athlete is about 800 N. Therefore, the athlete has a mass of about m=(800 N)/(9.8 m/s^2)=82 kg. The maximum impulsive force the athlete exerts on the platform during landing is about F_3=1000 N which reduces to zero (in the normalized scale) in a time interval of about Δt_3=0.4 s.

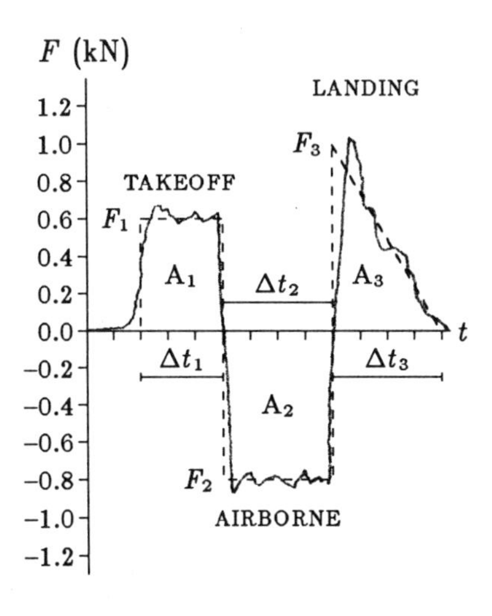

Figure 12.9 *Normalized force versus time plot for the athlete.* (1 kN = 1000 N.)

Determine an approximate takeoff velocity of the center of gravity of the athlete, calculate the height of jump, and determine the impulse and momentum of the athlete during landing by using the approximated areas under the force versus time curve.

Solution: Under the suggested assumptions, the analysis of the problem is quite straightforward. All we need to do is to utilize the impulse-momentum equation. The rectangular area designated by A_1 under the force versus time curve is equal to the impulse during takeoff:

$$A_1 = F_1\,\Delta t_1 = (600)(0.3) = 180 \text{ N-s}$$

The velocity of the athlete before takeoff is zero, and therefore, the impulse is converted into momentum at the instant of takeoff. If v_T is the speed of takeoff, then the impulse-momentum equation (Eq. 12.7) takes the following simplified form:

$$F_1\,\Delta t_1 = \Delta p = m\,v_T$$

Solving this equation for the takeoff speed will yield:

$$v_T = \frac{F_1\,\Delta t_1}{m} = \frac{180}{82} = 2.2 \text{ m/s}$$

To determine the height of the jump, we can utilize the conservation of energy principle. At the instant of takeoff, the athlete has a kinetic energy of $mv_t{}^2/2$ and zero potential energy. As the athlete ascends, the kinetic energy is converted to potential energy. At the instant when the athlete reaches the maximum height h, the kinetic energy of the athlete becomes zero and the potential energy is mgh. Therefore:

$$\frac{1}{2}\, m\, v_t{}^2 = m\, g\, h$$

Solving this equation for the maximum elevation of athlete's center of gravity will yield:

$$h = \frac{v_2{}^2}{2\,g} = \frac{(2.2)^2}{2\,(9.8)} = 0.25 \text{ m}$$

The impulse applied by the athlete on the force platform during landing is equal to the triangular area, A_3, under the force versus time curve:

$$A_3 = \frac{F_3\,\Delta t_3}{2} = \frac{(1000)(0.4)}{2} = 200 \text{ N-s}$$

During the course of landing the downward velocity of the athlete is reduced to zero and the impulse calculated during landing is essentially equal to the momentum, p_L, of the athlete at the instant of landing:

$$p_L = A_3 = 200 \text{ N-s}$$

Notice that we can also calculate the landing speed of the athlete as:

$$v_L = \frac{p_L}{m} = \frac{200}{82} = 2.4 \text{ m/s}$$

Example 12.4 A laboratory crash test is set up to measure the endurance of seat belts for automobile passengers (Figure 12.10). The initial horizontal speed of the test vehicle carrying a 70 kilogram dummy is set to 100 kilometers per hour. The speed of the vehicle and the dummy is brought to zero in a time interval of 0.1 seconds.

Assuming that the frictional effects are negligibly small, determine an average horizontal force applied by the dummy on the seat belt.

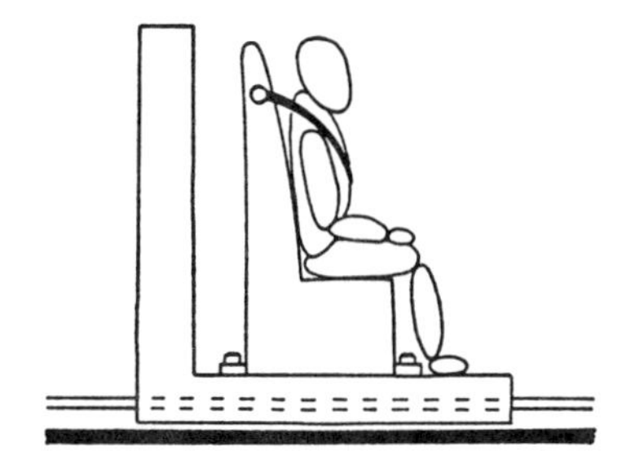

Figure 12.10 *A crash test.*

Solution: In the absence of frictional forces, the dummy would have continued moving along the positive x direction (towards the right) if the seat belt were not constraining its motion. The seat belt is applying a force on the dummy in the negative x direction, which brings the dummy to rest in a time interval of Δt=0.1 s. At the instant when the brakes are applied, the speed of the vehicle and the dummy is v_1=100 km/h or 27.8 m/s. Therefore, the momentum of the dummy at the same instant is:

$$p_1 = m\,v_1 = (70)(27.8) = 1946 \text{ kg-m/s}$$

In vector form, $\underline{p}_1 = 1946\,\underline{i}$ kg-m/s. In a time interval of Δt=0.1 s, the speed of the dummy is brought to zero. Therefore, the momentum of the dummy at the end of 0.1 seconds is zero:

$$\underline{p}_2 = 0$$

According to the impulse-momentum theorem, the change in momentum of the dummy must be equal to the impulse applied by the seat belt on the dummy:

$$\underline{F}\,\Delta t = \underline{p}_2 - \underline{p}_1$$

Or:

$$\underline{F} = -\frac{\underline{p}_1}{\Delta t} = -\frac{1946\,\underline{i}}{0.1} = -19460\,\underline{i} \quad \text{(N)}$$

Therefore, the seat belt applied a force of F=19460 N on the dummy in the negative x direction. Since action and reaction must have equal magnitudes, F is also the magnitude of the force applied by the dummy on the seat belt. If the objective was to design a proper seat belt, this result suggests that the seat belt material must be chosen to withstand a force of more than about 19.5 kN.

12.4 Conservation of Linear Momentum

The equation of motion (Eq. 12.4) relates the resultant of forces applied on an object and the time rate of change of momentum of the object. When the resultant force on an object is zero (i.e., when the object is in equilibrium), then the time rate of change of momentum is also zero, and the momentum of the object is constant. This condition is known as the *principle of conservation of linear momentum.* Conservation of momentum is particularly useful for impact and collision analyses.

Consider two objects that interact with each other. Assume that these objects are isolated from their surroundings so that there are no external forces present except for the forces they exert onto each other. In other words, the effects of external forces are negligibly small as compared to the forces they exert onto each other, which is particularly true in the case of impact and collision. Suppose that at some time t, the two objects have momenta $\underline{p}_1$ and $\underline{p}_2$, respectively. Let $\underline{F}_{12}$ be the force on object 1 applied by object 2, and $\underline{F}_{21}$ the force on object 2 applied by object 1. Using Eq. (12.4):

$$\underline{F}_{12} = \frac{d\underline{p}_1}{dt} \qquad \underline{F}_{21} = \frac{d\underline{p}_2}{dt}$$

According to Newton's third law, the action and reaction must be equal in magnitude and opposite in direction ($\underline{F}_{12}$=$-\underline{F}_{21}$), or:

$$\underline{F}_{12} + \underline{F}_{21} = \frac{d\underline{p}_1}{dt} + \frac{d\underline{p}_2}{dt} = \frac{d}{dt}(\underline{p}_1 + \underline{p}_2) = 0$$

This condition of equilibrium is equivalent to:

$$\underline{p}_1 + \underline{p}_2 = \text{ constant} \qquad (12.9)$$

Equation (12.9) represents the principle of conservation of linear momentum for two interacting bodies that form an isolated system. It states that whenever two objects collide their total momentum will remain constant, regardless of the nature of the forces involved between the two.

12.5 Impact and Collisions

When two bodies collide, they deform to a certain extent because of the forces involved. It will be discussed in Part III of this text that the amount of deformation depends on the objects' material properties, the extent and duration of applied forces, and other conditions such as temperature. In general, an object may undergo an *elastic* deformation, a *plastic* deformation, or both. The elastic deformation is recoverable upon release of the force causing the deformation, whereas plastic deformations are permanent.

There are several types of collisions, which may be distinguished either with respect to the orientations of the impact velocities of the two objects, or according to the nature of deformations they undergo. As illustrated in Figure 12.11, two objects may collide "head-on" which is known as the *direct central impact*. In this case, the velocities of the objects are collinear with the line of impact. The *line of impact* (AA) is a fictitious line passing through the mass centers of the colliding objects, and it is perpendicular to the *plane of contact* (BB) which is tangent to the contacting surfaces. Figure 12.12 illustrates the *oblique central impact* of two objects. In this case, the impact velocities of the objects are not collinear with the line of impact.

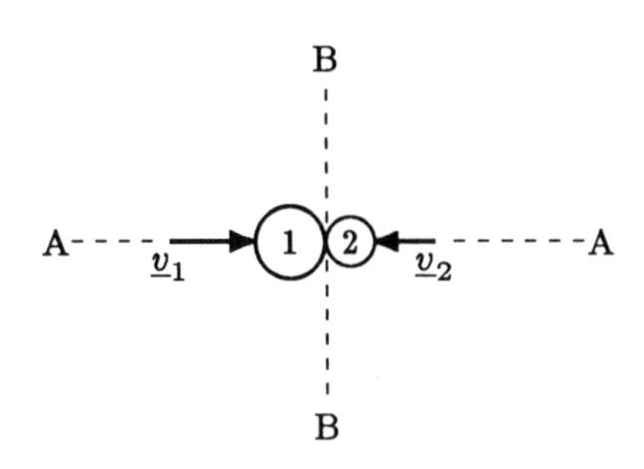

Figure 12.11 *Direct central impact.*

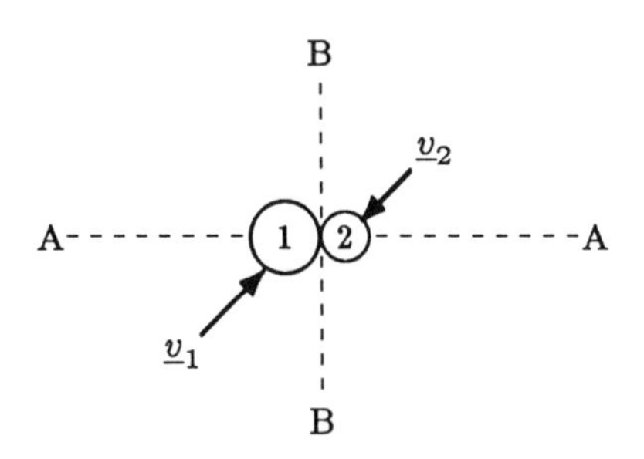

Figure 12.12 *Oblique central impact.*

Collisions can also be distinguished based on the nature of the deformations occurring during the course of a collision. An *elastic* or *perfectly elastic collision* is defined as a collision in which both the total momentum and total kinetic energy of the system (i.e., the two objects) are conserved. For an *inelastic* or *plastic collision*, on the other hand, only the total momentum of the system is conserved. During an inelastic collision, some of the kinetic energies of the colliding objects are dissipated as heat. The extreme case of inelastic collision is called the *perfectly inelastic collision*. This is a collision in which the objects stick together after the collision and move with a common velocity.

Whether a collision is an elastic or an inelastic one, the total momentum of the system is conserved. Therefore, the analyses of impact and collision problems are based on the conservation of momentum principle. Consider the collision of two objects with masses m_1 and m_2. Let $\underline{v}_1$ and $\underline{v}_2$ refer to the velocities of objects 1 and 2, respectively. Also let subscripts "i" and "f" denote the instants immediately before and after the collision. In the time interval between t_i and t_f, the principle of conservation of momentum (Eq. 12.9) can be expressed as:

$$m_1\,\underline{v}_{1i} + m_2\,\underline{v}_{2i} = m_1\,\underline{v}_{1f} + m_2\,\underline{v}_{2f} \tag{12.10}$$

Characteristics of different types of collisions will be provided by considering one-dimensional cases first. These concepts will then be expanded to analyze two-dimensional collisions by utilizing the vectorial properties of the parameters involved.

12.6 One-Dimensional Collisions

For a collision along a straight line (i.e., direct central impact), Eq. (12.10) can be simplified in the following scalar form:

$$m_1\,v_{1i} + m_2\,v_{2i} = m_1\,v_{1f} + m_2\,v_{2f} \tag{12.11}$$

This equation is valid both for elastic and inelastic collisions.

Before collision:

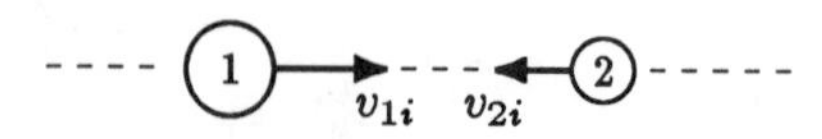

After collision:

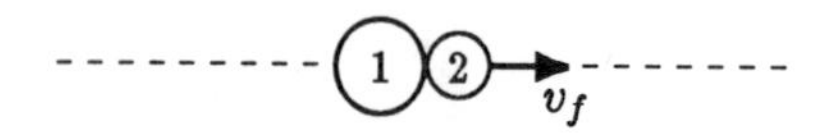

Figure 12.13 *Perfectly inelastic collision.*

12.6.1 Perfectly inelastic collision

A perfectly inelastic collision is one in which the objects stick together and move with a common velocity v_f after the collision (Figure 12.13). To determine v_f, it is sufficient to consider the conservation of momentum principle during the collision. Substituting $v_{1f}=v_{2f}=v_f$ into Eq. (12.11), and solving it for v_f will yield:

$$v_f = \frac{m_1\, v_{1i} + m_2\, v_{2i}}{m_1 + m_2} \tag{12.12}$$

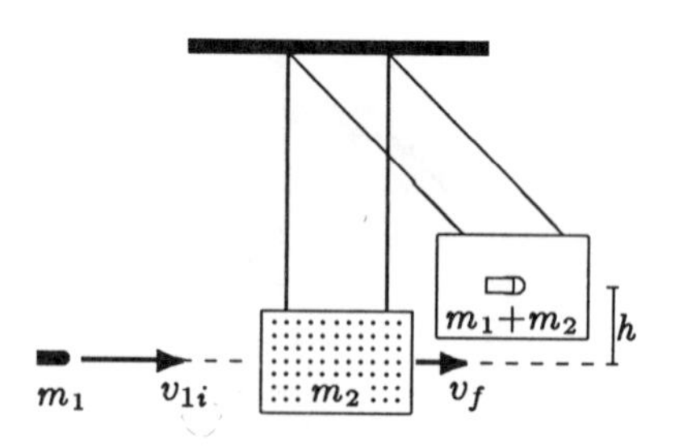

Figure 12.14 *Ballistic pendulum.*

Example 12.5 Figure 12.14 illustrates a *ballistic pendulum* which may be a block of wood suspended by light wires. This simple device can be used to measure the velocity of a bullet. A bullet of mass m_1 fired at a stationary block of mass m_2 will penetrate the block, and the bullet-block system with mass m_1+m_2 will swing to a height h.

Assuming that the bullet remains in the block, determine an expression for the initial speed v_{1i} of the bullet immediately before impact in terms of m_1, m_2, and h.

Solution: This is a typical example of perfectly inelastic collision. The velocity of the block before impact is zero. Immediately after the collision, the block gains a kinetic energy. Designating the speed immediately after impact by v_f, then the bullet-block system has a kinetic energy of $(m_1+m_2)v_f^{\,2}/2$ which is converted to a potential energy of $(m_1 + m_2)gh$ as the block swings to a height h. Applying the conservation of energy principle:

$$\frac{1}{2}\,(m_1 + m_2)\, v_f^{\,2} = (m_1 + m_2)\, g\, h$$

Soving this equation for the common speed v_f of the bullet and the block right after impact:

$$v_f = \sqrt{2\, g\, h} \tag{i}$$

During the course of impact the momentum is conserved. Since the velocity of the block before impact is zero, Eq. (12.12) takes the following simpler form:

$$v_f = \frac{m_1\, v_{1i}}{m_1 + m_2}$$

Solving this equation for v_{1i} and using Eq. (i) will yield:

$$v_{1i} = \left(\frac{m_1 + m_2}{m_1}\right) \sqrt{2\, g\, h}$$

12.6.2 Perfectly elastic collision

A perfectly elastic collision is one in which the total kinetic energy is conserved as well as the total momentum of the objects involved (Figure 12.15). The condition of conservation of total kinetic energy during the course of collision can be written as:

$$\frac{1}{2}\, m_1\, {v_{1i}}^2 + \frac{1}{2}\, m_2\, {v_{2i}}^2 = \frac{1}{2}\, m_1\, {v_{1f}}^2 + \frac{1}{2}\, m_2\, {v_{2f}}^2 \qquad (12.13)$$

In this case, Eqs. (12.11) and (12.13) must be solved simultaneously for the unknowns v_{1f} and v_{2f}. This will yield:

$$v_{1f} = \left(\frac{m_1 - m_2}{m_1 + m_2}\right) v_{1i} + \left(\frac{2\, m_2}{m_1 + m_2}\right) v_{2i} \qquad (12.14)$$

$$v_{2f} = \left(\frac{2\, m_1}{m_1 + m_2}\right) v_{1i} + \left(\frac{m_2 - m_1}{m_1 + m_2}\right) v_{2i} \qquad (12.15)$$

At this point, it should be remembered that velocity is a vector quantity and that appropriate signs for all velocities involved must be included in Eqs. (12.12), (12.14), and (12.15). For this purpose, a positive direction of motion must be chosen. Velocities acting in the same direction must be considered to be positive, and those acting in the opposite direction must be taken to be negative.

Before collision:

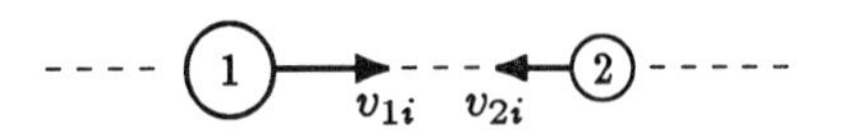

After collision:

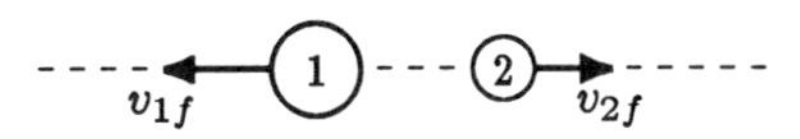

Figure 12.15 *Perfectly elastic collision.*

Example 12.6 As illustrated in Figure 12.16, consider a perfectly elastic collision of two billiard balls with equal masses m. Before the collision, ball 1 has a speed of v_{1i} and ball 2 is stationary.

Assuming a direct central impact, determine the velocities of the billiard balls immediately after the collision.

Solution: The fact that the masses of the balls are equal (i.e., $m_1{=}m_2{=}m$) and that one of the balls was stationary before the collision ($v_{2i}{=}0$) simplifies the problem considerably. Since the masses are equal, $m_1{-}m_2{=}m_2{-}m_1{=}0$ and $2m_1/(m_1{+}m_2){=}1$. Therefore, the first term on the right-hand side of Eq. (12.14) and the second term on the right-hand side of Eq. (12.15) are zero. The second term on the right-hand side of Eq. (12.14) is zero as well because $v_{2i}{=}0$. Hence, choosing the right to be the positive direction of motion, Eqs. (12.14) and (12.15) yield:

$$v_{1f} = 0$$
$$v_{2f} = v_{1i} \quad (\rightarrow)$$

Notice that the kinetic energy and momentum of ball 1 immediately before the collision is totally transferred into kinetic energy and momentum for ball 2 during the collision.

Before collision:

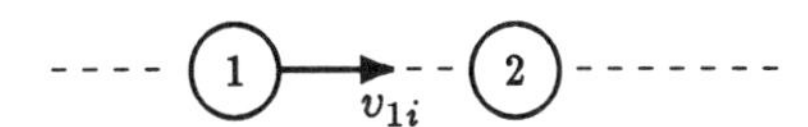

After collision:

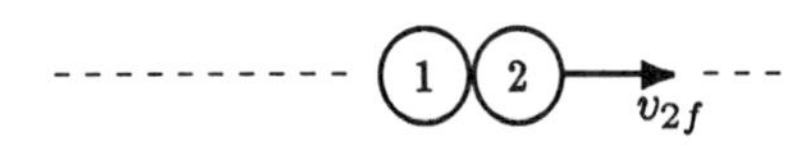

Figure 12.16 *Perfectly elastic collision of two identical billiard balls.*

12.6.3 Elasto-plastic collision

In reality, a material can undergo both elastic and plastic deformations. To facilitate the analyses of the impact characteristics of objects made up of such materials, a concept called the *coefficient of restitution* is developed which is defined as the ratio of the relative velocity of separation and the relative velocity of approach. If e represents the coefficient of restitution between two materials, then:

$$e = \frac{\text{relative velocity of separation}}{\text{relative velocity of approach}} \tag{12.16}$$

Velocities of approach:

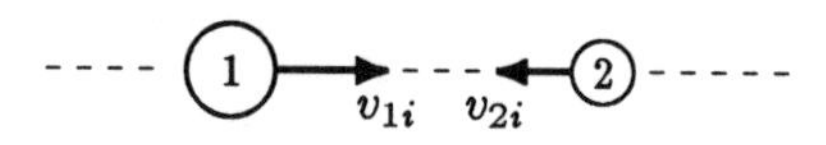

Velocities of separation:

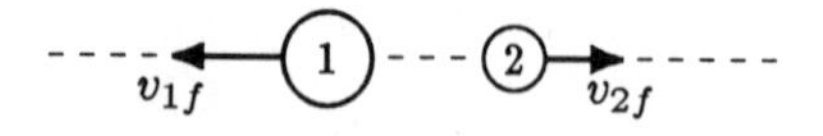

Figure 12.17 *The coefficient of restitution between two colliding objects is defined as the ratio of relative velocities of separation and approach.*

The coefficient of restitution can take values between 0 and 1, such that e=0 for perfectly inelastic impact and e=1 for perfectly elastic impact. The coefficient of restitution is a positive number, and while calculating e the absolute values of the relative velocities of separation and approach must be considered. Consider the two objects shown in Figure 12.17. Before the collision, object 1 is moving towards the right with velocity v_{1i} and object 2 is moving towards the left with velocity v_{2i}. Therefore, the relative velocity of approach is $v_{1i} + v_{2i}$. Assume that after the collision, the objects move in opposite directions with velocities v_{1f} and v_{2f}. Then the relative velocity of separation is $v_{1f}+v_{2f}$. If the objects were moving in the same direction before the collision, then the relative velocity of approach is $v_{1i} - v_{2i}$. If they move in the same direction after the collision, then the relative velocity of separation is $v_{1f} - v_{2f}$. If one of the objects is stationary before and after the collision, then the coefficient of restitution can be determined by considering the ratio of the final and initial velocities of the object. In fact, this is the easiest to measure the coefficient of restitution between two materials.

The coefficient of restitution depends on the material properties of the objects involved in a collision. Temperature is also a factor that can influence the coefficient of restitution, because temperature can alter the mechanical properties of materials. For example, a ball will bounce better after being heated. If the ball is air-filled, such as a tennis ball, heat can also increase the internal pressure of the ball. A highly inflated ball will bounce better than a flat ball. When a ball is deformed during impact, some of the energy is dissipated as heat. The rise in temperature is important in games like tennis and squash where the ball is impacted at a relatively high rate. Another factor that may affect the coefficient of restitution between two materials is the relative velocity of approach. The higher the velocity of approach is, the lower the coefficient of restitution will be.

Example 12.7 Consider the ball analyzed in Example 12.1, which has a mass m=0.25 kg. As illustrated in Figure 12.18, the ball is dropped from a height h_0=1 m. After hitting the floor, the ball bounces and reaches a height h_3=0.75 m. Using the conservation of energy principle between 0 and 1, and between 2 and 3, we calculated that the ball has speeds v_1=4.43 m/s and v_2=3.83 m/s immediately before and after impact.

What is the coefficient of restitution between the ball and the floor, and how much energy is lost during the course of impact?

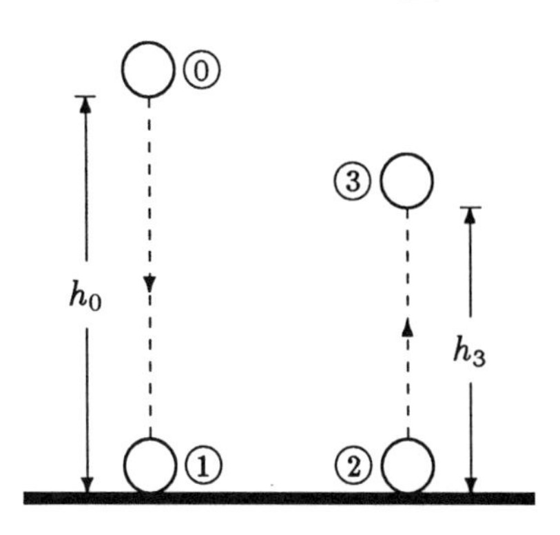

Figure 12.18 *The ball hits the floor and bounces.*

Solution: By definition, the coefficient of restitution is equal to the ratio of the velocities of separation and approach. In this case, the floor is stationary before and after the impact. Therefore, the speeds v_1 and v_2 of the ball before and after impact are also the magnitudes of the relative velocities of approach and separation, respectively. Therefore, the coefficient of restitution between the ball and the floor is:

$$e = \frac{v_2}{v_1} \qquad (i)$$

Substituting the numerical values and carrying out the calculations:

$$e = \frac{3.83}{4.43} = 0.86$$

The amount of energy lost (converted into heat, sound, or internal potential energy) during the course of impact can be determined by calculating the difference in the kinetic energies of the ball before and after the impact:

$$\mathcal{E}_{K2} - \mathcal{E}_{K1} = \frac{1}{2}\, m \left({v_2}^2 - {v_1}^2\right) = -0.62 \text{ J}$$

The coefficient of restitution (represented by Eq. (i) for the ball hitting a stationary surface and bouncing back) can alternatively be written in terms of the initial height h_0 from which the ball is dropped and the final height h_3 to which the ball bounced. In Example 12.1, using the conservation of energy principle, we determined that $v_1=\sqrt{2gh_0}$ and $v_2=\sqrt{2gh_3}$. Substituting these into Eq. (i) will yield:

$$e = \sqrt{\frac{h_3}{h_0}} \qquad (ii)$$

The U.S. National Collegiate Athletic Association (NCAA) rule requiring a basketball dropped from a height h_0=1.8 m to bounce a height of about h_3=1.0 m is equivalent to requiring a coefficient of restitution of about $e=\sqrt{1.0/1.8}=0.75$.

12.7 Two-Dimensional Collisions

As discussed in Section 12.4, if two interacting objects are isolated from their surroundings (if the effects of external forces are negligible as compared to the impulsive forces present), then the total momentum of the system is conserved. This principle of conservation of momentum is applied to analyze direct central impact, or one-dimensional collision, of some simple systems. By considering the vectorial properties of the parameters involved, the concepts developed for one-dimensional collisions can be expanded to analyze the oblique central impact or two-dimensional collision of two objects.

For a two-dimensional collision problem, the conservation of linear momentum principle must be applied in two coordinate directions. In addition, the nature of the collision must be known. For example, if the collision is perfectly elastic, then the total kinetic energy of the system is also conserved. If the collision is perfectly inelastic, then the velocities of the objects after the collision are equal.

The following example will illustrate some of the concepts involved in two-dimensional collision problems.

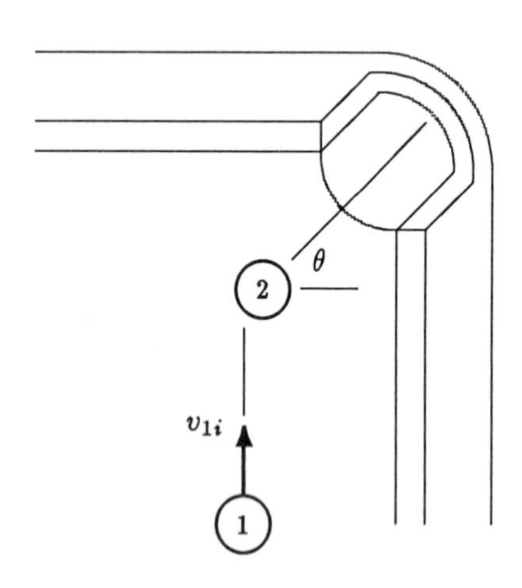

Figure 12.19 *Oblique central impact of two pool balls.*

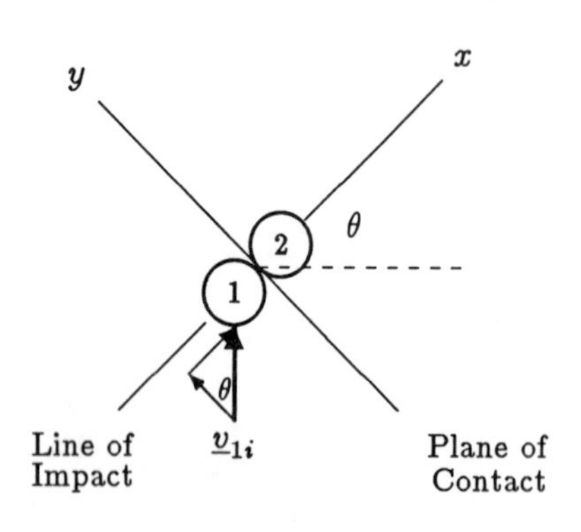

Figure 12.20 *Before collision.*

Example 12.8 Figure 12.19 illustrates an instant during a pool game. What the pool player wishes to do is to hit the stationary target ball (ball 2) by the cue ball (ball 1) so as to move the target ball towards and into the corner pocket. Consider that the cue ball is given a velocity of v_{1i}=5 m/s towards the target ball, and that a line connecting the center of mass and the geometric center of the corner pocket make an angle θ=45°, as shown in Figure 12.20. The rectangular coordinates x and y are chosen in such a way that they respectively coincide with the line of impact (perpendicular to the contacting surfaces), and the plane of contact (tangential to the contacting surfaces). Assume that the cue ball hits the target ball at a point along the line of impact, and that the balls have equal mass.

Neglecting the effects of friction, rotation, and gravity, determine the velocities of the cue and target balls immediately after collision if the coefficient of restitution between them is e=0.8.

Solution: This problem can be analyzed individually along the x and y directions first, and then by combining the results by utilizing the vectorial properties of velocity. Before impact, the target ball is stationary, and the cue ball has velocity components both in the x and y directions. Therefore:

$$(v_{1i})_x = v_{1i}\ \sin\theta = (5)(\sin 45°) = 3.54 \text{ m/s}$$
$$(v_{1i})_y = v_{1i}\ \cos\theta = (5)(\cos 45°) = 3.54 \text{ m/s}$$

$$(v_{2i})_x = 0$$
$$(v_{2i})_y = 0$$

Since the frictional effects are negligible, the motion along the y direction is equivalent to two balls moving along parallel lines without a chance of collision. The motion along the x direction, on the other hand, is equivalent to the direct central impact of the two balls.

(a) Motion along the y direction:

In the y direction, there is no collision. Therefore, the momentum of ball 1 in the y direction is conserved:

$$m\,(v_{1i})_y = m\,(v_{1f})_y$$

$$(v_{1f})_y = (v_{1i})_y = 3.54 \text{ m/s} \qquad (i)$$

Utilizing the conservation of momentum principle for ball 2 in the y direction:

$$m\,(v_{2i})_y = m\,(v_{2f})_y$$

$$(v_{2f})_y = (v_{2i})_y = 0 \qquad (ii)$$

(b) Motion along the x direction:

In the x direction, the balls undergo a direct central impact, and therefore, the total momentum of the system is conserved. Assume that the positive direction of motion is along the positive x axis (i.e., towards the corner pocket), and that after collision, both balls move along the positive x direction. In other words, both $(v_{1f})_x$ and $(v_{2f})_x$ are positive. Since, the balls have equal masses, the conservation of linear momentum along the x direction can be written as:

$$m\,(v_{1i})_x + m\,(v_{2i})_x = m\,(v_{1f})_x + m\,(v_{2f})_x$$

Eliminating the masses, and since $(v_{2i})_x$=0:

$$(v_{1f})_x + (v_{2f})_x = (v_{1i})_x \qquad (iii)$$

The coefficient of restitution for the collision is given as e=0.8. By definition, e is equal to the ratio of the relative velocities of separation and approach. Before impact, ball 2 is stationary. Hence the relative velocity of approach in the x direction is equal to the initial speed of ball 1 in the x direction. Because of the assumed directions of motion after the collision, the relative velocity of separation in the x direction is equal to the difference of the velocity components of the balls along the x direction. Assuming that $(v_{2f})_x$ is larger than $(v_{1f})_x$:

$$e = \frac{(v_{2f})_x - (v_{1f})_x}{(v_{1i})_x}$$

This equation can also be written as:

$$(v_{2f})_x - (v_{1f})_x = e\,(v_{1i})_x \qquad (iv)$$

We have two equations, Eqs. (*iii*) and (*iv*), for two unknowns, $(v_{2f})_x$ and $(v_{1f})_x$. Solving these equations simultaneously will yield:

$$(v_{1f})_x = \frac{(e-1)}{2}\,(v_{1i})_x \qquad (v)$$

$$(v_{2f})_x = \frac{(e+1)}{2}\,(v_{1i})_x \qquad (vi)$$

Substituting $(v_{1i})_x$=3.54 m/s and e=0.8 into these equations will yield:

$$(v_{1f})_x = -0.35$$
$$(v_{2f})_x = 3.19$$

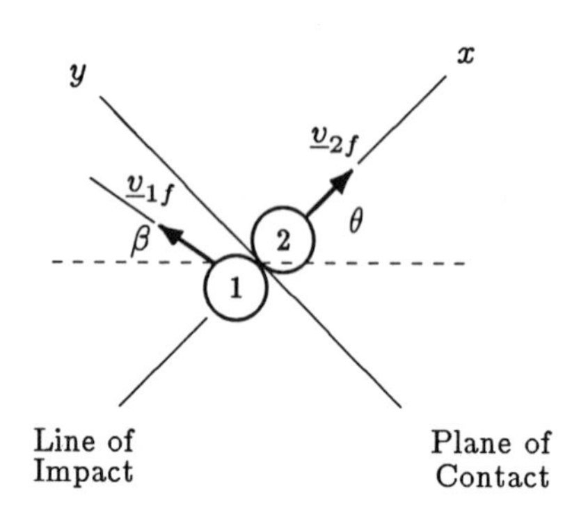

Figure 12.21 *After the collision.*

Hence, the velocities of the balls after the collision are:

$$\underline{v}_{1f} = -0.35\,\underline{i} + 3.54\,\underline{j} \quad (\text{m/s})$$
$$\underline{v}_{2f} = 3.19\,\underline{i} \quad (\text{m/s})$$

Immediately after the collision, the target ball will move with a speed of 3.19 m/s towards the corner pocket (i.e., along the positive x direction). As illustrated in Figure 12.21, the cue ball will move with a speed of $v_{1f}=\sqrt{(0.35)^2+(3.54)^2}$=3.56 m/s along a direction which makes an angle arctan (3.54/0.35)=84° with the negative x direction, or at an angle β=84°–45°=39° with the horizontal.

While utilizing the equations for the coefficient of restitution and conservation of momentum, we assumed that the velocity components of both balls would be in the positive x direction. As a result of our computations, we determined a negative value for $(v_{1f})_x$ which means that it is acting along the negative x direction.

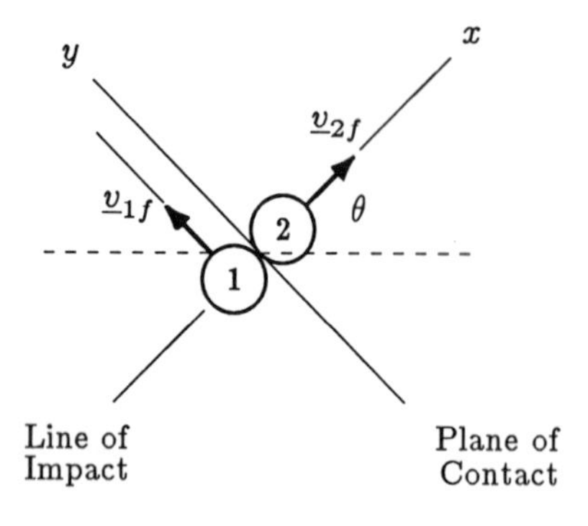

Figure 12.22 *Perfectly elastic collision of two pool balls.*

If the collision were a perfectly elastic one (i.e., e=1), then Eqs. (*v*) and (*vi*) would yield $(v_{1f})_x$=0 and $(v_{1f})_y=(v_{1i})_x$. Therefore, the velocity vectors for the balls after the collision would be:

$$\underline{v}_{1f} = 3.54\,\underline{j} \quad (\text{m/s})$$
$$\underline{v}_{2f} = 3.54\,\underline{i} \quad (\text{m/s})$$

Immediately after the collision, the target ball would move with a speed of 3.54 m/s along the positive x direction (towards the corner pocket) and the cue ball would move with the same speed along the positive y direction, as illustrated in Figure 12.22. The balls would move at right angles to each other after the collision.

12.8 Angular Impulse and Momentum

The rotational analogue of linear momentum is called *angular momentum.* Angular momentum is defined as the product of the mass moment of inertia and the angular velocity of the body undergoing rotational motion and is commonly denoted with L:

$$L = I\,\omega \tag{12.17}$$

The *impulse-momentum theorem for rotational motion* relates applied torque and change in angular momentum. If a torque with magnitude M is applied to a rotating body in the time interval between t_1 and t_2 so that the angular momentum of the body is changed from L_1 to L_2, then the impulse-momentum theorem for rotational motion states that:

$$\int_{t_1}^{t_2} M\,dt = L_2 - L_1 \tag{12.18}$$

The left-hand side of Eq. (12.18) is the *angular impulse*, and M is the *impulsive torque.* If the torque is constant, then the integral in Eq. (12.18) can be evaluated to yield:

$$M\,\Delta t = \Delta L = I\,\Delta\omega \tag{12.19}$$

That is, a constant impulsive torque M applied on a body in the time interval $\Delta t = t_2 - t_1$ will change the angular velocity of the body from ω_1 to ω_2. Consequently, the angular momentum of the body will change from L_1 to L_2. The angular velocity and momentum of the body will increase if the torque is applied in the direction of motion.

Notice that rotational kinetic energy, angular work done, and power have the same dimensions and units as their linear counterparts. On the other hand, the dimension of angular momentum is $[M][L^2]/[T]$ and has the unit of kg-m^2/s in SI.

12.9 Summary of Basic Equations

Table 12.2 provides a list of the basic equations necessary for rotational motion analyses about a fixed axis (circular motion) along with the equations for one-dimensional translational motion analyses. Note that the linear and angular quantities are analogous to each other in such pairs as x and θ, v and ω, a and α, m and I, F and M, and p and L.

TRANSLATIONAL MOTION	ROTATIONAL MOTION (CIRCULAR)
VELOCITY	
$v = \frac{dx}{dt}$	$\omega = \frac{d\theta}{dt}$
ACCELERATION	
$a = \frac{dv}{dt}$	$\alpha = \frac{d\omega}{dt}$
KINEMATIC RELATIONS FOR CONSTANT ACCELERATION	
$x = x_0 + v_0\, t + \frac{1}{2}\, a_0\, t^2$ $v = v_0 + a_0\, t$ $v^2 = {v_0}^2 + 2\, a_0\, (x - x_0)$	$\theta = \theta_0 + \omega_0\, t + \frac{1}{2}\, \alpha_0\, t^2$ $\omega = \omega_0 + \alpha_0\, t$ $\omega^2 = {\omega_0}^2 + 2\, \alpha_0\, (x - x_0)$
EQUATION OF MOTION	
$F = m\, a$	$M = I\, \alpha$
WORK DONE	
$\mathcal{W} = \int_{x_1}^{x_2} F\, dx$	$\mathcal{W} = \int_{\theta_1}^{\theta_2} M\, d\theta$
KINETIC ENERGY	
$\mathcal{E}_K = \frac{1}{2}\, m\, v^2$	$\mathcal{E}_K = \frac{1}{2}\, I\, \omega^2$
WORK-ENERGY	
$\mathcal{W} = \frac{1}{2}\, m\, ({v_2}^2 - {v_1}^2)$	$\mathcal{W} = \frac{1}{2}\, I\, ({\omega_2}^2 - {\omega_1}^2)$
POWER	
$\mathcal{P} = F\, v$	$\mathcal{P} = M\, \omega$
MOMENTUM	
$p = m\, v$	$L = I\, \omega$
IMPULSE-MOMENTUM	
$\int_{t_1}^{t_2} F\, dt = p_2 - p_1$	$\int_{t_1}^{t_2} M\, dt = L_2 - L_1$

Table 12.2 *Equations of translational and rotational motion.*

12.10 Kinetics of Rigid Bodies in Plane Motion

In most situations (e.g., when the effects due to air resistance are neglected), the size and shape of an object do not affect its translational motion characteristics. The size and shape (i.e., inertial effects) must be taken into consideration if the object is undergoing a rotational motion. Now that we have defined the basic concepts behind the rotational motion of rigid bodies, we can integrate our knowledge about translational and rotational motions to investigate their general motion characteristics.

Consider the rigid body illustrated in Figure 12.23. Let m be the total mass of the body, and C be the location of its mass center.

There are three coplanar forces acting on the body. Force F_3 is not producing any torque about C because its line of action is passing through C (i.e., its moment arm is zero). Forces $\underline{F}_1$ and $\underline{F}_2$ are producing clockwise moments about C with magnitudes $M_1{=}d_1F_1$ and $M_2{=}d_2F_2$, respectively. As illustrated in Figure 12.24, the three-force system can be reduced to a one-force and one-moment system such that $\sum \underline{F}{=}\underline{F}_1{+}\underline{F}_2{+}\underline{F}_3$ is the net or the resultant force acting on the body and $\underline{M}_c{=}\underline{M}_1{+}\underline{M}_2$ is the resultant of the couple-moments as measured about C. $\sum \underline{F}$ causes the body to translate and $\underline{M}_c$ causes it to rotate about C.

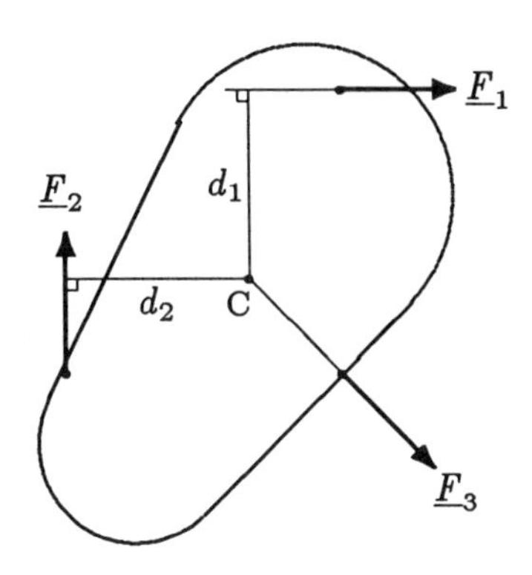

Figure 12.23 *A system of three forces acting on the body.*

Recall that the translational motion of a body depends on its mass and the net force applied on it. Newton's second law of motion states that:

$$\sum \underline{F} = m\,\underline{a}_c \qquad (12.20)$$

$\underline{a}_c$ is the acceleration of the mass center of the body, and Eq. (12.20) accounts for its translational motion. The rotational motion of the body depends on its mass moment of inertia and the net torque applied on it:

$$\sum \underline{M}_c = I_c\,\underline{\alpha} \qquad (12.21)$$

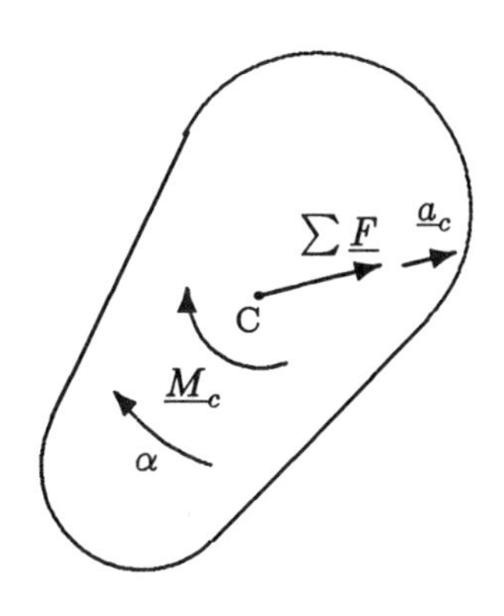

Figure 12.24 *The multi-force system can be reduced to a one-force and one-moment system.*

In Eq. (12.21), I_c is the mass moment of inertia of the body about the axis perpendicular to the plane of rotation and passing through the mass center of the body, and α is the angular acceleration. Notice that Eq. (12.20) is valid for any point within the body, but Eq. (12.21) is correct only about the mass center at C. For two-dimensional motion analyses in the xy plane, Eqs. (12.20) and (12.21) will yield three scalar equations:

$$\sum F_x = m\,a_{cx} \qquad (12.22)$$

$$\sum F_y = m\,a_{cy} \qquad (12.23)$$

$$\sum M_c = I_c\,\alpha \qquad (12.24)$$

These three equations are the governing equations of motion for studying the two-dimensional general motion characteristics of bodies.

Notice that if $M_c{=}0$ then the body is in pure translation, the body is in pure rotation when $F_x{=}0$ and $F_y{=}0$, and the body is said to be in equilibrium when $F_x{=}0$, $F_y{=}0$, and $M_c{=}0$.

12.11 Discussion

With this chapter, we conclude the second part of this text, which was devoted to motion analyses. We have introduced the basic concepts of dynamics along with the mathematics necessary to explain the relationships between these concepts. We have made a distinction between two general methods of motion analysis as kinetics and kinematics. We have defined various forms of motion as translational, rotational, and general, and provided various techniques for their analyses. We have paid particular attention to the applications of these techniques by providing many solved example problems.

The following five chapters constitute the last part of this text. In these chapters, the emphasis will be placed upon the deformability of bodies under consideration. Concepts such as stress, strain, elasticity, plasticity, viscoelasticity, torsion, bending, material failure, fatigue, and endurance will be explained in these chapters. These concepts will be applied to orthopaedic biomechanics by analyzing deformation characteristics of biomaterials.

Chapter 13

INTRODUCTION TO THE MECHANICS OF DEFORMABLE BODIES

13.1 Rigid and Deformable Bodies

The first two parts of this text introduced the fundamental concepts of statics and dynamics along with some of their applications. The fields of statics and dynamics are based on Newton's laws (Newtonian mechanics), and they constitute the main branches of the more general field of rigid body mechanics. The basic assumption in rigid body mechanics is that the bodies involved do not deform under applied loads. This idealization is necessary to simplify the problems under investigation for the sake of analyzing the external forces acting on bodies and their general dynamic behavior. The field of deformable body mechanics, on the other hand, does not treat the bodies as rigid, but incorporates the deformability (shape change) and the material properties of the bodies into the analyses. This field of applied mechanics utilizes the experimentally determined relationships between applied forces and corresponding deformations.

The mechanics of deformable bodies have been studied under various titles such as mechanics of materials, strength of materials and the mechanics of solids. The subjects covered within deformable body mechanics form the basis for the study of more advanced topics in elasticity, inelasticity, and continuum mechanics.

13.2 Why Deformable Body Mechanics?

Rigid body mechanics has its limitations. One of these limitations was mentioned in Section 5.15 in which the distinction between statically determinant and statically indeterminant systems was introduced. A system for which the equations of equilibrium are not sufficient to determine the unknown forces is called *statically indeterminate*. For the analyses of such systems, there is a need for equations in addition to those provided by the conditions of static equilibrium. These additional equations can be derived by considering the material properties of the parts constituting a system and by relating forces to deformations, which is the focus of the deformable body mechanics.

The desire to analyze statically indeterminate systems is only one of the reasons why deformable body mechanics is important. The applications of this field extend to almost all branches of engineering by providing essential design and analysis tools. The task of an engineer – mechanical, civil, electrical, or biomedical – is to determine the safest and most efficient operating condition for a machine, a structure, a piece of equipment, or a prosthetic device. A design engineer can accomplish this task by first assessing the proper operational environment through force analyses, making the correct structural design, and choosing the material to sustain the forces involved in that environment. The primary concern of a design engineer is to make sure that when loaded, a machine part, a structure, a piece of equipment, or a device will not break or deform excessively.

13.3 Applied Forces and Deformations

Mechanics is concerned with forces and motions. It is possible to distinguish two types of motions. If the resultants of external forces and/or moments applied on a body are not zero, then the body will undergo gross overall motion (translation and/or rotation). In other words, as studied in dynamics, the position of the body as a whole will change over time. The second type of motion involves local changes of shape within a body, called *deformations*, which are the primary concern of the field of deformable body mechanics. If a body is subjected to externally applied forces and moments but remains in static equilibrium, then it is most likely that there is some local shape change within the body. The extent of the shape change depends on the magnitude of the applied forces and the material properties of the body. Deformation and overall motion can also occur simultaneously. Here, we are mainly concerned with deformations under static conditions.

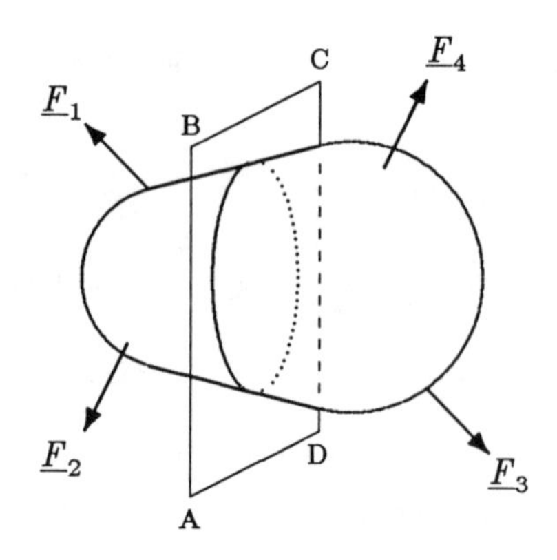

Figure 13.1 *An object with arbitrary shape.*

13.4 Internal Forces and Moments

Consider the arbitrarily shaped object illustrated in Figure 13.1, which is subjected to a number of externally applied forces. Assume that the resultant of these forces and the net moment acting on the object are equal to zero. That is, the object is in static equilibrium. Also assume that the object is fictitiously separated into two parts by passing an arbitrary plane ABCD through the object. If the object as a whole is in equilibrium, then its individual parts must be in equilibrium as well. If one of these two parts is considered, then the equilibrium condition requires that there is a force vector and/or a moment vector acting on the cut section to counterbalance the effects of the external forces and moments applied on that part. These are called the *internal* force and moment vectors. Of course, the same argument is true for the other part of the object. Furthermore, for the overall equilibrium of the object, the force vectors and moment vectors on either surface of the cut section must have equal magnitudes and opposite directions (Figure 13.2).

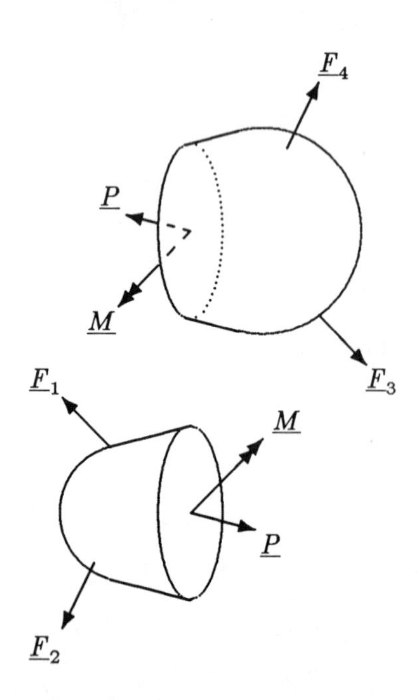

Figure 13.2 *The method of sections.*

For a three-dimensional object, the internal forces and moments can be resolved into their components along three mutually perpendicular directions, as illustrated in Figure 13.3. The force and moment components measured at the cut sections take special names reflecting their orientation and effects on the cut sections. The force component P_x in Figure 13.3 is called the *axial* or the *normal* force, and it is a measure of the pulling or pushing action of the externally applied forces in a direction perpendicular to the cut section. It is called a *tensile* force if it has a pulling action trying to elongate the part, or a *compressive* force if it has a pushing action tending to shorten the part. The force components P_y and P_y are called *shear* forces, and they are measures of resistance to the sliding action of one cut section over the other. Their subscripts indicate their lines of action. The moment component M_x is also called *torque*, and it is a measure of the twisting action of the externally applied forces along an axis normal to the plane of the cut section (in this case, in the x direction). The components M_y and M_z of the moment vector are called the *bending moments*, and they respectively indicate the extent of bending action to which the cut part is subjected in the y and z directions.

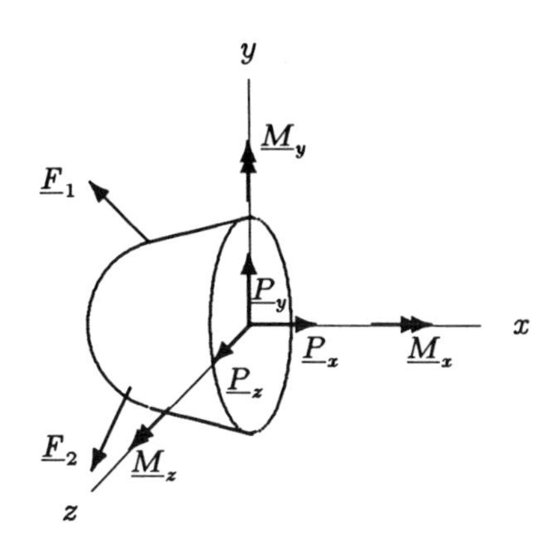

Figure 13.3 *Internal forces and moments.*

Note here that it may be more informative to refer to forces and moments with double subscripted symbols. For example, using P_{xy} instead of P_y indicates that the force component is acting in the y direction (second subscript) on a section whose normal is in the x direction (first subscript). Similarly, M_{xz} would refer to the component of the moment vector in the z direction measured on the same section.

13.5 Stress and Strain

The purpose of studying the mechanics of deformable bodies or strength of materials is to make sure that the design of a structure is safe against the combined effects of applied forces and moments. The idea is to select the proper material for the structure, or if there is an existing structure, to determine the loading conditions under which the structure can operate safely and efficiently. To make a selection, however, one needs to know the mechanical properties of materials under different loading conditions.

Consider the two suspended bars shown in Figure 13.4, which are made of the same material, and have the same length but different sizes. The cross-sectional area A_1 of bar 1 is less than the cross-sectional area A_2 of bar 2. Assume that these bars are subjected to successively increasing forces until they break. If the forces F_1 and F_2 at which the bars 1 and 2 break were recorded, it would be observed that the force F_2 required to break bar 2 is greater than the force F_1 required to break bar 1. These forces might be an indication of the strength of the bars. However, the fact that the force-to-failure depends on the cross-sectional area of the specimen in addition to some other factors makes force an impractical measure of the strength of a material. To eliminate this inconvenience, a concept called *stress* is defined by dividing force with the cross-sectional area:

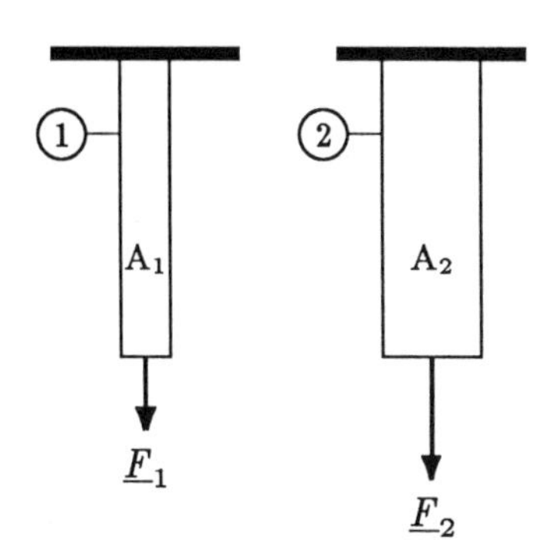

Figure 13.4 *Two bars with different cross-sectional areas are subjected to tensile forces.*

$$\text{STRESS} = \frac{\text{FORCE}}{\text{AREA}}$$

Although the bars in Figure 13.4 have different cross-sectional areas and require different forces-to-failure, since they are made of the same material, their stress measurement at failure would be equal.

As stated earlier, mechanics of deformable bodies is concerned with applied forces and their internal effects on bodies. One of these effects is shape change or deformation. The amount of deformation an object will undergo depends on its dimensions, material properties, and the magnitude and duration of applied forces. Consider the two bars shown in Figure 13.5, which are made of the same material, have the same cross-sectional area, but different lengths. The length ℓ_1 of bar 1 is less than the length ℓ_2 of bar 2. Assume that the same force F is applied to both bars, and the elongation of each bar is measured. It would be observed that the increase of length in bar 2 would be greater than the increase of length in bar 1, indicating that the amount of elongation depends on the original length of the specimen. To eliminate size dependence of deformation measurements, another concept called *strain* is defined by dividing the amount of elongation with the original length of the specimen in the direction of elongation:

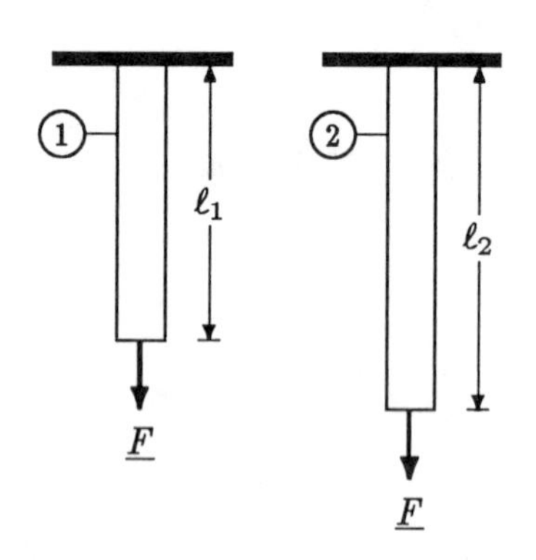

Figure 13.5 *Two bars with different lengths are subjected to tensile forces.*

$$\text{STRAIN} = \frac{\text{AMOUNT OF ELONGATION}}{\text{ORIGINAL LENGTH}}$$

Broad definitions of stress and strain are introduced here. More detailed descriptions of these concepts will be provided in the following chapters.

13.6 General Procedure

An overall procedure for analyzing problems in mechanics of deformable bodies, including the purpose of these analyses, is listed below.

- Static analyses. At this first stage, the analytical methods of statics are employed to determine the external reaction forces. This stage involves drawing free-body diagrams and applying the conditions of equilibrium to determine the unknown reaction forces by utilizing concepts such as equivalent force systems.

- Analyses of internal forces and moments. The internal forces and moments can be determined by the method of sections. As discussed briefly in Section 13.4, this can be done by separating the body into two sections at the location where the forces and moments need to be calculated. Here, the concern is to determine critical load conditions which correspond to maximum stress levels. These critical loads can be determined by drawing the shear and bending diagrams of the body, which essentially requires the application of the method of sections throughout the body.

• Stress analyses. This stage involves the conversion of internal forces and moments, in particular the critical forces and moments, into corresponding stresses by using the formulas which incorporate the material and geometric properties of the problem into the analyses.

• Selection of material. Materials can be distinguished by their physical and mechanical properties. At this final stage of analysis, a material must be selected for the safe operation of the structure based on the maximum stresses calculated. If the material is already selected and the design is already made, then the maximum stresses calculated are used to set the allowable load conditions.

Notice that the prerequisite for deformable body mechanics is statics. It is important to note here that some bodies may be in dynamic equilibrium, rather than static equilibrium. Such a problem can also be analyzed with the same procedure outlined above by reducing it to a problem of static equilibrium. This can be achieved by utilizing the *d'Alembert principle.* This principle can be applied by computing the acceleration of the body, calculating the inertial force by multiplying the acceleration and the mass of the body, and treating this force as another external force acting at the mass center of the body in a direction opposite to the direction of acceleration.

13.7 Mathematics

The analyses in this part of the text will utilize vector algebra and differential and integral calculus as computational tools. Therefore, the reader is advised to review Chapter 2 and Appendices A through C. Some of the analyses may also require familiarity with ordinary differential equations.

13.8 Topics to Be Covered

At the beginning of Chapter 14, detailed definitions of stress and strain will be provided. Based on the stress-strain diagrams, material properties such as ductility, stiffness, and brittleness will be discussed. Elastic and plastic deformations, Hooke's law, the necking phenomena, and the concepts of work and strain energy will be explained. The analyses in Chapter 14 will be limited to uniaxial deformations. The concepts introduced will be applied to analyze statically indeterminate systems, and special applications to orthopaedic biomechanics will be provided.

In Chapter 15, more advanced topics in stress-strain analyses will be introduced along with analyses of bodies subjected to torsion, bending, and combined loading. Again, the emphasis will be placed upon applications to orthopaedic mechanics. In Chapter 16, techniques of transforming stresses from one plane to another are explained along with methods for finding maximum stresses. Also introduced are some failure theories, fatigue, endurance, and other factors affecting the strength of materials. Chapter 16 will illustrate some of the reasons why stress analyses are important in the design of structures. In the final chapter, the viscoelastic behavior of materials are studied. The concept of viscoelasticity has utmost importance in biomechanics because most biological materials possess viscoelastic properties.

13.9 Suggested Reading

The following books can be reviewed to gain more detailed information on the principles of mechanics of deformable bodies:

Crandall, S.H., Dahl, N.C. and Lardner, T.J. 1978. *An Introduction to the Mechanics of Solids.* 2nd ed. New York: McGraw-Hill.

Popov, E.P. 1978. *Mechanics of Materials.* 2nd ed. Englewood Cliffs, NJ: Prentice-Hall.

Pytel, A., and Singer, F.L. 1987. *Strength of Materials.* 4th ed. New York: Harper & Row.

Chapter 14

STRESS AND STRAIN

14.1 Basic Loading Configurations

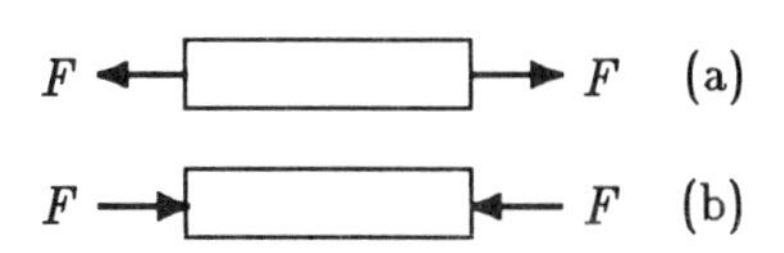

Figure 14.1 *Axial loading: (a) tension and (b) compression.*

An object subjected to an external force will move in the direction of the applied force. The object will deform if its motion is constrained in the direction of the applied force. Deformation implies relative displacement of any two points within the body. The extent of deformation an object undergoes is dependent on many factors including the material properties of the object, the geometry of the object, environmental factors such as heat and humidity, and the magnitude, duration, and direction of the applied forces.

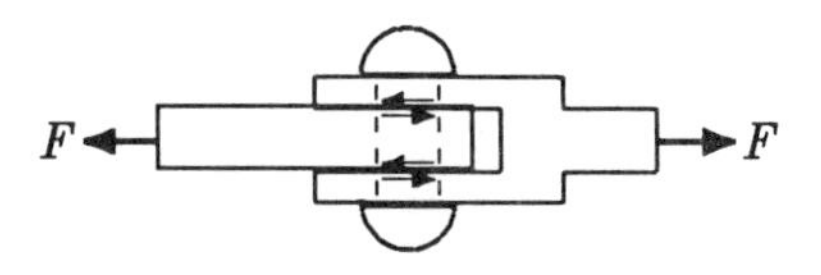

Figure 14.2 *Shearing effects.*

External forces can be classified in various ways. Forces can be distinguished by the way they tend to deform the body upon which they are exerted. Figure 14.1 illustrates *axial loading* which can occur in two ways. The body is said to be in *tension* if the body tends to elongate, and in *compression* if it tends to shrink in the direction of the applied force. Figure 14.2 illustrates an example of *shear loading*. Shearing differs from tension and compression in that it is caused by parallel forces acting in opposite directions tangent to the area resisting the forces causing shear, whereas both tension and compression are caused by collinear forces applied perpendicular to the areas on which they act. It is common to call tensile and compressive forces *normal forces*, whereas shearing forces are *tangential forces*.

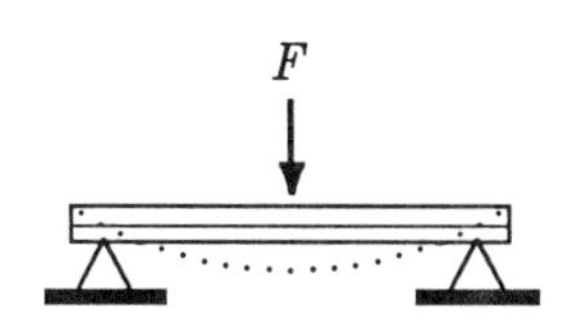

Figure 14.3 *Bending.*

Objects also deform when they are subjected to forces which cause bending and torsion. *Bending* and *torsion* are related to the moment and torque actions of applied forces. As illustrated in Figure 14.3, when a downward force is applied on a simply supported beam, the beam will deflect or bend. The force applied at the free-end (C) of the L-shaped cantilever beam shown in Figure 14.4 has both bending and torsional effects on the beam.

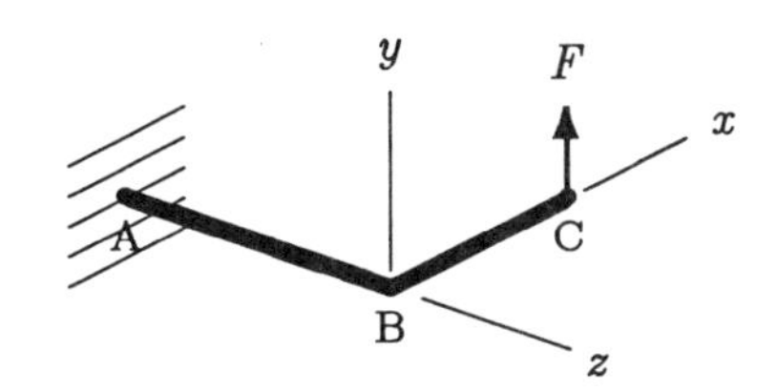

Figure 14.4 *Torsional and bending effects.*

14.2 Uniaxial Tension Test

It is important to point out that a material may respond differently to different loading configurations. For a given material, there are different physical properties that must be considered while analyzing the response of that material to tensile loading as compared to loading which may cause shear, bending, or torsion. The physical properties of materials are established by subjecting them to various experiments. The mechanical response of materials under tensile forces are analyzed by the *uniaxial* or *simple tension test*.

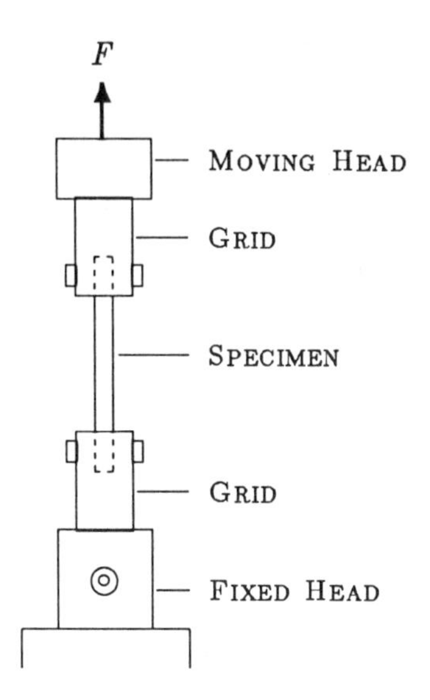

Figure 14.5 *Uniaxial tension test.*

The experimental setup for the uniaxial tension test is illustrated in Figure 14.5. It consists simply of one fixed and one moving head with attachments to grip the test specimen. A specimen is placed and firmly fixed in the equipment, a tensile force of known magnitude is applied through the moving head, and the

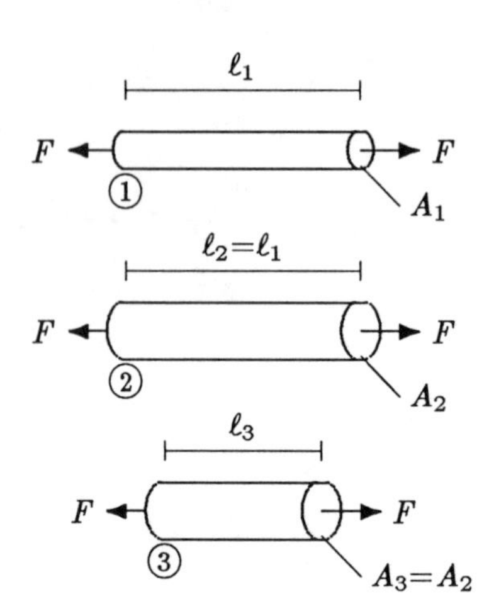

Figure 14.6 *Specimens.*

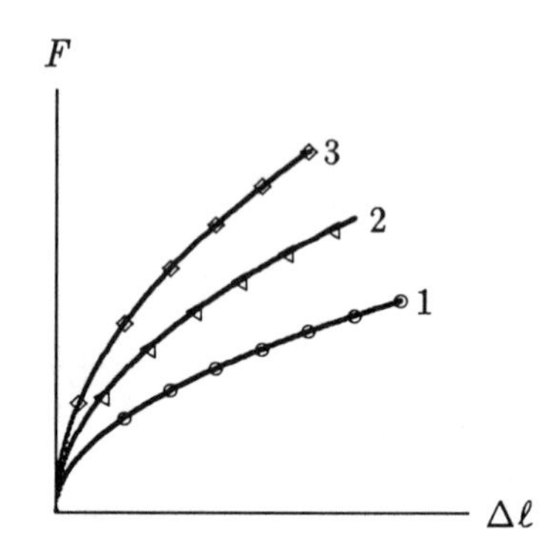

Figure 14.7 *Load-elongation diagrams.*

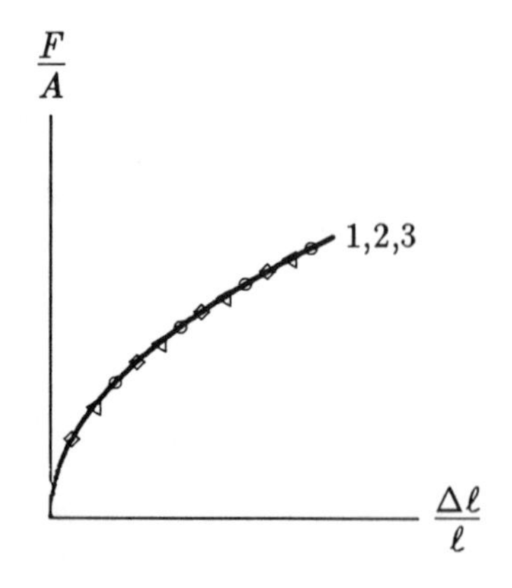

Figure 14.8 *Load over area versus elongation over length diagram.*

corresponding elongation of the specimen is measured. A general understanding of the response of the material to tensile loading is obtained by repeating this test for a number of specimens made of the same material, but with different lengths, cross-sectional areas, and under forces of varying magnitudes.

14.3 Load-Elongation Diagrams

Consider the three bars shown in Figure 14.6. Assume that these bars are made of the same material. The first and the second bars have the same length but different cross-sectional areas, and the second and third bars have the same cross-sectional area but different lengths. Each of these bars can be subjected to a series of uniaxial tension tests by gradually increasing the applied forces and measuring corresponding increases in their lengths. If F is the applied force and $\Delta\ell$ is the increase in length, then the data collected can be plotted to obtain a load versus elongation diagram for each specimen. Effects of geometric parameters (cross-sectional area and length) on the load-bearing ability of the material can be judged by drawing the curves obtained for each specimen on a single graph (Figure 14.7) and comparing them. For a given force, the comparison of curves 1 and 2 indicate that the larger the cross-sectional area, the more difficult it is to deform the specimen in a simple tension test, and curves 2 and 3 indicate that the longer the specimen, the larger the deformation in tension.

Note that instead of applying a series of tensile forces to a single specimen, it is more realistic to have a number of specimens with almost identical geometries and apply one force to one specimen only once. We shall observe later in this and following chapters that a force applied on an object may alter its physical properties, and using one specimen for more than one test may cause discrepancies in the measurements.

Another method of representing the results obtained in a uniaxial tension test is by first dividing the applied force with the cross-sectional area of the specimen, normalizing the amount of deformation by dividing the measured elongation with the original length of the specimen, and then plotting the data on a F/A versus $\Delta\ell/\ell$ graph as shown in Figure 14.8. The three curves in Figure 14.7, representing three specimens made of the same material, are represented by a single curve in Figure 14.8. It is obvious that some of the information provided in Figure 14.7 is lost in Figure 14.8. That is why the representation in Figure 14.8 is more advantageous than that in Figure 14.7. The single curve in Figure 14.8 is unique for a particular material, independent of the geometries of the specimens used during the experiments. This type of representation eliminates geometry as one of the variables, and makes it possible to focus attention on the physical properties of different materials. For example, consider the curves in Figure 14.9, representing the mechanical behavior of materials A and B in simple tension. It is clear that material B can be deformed more easily than material A in a uniaxial tension test, or in other words, that material A is "stiffer" than material B.

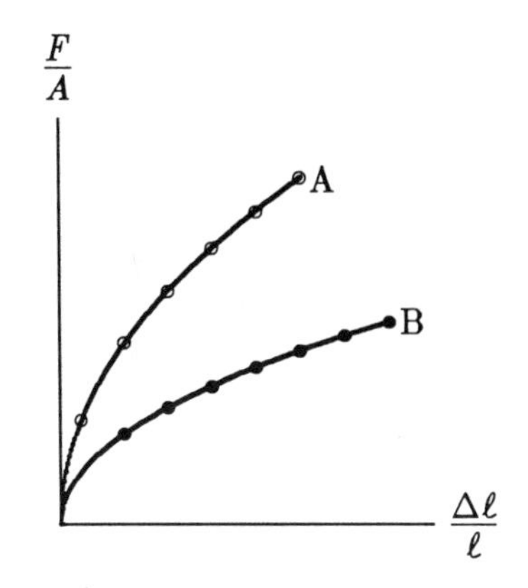

Figure 14.9 *Material A is stiffer than material B.*

14.4 Simple Stress

Consider the bar shown in Figure 14.10(a) which has a length ℓ, a cross-sectional area A, and is subjected to a pair of tensile forces of magnitude F. The bar is in static equilibrium. To analyze the forces induced within the bar, the method of sections can be applied by hypothetically cutting the bar into two pieces through a plane perpendicular to the long axis of the bar. Since the bar as a whole is in equilibrium, the two pieces must individually be in equilibrium as well. This requires that at the cut section of each piece there is an internal force which is equal in magnitude but opposite in direction to the externally applied force (Figure 14.10b). Note that the internal force is distributed over the entire cross-sectional area (Figure 14.10c) of the cut section, and F represents the resultant of the distributed internal force. The intensity or force per unit area of this distributed force may or may not be uniform (constant) throughout the cut section. The intensity of the internal force over the cut section is known as the *stress*, which is defined as the force per unit area over which the force is acting. For the case shown in Figure 14.10, since the force resultant at the cut section is perpendicular to the plane of the cut, the corresponding stress is called a *normal stress*. It is customary to use the symbol σ (sigma) to refer to normal stresses. Assuming that the intensity of the distributed force at the cut section is uniform over the cross-sectional area A, the normal stress can be calculated using:

$$\sigma = \frac{F}{A} \tag{14.1}$$

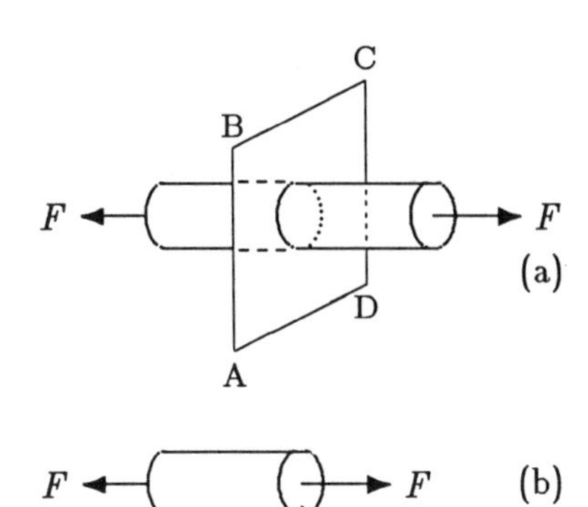

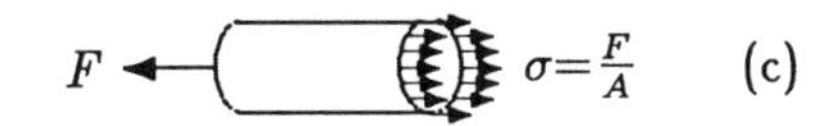

Figure 14.10 *Normal (axial) stress.*

If the intensity of the stress distribution over the area is not uniform, then Eq. (14.1) will yield an *average normal stress*. It is also customary to refer to a normal stress that causes tension on the cut surface as a *tensile stress*. On the other hand, a normal stress that causes compression is a *compressive stress*.

The other form or component of stress is called *shear stress*, which is a measure of the intensity of internal forces acting parallel or tangent to a plane of cut. Consider Figure 14.11(a) in which two rectangular beams are fastened by a rivet, and a pair of forces with magnitude F are applied to the beams. The method of sections is applied on the rivet by cutting it at a plane perpendicular to the axis of the rivet. The free-body diagram of one of the two pieces thus obtained is shown in Figure 14.11(b). To maintain the static equilibrium of the members, there has to be

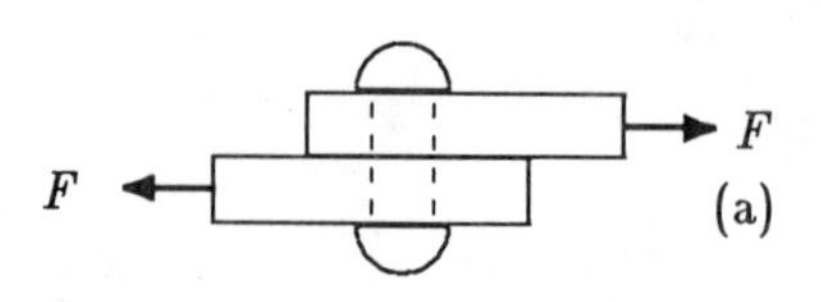

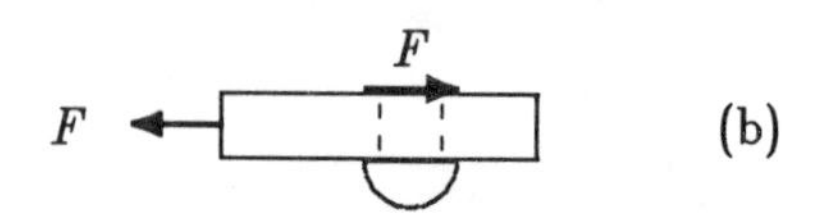

$\tau = \frac{F}{A}$ (c)

A

Figure 14.11 *Shear (tangential) stress.*

a force over the cut area of the rivet with magnitude F acting parallel to the cut surface in the direction opposite to that of the external force. This is known as the *internal shearing force* and is the resultant of a distributed load over the cut surface of the rivet (Figure 14.11c). The intensity of the shearing force over the surface is known as the *shear stress*, and is commonly denoted with the symbol τ (tau). If the area of this surface (in this case, the cross-sectional area of the rivet) is A, then the shear stress is:

$$\tau = \frac{F}{A} \tag{14.2}$$

It is important to reiterate that in Eq. (14.2), F is the magnitude of the tangential force acting over the cut surface which has a cross-sectional area A. Shear stresses are also known as *tangential stresses*. The underlying assumption in Eq. (14.2) is that the shear stress is distributed uniformly over the area. For most cases, however, this assumption may not be true. In such cases, the shear stress calculated by Eq. (14.2) will represent an average value.

The dimension of stress can be determined by dividing the dimension of force [F]=[M][L]/[T^2] with the dimension of area [L^2]. Therefore, stress has the dimension of [M]/[L][T^2]. The units of stress in different unit systems are listed in Table 14.1.

SYSTEM	UNITS OF STRESS	SPECIAL NAME
SI	N/m^2	Pascal (Pa)
c-g-s	dyn/cm^2	
British	lb/ft^2 or lb/in^2	psf or psi

Table 14.1 *Units of stress.*

14.5 Simple Strain

A body may deform when it is subjected to a change in temperature or to an externally applied force. Deformation implies change in shape. *Strain*, which is also known as *unit deformation*, is a measure of the degree or intensity of deformation.

Consider the bar in Figure 14.12. Let A and B be two points on the bar located at a distance ℓ_1, and C and D be two other points located at a distance ℓ_2 from one another, such that $\ell_1 > \ell_2$. ℓ_1 and ℓ_2 are called *gage lengths*. The bar will elongate when it is subjected to a tensile load. Let $\Delta\ell_1$ be the amount of elongation measured between A and B, and $\Delta\ell_2$ be the increase in length between C and D. $\Delta\ell_1$ and $\Delta\ell_2$ are certainly some measures of deformation. However, they depend on the respective gage lengths, such that $\Delta\ell_1 > \Delta\ell_2$. On the other hand, if the ratio of the amount elongation to the gage length is calculated for each case and compared, it would be observed that $(\Delta\ell_1/\ell_1) \simeq (\Delta\ell_2/\ell_2)$. Elongation per unit gage length is known as strain and is a more fundamental means of measuring deformation.

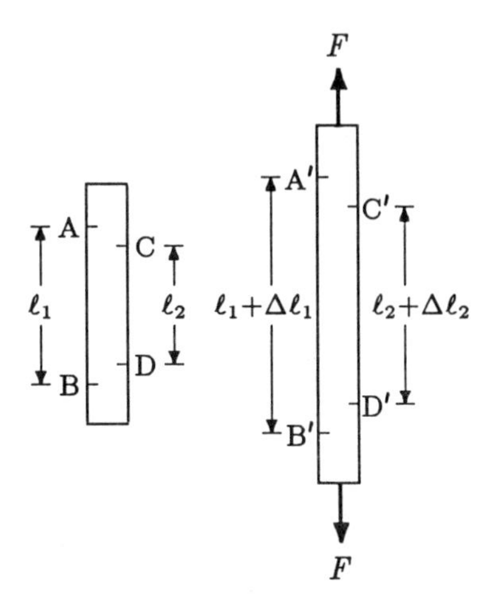

Figure 14.12 *Normal strain.*

As in the case of stress, two types of strains can be distinguished. The ***normal*** or ***axial strain*** is associated with axial forces and defined as the ratio of the change (increase or decrease) in length, $\Delta\ell$, to the original gage length, ℓ, and is denoted with the symbol ϵ (epsilon):

$$\epsilon = \frac{\Delta\ell}{\ell} \tag{14.3}$$

When a body is subjected to tension its length increases, and both $\Delta\ell$ and ϵ are positive. The length of a specimen under compression decreases, and both $\Delta\ell$ and ϵ become negative.

The second form of strain is related to distortions caused by shearing forces. Consider the rectangle (ABCD) shown in Figure 14.13, which is acted upon by a pair of shearing forces. Shear forces deform the rectangle into a parallelogram (AB′C′D). If the relative horizontal displacement of the top and the bottom of the rectangle is d and the height of the rectangle is ℓ, then the ***average shear strain*** is defined as the ratio of d and ℓ which is equal to the tangent of angle γ (gamma). The angle γ is usually very small. For small angles, the tangent of the angle is approximately equal to the angle itself. Hence, the average shear strain is equal to angle γ (measured in radians), which can be calculated using:

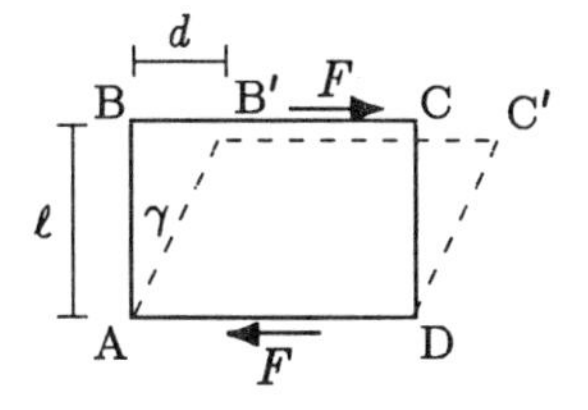

Figure 14.13 *Shear strain.*

$$\gamma = \frac{d}{\ell} \tag{14.4}$$

Notice that the effects of shearing forces can be observed using a stack of paper. Place a stack of paper on a flat horizontal surface, press your hand gently on the top of the stack so as to slide pages over one another by applying a horizontal (shear) force on the stack.

Strains are calculated by dividing two quantities having the dimension of length. Therefore, they are dimensionless quantities and there is no unit associated with them. For most applications, the deformations and consequently the strains involved are very small, and the precision of the measurements taken is very important. To indicate the type of measurements taken, it is not unusual to attach units such as cm/cm or mm/mm next to a strain value. Strains can also be given in percent. In engineering applications, the strains involved are of the order of magnitude 0.1 percent.

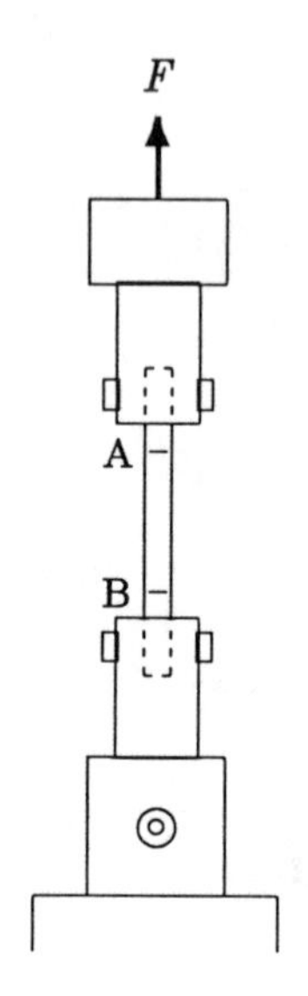

Figure 14.14 *Example 14.1.*

Example 14.1 A circular cylindrical rod with radius r=1.26 cm is tested in a uniaxial tension test (Figure 14.14). Before applying a tensile force of F=1000 N, two points A and B which are at a distance ℓ_1=30 cm (gage length) are marked on the rod. After the force is applied, the distance between A and B is measured as ℓ_1=31.5 cm.

Determine the strain and stress generated in the rod.

Solution: By definition, the tensile strain is equal to the ratio of the amount of elongation to the original length. The amount of elongation $\Delta\ell$ is the difference between the gage lengths before and after deformation:

$$\Delta\ell = \ell_1 - \ell_0 = 31.5 - 30.0 = 1.5 \text{ cm}$$

Therefore, the strain ϵ generated in the rod is:

$$\epsilon = \frac{\Delta\ell}{\ell_0} = \frac{1.5}{30} = 0.05 \text{ cm/cm}$$

The tensile stress is defined as the applied force per unit cross-sectional area. Since the bar has a circular cross-section with radius r=1.26 cm, the cross-sectional area A of the bar is:

$$A = \pi\, r^2 = (3.1416)(1.26)^2 = 5 \text{ cm}^2$$

Before calculating the stress generated in the bar, we must convert the unit of the cross-sectional area A=5 cm^2 to be consistent with SI. Since 1 m is equal to 100 cm, this can be achieved by dividing 5 cm^2 by 10,000. Therefore:

$$A = 0.0005 = 5 \times 10^{-4} \text{ m}^2$$

Hence, the tensile stress σ generated in the rod is:

$$\sigma = \frac{F}{A} = \frac{1000}{0.0005} = 2{,}000{,}000 = 2 \times 10^6 \text{ Pa}$$

14.6 Stress-Strain Diagrams

It was demonstrated in Section 14.3 that the results of uniaxial tension tests can be used to obtain a unique curve representing the relationship between the applied load and corresponding deformation for a material. This can be achieved by dividing the applied load with the cross-sectional area (F/A) of the specimen, dividing the amount of elongation measured with the gage length ($\Delta\ell/\ell$), and plotting a F/A versus $\Delta\ell/\ell$ graph. Notice however that for a specimen under tension, F/A is the average tensile stress σ and $\Delta\ell/\ell$ is the average tensile strain ϵ. Therefore, the F/A versus $\Delta\ell/\ell$ graph of a material is essentially the *stress-strain diagram* of that material.

Different materials demonstrate different stress-strain relationships, and the stress-strain diagrams of two or more materials can be compared to determine which material is relatively stiffer, harder, tougher, more ductile, and/or more brittle. Before explaining these concepts related to the strength of materials, it is more appropriate to first analyze a typical stress-strain diagram in detail.

Consider the stress-strain diagram shown in Figure 14.15. There are six distinct points on the curve, which are labeled O, P, E, Y, U, and R. Point O is the origin of the σ-ϵ diagram, which corresponds to the initial no load, no deformation stage. Point P represents the *proportionality limit.* Between O and P, stress and strain are linearly proportional, and the σ-ϵ curve is a straight line. Point E represents the *elastic limit.* The stress corresponding to the elastic limit is the greatest stress that can be applied to the material without causing any permanent deformation within the material. The material will not take its original size and shape upon unloading if it is subjected to stress levels beyond the elastic limit. Point Y is the *yield point*, and the stress σ_y corresponding to the yield point is called the *yield strength* of the material. At this stress level considerable elongation (yielding) can occur without a corresponding increase of load. U is the highest stress point on the σ-ϵ curve. The stress σ_u is the *ultimate strength* of the material. For some materials, once the ultimate strength is reached, the applied load can be decreased and continued yielding may be observed. This is due to the phenomena called *necking* which will be discussed in Section 14.10. The last point on the σ-ϵ curve is R which represents the *rupture* or *failure point.* The stress at which the rupture occurs is called the *rupture strength* of the material.

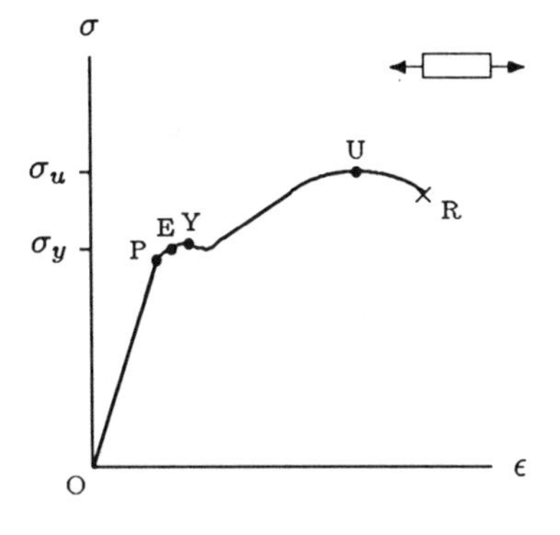

Figure 14.15 *Stress-strain diagram for axial loading.*

For some materials, it may not be easy to determine or distinguish the elastic limit and the yield point. The yield strength of such materials is determined by the *offset method*, illustrated in Figure 14.16. The offset method is applied by drawing a line parallel to the linear section of the stress-strain diagram and passing

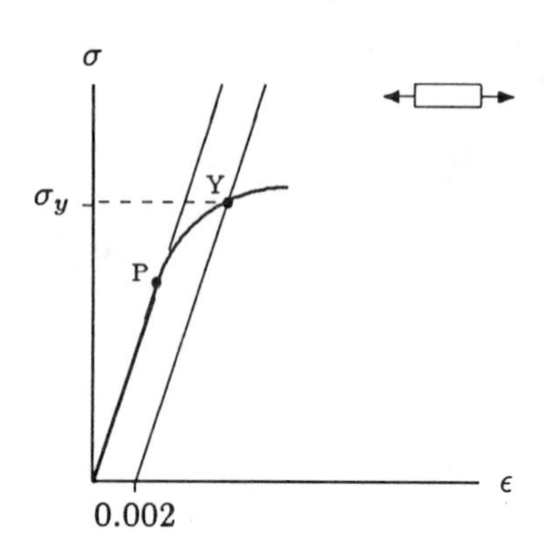

Figure 14.16 *Offset method.*

through a strain level of about 0.2 percent (0.002). The intersection of this line with the σ-ϵ curve is taken to be the yield point, and the stress corresponding to this point is called the *apparent yield strength.*

Note that a given material may behave differently under different load and environmental conditions. If the curve shown in Figure 14.15 represents the stress-strain relationship for a material under tensile loading, there may be a similar but different curve representing the stress-strain relationship for the same material under compressive or shear loading. Also, temperature is known to alter the relationship between stress and strain. For a given material and fixed mode of loading, different stress-strain diagrams may be obtained under different temperatures. Furthermore, the data collected in a particular tension test may depend on the rate at which the tension is applied on the specimen. Some of these factors affecting the relationship between stress and strain will be discussed later.

14.7 Elastic Deformations

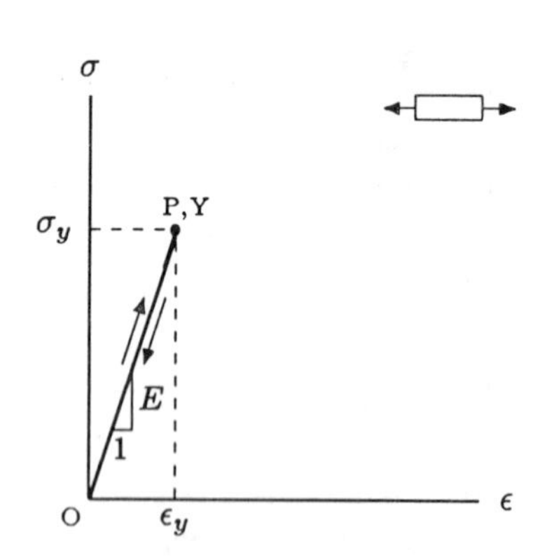

Figure 14.17 *Stress-strain diagram for a linearly elastic material. (↗: loading, ↙: unloading)*

Consider the stress-strain diagram shown in Figure 14.17. Y is the yield point, and in this case, it also represents the proportionality and elastic limits. σ_y is the yield strength and ϵ_y is the corresponding strain. (The σ-ϵ curve beyond the elastic limit is not shown.) The straight line in Figure 14.17 represents the stress-strain relationship in the elastic region. *Elasticity* is defined as the ability of a material to resume its original (stress free) size and shape upon removal of applied loads. In other words, if a load is applied on a material such that the stress generated in the material is equal to or less than σ_y, then the deformations that took place in the elastic material will be completely recovered once the applied loads are removed (the material is unloaded).

An elastic material whose σ-ϵ diagram is a straight line is called a *linearly elastic* material. For such a material, the stress is linearly proportional to strain, and the constant of proportionality is called the *elastic* or *Young's modulus* of the material. Denoting the elastic modulus with E:

$$\sigma = E\,\epsilon \tag{14.5}$$

The elastic modulus, E, is equal to the slope of the σ-ϵ diagram in the elastic region, which is constant for a linearly elastic material. Physically E represents the *stiffness* of a material, such that the larger the elastic modulus, the stiffer the material.

In general, the equation relating the stress and strain for a given material is known as the *constitutive equation* of that material. Eq. (14.5) is the constitutive equation for a linearly elastic material with elastic modulus E when it is subjected to tensile loads.

For a given material, different constitutive equations may exist for different modes of deformation.

The distinguishing factor in linearly elastic materials is their elastic moduli. That is, different materials have different elastic moduli. If the elastic modulus of a material is known, then the mathematical definitions of stress and strain ($\sigma = F/A$ and $\epsilon = \Delta\ell/\ell$) can be substituted into Eq. (14.5) to derive a relationship between the applied load and corresponding deformation. This yields:

$$\Delta\ell = \frac{F\,\ell}{E\,A} \tag{14.6}$$

For a given specimen and applied load, Eq. (14.6) can be used to calculate the corresponding deformation. This equation can be used when the object is under a tensile or compressive force.

Not all elastic materials demonstrate linear behavior. As illustrated in Figure 14.18, the stress-strain diagram of a material in the elastic region may be a straight line up to the proportionality limit followed by a curve. A curve implies varying slope and nonlinear behavior. Therefore, such materials are known as ***nonlinear elastic materials.*** For a nonlinear elastic material, there is not a single elastic modulus because the slope of the σ-ϵ curve is not constant throughout the elastic region. Therefore, the stress-strain relationships for nonlinear materials take more complex forms. Note however that even nonlinear materials may have a linear elastic region in their σ-ϵ diagrams at low stress levels (region between O and P in Figure 14.18).

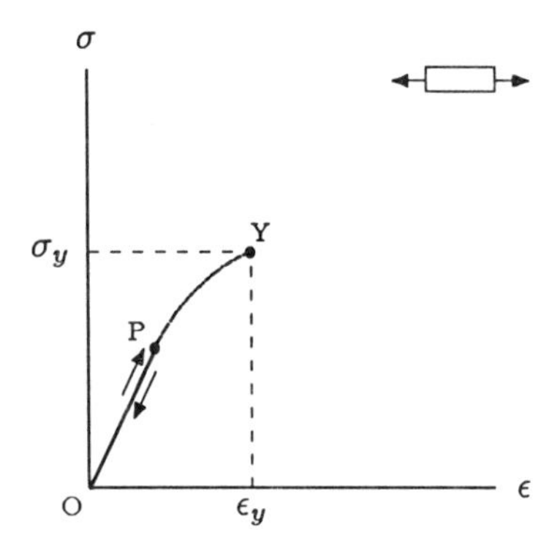

Figure 14.18 *Stress-strain diagram for a nonlinearly elastic material.*

Some materials may exhibit linearly elastic behavior when they are subjected to shear loading (Figure 14.19). For such materials, the shear stress, τ, is linearly proportional to the shear strain, γ, and the constant of proportionality is called the ***shear modulus*** or the ***modulus of rigidity***, commonly denoted as G:

$$\tau = G\,\gamma \tag{14.7}$$

The shear modulus of a given linear material is equal to the slope of the τ-γ curve in the elastic region.

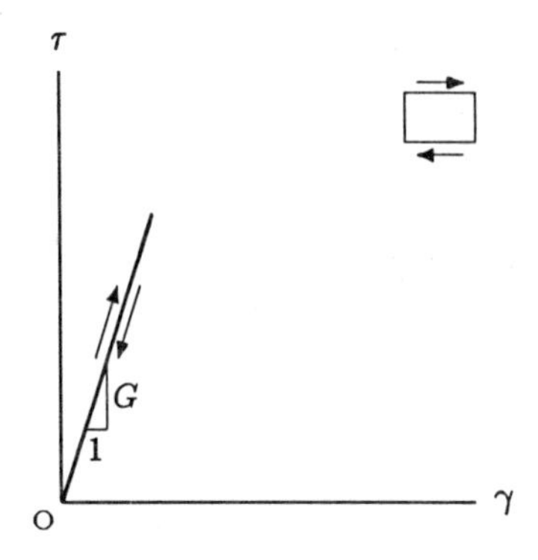

Figure 14.19 *Shear stress versus shear strain diagram for a linearly elastic material.*

14.8 Hooke's Law

The load-bearing characteristics of elastic materials are similar to those of springs, which was first noted by Robert Hooke. Like springs, elastic materials have the ability to store potential energy when they are subjected to externally applied loads. During unloading, it is the release of this energy that causes the material to resume its undeformed configuration. As discussed in Section 11.7, a linear spring subjected to a tensile load will elongate, the amount of elongation being linearly proportional to the applied load (Figure 14.20). The constant of proportionality between

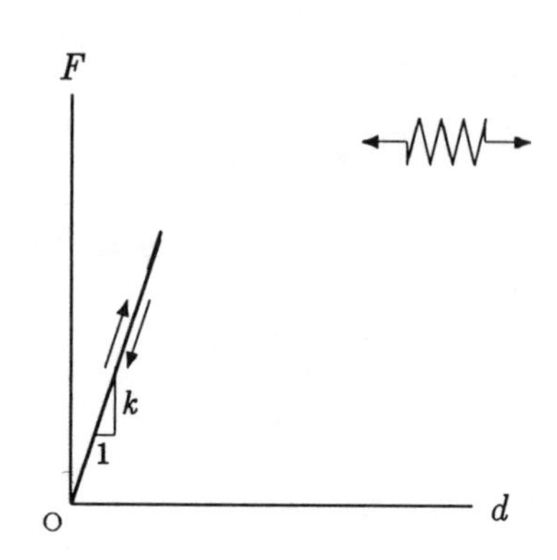

Figure 14.20 *Load-elongation diagram for a linear spring.*

the load and the deformation is usually denoted with k, which is called the ***spring constant*** or ***stiffness*** of the spring. For a linear spring with a spring constant k, the relationship between the applied load F and the amount of elongation d is:

$$F = k\,d \tag{14.8}$$

By comparing Eqs. (14.5) and (14.8), it can be observed that stress in an elastic material is analogous to the force applied to a spring, strain in an elastic material is analogous to the amount of deformation of a spring, and the elastic modulus of an elastic material is analogous to the spring constant of a spring. This analogy between elastic materials and springs is known as *Hooke's Law*.

14.9 Plastic Deformations

We have defined elasticity as the ability of a material to regain completely its original dimensions upon removal of the applied force. Elastic behavior implies the absence of any permanent deformation. On the other hand, ***plasticity*** implies permanent deformations. A material will undergo a plastic deformation following an elastic deformation when it is loaded beyond its elastic limit (or yield point).

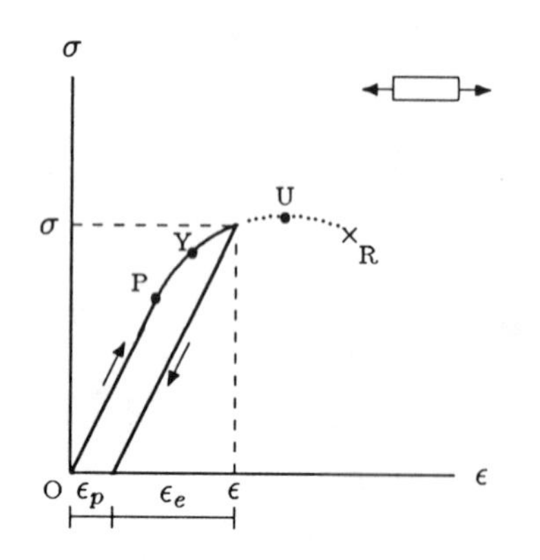

Figure 14.21 *Plastic deformation.*

Consider the stress-strain diagram of a material shown in Figure 14.21. Assume that a specimen made of the same material is subjected to a tensile load and the stress σ in the specimen is brought to such a level that $\sigma > \sigma_y$. The corresponding strain in the specimen is measured as ϵ. Upon removal of the applied load, the material will recover the elastic deformation which had taken place by following an unloading path parallel to the straight line between O and P (the initial linearly elastic region). The point where this path cuts the strain axis is called the plastic strain, ϵ_p. ϵ_p signifies the extent of permanent (unrecoverable) shape change that has taken place in the specimen.

The difference in strains between when the specimen is loaded and unloaded (i.e., $\epsilon - \epsilon_p$) is equal to the amount of elastic strain ϵ_e which had taken place in the specimen and which was recovered upon unloading. Therefore, for a material loaded to a stress level beyond its elastic limit, the total strain is equal to the sum of the elastic and plastic strains:

$$\epsilon = \epsilon_e + \epsilon_p \tag{14.9}$$

The elastic strain, ϵ_e, is completely recoverable upon unloading, whereas the plastic strain, ϵ_p, is a permanent residue of the deformations.

Example 14.2 Two specimens made of two different materials are tested in a uniaxial tension test by applying a force of F=20 kN (20×10^3 N) on each specimen (Figure 14.22). Specimen 1 is an aluminum bar with elastic modulus E_1=70 MPa (70×10^9 Pa) and a rectangular cross-section (1 cm by 2 cm). Specimen 2 is a steel rod with elastic modulus E_2=200 MPa (200×10^9 Pa) and a circular cross-section (radius 1 cm).

Calculate the stresses generated in each specimen. Assuming that the stress in each specimen is below the proportionality limit of the material, calculate the amount of strain in each specimen.

SPECIMEN 1: ALUMINUM BAR

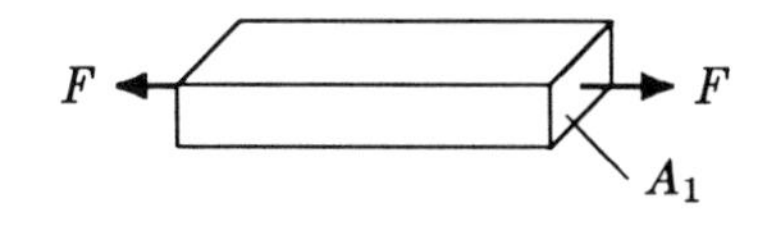

SPECIMEN 2: STEEL ROD

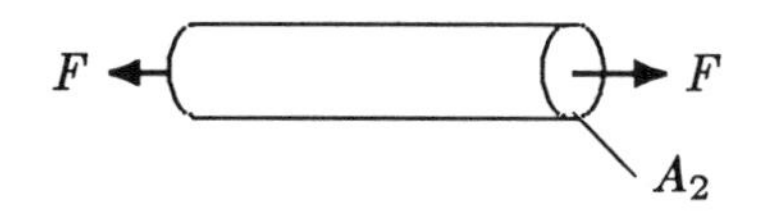

Figure 14.22 *Example 14.2.*

Solution: The tensile stress is equal to the ratio of the applied force and the cross-sectional area of the specimen. The cross-sectional areas of the specimens are:

$$A_1 = (1\text{ cm})(2\text{ cm}) = 2\text{ cm}^2 = 2\times10^{-4}\text{ m}^2$$
$$A_2 = \pi(1\text{ cm})^2 = 3.14\text{ cm}^2 = 3.14\times10^{-4}\text{ m}^2$$

Therefore, the tensile stresses developed in each specimen are:

$$\sigma_1 = \frac{F}{A_1} = \frac{20\times10^3}{2\times10^{-4}} = 100\times10^6\text{ Pa} = 100\text{ MPa}$$
$$\sigma_2 = \frac{F}{A_2} = \frac{20\times10^3}{3.14\times10^{-4}} = 63.7\times10^6\text{ Pa} = 63.7\text{ MPa}$$

That is, the aluminum bar (specimen 1) is stressed more than the steel rod (specimen 2). Note that the smaller the cross-sectional area, the greater the tensile stress.

To calculate the strains corresponding to the stresses σ_1 and σ_2, we can assume that the deformations are elastic and that the stresses σ_1 and σ_2 are below the proportionality limits for aluminum and steel. In other words, the stresses are linearly proportional to strains and the elastic moduli E_1 and E_2 are the constants of proportionality. Hence:

$$\epsilon_1 = \frac{\sigma_1}{E_1} = \frac{100\times10^6}{70\times10^9} = 1.43\times10^{-3}$$
$$\epsilon_1 = \frac{\sigma_2}{E_2} = \frac{63.7\times10^6}{200\times10^9} = 0.32\times10^{-3}$$

These results suggest that the aluminum bar is stretched more than the steel rod.

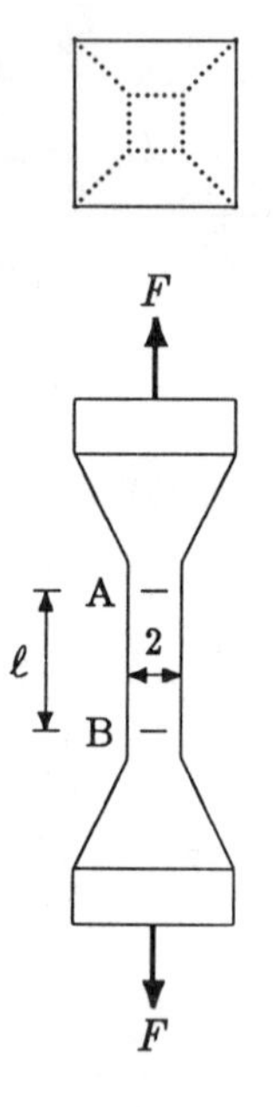

Figure 14.23 *The top and side views of the specimen.*

Example 14.3 An experiment was designed to determine the elastic modulus of the human bone (cortical) tissue. Three almost identical bone specimens were prepared. The specimen size and shape used is shown in Figure 14.23, which has a square (2×2 mm) cross-section. Two sections A and B are marked on each specimen at a fixed distance apart. Each specimen was then subjected to tensile loading of varying magnitudes, and the lengths between the marked sections were again measured electronically. The following data was obtained:

Applied Force, F (N)	Measured Gage Length, ℓ (mm)
0	5.000
240	5.017
480	5.033
720	5.050

Determine the stresses and strains developed in each specimen, plot a stress-strain diagram for the bone, and determine the elastic modulus (E) for the bone.

Solution: The cross-sectional area of each specimen is A=4 mm^2 or 4×10^{-6} m^2. When the applied load is zero, the gage length is 5 mm, which is the original (undeformed) gage length ℓ_0. Therefore, the stress and strain developed in each specimen can be calculated using:

$$\sigma = \frac{F}{A} \qquad \epsilon = \frac{\ell - \ell_0}{\ell_0}$$

The following table lists stresses and strains calculated using the above formulas:

F (N)	$\sigma \times 10^6$ (Pa)	ℓ (mm)	ϵ (mm/mm)
0	0	5.000	0.0
240	60	5.017	0.0034
480	120	5.033	0.0066
720	180	5.050	0.0100

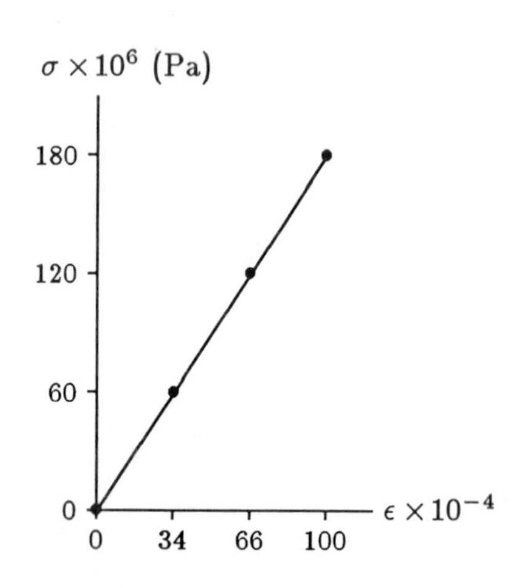

Figure 14.24 *Stress-strain diagram for the bone.*

In Figure 14.24, the stress and strain values computed are plotted to obtain a σ-ϵ graph for the bone. Notice that the relationship between the stress and strain is almost linear, which is indicated in Figure 14.24 by a straight line.

Recall that the elastic modulus of a linearly elastic material is equal to the slope of the straight line representing the σ-ϵ relationship for that material. Therefore, the elastic modulus E for the bone is equal to the slope of the straight line in Figure 14.24, which can be calculated as:

$$E = \frac{\sigma}{\epsilon} = \frac{180 \times 10^6}{0.0100} = 18 \times 10^9 \text{ Pa} = 18 \text{ MPa}$$

Example 14.4 Figure 14.25 illustrates a fixation device consisting of a plate and two screws, which can be used to stabilize fractured bones. Also shown is a sectional view of the device and the bone around the fracture.

During a single leg stance, a person can apply his/her entire weight to the ground via a single foot. In such situations, the weight of the person is applied back on the person through the same foot, which has a compressive effect on the leg, its bones, and joints. In the case of a patient with a fractured leg bone (in this case, the femur), this force is transferred from below to above the fracture through the screws of the fixation device.

If the diameter of the screws is D=5 mm and the weight of the patient is W=700 N, determine the shear stress exerted on the screws of a two-screw fixation device when the patient stands on the leg with a fractured bone.

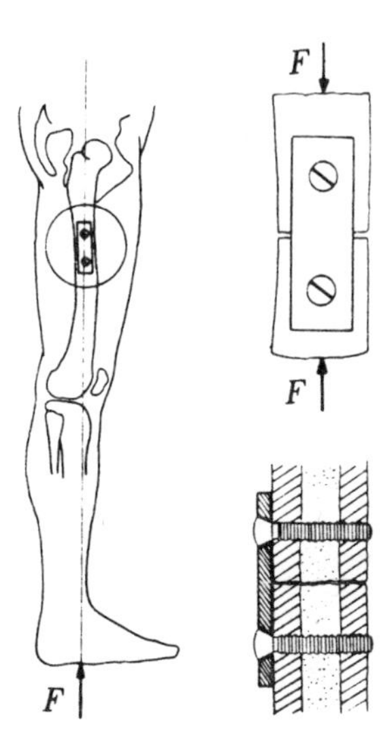

Figure 14.25 *Example 14.4.*

Solution: Free-body diagrams of the fixation device and the screws are shown in Figure 14.26. Note that the screw above the fracture is pushing the plate downward, whereas the screw below the fracture is pushing the plate upward. Each screw is applying a force on the plate equal to the weight of the person. The same magnitude force is also acting on the screws but in the opposite directions. For example, for the screw above the fracture, the plate is exerting an upward force on the head of the screw and the bone is applying a downward force. The effects of the forces applied on the screws are such that they are trying to shear the screws in a plane perpendicular to the lengths of the screws. That is, with respect to the cross-sectional areas of the screws, they are shearing forces.

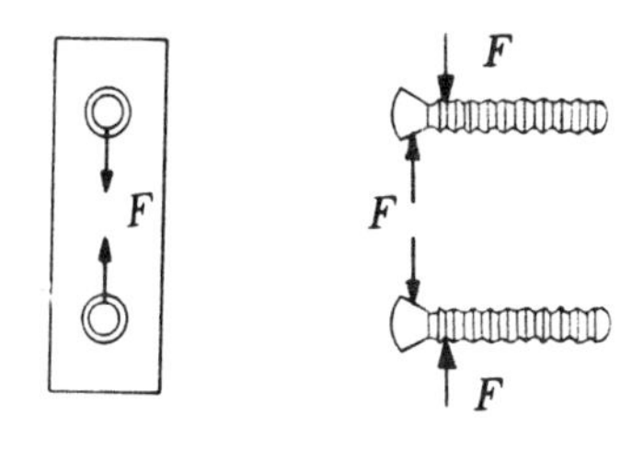

Figure 14.26 *Forces applied on the plate and screws.*

The shear stress τ generated in the screws can be calculated using the following relationship between the shear force F and area A over which the shear stress is desired:

$$\tau = \frac{F}{A}$$

In this case, F is equal to the weight (W=700 N) of the patient. Since the diameter of the screws are given, the cross-sectional area of each screw can be calculated as $A=\pi D^2/4=19.6$ mm^2 or $A=19.6\times10^{-6}$ m^2. Substituting the numerical value of A and $F=W=700$ N into the above formula, and carrying out the computation will yield $\tau=35.7\times10^6$ Pa.

Note that if we had a four-screw rather than a two-screw fixation device as shown in Figure 14.27, then each screw would be subjected to a shearing force equal to half of the total weight of the patient.

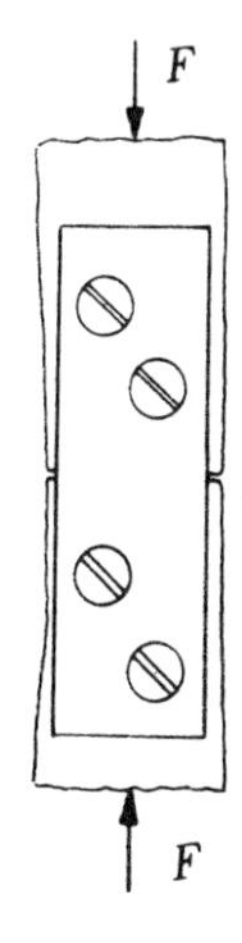

Figure 14.27 *A four-screw fixationdevice.*

14.10 Necking

As defined in Section 14.6, the largest stress a material can endure is called the ultimate strength of that material. Once a material is subjected to a stress level equal to its ultimate strength, an increased rate of deformation can be observed, and in most cases, continued yielding can occur even by reducing the applied load. The material will eventually fail to hold any load and rupture. The stress at failure is called the rupture strength of the material, which may be lower than its ultimate strength. Although this may seem to be unrealistic, the reason is due to a phenomenon called *necking*.

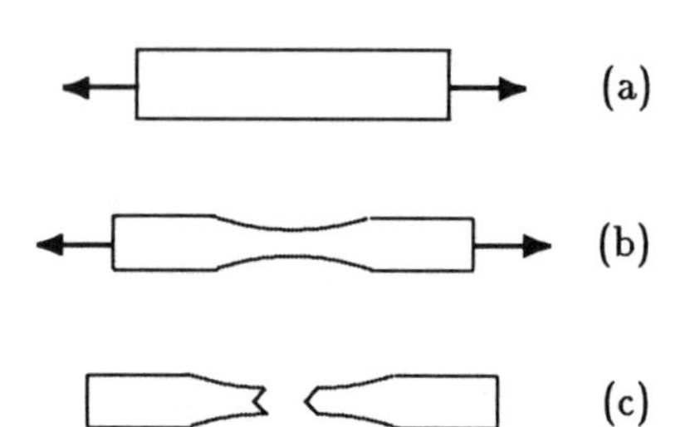

Figure 14.28 *Necking.*

Stresses are usually computed on the basis of the original cross-sectional area of the material. Such stresses are called *conventional stresses*. Calculating a stress by dividing the applied force with the original cross-sectional area is convenient but not necessarily correct. The *true* or *actual stress* calculations must be made by taking the cross-sectional area of the deformed material into consideration. As illustrated in Figure 14.28, under a tensile load a material may elongate in the direction of the applied load but contract in the transverse directions. At stress levels close to the breaking point, the elongation may occur very rapidly and the material may narrow simultaneously. The cross-sectional area at the narrowed section decreases, and although the force required to further deform the material may decrease, the force per unit area (stress) may increase. As illustrated with the dotted curve in Figure 14.29, the actual stress-strain curve may continue having a positive slope, which indicates increasing strain with increasing stress rather than a negative slope, which implies increasing strain with decreasing stress. Also the rupture point and the point corresponding to the ultimate strength of the material may be the same.

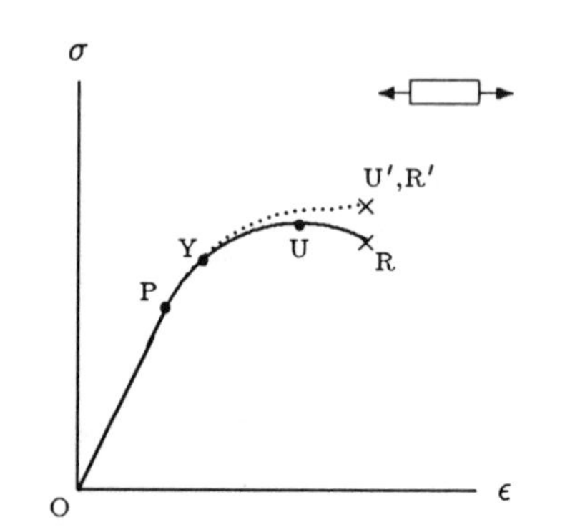

Figure 14.29 *Conventional (solid curve) and actual (dotted curve) stress-strain diagrams.*

14.11 Work and Strain Energy

In dynamics, *work done* is defined as the product of force and the distance traveled in the direction of applied force, and *energy* is the capacity of a system to do work. Stress and strain in deformable body mechanics are respectively related to force and displacement. Stress multiplied by area is equal to force, and strain multiplied by length is displacement. Therefore, the product of stress and strain is equal to the work done on a body per unit volume of that body, or the *internal work* done on the body by the externally applied forces. For an elastic body, this work is stored as an internal *elastic strain energy*, and it is the release of this energy which brings the body back to its original shape upon unloading. The maximum elastic strain energy (per unit volume) which can be stored in a body is equal to the total area under the σ-ϵ diagram in the elastic region (Figure 14.30).

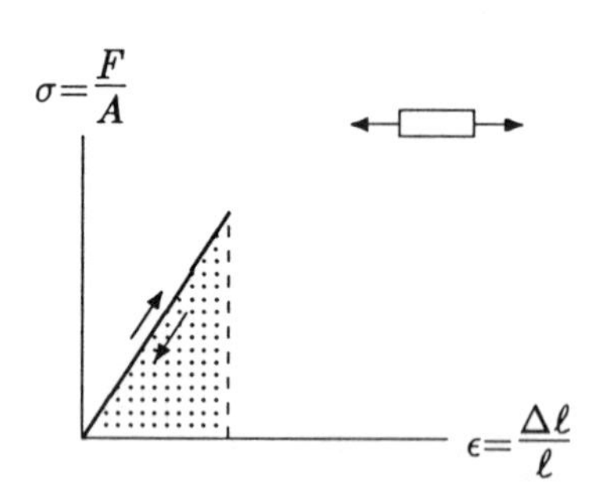

Figure 14.30 *Internal work done and elastic strain energy per unit volume.*

There is also a *plastic strain energy* which is dissipated as heat while deforming the body.

14.12 Strain Hardening

Consider a material whose σ-ϵ diagram in tension is as shown in Figure 14.31. Assume that the material is subjected to a tensile force such that the stress generated is beyond the elastic limit (yield point) of the material. The stress level in the material is designated with point A on the σ-ϵ diagram. Upon removal of the applied force, the material will follow the path designated as AB which is almost parallel to the initial linear section OP of the σ-ϵ diagram. The strain at B corresponds to the amount of plastic deformation in the material. If the material is reloaded, it will exhibit elastic behavior between B and A, the stress at A being the new yield strength of the material. This technique of changing the yield point of a material is called ***strain hardening***. Since the stress at A is greater than the original yield strength of the material, strain hardening increases the yield strength of the material. Upon reloading, if the material is stressed beyond A, then the material will deform according to the original σ-ϵ relationship for the material.

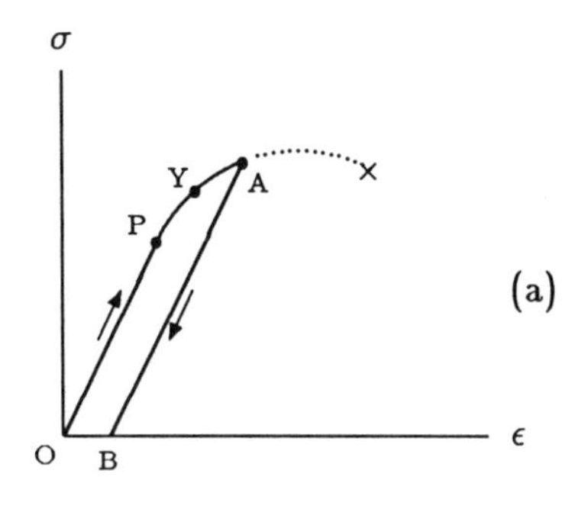

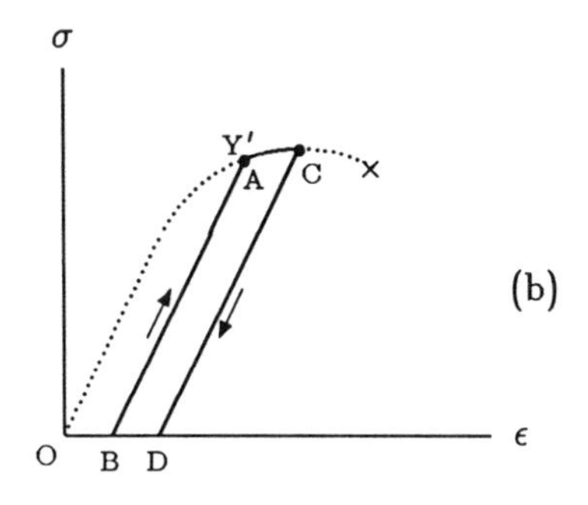

Figure 14.31 *Strain hardening.*

14.13 Hysteresis Loop

Consider the σ-ϵ diagram shown in Figure 14.32. Between O and A a tensile force is applied on the material and the material is deformed beyond its elastic limit. At A, the tensile force is removed, and the line AB represents the unloading path. At B the material is reloaded, this time with a compressive force. At C, the compressive force applied on the material is removed. Between C and O, second unloading occurs, and finally the material resumes its original shape. The loop OABCO is called the *hysteresis loop*, and the area enclosed by this loop is equal to the total strain energy dissipated to deform the body in tension and compression.

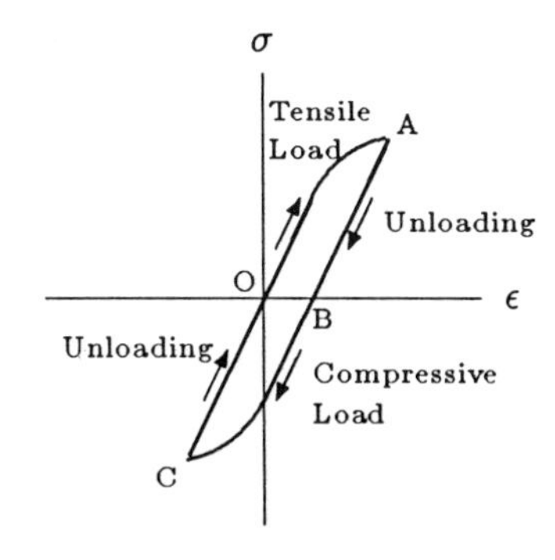

Figure 14.32 *Hysteresis loop.*

14.14 Material Properties Based on Stress-Strain Diagrams

As defined earlier, the elastic modulus of a material is equal to the slope of its stress-strain diagram in the elastic region. The elastic modulus is a relative measure of the *stiffness* of one material with respect to another. The higher the elastic modulus, the stiffer the material and the higher the resistance to deformation. For example, material 1 in Figure 14.33 is stiffer than material 2.

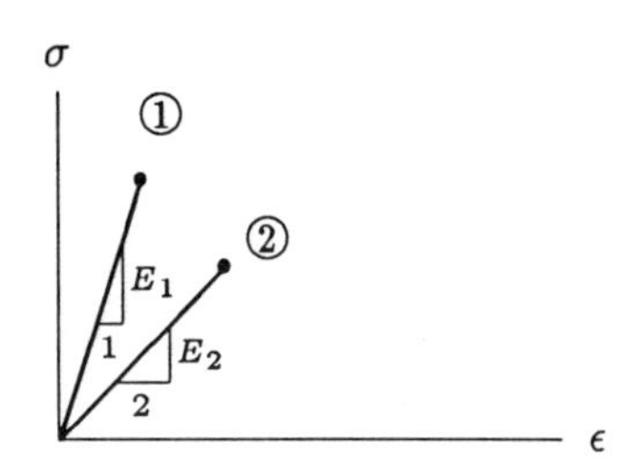

Figure 14.33 *Material 1 is stiffer than material 2.*

A *ductile* material is one that exhibits a large plastic deformation prior to failure. For example, material 1 in Figure 14.34 is more ductile than material 2. A *brittle* material, on the other hand,

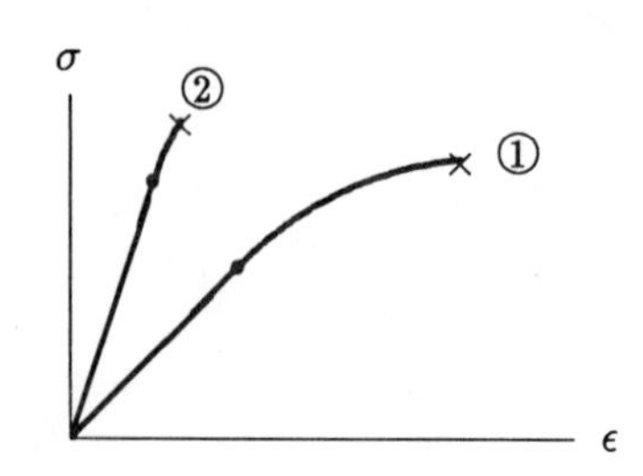

Figure 14.34 *Material 1 is more ductile and less brittle than material 2.*

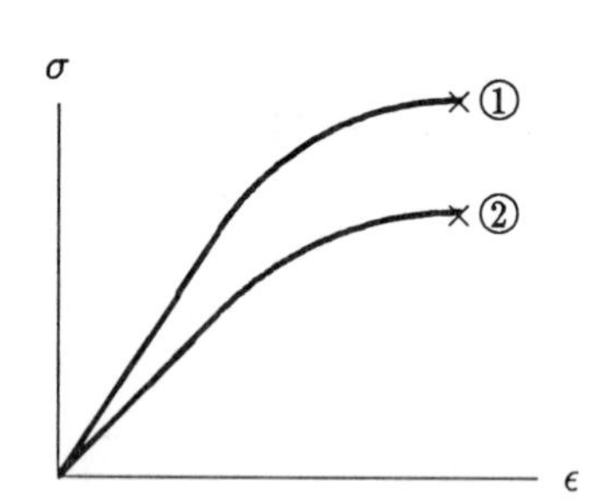

Figure 14.35 *Material 1 is tougher than material 2.*

shows a sudden failure (rupture) without undergoing a considerable plastic deformation. A typical example of a brittle material is glass.

Toughness is a measure of the capacity of a material to sustain permanent deformation. The toughness of a material is measured by considering the total area under its stress-strain diagram. The larger this area, the tougher the material. For example, material 1 in Figure 14.35 is tougher than material 2.

The ability of a material to store or absorb energy without permanent deformation is called the *resilience* of the material. The resilience of a material is measured by its *modulus of resilience* which is equal to the area under the stress-strain curve in the elastic region. The modulus of resilience is equal to $\sigma_y \epsilon_y / 2$ or ${\sigma_y}^2/2E$ for linearly elastic materials.

Although they are not directly related to the stress-strain diagrams, there are other important concepts used to describe material properties. A material is called *homogeneous* if its properties do not vary from location to location within the material. A material is called *isotropic* if its properties are independent of direction or orientation. A material is called *incompressible* if it has a constant density.

14.15 Idealized Models of Material Behavior

The stress-strain diagrams are most useful when they are represented by mathematical functions, called constitutive equations. The stress-strain diagrams of materials may come in various forms, and it may not be possible to find a single mathematical function to represent them. For the sake of mathematical modeling and the analytical treatment of material behavior, these diagrams can be simplified. Some of these diagrams representing certain idealized material behavior are illustrated in Figure 14.36.

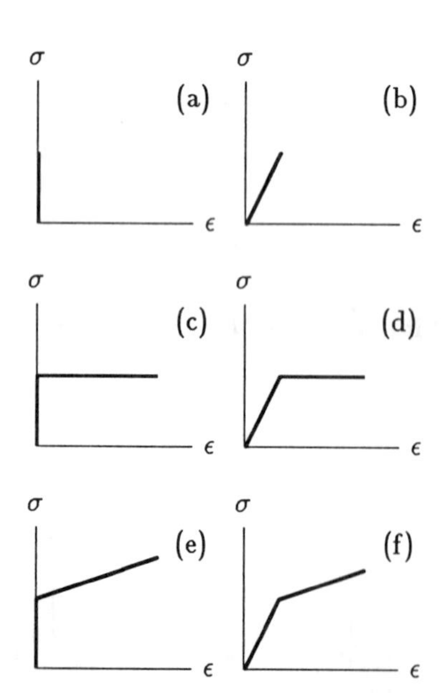

Figure 14.36 *Idealized models of material behavior.*

A *rigid* material is one that cannot be deformed even under very large loads (Figure 14.36a). A *linearly elastic* material is one for which the stress and strain are linearly proportional, with modulus of elasticity being the constant of proportionality (Figure 14.36b). A *rigid-perfectly plastic* material does not exhibit any elastic behavior, and once a critical stress level is reached (Figure 14.36c), it will deform continuously and permanently until failure. A *linearly elastic-perfectly plastic* material, after an elastic response, deforms at very large amounts at a constant stress (Figure 14.36d). Figure 14.36(e) represents the stress-strain diagram for *rigid-linearly plastic* behavior. The stress-strain diagram of a *linearly elastic-linearly plastic* material has two distinct regions with two different slopes (bilinear), in which stresses and strains are linearly proportional to one another (Figure 14.36f).

Example 14.5 Specimens of human cortical bone tissue are subjected to simple tension test until fracture. The test results reveal a stress-strain diagram shown in Figure 14.37, which has three distinct regions. An initial linearly elastic region (between O and A), an intermediate nonlinear elasto-plastic region (between A and B), and a final linearly plastic region (between B and C). The average stresses and corresponding strains at points O, A, B, and C are determined to be:

Point	Stress σ (MPa)	Strain ϵ (mm/mm)
O	0	0.0
A	85	0.005
B	114	0.010
C	128	0.026

Using this information, determine the elastic and strain hardening moduli of the bone tissue in the linear regions of its σ-ϵ diagram. Note that the ***strain hardening modulus*** is the slope of the σ-ϵ curve in the plastic region. Also, find mathematical expressions relating stresses to strains in the linearly elastic and plastic regions.

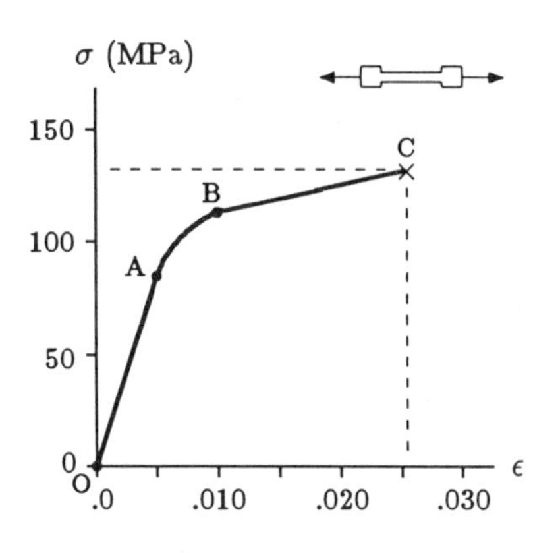

Figure 14.37 *Tensile stress-strain diagram for cortical bone (1 MPa = 10^6 Pa).*

Solution: The elastic modulus E is the slope of the σ-ϵ curve in the elastic region. Between O and A, the bone exhibits linearly elastic material behavior, and the σ-ϵ curve is a straight line. The slope of this straight line is:

$$E = \frac{\sigma_A - \sigma_O}{\epsilon_A - \epsilon_O} = \frac{85 \times 10^6 - 0}{0.005 - 0.0} = 17 \times 10^9 \text{ Pa} = 17 \text{ GPa}$$

The strain hardening modulus E' is equal to the slope of the σ-ϵ curve in the plastic region. Between B and C, the bone exhibits a linearly plastic material behavior, and its σ-ϵ curve is a straight line. Therefore:

$$E' = \frac{\sigma_C - \sigma_B}{\epsilon_C - \epsilon_B} = \frac{128 \times 10^6 - 114 \times 10^6}{0.026 - 0.010} = 0.875 \times 10^9 \text{ Pa}$$

The stress-strain relationship between O and A is:

$$\sigma = E\,\epsilon \quad \text{or} \quad \epsilon = \frac{\sigma}{E}$$

The relationship between σ and ϵ in the linearly plastic region between B and C can be expressed as:

$$\sigma = \sigma_B + E'\,(\epsilon - \epsilon_B) \quad \text{or} \quad \epsilon = \epsilon_B + \frac{1}{E'}(\sigma - \sigma_B)$$

For example, the strain corresponding to a tensile stress of σ=120 MPa can be calculated as:

$$\epsilon = 0.010 + \frac{120 \times 10^6 - 114 \times 10^6}{0.875 \times 10^9} = 0.017$$

14.16* Statically Indeterminate Systems

In Section 5.15 a distinction was made between statically determinate and indeterminate systems. It was stated that a mechanical system is statically determinate if the equations of equilibrium are sufficient to analyze the system. It was demonstrated that the translational and rotational conditions might not be sufficient to analyze some systems that are called statically indeterminate. For statically indeterminate systems, the number of unknown forces and moments exceed the number of equations available from statics. Such systems can be analyzed, however, by taking into consideration the deformability of the members constituting them, which will be demonstrated through the following example.

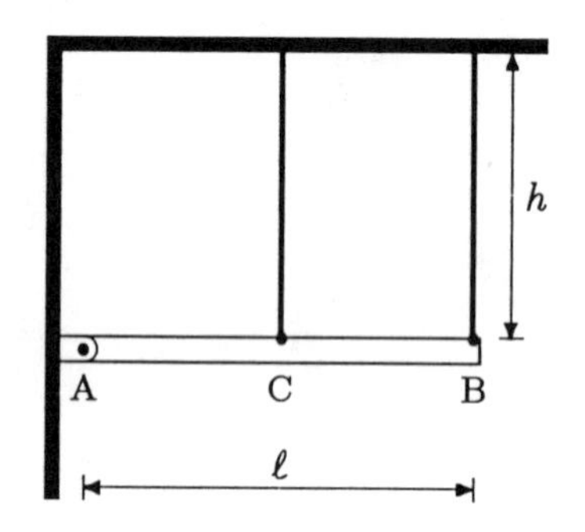

Figure 14.38 *A statically indeterminate system.*

Example 14.6 Consider the structure shown in Figure 14.38. The horizontal beam AB has a length ℓ, weight W, and is hinged to the wall at A. The beam is supported by two identical steel bars of length h, cross-sectional area A, and elastic modulus E. The steel bars are attached to the beam at B and C, where C is equidistant from both ends of the beam.

Taking ℓ=4 m, W=500 N, h=3 m, A=2 cm^2, and E=200 GPa (200×10^9 Pa), determine the forces applied by the beam on the steel bars, the amount of elongation of each bar, and the stresses generated in each bar. (Assume that the beam material is much stiffer than the steel bars.)

Solution: This problem will be analyzed in three stages.

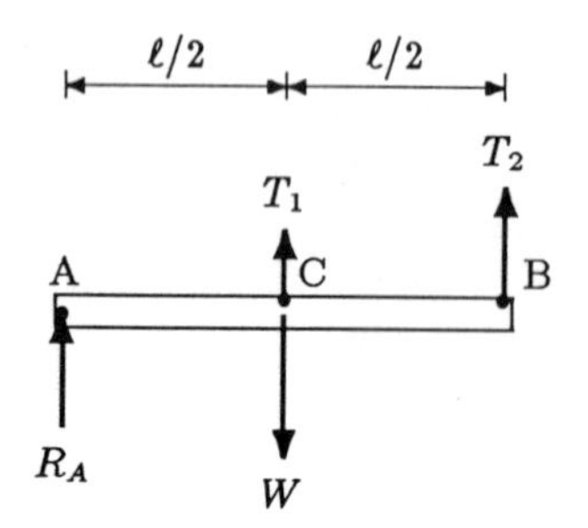

Figure 14.39 *The free-body diagram of the beam.*

Static analysis. The free-body diagram of the beam is shown in Figure 14.39. The total weight W of the beam is assumed to act at its geometric center located at C. R_A is the ground reaction force applied on the beam through the hinge joint at A, and T_1 and T_2 are the forces applied by the steel bars on the beam. Note that only a vertical reaction force is considered at A since there is no horizontal force acting on the beam.

This is a coplanar system, and there are three equations of equilibrium available from statics. Since there is no horizontal force, the horizontal equilibrium of the beam is automatically satisfied. The vertical equilibrium of the beam requires that:

$$\sum F_y = 0: \qquad T_1 + T_2 + R_A = W \qquad (i)$$

For the rotational equilibrium of the beam about point A:

$$\sum M_A = 0: \qquad \frac{1}{2}T_1 + T_2 = \frac{1}{2}W \qquad (ii)$$

We have two equations but three unknowns (R_A, T_1, and T_2). Since the number of unknowns exceeds the number of equations (a statically indeterminate problem), the equations of equilibrium are not sufficient to solve this problem. For the solution of the problem, we have to take into consideration the deformation characteristics of the steel bars.

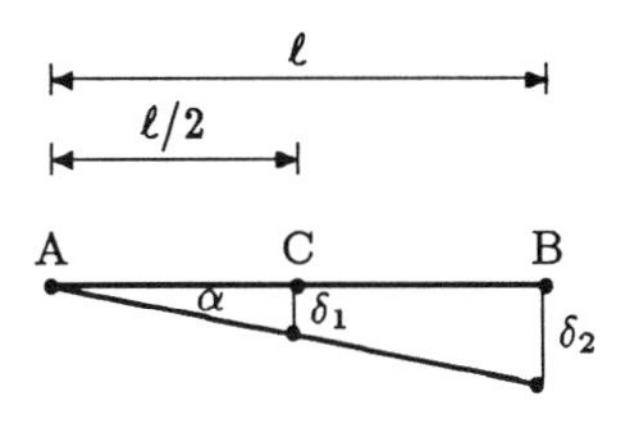

Figure 14.40 *Deflection of the beam.*

Geometric compatibility. The horizontal beam is hinged to the wall at A, and the weight of the beam tends to rotate the beam about A in the clockwise direction. Because of the weight of the beam, which is applied as tensile forces T_1 and T_2 on the bars, the steel bars will deflect (elongate) and the beam will slightly swing about A (Figure 14.40). Because of the forces (R_A, W, T_1, and T_2) acting on the beam, the beam may bend a little as well. Since it is stated that the beam material is much stiffer than the steel bars, we can ignore the deformability of the beam and assume that it maintains its straight configuration. If δ_1 and δ_2 refer to the amount of deflections in the steel bars, then from Figure 14.40:

$$\tan\alpha \simeq \frac{\delta_1}{\ell/2} \simeq \frac{\delta_2}{\ell} \qquad (iii)$$

Note that this relationship is correct when deflections (δ_1 and δ_2) are small for which angle α is small.

Next we need take into consideration the relationship between applied forces and corresponding deformations.

Stress-strain (force-deformation) analyses. The steel bars with length h elongate by δ_1 and δ_2. Therefore, the tensile strains in the bars are:

$$\epsilon_1 = \frac{\delta_1}{h} \qquad \epsilon_2 = \frac{\delta_2}{h} \qquad (iv)$$

The bars are subjected to tensile forces T_1 and T_2. The cross-sectional area of the steel bars is given as A. Therefore, the tensile stresses exerted by the beam on the steel rods are:

$$\sigma_1 = \frac{T_1}{A} \qquad \sigma_2 = \frac{T_2}{A} \qquad (v)$$

Assuming that the stresses involved are within the proportionality limit for steel, we can apply the Hooke's law to relate stresses to strains:

$$\sigma_1 = E\,\epsilon_1 \qquad \sigma_2 = E\,\epsilon_2 \qquad (vi)$$

Now, we can subtitute Eqs. (iv) and (v) into Eqs. (vi) so as to eliminate stresses and strains. This will yield:

$$\delta_1 = \frac{T_1\,h}{E\,A} \qquad \delta_2 = \frac{T_2\,h}{E\,A} \qquad (vii)$$

Note that from Eq. (iii):

$$\delta_2 = 2\,\delta_1 \qquad (viii)$$

Substituting Eqs. (*vii*) into Eq. (*viii*) will yield:

$$T_2 = 2\,T_1 \qquad (ix)$$

Now, we have a total of three equations, Eqs. (*i*), (*ii*), and (*ix*), with three unknowns, R_A, T_1, and R_2. Solving these equations simultaneously will yield:

$$T_1 = \frac{1}{5}W = 100 \text{ N}$$
$$T_2 = \frac{2}{5}W = 200 \text{ N}$$
$$R_A = \frac{2}{5}W = 200 \text{ N}$$

Once tensile forces T_1 and T_2 are computed, Eqs. (*vii*) can be used to calculate the amount of elongations, Eqs. (*v*) can be used to calculate the tensile stresses, and Eqs. (*iv*) can be used to calculate the tensile strains developed in the steel bars:

$$\sigma_1 = \frac{T_1}{A} = 0.5 \times 10^6 \text{ Pa} = 0.5 \text{ MPa}$$
$$\sigma_2 = \frac{T_2}{A} = 1.0 \times 10^6 \text{ Pa} = 1.0 \text{ MPa}$$
$$\epsilon_1 = \frac{\sigma_1}{E} = 2.5 \times 10^{-6}$$
$$\epsilon_2 = \frac{\sigma_2}{E} = 5.0 \times 10^{-6}$$

Note that the calculated strains are very small. Correspondingly, the deformations are very small as predicted earlier while deriving the relationship in Eq. (*iii*). Also note that the stresses developed in the steel bars are much lower than the proportionality limit for steel. Therefore, the assumption made to relate stresses and strains in Eqs. (*vi*) was correct as well.

Chapter 15

MULTIAXIAL DEFORMATIONS, TORSION, AND BENDING

15.1 Poisson's Ratio

When a structure is subjected to uniaxial tension, the transverse dimensions decrease (the structure undergoes lateral contractions) while simultaneously elongating in the direction of the applied load. This was illustrated in the previous chapter through the phenomenon called necking. For stresses within the proportionality limit, the results of uniaxial tension and compression experiments suggest that the ratio of deformations occurring in the axial and lateral directions is constant. For a given material, this constant is called the *Poisson's ratio* and is commonly denoted by the symbol ν (nu):

$$\nu = -\frac{\text{lateral strain}}{\text{axial strain}}$$

Consider the rectangular bar with dimensions a, b, and c shown in Figure 15.1. To be able to differentiate strains involved in different directions, a rectangular coordinate system is adopted. The bar is subjected to tensile forces of magnitude F_x in the x direction. The tensile force produces a tensile stress σ_x within the material (Figure 15.2). Assuming that this stress is uniformly distributed over the cross-sectional area (A=ab) of the bar, its magnitude can be determined using:

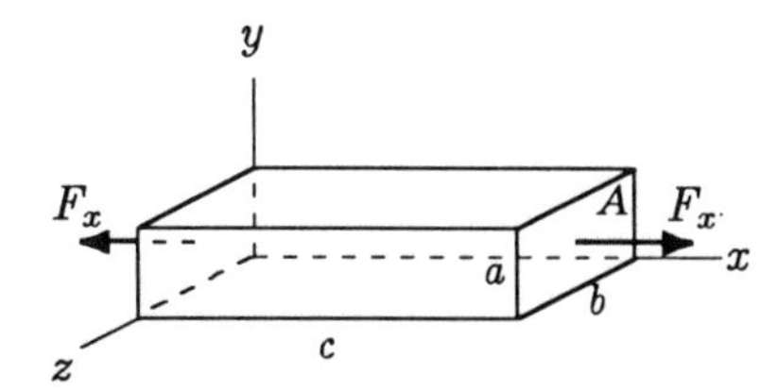

Figure 15.1 *A rectangular bar subjected to uniaxial tension.*

$$\sigma_x = \frac{F_x}{A} \tag{15.1}$$

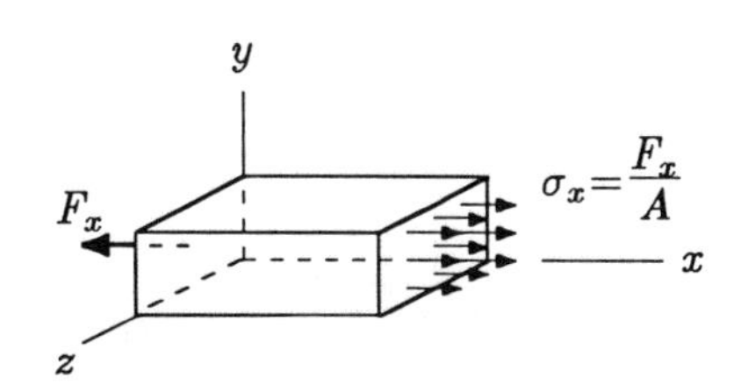

Figure 15.2 *Stress distribution is uniform over the cross-sectional area A=ab of the bar.*

The material elongates in the x direction, and contracts in the y and z directions. If the elastic modulus E of the bar material is known and the deformations involved are within the proportionality limit, then the stress and strain in the x direction are related through the Hooke's law which can be expressed as:

$$\epsilon_x = \frac{\sigma_x}{E} \tag{15.2}$$

Equation (15.2) yields the unit deformation of the bar in the direction of the applied force. Strains in the lateral directions can be determined by utilizing the definition of the Poisson's ratio. If ϵ_y and ϵ_z are the unit contractions in the y and z directions due to the uniaxial loading in the x direction, then the Poisson's ratio is:

$$\nu = -\frac{\epsilon_y}{\epsilon_x} = -\frac{\epsilon_z}{\epsilon_x} \tag{15.3}$$

In other words, if the Poisson's ratio of the bar material is known, then the strains in the lateral directions can be determined:

$$\epsilon_y = \epsilon_z = -\nu\,\epsilon_x = -\nu\,\frac{\sigma_x}{E} \tag{15.4}$$

The minus signs in Eqs. (15.3) and (15.4) indicate a decrease in the lateral dimensions when there is an increase in the longitudinal direction. Strains ϵ_y and ϵ_z are negative when ϵ_x is positive,

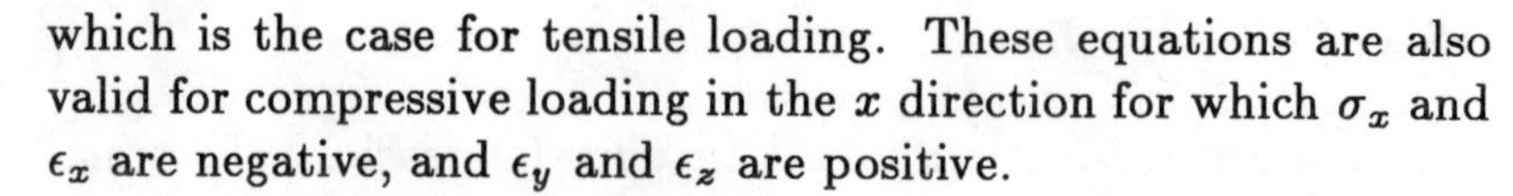

which is the case for tensile loading. These equations are also valid for compressive loading in the x direction for which σ_x and ϵ_x are negative, and ϵ_y and ϵ_z are positive.

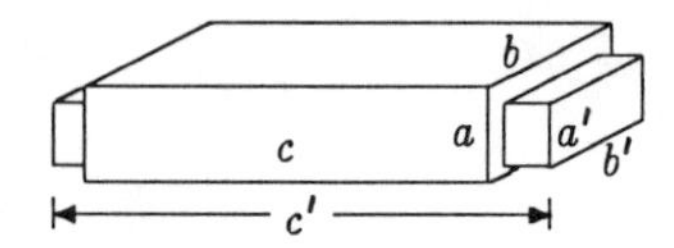

Figure 15.3 *The bar elongates and undergoes lateral contractions simultaneously.*

Notice that once the strains in all three directions are determined, then the deformed dimensions a', b' and c' (Figure 15.3) of the bar can also be calculated. By definition, strain is equal to the ratio of the change in length and the original length. Therefore:

$$\epsilon_x = \frac{c' - c}{c}$$

Solving this equation for the deformed length c' of the object in the x direction will yield $c'=(1+\epsilon_x)c$. Similarly, $a'=(1+\epsilon_y)a$ and $b'=(1+\epsilon_z)b$.

Also note that the stress-strain relationships provided here are valid only for linearly elastic materials, or within the proportionality limits of any elastic-plastic material.

For a given elastic material, the elastic modulus, shear modulus, and Poisson's ratio are related through the expression:

$$G = \frac{E}{2(1+\nu)} \tag{15.5}$$

This formula can be used to calculate the Poisson's ratio of a material whose elastic and shear moduli are known.

15.2 Biaxial and Triaxial Stresses

When an object is subjected to uniaxial loading, strains can occur in all three directions. The strains in the lateral directions can be calculated by utilizing the definition of Poisson's ratio. Poisson's ratio also makes it possible to analyze situations in which there is more than one normal stress acting in more than one direction.

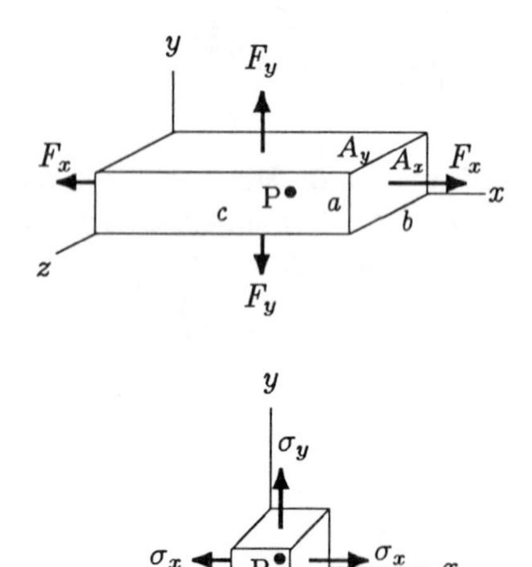

Figure 15.4 *A rectangular bar subjected to tensile forces in the x and y directions and a material element under biaxial stresses.*

Consider the rectangular bar shown in Figure 15.4. The bar is subjected to biaxial loading in the xy plane. Let P be a point in the bar. Stresses acting at point P can be analyzed by constructing a cubical material element around the point. A cubical material element whose sides are parallel to the sides of the bar itself is shown in Figure 15.4, along with the stresses acting on it. σ_x and σ_y are the magnitudes of the normal stresses due to the tensile forces applied on the bar in the x and y directions, respectively:

$$\sigma_x = \frac{F_x}{A_x} = \frac{F_x}{a\,b}$$

$$\sigma_y = \frac{F_y}{A_y} = \frac{F_y}{b\,c}$$

The effects of these biaxial stresses are illustrated graphically in Figure 15.5. Stress σ_x elongates the material in the x direction and causes a contraction in the y (also z) direction. Strains due to σ_x in the x and y directions are:

$$\epsilon_{x1} = \frac{\sigma_x}{E}$$

$$\epsilon_{y1} = -\nu\,\epsilon_{x1} = -\nu\,\frac{\sigma_x}{E}$$

Similarly, σ_y elongates the material in the y direction and causes a contraction in the x direction (Figure 15.5b):

$$\epsilon_{y2} = \frac{\sigma_y}{E}$$

$$\epsilon_{x2} = -\nu\,\epsilon_{y2} = -\nu\,\frac{\sigma_y}{E}$$

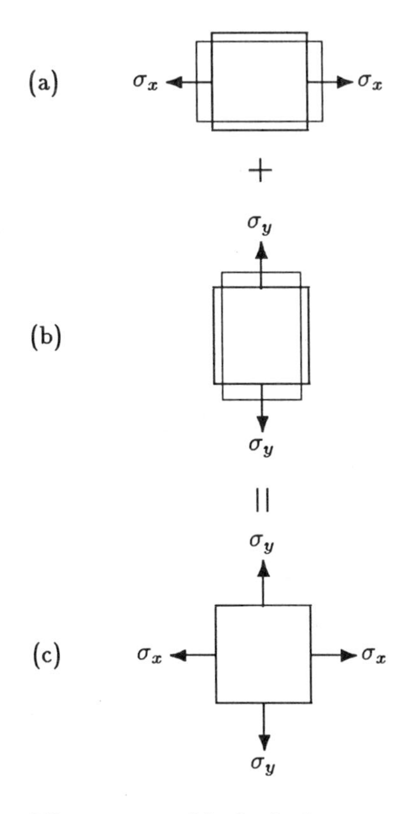

Figure 15.5 *Method of superposition.*

The combined effect of σ_x and σ_y on the plane material element is shown in Figure 15.5(c). The same effect can be represented mathematically by adding the individual effects of σ_x and σ_y. The resultant strains in the x and y directions can be determined as:

$$\begin{aligned} \epsilon_x &= \epsilon_{x1} + \epsilon_{x2} = \frac{\sigma_x}{E} - \nu\,\frac{\sigma_y}{E} \\ \epsilon_y &= \epsilon_{y1} + \epsilon_{y2} = \frac{\sigma_y}{E} - \nu\,\frac{\sigma_x}{E} \end{aligned} \tag{15.6}$$

If required, these equations can be solved simultaneously to express stresses in terms of strains:

$$\begin{aligned} \sigma_x &= \frac{(\epsilon_x + \nu\,\epsilon_y)\,E}{1-\nu^2} \\ \sigma_y &= \frac{(\epsilon_y + \nu\,\epsilon_x)\,E}{1-\nu^2} \end{aligned} \tag{15.7}$$

This discussion can be extended to derive the following stress-strain relationships for the case of triaxial loading (Figure 15.6):

$$\begin{aligned} \epsilon_x &= \frac{1}{E}\left[\sigma_x - \nu\,(\sigma_y + \sigma_z)\right] \\ \epsilon_y &= \frac{1}{E}\left[\sigma_y - \nu\,(\sigma_z + \sigma_x)\right] \\ \epsilon_z &= \frac{1}{E}\left[\sigma_z - \nu\,(\sigma_x + \sigma_y)\right] \end{aligned} \tag{15.8}$$

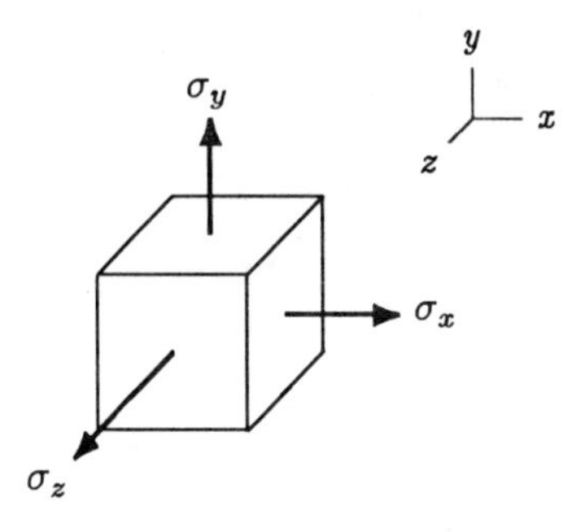

Figure 15.6 *Triaxial stresses.*

These formulas are valid for linearly elastic materials, and they can be used when the stresses induced are tensile or compressive, by adapting the convention that tensile stresses are positive and the compressive stresses are negative.

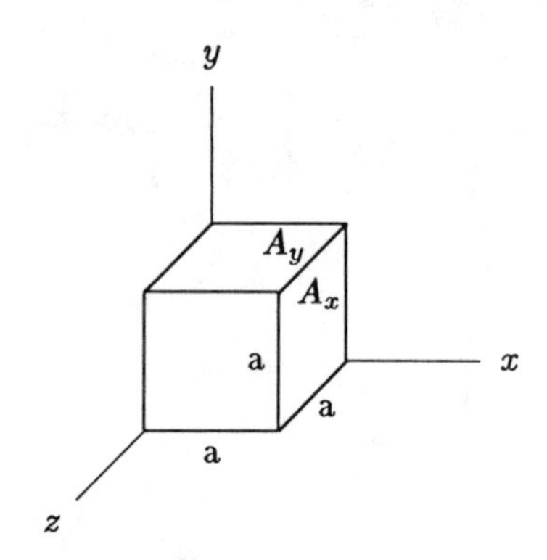

Figure 15.7 *Example 15.1.*

Example 15.1 Consider the cube with sides a=10 cm shown in Figure 15.7. This block is tested under biaxial forces that are applied in the x and y directions. Assume that the forces applied have equal magnitudes of F_x=F_y=F=2×10^6 N, and that the elastic modulus and Poisson's ratio of the block material are given as E=2×10^{11} Pa and ν=0.3.

Determine the strains in the x, y, and z directions, and the deformed dimensions of the block if:
(a) both F_x and F_y are tensile (Figure 15.8a),
(b) F_x is tensile and F_y is compressive (Figure 15.8b), and
(c) both F_x and F_y are compressive (Figure 15.8c).

Solution: To be able to compute stresses involved, we need to know the areas to which forces F_x and F_y are applied. Let A_x and A_y be the areas of the sides of the cube whose normals are in the x and y directions, respectively. Since the object is a cube, these areas are equal:

$$A_x = A_y = A = a^2 = 100\ \text{cm}^2 = 1 \times 10^{-2}\ \text{m}^2$$

We can now calculate the normal stresses in the x and y directions. Notice that the magnitudes of the applied forces in the x and y directions and the areas upon which they are applied are equal. Therefore, the magnitudes of the stresses in the x and y directions are equal as well:

$$\sigma_x = \sigma_y = \sigma = \frac{F}{A} = \frac{2 \times 10^6\ \text{N}}{1 \times 10^{-2}\ \text{m}^2} = 2 \times 10^8\ \text{Pa}$$

The normal stress component in the z direction is zero, since there is no force applied on the block in that direction.

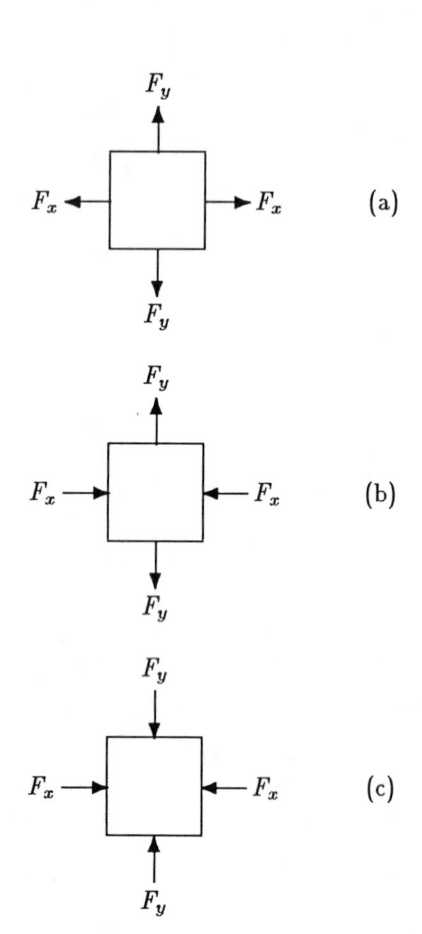

Figure 15.8 *Various types of biaxial loading of the block.*

Case (a) Both forces are tensile, and therefore, both σ_x and σ_y are tensile and positive (Figure 15.8a). We can utilize Eqs. (15.8) to calculate the strains involved. Since σ_x=σ_y=σ and σ_z=0, these equations can be simplified as:

$$\epsilon_x = \frac{1}{E}(\sigma_x - \nu\,\sigma_y) = \frac{1-\nu}{E}\,\sigma$$
$$\epsilon_y = \frac{1}{E}(\sigma_y - \nu\,\sigma_x) = \frac{1-\nu}{E}\,\sigma$$
$$\epsilon_z = -\frac{\nu}{E}(\sigma_x + \sigma_y) = -\frac{2\,\nu}{E}\,\sigma$$

Substituting the numerical values of E, ν, σ_x, and σ_y into these equations, and carrying out the computations will yield:

$$\epsilon_x = 0.7 \times 10^{-3}$$
$$\epsilon_y = 0.7 \times 10^{-3}$$
$$\epsilon_z = -0.6 \times 10^{-3}$$

Note that both ϵ_x and ϵ_y are positive, while ϵ_z is negative. As a result of the applied forces, the dimensions of the block in the x and y directions increase, while its dimension in the z direction decreases. If a_x, a_y, and a_z designate the new (deformed) dimensions of the block, then:

$$a_x = (1+\epsilon_x)\,a = (1+0.0007)(10\text{ cm}) = 10.007\text{ cm}$$
$$a_y = (1+\epsilon_y)\,a = (1+0.0007)(10\text{ cm}) = 10.007\text{ cm}$$
$$a_z = (1+\epsilon_z)\,a = (1-0.0006)(10\text{ cm}) = 9.994\text{ cm}$$

Case (b) In this case, σ_x is tensile and positive while σ_y is compressive and negative (Figure 15.8b). The stress-strain relationships take the following special forms:

$$\epsilon_x = \frac{1}{E}\,(\sigma_x + \nu\,\sigma_y) = \frac{1+\nu}{E}\,\sigma$$
$$\epsilon_y = \frac{1}{E}\,(-\sigma_y - \nu\,\sigma_x) = -\frac{1+\nu}{E}\,\sigma$$
$$\epsilon_z = -\frac{\nu}{E}\,(\sigma_x - \sigma_y) = 0$$

Substituting the numerical values of the parameters involved into these equations, and carrying out the computations will yield:

$$\epsilon_x = 1.3 \times 10^{-3}$$
$$\epsilon_y = -1.3 \times 10^{-3}$$
$$\epsilon_z = 0$$

The tensile force applied in the x direction and the compressive force applied in the y direction elongates the block in the x direction and reduces its dimension in the y direction. In the z direction, the contraction caused by the tensile stress σ_x is counterbalanced by the expansion caused by the compressive stress σ_y.

Case (c) In this case, both forces are compressive. Therefore, both σ_x and σ_y are compressive and negative (Figure 15.8c). The simplified equations relating normal strains to normal stresses for this case are:

$$\epsilon_x = \frac{1}{E}\,(-\sigma_x + \nu\,\sigma_y) = -\frac{1-\nu}{E}\,\sigma$$
$$\epsilon_y = \frac{1}{E}\,(-\sigma_y + \nu\,\sigma_x) = -\frac{1-\nu}{E}\,\sigma$$
$$\epsilon_z = -\frac{\nu}{E}\,(-\sigma_x - \sigma_y) = \frac{2\,\nu}{E}\,\sigma$$

Substituting the numerical values and carrying out the computations will yield:

$$\epsilon_x = -0.7 \times 10^{-3}$$
$$\epsilon_y = -0.7 \times 10^{-3}$$
$$\epsilon_z = 0.6 \times 10^{-3}$$

15.3 Stress and Strain Tensors

Further generalization of stress-strain relationships for linearly elastic materials should take into consideration the relationships between shear stresses and shear strains. As illustrated in Figure 15.9, the most general case of material loading occurs when an object is subjected to normal and shear stresses in three mutually perpendicular directions. Corresponding to these stresses there are normal and shear strains. Relationships between the normal stresses (σ_x, σ_y, σ_z) and normal strains (ϵ_x, ϵ_y, ϵ_z) are given in Eq. (15.8).

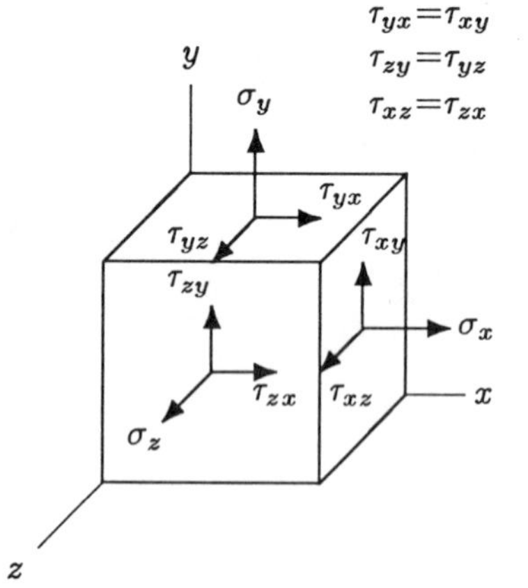

Figure 15.9 *Normal and shear components of the stress tensor.*

In Figure 15.9, shear stresses are identified by double subscripted symbols. The first subscript indicates the direction normal to the surface over which the shear stress is acting and the second subscript indicates the direction in which the stress is applied. A total of six shear stresses (τ_{xy}, τ_{yx}, τ_{yz}, τ_{zy}, τ_{zx}, τ_{xz}) are defined. However, the condition of static equilibrium requires that $\tau_{xy}=\tau_{yx}$, $\tau_{yz}=\tau_{zy}$, and $\tau_{zx}=\tau_{xz}$. Therefore, there are only three independent shear stresses (τ_{xy}, τ_{yz}, τ_{zx}), and three corresponding shear strains (γ_{xy}, γ_{yz}, γ_{zx}). For linearly elastic materials, shear stresses are linearly proportional to shear strains and the shear modulus G is the constant of proportionality:

$$\begin{aligned} \tau_{xy} &= G\,\gamma_{xy} \\ \tau_{yz} &= G\,\gamma_{yz} \\ \tau_{zx} &= G\,\gamma_{zx} \end{aligned} \tag{15.9}$$

As illustrated in Figure 15.10, for two-dimensional problems in the xy plane, only one shear stress (τ_{xy}) and two normal stresses (σ_x, σ_y) need to be considered.

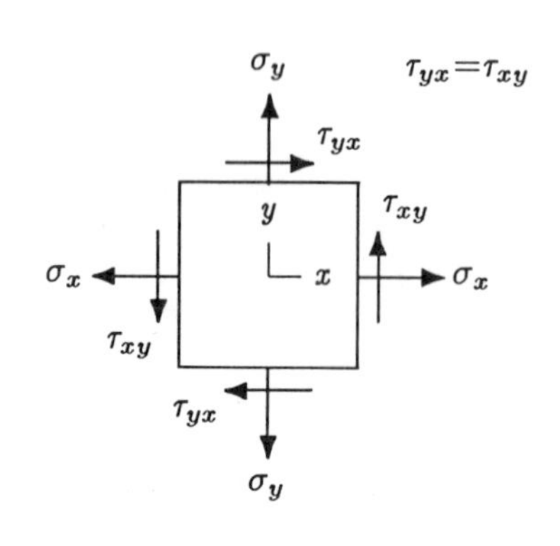

Figure 15.10 *Plain stress components.*

The above discussion indicates that stress and strain can have up to six independent components. Stress and strain are known as *second order tensors.* Recall that a vector quantity has three independent components, and a scalar quantity has one independent component, its magnitude. Vectors are also known as *first order tensors,* while scalars are *zero order tensors.*

It is important to know that components of the stress and strain tensors at a material point may vary with respect to the coordinate system adapted. However, if the state of stress (or strain) with respect to one coordinate frame is known, then the state of stress with respect to another coordinate frame can be determined through appropriate coordinate transformations. The transformation of stress components from one set of coordinates to another will be studied in the following chapter.

15.4 Torsion

Torsion is one of the fundamental modes of loading resulting from the twisting action of applied forces. Here, torsional analyses will be limited to circular shafts. The analyses of structures with noncircular cross-sections subjected to torsional loading are complex and beyond the scope of this text.

Consider the solid circular shaft shown in Figure 15.11. The shaft has a length ℓ and a radius r_o. AB represents a straight line on the outer surface of the shaft which is parallel to the centerline of the shaft. Note that a plane passing through the centerline and cutting the shaft into two semicylinders is called a ***longitudinal plane***, and a plane perpendicular to the longitudinal planes is called a ***transverse plane*** (plane abcd in Figure 15.13). In this case, line AB lies along one of the longitudinal planes. The shaft is mounted to the wall at one end, and a twisting torque with magnitude M is applied in the counterclockwise direction at the other end (shown in Figure 15.11 with a double-headed arrow). Due to the externally applied torque, the shaft deforms in such a way that the straight line AB is twisted into a helix AB′. The deformation at A is zero because the shaft is fixed at that end. The extent of deformation increases in the direction from the fixed end towards the free end. Angle γ in Figure 15.11 is a measure of the deformation of the shaft, and it represents the shear strain due to the shear stresses induced in the transverse planes. The tangent of angle γ is approximately equal to the ratio of the arc length BB′ and the length ℓ of the shaft. For small deformations, the tangent of this angle is approximately equal to the angle itself measured in radians. Therefore:

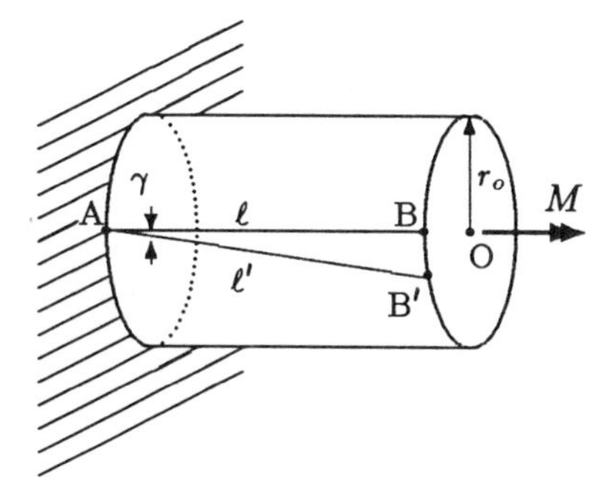

Figure 15.11 *A circular shaft subjected to torsion.*

$$\gamma = \frac{\text{arc length BB}'}{\ell} \tag{15.10}$$

As illustrated in Figure 15.12, the amount of deformation within the shaft also varies with respect to the radial distance r measured from the centerline of the shaft. This variation is such that the deformation is zero at the center, increases towards the rim, and reaches a maximum on the outer surface. Angle θ in Figure 15.12, which is called the ***angle of twist***, is a measure of the extent of the twisting action that the shaft suffers. From the geometry of the problem:

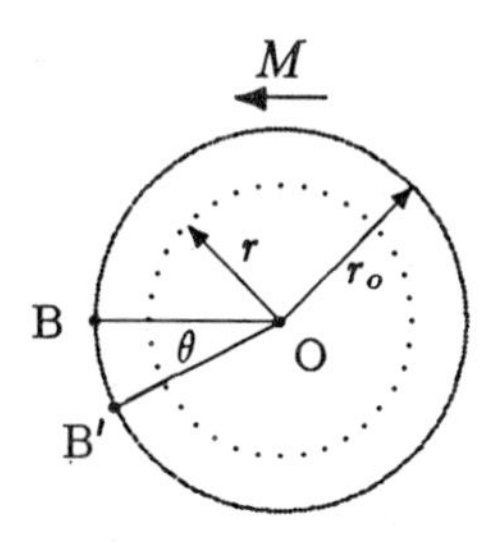

Figure 15.12 θ *is the angle of twist.*

$$\theta = \frac{\text{arc length BB}'}{r_o} \tag{15.11}$$

Equations (15.10) and (15.11) can be combined together by eliminating the arc length BB′. Solving the resulting equation for the angle of twist will yield:

$$\theta = \frac{\ell}{r_o}\gamma \tag{15.12}$$

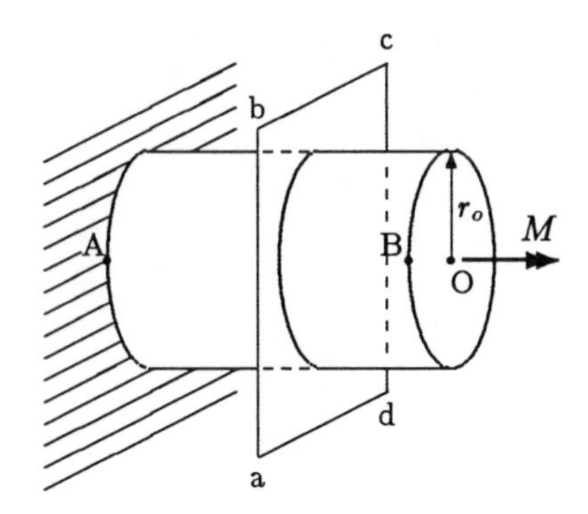

Figure 15.13 *A plane perpendicular to the centerline cuts the shaft into two.*

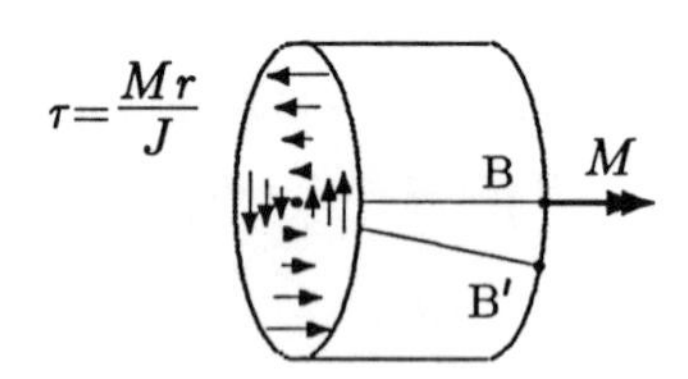

Figure 15.14 *Shear stress, τ, distribution over the cross-sectional area of the shaft.*

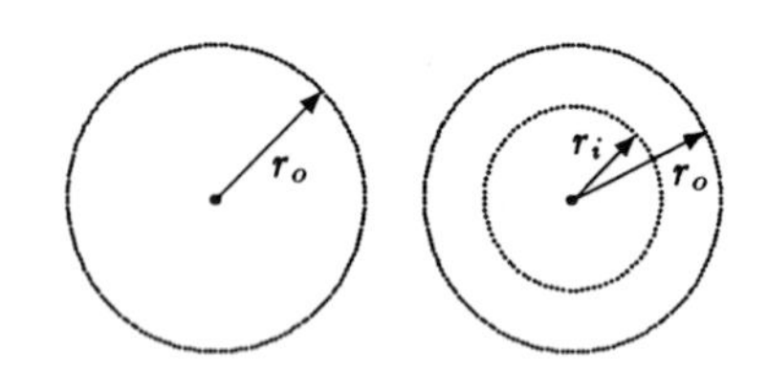

(a) Solid shaft (b) Hollow shaft

$J = \frac{\pi r_o^4}{2}$ $J = \frac{\pi (r_o^4 - r_i^4)}{2}$

Figure 15.15 *Polar moments of inertia for circular cross-sections.*

Now consider a plane perpendicular to the centerline of the shaft (plane abcd in Figure 15.13) which cuts the shaft into two segments. Since the shaft as a whole is in static equilibrium, its individual parts have to be in static equilibrium as well. This condition requires the presence of internal shearing forces distributed over the cross-sectional area (Figure 15.14). The intensity of these internal forces (force per unit area) is the shear stress τ. The magnitude of the shear stress is related to the magnitude M of the applied torque, the cross-sectional area of the shaft, and the radial distance r between the centerline and the point at which the shear stress is to be determined. This relationship can be determined by satisfying the static equilibrium of either the right-hand or the left-hand segment of the shaft, which will yield:

$$\tau = \frac{M\,r}{J} \tag{15.13}$$

This is known as the *torsion formula*. In Eq. (15.13), J is the *polar moment of inertia* of the cross-sectional area about the centerline of the shaft. The polar moment of inertia for a solid circular shaft with radius r_o (Figure 15.15a) about its centerline is:

$$J = \frac{\pi\, r_o^{\,4}}{2}$$

The polar moment of inertia for a hollow circular shaft (Figure 15.15b) with inner radius r_i and outer radius r_o is:

$$J = \frac{\pi}{2}\,(r_o^{\,4} - r_i^{\,4})$$

The polar moment of inertia has the dimension of length to the power four, and therefore, has a unit m^4 in SI.

If the shaft material is linearly elastic, or the deformations are within the proportionality limit, then the stress and strain must be linearly proportional. In the case of shear loading, the constant of proportionality is the shear modulus of elasticity G of the material:

$$\tau = G\,\gamma \tag{15.14}$$

Solving Eq. (15.14) for γ and substituting Eq. (15.13) into the resulting equation will yield:

$$\gamma = \frac{\tau}{G} = \frac{M\,r}{G\,J} \tag{15.15}$$

On the circumference of the shaft, r in Eq. (15.15) is equal to r_o. Substituting Eq. (15.15) into Eq. (15.12), a more useful expression for the angle of twist can be obtained:

$$\theta = \frac{M\,\ell}{G\,J} \tag{15.16}$$

15.5 Some Remarks About Torsional Stresses

• To derive the torsion formula several assumptions and idealizations were made. For example, it is assumed that the material is isotropic, homogeneous, and linearly elastic.

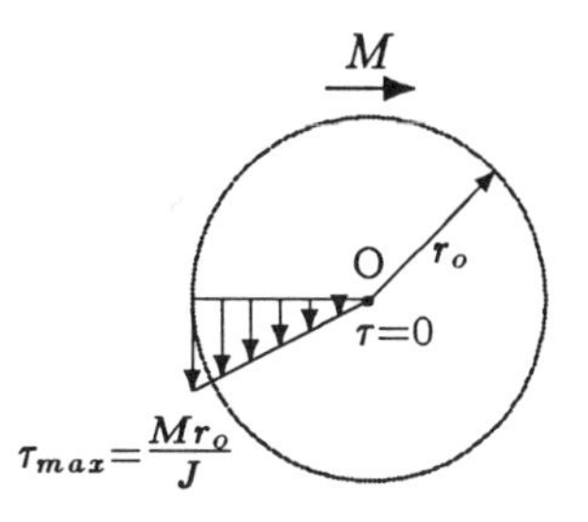

Figure 15.16 *Shear stress varies linearly with radial distance.*

• For a given shaft and applied torque, the torsional shear stress τ is a linear function of the radial distance r measured from the center of the shaft. The shear stress is distributed nonuniformly over the cross-sectional area of the shaft. At the center of the shaft, r=0 and τ=0. The stress-free centerline of the solid circular shaft is called the *neutral axis*. The magnitude of the torsional shear stress increases in the direction from the center towards the rim, and reaches a maximum on the circumference of the shaft where $r=r_o$ and $\tau=Mr_o/J$ (Figure 15.16).

• Torsion formula takes a special form, $\tau=2M/\pi r_o{}^3$, at the rim of a solid circular shaft for which $J=\pi r_o{}^4/2$. This equation indicates that the larger the radius of the shaft, the harder it is to deform it in torsion.

• Since $\gamma=\tau/G=Mr/GJ$, the greater the magnitude of the applied torque, the larger the shear stress and shear deformation. The greater the shear modulus of the shaft material, the stiffer the material and the more difficult to deform it in torsion.

• The shear stress discussed herein is that induced in the transverse planes. For a shaft subjected to torsional loading, shear stresses are also developed along the longitudinal planes. This is illustrated in Figure 15.17 on a material element which is obtained by cutting the shaft with two transverse and two longitudinal planes. The transverse and longitudinal stresses are denoted with τ_t and τ_ℓ, respectively, and the equilibrium of the material element requires that τ_t and τ_ℓ are numerically equal.

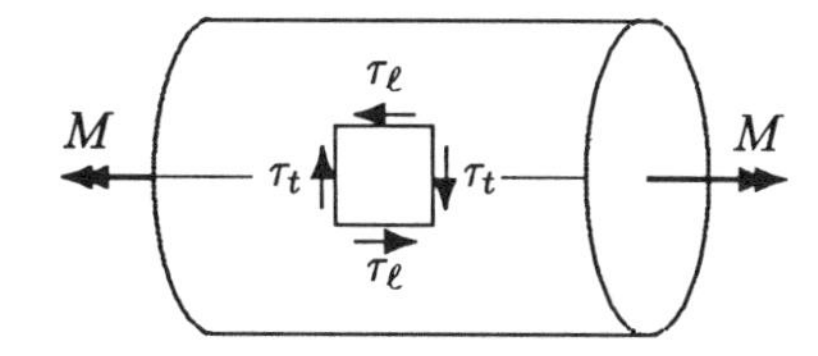

Figure 15.17 *Transverse (τ_t) and longitudinal (τ_ℓ) stresses.*

• A shaft subjected to torsion not only deforms in shear but is also subjected to normal stresses. This can be explained by the fact that the straight line AB deforms into a helix AB′, as illustrated in Figure 15.11. The length ℓ before deformation is increased to length ℓ' after the deformation, and an increase in length indicates the presence of tensile stresses along the direction of elongation.

• Consider the material element in Figure 15.18. The normals of the sides of this material element make an angle 45° with the centerline of the shaft. It can be illustrated by proper coordinate transformations that the only stresses induced on the sides of such an element are normal stresses (tensile stress σ_1 and compressive stress σ_2). The absence of shear stresses on a material element indicates that the normal stresses present are the principal (maximum and minimum) stresses, and that the planes on which these stresses act are the principal planes. (These concepts

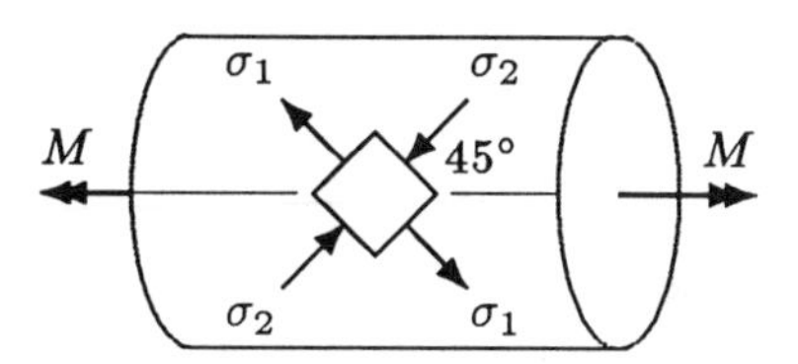

Figure 15.18 *Under pure torsion, principal normal stresses (σ_1, σ_2) occur on planes whose normals are at 45° with the centerline.*

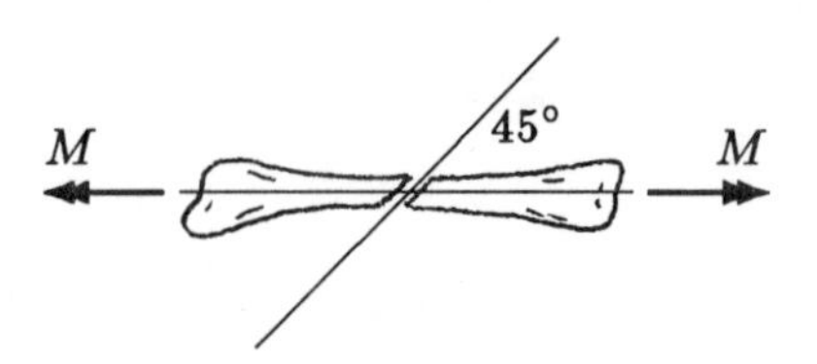

Figure 15.19 *Spiral fracture pattern for a bone subjected to pure torsion.*

will be discussed in the following chapter.) For structures subjected to pure torsion, material failure occurs along one of the principal planes. This can be demonstrated by twisting a piece of chalk until it breaks into two pieces. A careful examination of the chalk will reveal the occurrence of the fracture along a spiral line normal to the direction of maximum tension. For circular shafts, the spiral lines make an angle of 45° with the neutral axis (centerline). The same fracture pattern has been observed for bones subjected to pure torsion (Figure 15.19).

15.6 Torsion Test

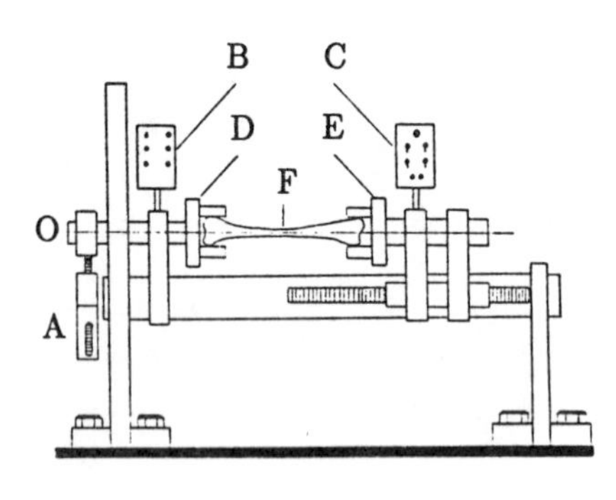

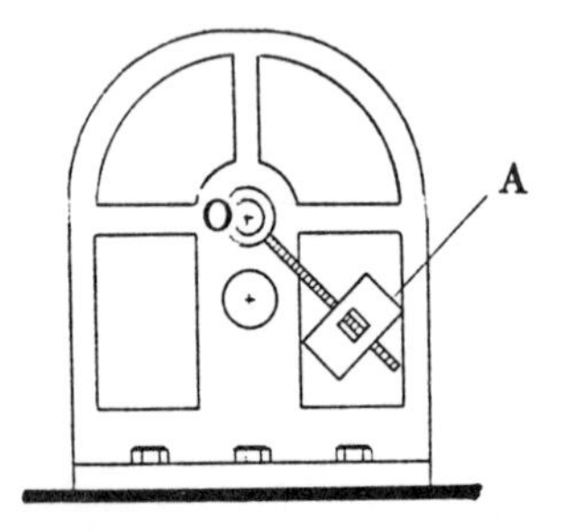

Figure 15.20 *Standard torsion testing machine.*

Figure 15.20 shows a simplified, schematic drawing of a standard torsion testing machine*. The important components of this machine are: an adjustable pendulum (A), an angular displacement transducer (B), a torque transducer (C), a rotating grip (D), and a stationary grip (E). This machine can be used to determine the torsional characteristics of specimens, in this case, of bone (F).

The pendulum generates a twisting torque about its shaft, which is also connected to the rotating grip of the machine, and forces an angular deformation in the specimen. The magnitude of the torque applied on the specimen can be controlled by adjusting the position of the mass of the pendulum relative to its center of rotation (O). The closer the mass to the center of rotation of the pendulum, the smaller the length of the moment arm as measured from the center and therefore the smaller the torque generated. Conversely, the more distal the mass of the pendulum from the center, the larger the length of the moment arm and the larger the twisting action (torque) of the weight of the pendulum. The torque generated by the weight of the pendulum is transmitted through a shaft which is connected to the rotating grip, thereby applying the same torque to the specimen firmly placed between the two grips. The torque and angular displacement transducers measure the amount of torque applied on the specimen and the corresponding angular deformation of the specimen. The fracture occurs when the torque applied is sufficiently high so that the stresses generated in the specimen are beyond the ultimate strength of the material.

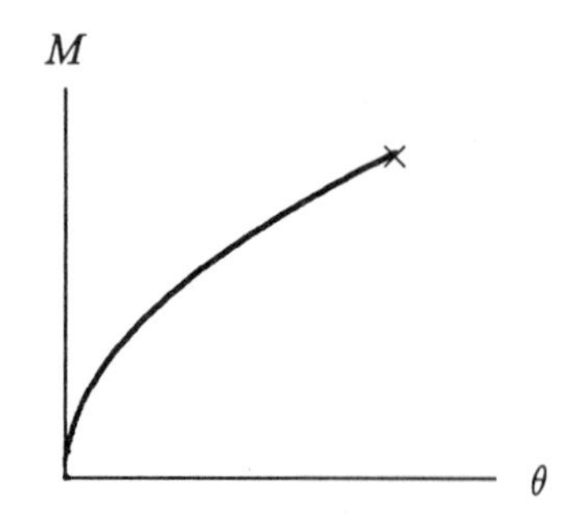

Figure 15.21 *Applied torque versus angular displacement.*

The data collected by the transducers of the torsion test machine can be plotted on a torque (M) versus angular deformation (angle of twist, θ) graph. A typical M-θ graph is shown in Figure 15.21. This graph can be analyzed to gather information about the material properties of the specimen.

* Adapted from Burstein, A.H., and Frankel, V.H. 1971. A standard test for laboratory animal bone. *J. Biomechanics* 4:155–158.

Example 15.2 A human femur is mounted in the grips of the torsion testing machine (Figure 15.22). The length of the bone at sections between the rotating (D) and stationary (E) grips is measured as L=37 cm. The femur is subjected to a torsional loading until fracture, and the applied torque versus angular displacement (deflection) graph shown in Figure 15.23, is obtained. The femur is fractured at a section (section aa in Figure 15.22) which is ℓ=25 cm distance away from the stationary grip. The geometry of the bony tissue at the fractured section is observed to be a circular ring with inner radius r_i=7 mm and outer radius r_o=13 mm (Figure 15.22).

Calculate the maximum shear strain and shear stress at the fractured section of the femur, and determine the shear modulus of elasticity of the femur.

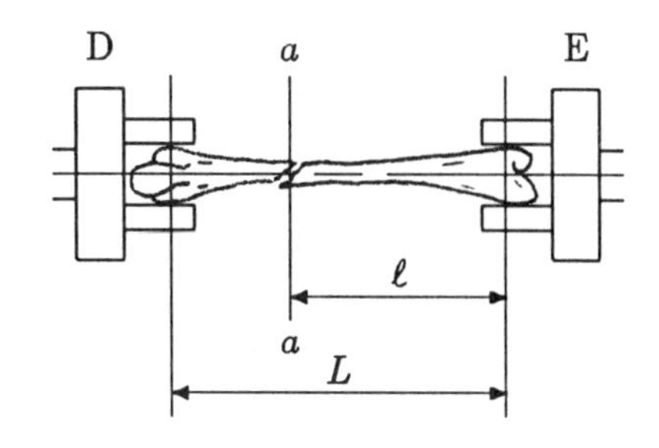

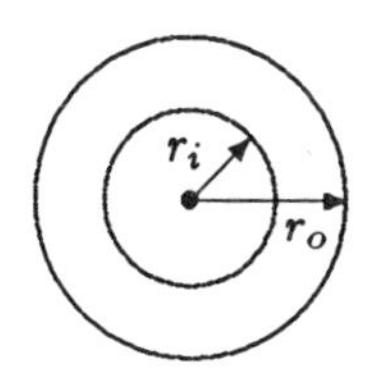

Section aa

Figure 15.22 *Fractured bone and its cross-sectional geometry.*

Solution: In Figure 15.22, θ=20° is the maximum angle of deformation (angle of twist) measured at the rotating grip at the instant when fracture occurred. The total length of the bone between the rotating and stationary grips is measured as L=37 cm. Therefore, the angular deformation is (20°)/(37 cm)=0.54 degrees per unit centimeter of bone length as measured from the stationary grip. The fracture occurred at section aa which is ℓ=25 cm away from the stationary grip. Therefore, the angular deflection θ_{aa} at section aa just before fracture is (0.54°/cm)(25 cm)=13.5° or (13.5°)(π/180°)=0.236 radian.

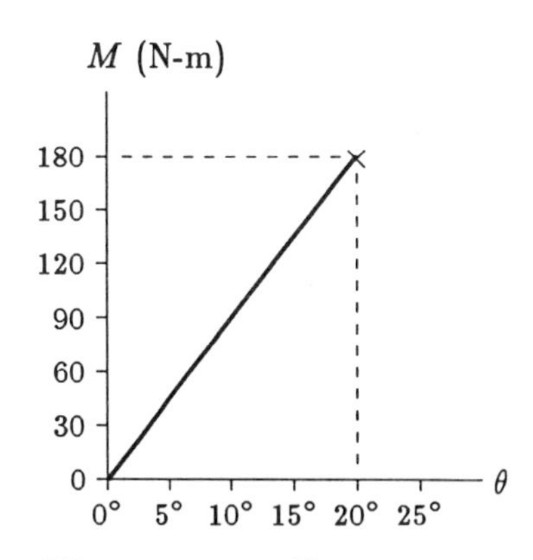

Figure 15.23 *Torque versus angle of twist diagram.*

The shear strain γ on the surface of the bone at section aa can be determined from Eq. (15.12). In this case, ℓ=0.25 m is the distance between section aa and the section of the femur where the stationary grip holds the bone, r_o=0.013 m is the outer radius of the cross-section of the bone at the fracture, and $\theta=\theta_{aa}$=0.236 is the angle of twist measured in radians. Therefore:

$$\gamma = \frac{r_o}{\ell}\,\theta_{aa} = \left(\frac{0.013}{0.25}\right)(0.236) = 0.0123 \text{ rad}$$

The torsion formula (Eq. 15.13) relates the applied torque M, radial distance r, polar moment of inertia J of the cross-section, and the shear stress τ. The cross-sectional geometry of the bony structure of the femur at section aa is a ring with inner radius r_i=0.007 m and outer radius r_o=0.013 m. Therefore, the polar moment of inertia of the cross-section of the femur at section aa is:

$$J = \frac{\pi}{2}\,(r_o^4 - r_i^4) = \frac{3.14}{2}\,[(0.013)^4 - (0.007)^4] = 41.1 \times 10^{-9} \text{ m}^4$$

In Figure 15.23, the magnitude of the applied torque at the instant when the fracture occurred (maximum torque) is M=180

N-m. Hence, the maximum shear stress on the surface of the bone at section *aa* can be determined using the torsion formula:

$$\tau = \frac{M\, r_o}{J} = \frac{(180)\,(0.013)}{41.1 \times 10^{-9}} = 56.9 \times 10^6 \text{ Pa} = 56.9 \text{ MPa}$$

Assuming that the deformations are elastic and the relationship between the shear stress and shear strain is linear, the shear modulus G of the bone can be determined from Eq. (15.14):

$$G = \frac{\tau}{\gamma} = \frac{56.9 \times 10^6}{0.0123} = 4.6 \times 10^9 \text{ Pa} = 4.6 \text{ GPa}$$

Remarks:

- For a specimen subjected to torsion, the maximum shear stress before fracture is the ***torsional strength*** of that specimen. In this case, the torsional strength of the femur is 56.9 MPa.

- The ***torsional stiffness*** is the ratio of the applied torque and the resultant angular deformation. In this case, the torsional stiffness of the femur is (180 N-m)/($20° \times \pi/180°$)=515.7 N-m/rad.

- The ***torsional rigidity*** is the torsional stiffness multiplied by the length of the specimen. In this case, the torional rigidity of the femur is (509.9 N-m/rad)(0.37 m)=190.8 N-m^2/rad.

- The maximum amount of torque applied to a specimen before fracture is defined as the ***torsional load capacity*** of the specimen. In this case, the torsional load capacity of the femur is 180 N-m.

- The total area under the torque versus angular displacement diagram represents the ***torsional energy storage capacity*** of the specimen or the ***torsional energy absorbed*** by the specimen. In this case, the torsional energy storage capacity of the femur is $\frac{1}{2}$(180 N-m)($20° \times \pi/180°$)=31.4 N-m-rad.

- Torsional fractures are usually initiated at regions of the bones where the cross-sections are the smallest. Some particularly weak sections of human bones are the upper and lower thirds of the humerus, femur, and fibula; the upper third of the radius; and the lower fourth of the ulna and tibia.

15.7 Bending

The long bones of the human body are slender, slightly curved and are loaded primarily by compressive forces applied at the joints. For example, during ordinary standing, a pair of compressive forces are applied on the femur through the hip and knee joints (Figure 15.24). Because of its curved shape, the femur is not only subjected to compressive loading but also to bending. The compressive forces applied at the joints tend to elongate the convex surface of the femur, while shortening its concave surface. Elongation is associated with tensile stresses, whereas shortening indicates the presence of compressive stresses.

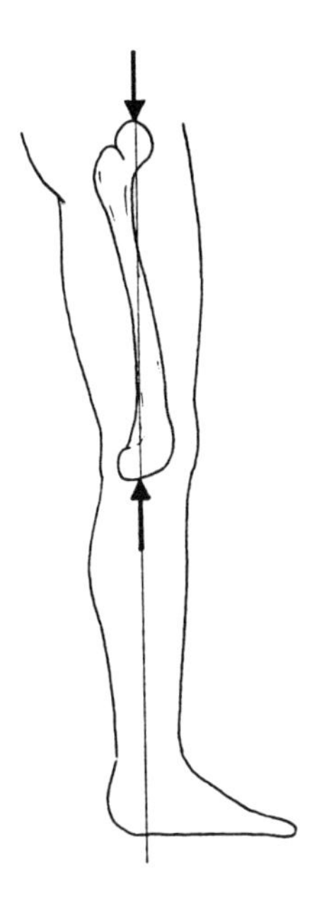

Figure 15.24 *Compressive forces applied on the femur.*

Consider the simply supported, originally straight beam shown in Figure 15.25. A force with magnitude F applied downward bends the beam, subjecting parts of the beam to shear, tension, and compression. The stress analyses of structures subjected to bending start with static analyses. The static analyses can be performed to determine both externally applied forces and internal resistances. While applying the equations of equilibrium to determine the unknown external reactions, the deformation characteristics of the structures can be neglected. For two-dimensional (plane) problems, the number of equations available to us from statics is three. Two of these equations are translational equilibrium conditions ($\sum F_x=0$ and $\sum F_y=0$), and the third equation guarantees the rotational equilibrium ($\sum M=0$).

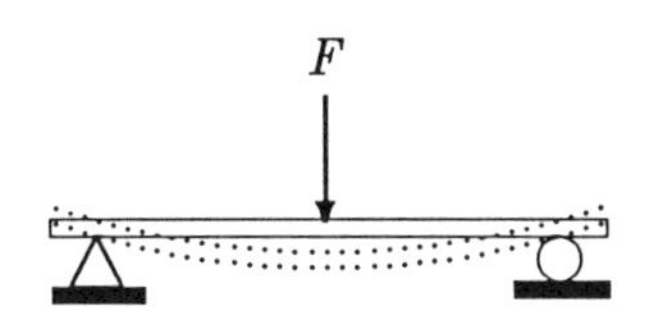

Figure 15.25 *Bending.*

The internal resistance of structures to externally applied loads can be determined by applying the *method of sections*, which is based on the concept that the individual parts of a structure which is itself in static equilibrium must also be in equilibrium. This concept makes it possible to utilize the equations of equilibrium for computing internal forces and moments, which can then be used to determine stresses. Stresses in a structure subjected to bending may vary from one section to another and from one point to another over a given section. For design and failure analyses, maximum normal and shear stresses must be considered. These critical stresses can be determined by the repeated application of the method of sections throughout the structure.

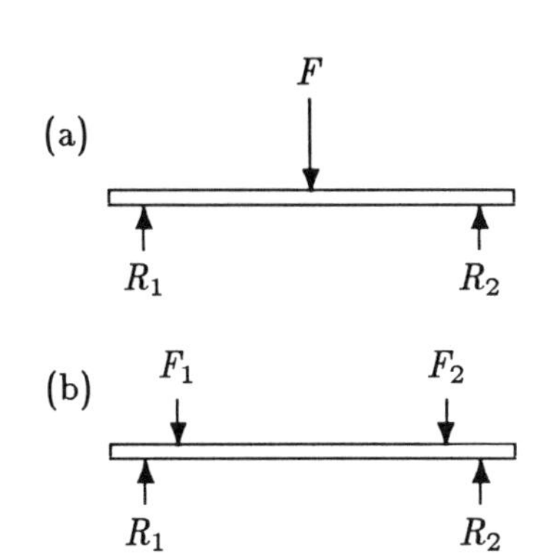

Figure 15.26 *Three- and four-point bending.*

There are several ways to subject structures to bending, two of which are illustrated in Figures 15.24 and 15.25. The free-body diagram of the beam in Figure 15.25 is shown in Figure 15.26(a). The number of parallel forces acting on the beam are three. F is the applied load, and R_1 and R_2 are the reaction forces. The type of bending to which this beam is subjected is called a *three-point bending*. The beam in Figure 15.26(b), on the other hand, is subjected to a *four-point bending*.

15.8 Internal Shear Force and Bending Moment

At a given section of a structure, depending on the loading configuration, there may be an axial force, a shear force, and a bending moment. These internal reactions to externally applied loads have different effects on the deformation characteristics of the structure. The response of a structure to axially applied loads (tension and compression) was studied in the previous chapter. Here, the loading configurations that primarily cause internal shear forces and bending moments will be discussed.

Consider the simply supported beam shown in Figure 15.27(a). Assume that the weight of the beam is negligible. The length of the beam is ℓ, and the two ends of the beam are labeled A and B. The beam is subjected to a concentrated load with magnitude F, which is applied vertically downward at point C. The distance between A and C is ℓ_1. The free-body diagram of the beam is shown in Figure 15.27(b). By applying the equations of equilibrium, the reaction forces at A and B can be determined as $R_A=(1-\ell_1/\ell)F$ and $R_B=(\ell_1/\ell)F$.

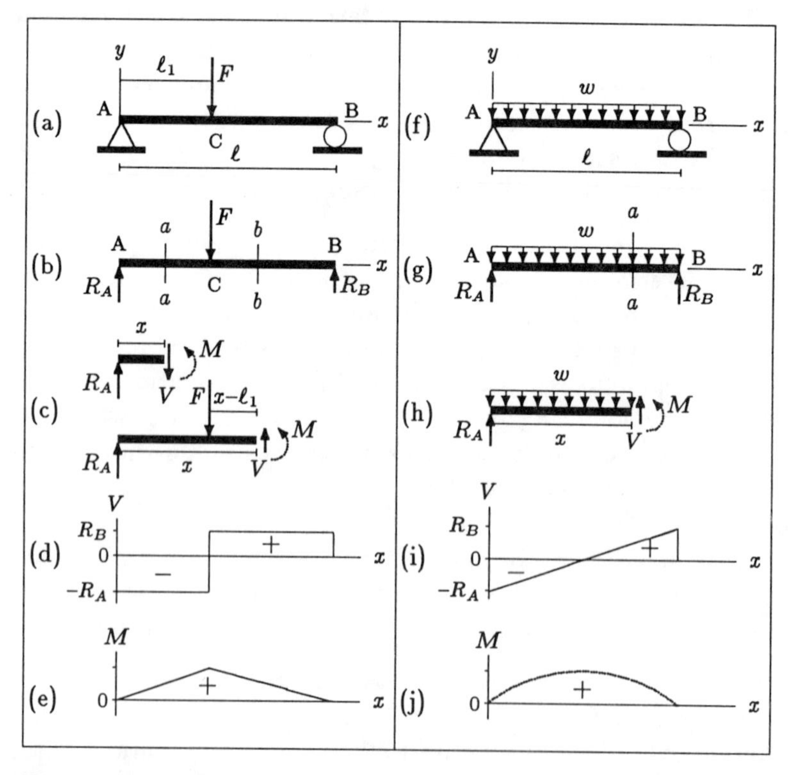

Figure 15.27 *Free-body, shear, and moment diagrams for simply supported beams subjected to concentrated and distributed loads.*

To determine the internal reactions, the method of sections must be applied. As shown in Figure 15.27(c), this method is applied at sections *aa* and *bb* because the nature of the internal reactions on the left-hand and right-hand sides of the concentrated load are different. Once a hypothetical cut is made, either the left-hand or the right-hand segment of the beam can be analyzed for the internal reactions. In Figure 15.27(c), the free-body diagrams of the left-hand segments of the beam are shown. For the vertical equilibrium of each of these segments, there must be an internal shear force at the cut. The magnitude of this force at section *aa* can be determined as $V=R_A$ by applying the equilibrium condition $\sum F_y=0$. This force acts vertically downward and is constant between A and C. The magnitude of the shear force at section *bb* is $V=F-R_A=R_B$, and since F is greater than R_A, it acts vertically upward. The sign convention adopted in this text for the shear force is illustrated in Figures 15.28 and 15.29. An upward internal force on the left-hand segment (or the downward internal force on the right-hand segment) of a cut is positive. Otherwise the shear force is negative. Therefore, as illustrated in Figure 15.27(d), the shear force is positive between A and C, and negative between C and B.

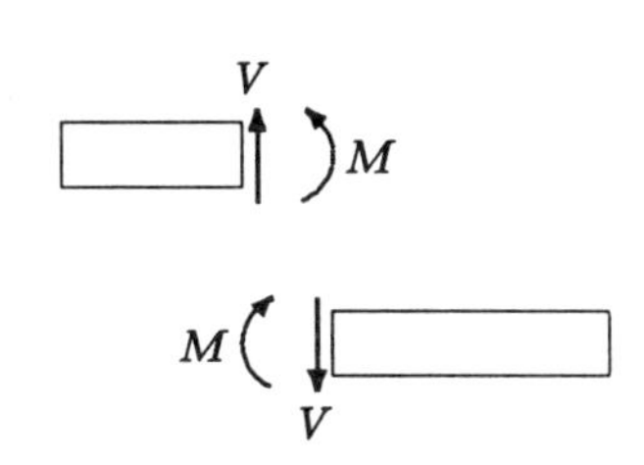

Figure 15.28 *Positive shear force and bending moment.*

In addition to the vertical equilibrium, the segments must also be checked for rotational equilibrium. As illustrated in Figure 15.27(c), this condition is satisfied if there are internal resisting moments at sections *aa* and *bb*. By utilizing the equilibrium condition $\sum M=0$, the magnitudes of these counterclockwise moments can be determined for sections *aa* and *bb* as $M=xR_A$ and $M=xR_A-(x-\ell_1)F$, respectively. Note that M is a function of the axial distance x measured from A. The function relating M and x between A and C is different than that between C and B. These functions are plotted on an M versus x graph in Figure 15.27(e). The moment is maximum at C where the load is applied. As illustrated in Figures 15.28 and 15.29, the sign convention adopted for the moment is such that a counterclockwise moment on the left-hand segment (or the clockwise moment on the right-hand segment) of a cut is positive.

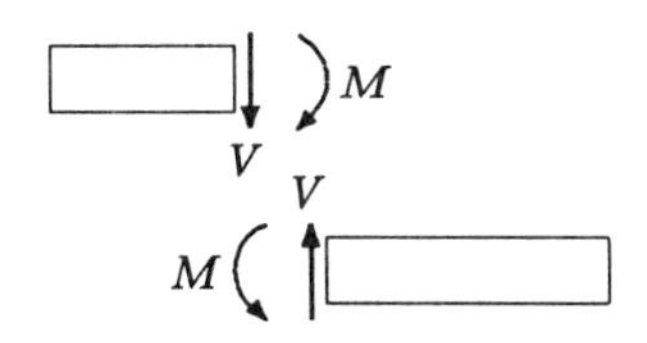

Figure 15.29 *Negative shear force and bending moment.*

It can also be shown that shear force and bending moment are related through the equation:

$$V=-\frac{dM}{dx}$$

If the variation of M along the length of the structure is known, then the shear force at a given section of the structure can also be determined.

The procedure outlined above is also applied to analyze a simply supported beam subjected to a distributed load of w (force per unit length of the beam), which is illustrated in Figure 15.27(f) through (j). The procedure for determining the internal shear forces and resisting moments in cantilever beams subjected to concentrated and distributed loads is outlined graphically in Figure 15.30.

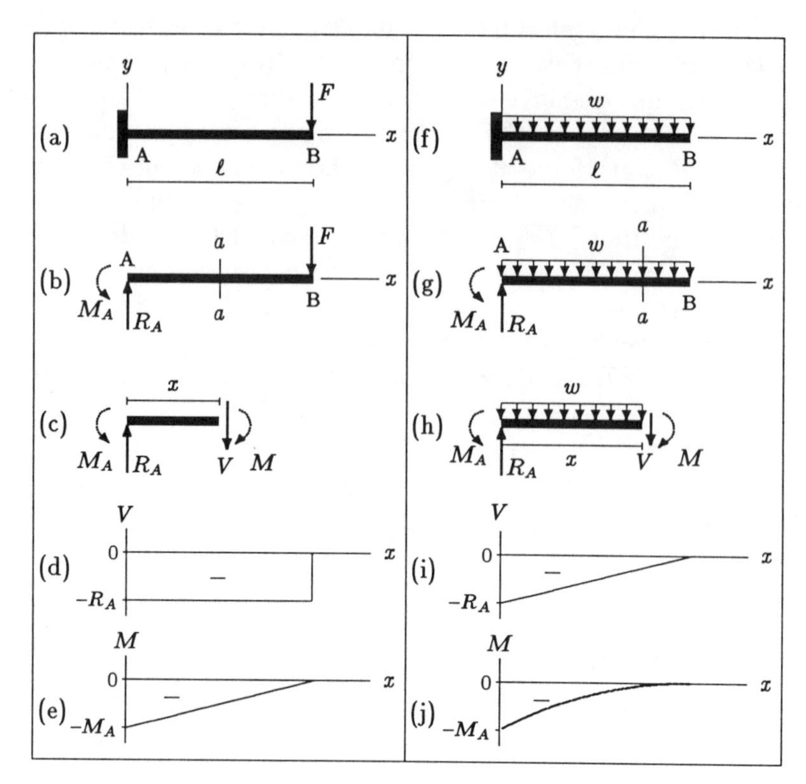

Figure 15.30 *Free-body, shear, and moment diagrams for cantilever beams subjected to concentrated and distributed loads.*

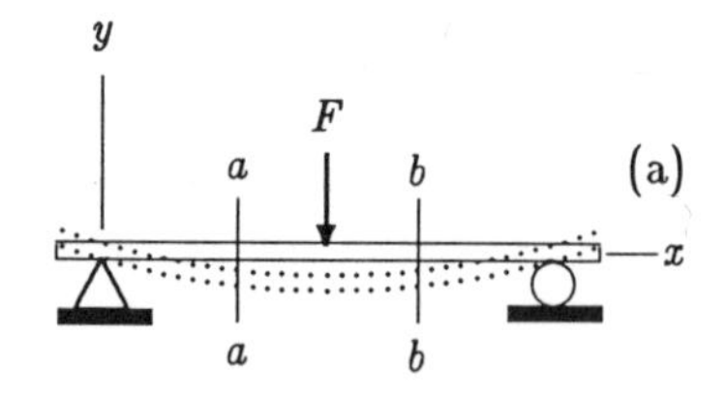

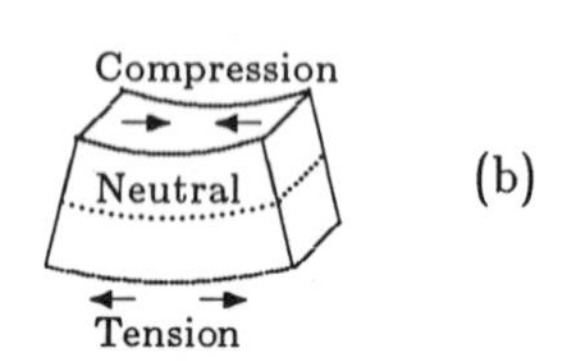

Figure 15.31 *Tension and compression in bending.*

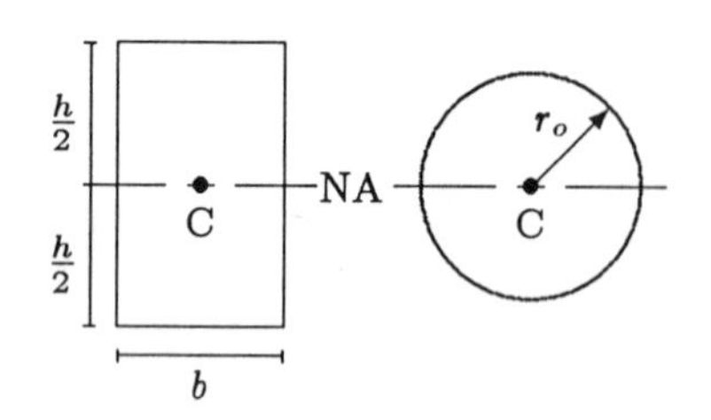

Figure 15.32 *Centroids (C) and neutral axes (NA) for a rectangle and a circle.*

15.9 Normal Stresses in Bending

Consider the beam shown in Figure 15.31. The beam is bent by a downward force. If the beam is assumed to consist of layers of material, then the upper layers of the beam are compressed while the layers on the lower portion of the beam are subjected to tension. The extent of compression or the amount of contraction is maximum at the uppermost layer, and the amount of elongation is maximum on the bottom layer. There is a layer somewhere in the middle of the beam, where a transition from tension to compression occurs. For such a layer, there is no tension or compression, and therefore, no deformation in the longitudinal direction. This stress-free layer, which separates the zones of tension and compression, constitutes the ***neutral plane*** of the beam. The line of intersection of the neutral plane with a plane (transverse) cutting the longitudinal axis of the beam in right angles is called the ***neutral axis***. The neutral axis passes through the centroid. Centroids of symmetrical cross-sections are located at their geometric centers (Figure 15.32).

The above discussion indicates that when a beam is bent, it is subjected to stresses occurring in the longitudinal direction or in a direction normal to the cross-section of the beam. Furthermore, for the loading configuration shown in Figure 15.31, the distribution of these normal stresses over the cross-section of the beam is such that it is zero on the neutral axis, negative (compressive) above the neutral axis, and positive (tensile) below the neutral axis. For a beam subjected to pure bending, the following equation can be derived from the equilibrium considerations of a beam segment:

$$\sigma_x = -\frac{M\,y}{I} \tag{15.17}$$

This equation is known as the ***flexure formula***, and stress σ_x is called the ***flexural stress***. In Eq. (15.17), M is the bending moment, y is the vertical distance between the neutral axis and the point at which the stress is sought, and I is the ***area moment of inertia*** of the cross-section of the beam about the neutral axis. The area moments of inertia for a number of simple geometries are listed in Table 15.1.

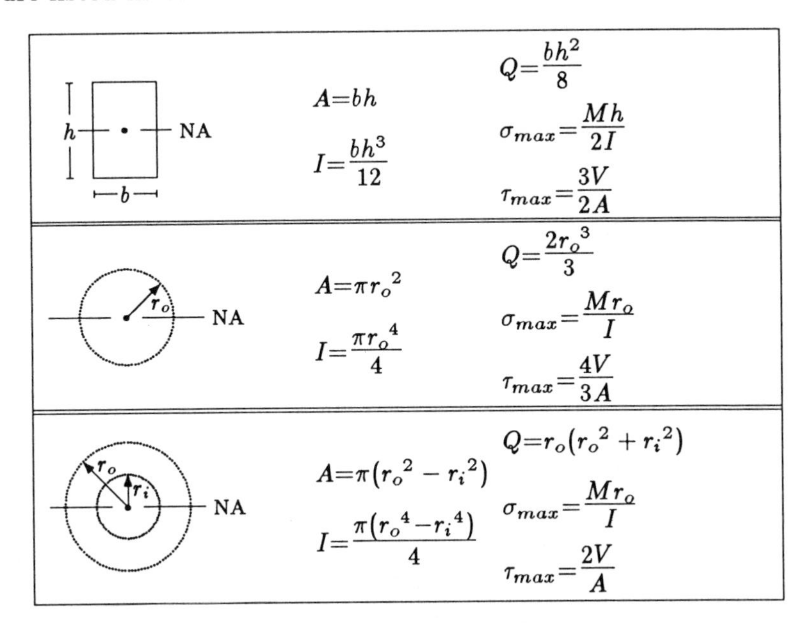

Cross-section	A, I	Q, σ_{max}, τ_{max}
Rectangle (h, b), NA	$A=bh$ $I=\frac{bh^3}{12}$	$Q=\frac{bh^2}{8}$ $\sigma_{max}=\frac{Mh}{2I}$ $\tau_{max}=\frac{3V}{2A}$
Circle (r_o), NA	$A=\pi r_o^2$ $I=\frac{\pi r_o^4}{4}$	$Q=\frac{2r_o^3}{3}$ $\sigma_{max}=\frac{Mr_o}{I}$ $\tau_{max}=\frac{4V}{3A}$
Hollow circle (r_o, r_i), NA	$A=\pi(r_o^2 - r_i^2)$ $I=\frac{\pi(r_o^4-r_i^4)}{4}$	$Q=r_o(r_o^2+r_i^2)$ $\sigma_{max}=\frac{Mr_o}{I}$ $\tau_{max}=\frac{2V}{A}$

Table 15.1 *Area (A), area moment of inertia (I) about the neutral axis (NA), first moment (Q) of the area about the neutral axis, and maximum normal (σ_{max}) and shear (τ_{max}) stresses for beams subjected to bending and with cross-sections as shown.*

The stress distribution at a section of a beam subjected to pure bending is shown in Figure 15.33. At a given section of the beam, both the bending moment and the area moment of inertia of the cross-section are constant. According to the flexure formula, the flexural stress σ_x is a linear function of the vertical distance y measured from the neutral axis, which can take both positive and negative values. At the neutral axis, y=0 and σ_x is zero. For points above the neutral axis, y is positive and σ_x is negative, indicating compression. For points below the neutral axis, y is negative and σ_x is positive and tensile. At a given section, the stress reaches its absolute maximum value either at the top or

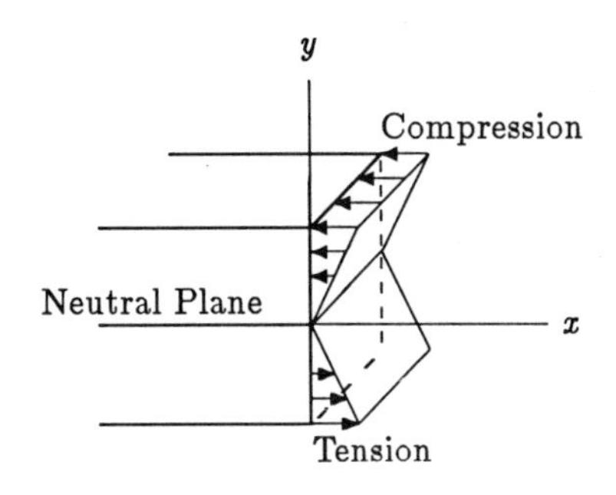

Figure 15.33 *Normal (flexural) stress distribution over the cross-section of the beam.*

the bottom of the beam where y is maximum. It is a common practice to indicate the maximum value of y with c, eliminate the negative sign indicating direction (which can be found by inspection), and write the flexure formula as:

$$\sigma_{max} = \frac{M\,c}{I} \tag{15.18}$$

For example, for a beam with a rectangular cross-section, width b, height h, and bending moment M, $c=h/2$ and the magnitude of the maximum flexural stress is:

$$\sigma_{max} = \frac{M}{I}\frac{h}{2} \qquad \text{with} \quad I = \frac{b\,h^3}{12}$$

Note that while deriving the flexure formula, a number of assumptions and idealizations are made. For example, it is assumed that the beam is subjected to pure bending. That is, it is assumed that shear, torsional, or axial forces are not present. The beam is initially straight with a uniform, symmetric cross-section. The beam material is isotropic and homogeneous, and linearly elastic. Therefore, Hooke's law ($\sigma_x = E\epsilon_x$) can be used to determine the strains due to flexural stresses. Furthermore, Poisson's ratio of the beam material can be used to calculate lateral contractions and/or elongations.

15.10 Shear Stresses in Bending

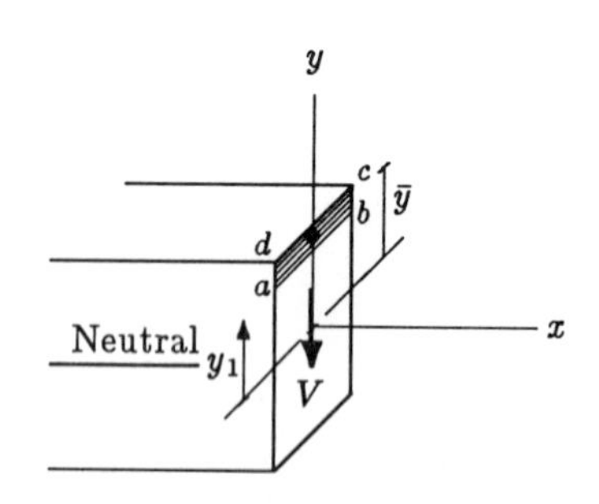

Figure 15.34 *Definitions of the parameters involved.*

When the internal shear force in a beam subjected to bending is not zero, a shear stress is also developed in the beam. The distribution of this shear stress on the cross-section of the beam is such that it is maximum on the neutral axis and zero on the top and bottom surfaces of the beam. The following formula is established to calculate shear stresses in bending:

$$\tau_{xy} = \frac{V\,Q}{I\,b} \tag{15.19}$$

In Eq. (15.19), V is the shear force at a section where the shear stress τ_{xy} is sought, I is the moment of inertia of the cross-sectional area about the neutral axis, and b is the width of the cross-section. As shown in Figure 15.34, let y_1 be the vertical distance between the neutral axis and the point at which τ_{xy} is to be determined. Then Q is the *first moment* of the area $abcd$ about the neutral axis. The first moment of area $abcd$ can be calculated as $Q = A\bar{y}$ where A is the area enclosed by $abcd$ and $\bar{y}$ is the distance between the neutral axis and the centroid of area $abcd$. Note that both A and $\bar{y}$ are maximum at the neutral axis, and A is zero at the top and bottom surfaces. Therefore, Q is maximum at the neutral axis, and zero at the top and bottom surfaces. Maximum Q's for different cross-sections are listed in Table 15.1.

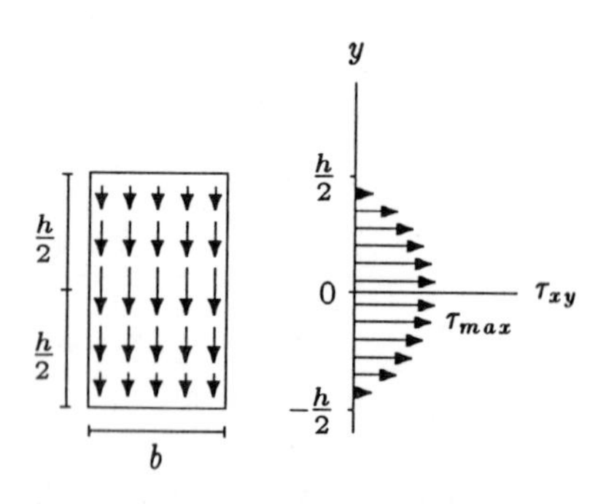

Figure 15.35 *Shear stress distribution over a section of the beam.*

The shear stress distribution over the cross-section of a beam is shown in Figure 15.35. The shear stress is constant along lines parallel to the neutral axis. The shear stress is maximum at the neutral axis where $y_1=0$ and Q is maximum. Maximum shear stresses for different cross-sections can be obtained by substituting the values of I and Q into Eq. (15.19), which are listed in Table 15.1. For example, for a rectangular cross-section:

$$\tau_{max} = \frac{3\,V}{2\,b\,h}$$

Consider a cubical material element in a beam subjected to shear force and bending moment as illustrated in Figure 15.36. On this material element, the effect of bending moment M is represented by a normal (flexural) stress σ_x, and the effect of shear force V is represented by the shear stress τ_{xy} acting on the surfaces whose normals are in the positive and negative x (longitudinal) directions. As shown in Figure 15.36(b), for the rotational equilibrium of this material element, there have to be additional shear stresses (τ_{yx}) on the upper and lower faces of the cube (whose normals are in the positive and negative y directions) such that numerically $\tau_{yx}=\tau_{xy}$. The occurrence of τ_{yx} can be understood by assuming that the beam is made of layers of material, and that these layers tend to slide over one another when the beam is subjected to bending (Figure 15.37).

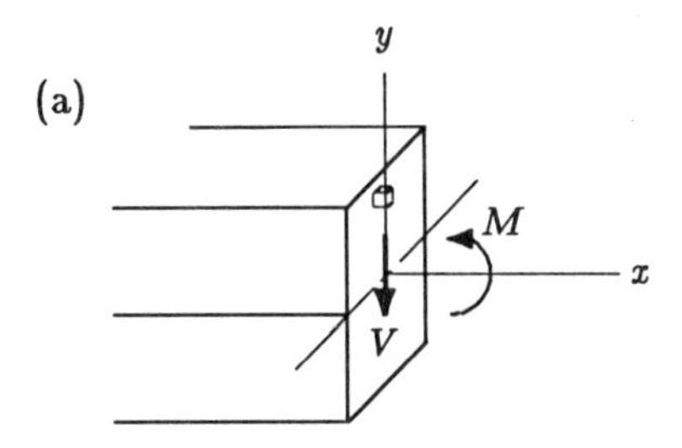

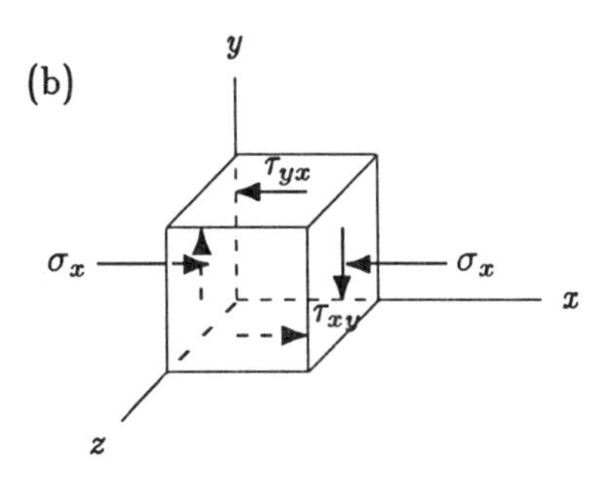

Figure 15.36 *Both normal and shear stresses occur on a material element subjected to bending.*

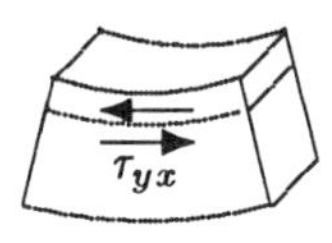

Figure 15.37 *Shear stresses in the longitudinal direction.*

15.11 Bending Tests

There are various experiments that may be conducted to analyze the behavior of specimens subjected to bending forces. Some of these experiments will be introduced within the context of the following examples.

Example 15.3 Figure 15.38 illustrates an apparatus that may be used to conduct three-point bending experiments. This apparatus consists of a stationary head (A) to which the specimen (B) is attached, two rings (C and D), and a mass (E) whose weight W is applied to the specimen through the rings.

For a weight W=1000 N applied to the middle of the specimen and for a support length of ℓ=16 cm (the distance between the left and the right supports), determine the maximum flexural and shear stresses generated at section bb of a specimen. The specimen has a square (a=1 cm) cross-section, and the distance between the left support and section bb is d=4 cm (Figure 15.39a).

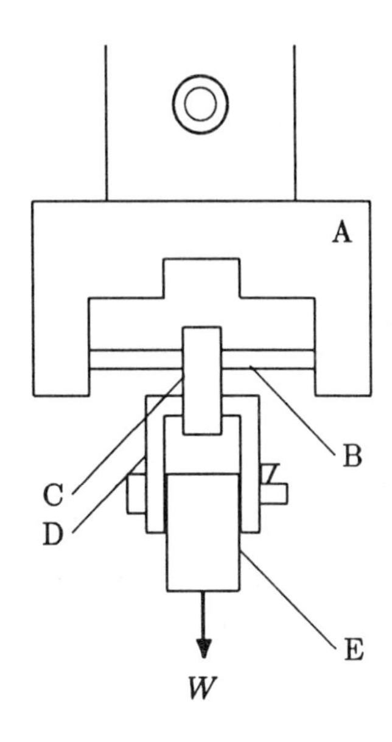

Figure 15.38 *Three-point bending apparatus.*

Solution: The free-body diagram of the specimen is shown in Figure 15.39(a). The force (W) is applied to the middle of the specimen. The rotational and translational equilibrium of the

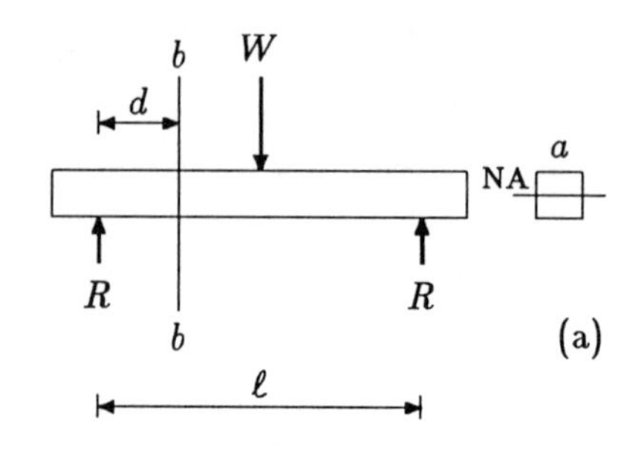

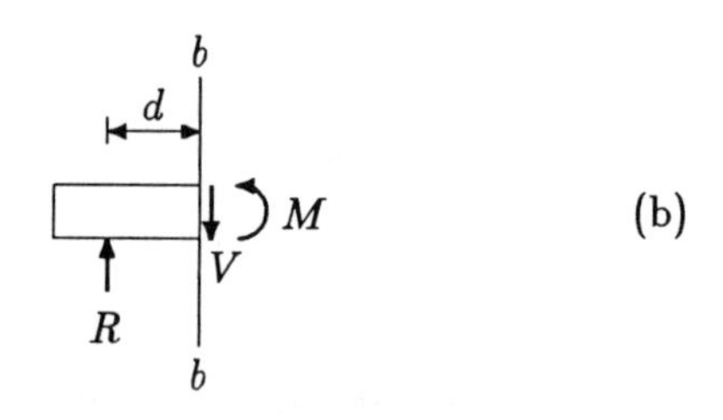

Figure 15.39 *The free-body diagrams.*

specimen require that the magnitude R of the reaction forces generated at the supports must be equal to half of W. That is, R=500 N.

The specimen has a square cross-section, and its neutral axis is located at a vertical distance $a/2$ measured from both the top and bottom surfaces of the specimen. The normal (flexural) stresses generated at section bb of the specimen depend on the magnitude of the bending moment M at section bb and the area moment of inertia I of the specimen at section bb about the neutral axis. At section bb, the magnitude of the flexural stress is maximum (σ_{max}) at the top (compressive) and the bottom (tensile) surfaces of the specimen:

$$\sigma_{max} = \frac{M}{I}\frac{a}{2}$$

The internal resistances at section bb of the specimen are shown in Figure 15.39(b). For the rotational equilibrium of the specimen:

$$M = d\,R = (0.04)(500) = 20 \text{ N-m}$$

The area moment of inertia of a square with sides a is:

$$I = \frac{a^4}{12} = \frac{(0.01)^4}{12} = 8.33 \times 10^{-10} \text{ m}^4$$

Substituting M and I into the flexure formula will yield:

$$\sigma_{max} = \left(\frac{20}{8.33 \times 10^{-10}}\right)\frac{0.01}{2} = 120 \times 10^6 \text{ Pa} = 120 \text{ MPa}$$

The shear stress generated at section bb of the specimen is a function of the shear force V at section bb, and the first moment Q and the area moment of inertia I of the cross-section of the specimen at section bb. The shear stress is maximum (τ_{max}) along the neutral axis, such that:

$$\tau_{max} = \frac{V\,Q}{I\,a}$$

For the vertical equilibrium of the specimen (Figure 15.39b), the magnitude V of the internal shear force at section bb must be equal to the magnitude R of the reaction force. That is, V=R=500 N. The first moment of the cross-sectional area of the specimen about the neutral axis is:

$$Q = \frac{a^3}{8} = \frac{(0.01)^3}{8} = 0.125 \times 10^{-6} \text{ m}^3$$

Therefore, the maximum shear stress occurring at section bb along the neutral axis is:

$$\tau_{max} = \frac{(500)(0.125 \times 10^{-6})}{(8.33 \times 10^{-10})(0.01)} = 7.5 \times 10^6 \text{ Pa} = 7.5 \text{ MPa}$$

Example 15.4 Figure 15.40 illustrates a bench test which may be used to subject bones to bending. In the case shown, the distal end of a human femur is firmly clamped to the bench and a horizontal force with magnitude F=500 N is applied to the head of the femur at point P.

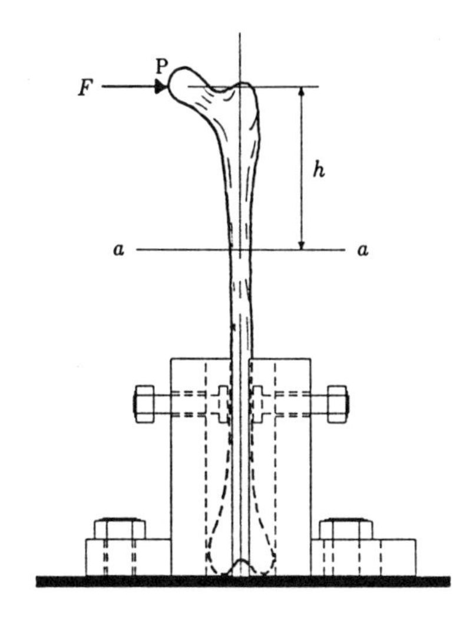

Figure 15.40 *A bench test.*

Determine the maximum normal and shear stresses generated at section aa of the femur which is located at a vertical distance h=16 cm measured from point P. Assume that the geometry of the femur at section aa is circular with inner radius r_i=6 mm and outer radius r_o=13 mm.

Solution: The femur is hypothetically cut into two parts by a plane passing through section aa, and the free-body diagram of the proximal part of the femur is shown in Figure 15.41. The internal resistances at section aa are the shear force V and bending moment M. The translational and rotational equilibrium of the proximal part of the femur require that:

$$V = F = 500 \text{ N}$$
$$M = h\,F = (0.16)(500) = 80 \text{ N-m}$$

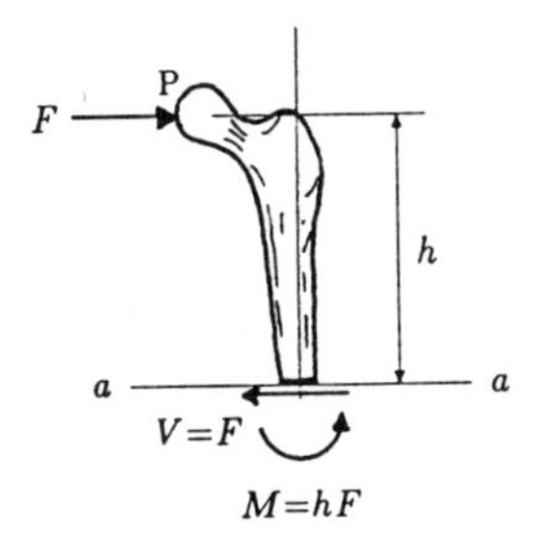

Figure 15.41 *The free-body diagram.*

Formulas to determine cross-sectional area A, area moment of inertia I, and first moment Q of cross-sectional area about the neutral axis of a structure with a hollow, circular cross-section are provided in Table 15.1. Accordingly:

$$A = \pi(r_o^2 - r_i^2) = 4.18 \times 10^{-4} \text{ m}^2$$
$$I = \frac{\pi}{4}(r_o^4 - r_i^4) = 2.14 \times 10^{-8} \text{ m}^4$$
$$Q = r_o(r_o^2 + r_i^2) = 2.67 \times 10^{-6} \text{ m}^3$$

Formulas to determine the maximum normal stress σ_{max} and maximum shear stress τ_{max} for a structure with hollow, circular cross-section and subjected to bending forces are also provided in Table 15.1:

$$\sigma_{max} = \frac{Mr_o}{I} = \frac{(80)(0.013)}{2.14 \times 10^{-8}} = 48.6 \times 10^6 \text{ Pa} = 48.6 \text{ MPa}$$

$$\tau_{max} = \frac{2V}{A} = \frac{2(500)}{4.18 \times 10^{-4}} = 2.4 \times 10^6 \text{ Pa} = 2.4 \text{ MPa}$$

Due to the orientation of the applied force, the flexural stress is maximum on the medial and lateral sides of the femur. The flexural stress is tensile on the medial side and compressive on the lateral side. The shear stress is maximum along the inner surface of the bony structure of the femur on the ventral and dorsal sides. The loading configuration of the bone is such that it behaves like a cantilever beam.

15.12 Combined Loading

The stress analyses discussed so far were concerned with axial (tension or compression), pure shear, torsional, and flexural (bending) loading of structures based on the assumption that these loads were applied on a structure one at a time. The stresses due to these basic types of loading configurations can be calculated using the following formulas:

$$\text{Axial loading:} \quad \sigma_a = \frac{F_a}{A_a}$$

$$\text{Pure shear loading:} \quad \tau_s = \frac{F_s}{A_s}$$

$$\text{Torsional loading:} \quad \tau_t = \frac{M_t\, r}{J}$$

$$\text{Flexural loading:} \quad \sigma_b = \frac{M_b\, y}{I}$$

Here, σ_a is the normal stress due to axial load F_a applied on an area A_a. τ_s is the shear stress due to shear load F_s applied on an area A_s. τ_t is the shear stress due to applied twisting torque M_t at a point r from the centerline of a cylindrical shaft and at a section whose polar moment of inertia is J. σ_b is the normal stress due to bending moment M_b at a distance y from the neutral axis of the structure at a section with area moment of inertia of I.

A structure may be subjected to two or more of these loads simultaneously. To analyze the overall effects of such combined loading configurations, first stresses generated at a given section of the structure due to each load are determined individually. Next the normal stresses are combined (added or subtracted) together, and the shear stresses are combined together. The following example will illustrate how combined stresses are handled.

Example 15.5 Figure 15.42 illustrates a bench test performed on an intertrochanteric nail. The nail is firmly clamped to the bench and a downward force with magnitude F=1000 N is applied.

Determine the stresses generated at section bb of the nail which is located at a horizontal distance d=6 cm measured from the point of application of the force on the nail. The geometry of the nail at section bb is a square with sides a=15 mm.

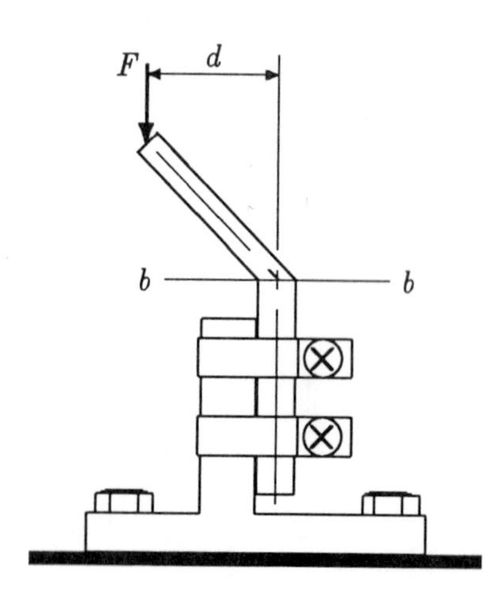

Figure 15.42 *A bench test.*

Solution: The nail is hypothetically cut into two parts by a plane passing through section bb, and the free-body diagram of the proximal part of the nail is shown in Figure 15.43. The translational equilibrium of the nail in the vertical direction requires the presence of a compressive force at section bb with magnitude equal to the magnitude F=1000 N of the external force applied on the nail. The rotational equilibrium condition requires that there is a clockwise internal bending moment at section bb whose magnitude is:

$$M = d\,F = (0.06)(1000) = 60 \text{ N-m}$$

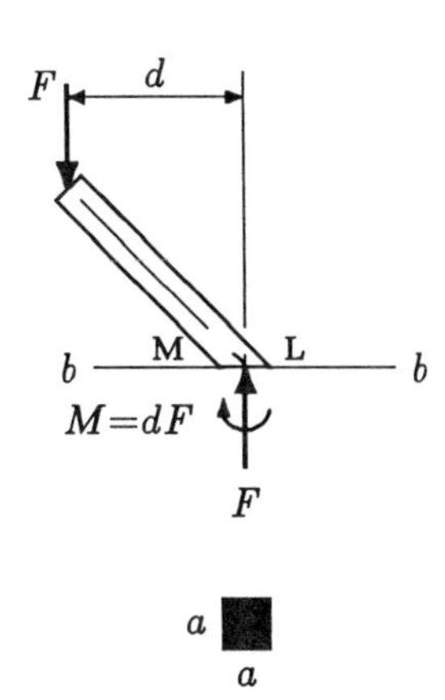

Figure 15.43 *The free-body diagram.*

The compressive force at section bb gives rise to an axial stress σ_a. The nail has a square geometry at section bb, and its cross-sectional area is:

$$A = a^2 = (0.015)^2 = 2.25 \times 10^{-4} \text{ m}^2$$

Therefore, the magnitude of the axial stress at section bb due to the compressive force is:

$$\sigma_a = \frac{F}{A} = \frac{1000}{2.25 \times 10^{-4}} = 4.4 \times 10^6 \text{ Pa} = 4.4 \text{ MPa}$$

The area moment of inertia I of the nail at section bb is:

$$I = \frac{a^4}{12} = \frac{(0.015)^4}{12} = 4.2 \times 10^{-9} \text{ m}^4$$

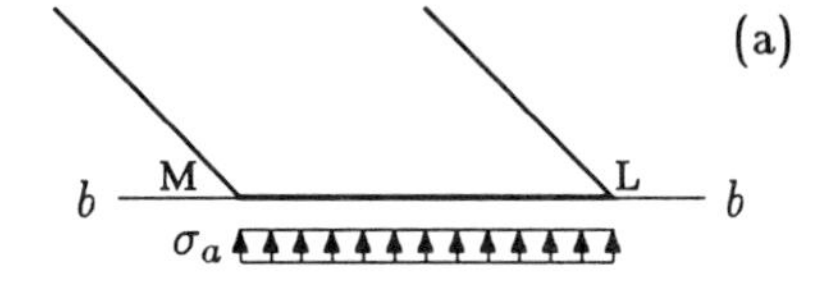

The bending moment M at section bb gives rise to a flexural stress σ_b. The flexural stress is maximum on the medial and lateral sides of the nail, which are indicated as M and L in Figure 15.43. The maximum flexural stress is:

$$\sigma_{b_{max}} = \frac{M\,a}{2\,I} = \frac{(60)(0.015)}{2(4.2 \times 10^{-9})} = 107.1 \times 10^6 \text{ Pa} = 107.1 \text{ MPa}$$

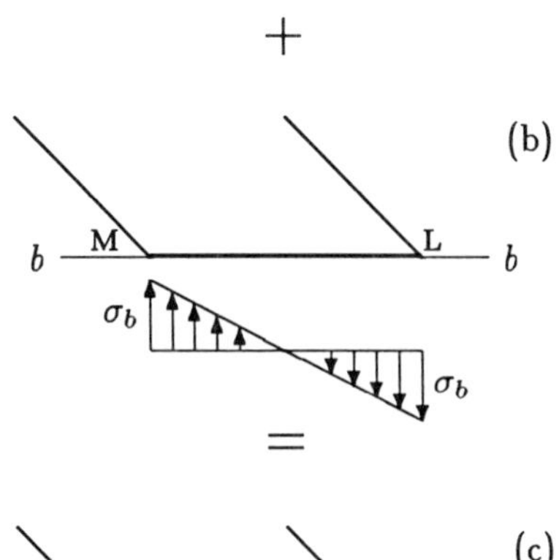

The flexural stress σ_b varies linearly over section bb. It is compressive on the medial half of the nail, zero in the middle, and tensile on the lateral half of the nail. The distribution of normal stresses σ_a and σ_b due to the compressive force F and bending moment M at section bb are shown in Figure 15.44(a) and (b), respectively. The combined effect of these stresses is shown in Figure 15.44(c). Note from Figure 15.44(c) that the resultant normal stress generated at section bb is maximum at the medial (M) side of the nail. This maximum stress is compressive and has a magnitude:

$$\sigma_{max} = \sigma_a + \sigma_{b_{max}} = 111.5 \text{ MPa}$$

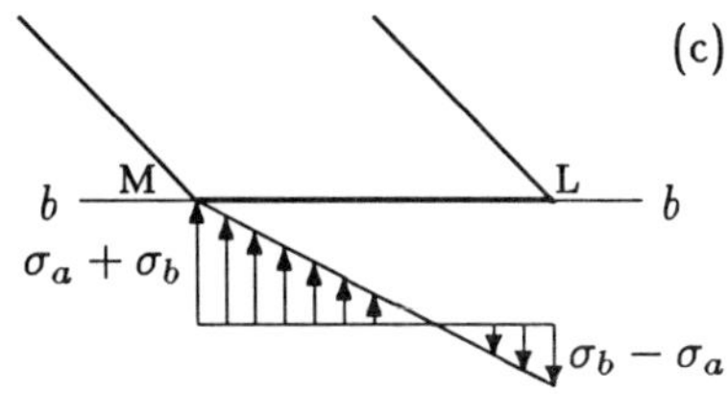

Figure 15.44 *Combined stresses.*

Since $\sigma_{b_{max}}$ (tensile) is greater than σ_a (compressive), the resultant stress σ_L on the lateral end of section bb is tensile, and its magnitude is equal to:

$$\sigma_L = \sigma_{b_{max}} - \sigma_a = 102.7 \text{ MPa}$$

Chapter 16

STRESS ANALYSIS AND FAILURE THEORIES

16.1 Stress Transformation

It was demonstrated in the previous chapter that there may be both normal and shear stresses occurring in structures that are subjected to bending or torsion. There are many other situations in which structures may be subjected to various modes of loading simultaneously. An important property of stress analyses is that the state of stress at a point within a structure is dependent upon the orientation of the material element constructed around that point.

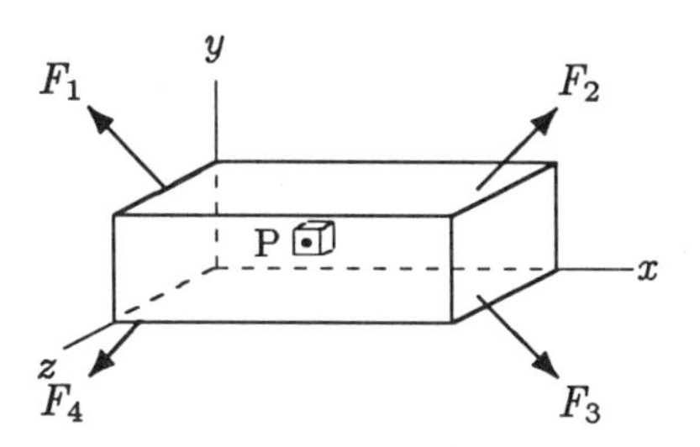

Figure 16.1 *A bar subjected to external forces applied in the xy plane.*

Consider the rectangular bar shown in Figure 16.1. The bar is subjected to externally applied forces that cause various modes of deformation within the bar. Let P be a point within the structure. Assume that a small cubical material element at point P whose sides lie parallel to the sides of the bar is cut out and analyzed. As illustrated in Figure 16.2, this material element is subjected to a combination of normal (σ_x and σ_y) and shear (τ_{xy}) stresses in the xy plane. Now, consider a second element at the same material point but with a different orientation than the first element (Figure 16.3). One can assume that the second material element is obtained simply by rotating the first in the counterclockwise direction through an angle θ. Let x' and y' be two mutually perpendicular directions representing the normals to the surfaces of the transformed material element. The stress distribution on the transformed material element would be different than that of the first. In general, the second element may be subjected to two normal stresses ($\sigma_{x'}$ and $\sigma_{y'}$) and one shear stress ($\tau_{x'y'}$) as well. If stresses σ_x, σ_y, and τ_{xy}, and the angle of orientation θ are given, then stresses $\sigma_{x'}$, $\sigma_{y'}$, and $\tau_{x'y'}$ can be calculated using the following formulas:

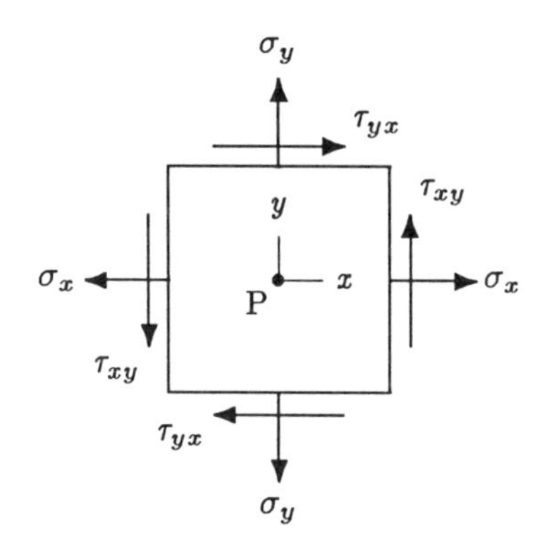

Figure 16.2 *Stress components on a plain element.*

$$\sigma_{x'} = \frac{\sigma_x + \sigma_y}{2} + \frac{\sigma_x - \sigma_y}{2}\cos(2\theta) + \tau_{xy}\sin(2\theta) \qquad (16.1)$$

$$\sigma_{y'} = \frac{\sigma_y + \sigma_x}{2} - \frac{\sigma_x - \sigma_y}{2}\cos(2\theta) + \tau_{xy}\sin(2\theta) \qquad (16.2)$$

$$\tau_{x'y'} = -\frac{\sigma_x - \sigma_y}{2}\sin(2\theta) + \tau_{xy}\cos(2\theta) \qquad (16.3)$$

These equations can be used for transforming stresses from one set of coordinates (x, y) to another (x', y'). A particularly useful application of these equations will be provided next.

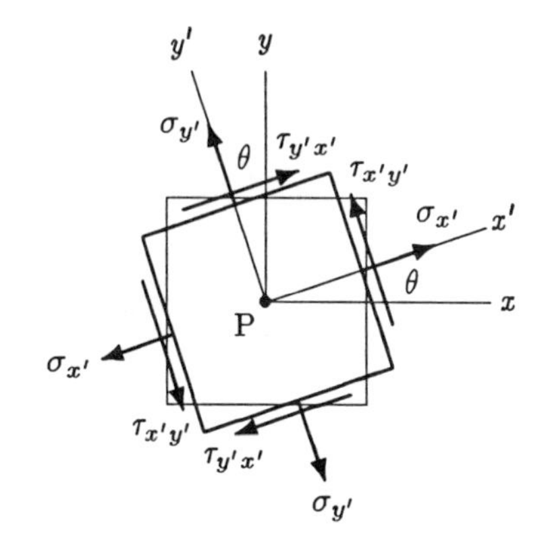

Figure 16.3 *Transformation of stress components.*

16.2 Principal Stresses

As discussed in the preceding section, the stress distribution at a given point within a structure may vary depending on the orientation of the element constructed around that point. There are infinitely many possibilities of constructing elements around a given point within a structure. Among these possibilities, there

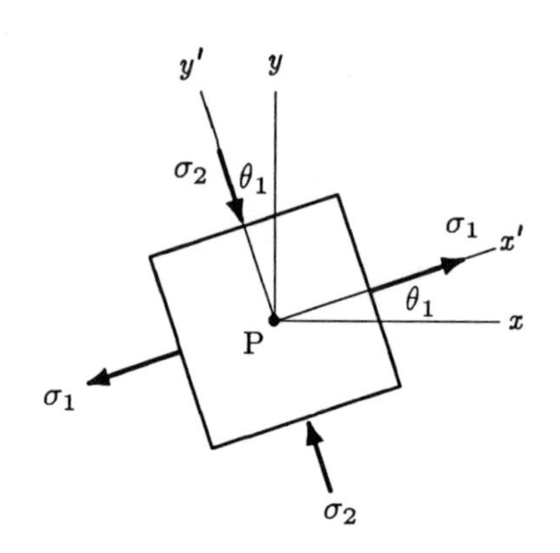

Figure 16.4 *Principal stresses.*

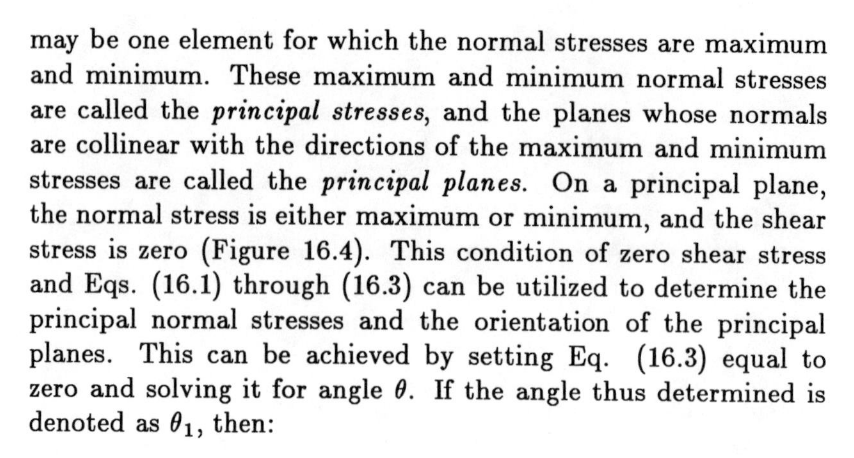

may be one element for which the normal stresses are maximum and minimum. These maximum and minimum normal stresses are called the ***principal stresses***, and the planes whose normals are collinear with the directions of the maximum and minimum stresses are called the ***principal planes***. On a principal plane, the normal stress is either maximum or minimum, and the shear stress is zero (Figure 16.4). This condition of zero shear stress and Eqs. (16.1) through (16.3) can be utilized to determine the principal normal stresses and the orientation of the principal planes. This can be achieved by setting Eq. (16.3) equal to zero and solving it for angle θ. If the angle thus determined is denoted as θ_1, then:

$$\theta_1 = \frac{1}{2} \tan^{-1} \left(\frac{2\,\tau_{xy}}{\sigma_x - \sigma_y} \right) \tag{16.4}$$

Here, θ_1 represents the angle of orientation of the principal planes relative to the x and y axes. If θ in Eqs. (16.1) and (16.2) is replaced by θ_1, then the following expressions can be derived for the principal stresses σ_1 and σ_2:

$$\sigma_1 = \frac{\sigma_x + \sigma_y}{2} + \sqrt{\left(\frac{\sigma_x - \sigma_y}{2} \right)^2 + {\tau_{xy}}^2} \tag{16.5}$$

$$\sigma_2 = \frac{\sigma_x + \sigma_y}{2} - \sqrt{\left(\frac{\sigma_x - \sigma_y}{2} \right)^2 + {\tau_{xy}}^2} \tag{16.6}$$

The concept of principal stresses is important in stress analyses. It is known that fracture or material failure occurs along the planes of maximum stresses, and structures must be designed by taking into consideration the maximum stresses involved.

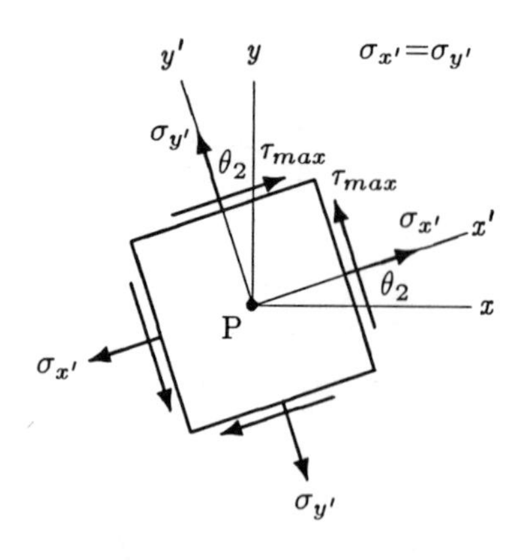

Figure 16.5 *Maximum shear stress.*

Remember that the response of a material to different modes of loading are different, and different physical properties of a given material must be considered while analyzing its behavior under shear, tension, and compression. Notice that Eqs. (16.5) and (16.6) are useful for calculating the maximum and minimum normal stresses. For a given structure and loading conditions, the maximum normal stress computed using Eq. (16.5) may be well within the limits of operational safety. However, the structure must also be checked for a critical shearing stress, called the ***maximum shear stress***. The maximum shear stress, τ_{max}, occurs on a material element for which the normal stresses are equal (Figure 16.5). Therefore, Eqs. (16.1) and (16.2) can be set equal, and the resulting equation can be solved for the angle of orientation, θ_2, of the element on which the shear stress is maximum. This will yield:

$$\theta_2 = \frac{1}{2} \tan^{-1} \left(\frac{\sigma_y - \sigma_x}{2\,\tau_{xy}} \right) \tag{16.7}$$

An expression for the maximum shear stress, τ_{max}, can then be derived by replacing θ in Eq. (16.3) with θ_2:

$$\tau_{max} = \sqrt{\left(\frac{\sigma_x - \sigma_y}{2}\right)^2 + \tau_{xy}{}^2} \tag{16.8}$$

A graphical method of finding principal stresses will be discussed in the following section.

16.3 Mohr's Circle

An effective way of visualizing the state of stress at a material point and calculating the principal stresses can be achieved by means of *Mohr's circle.* A typical Mohr's circle is drawn in Figure 16.7 for the plane stress element shown in Figure 16.6. The procedure for constructing such a diagram and finding the maximum and minimum stresses is outlined below.

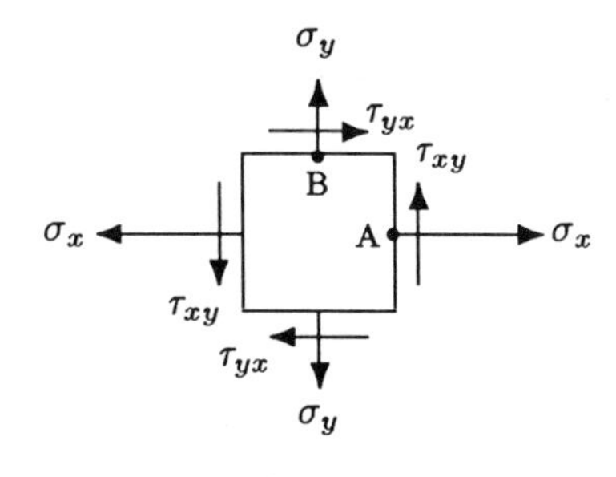

Figure 16.6 *Positive stresses.*

- As in Figure 16.6, make a sketch of the element for which the stresses are known and indicate on this element the proper directions of the stresses involved. The sign convention for positive and negative stresses is such that a tensile stress is positive while a compressive stress is negative. A shear stress on the right-hand surface (whose normal is in the positive y direction), which tends to rotate the material element in the counterclockwise direction, and a shear stress on the upper surface (whose normal is in the positive x direction), which tends to rotate the element in the clockwise direction, are positive. Note that the stresses shown in Figure 16.6 are all positive.

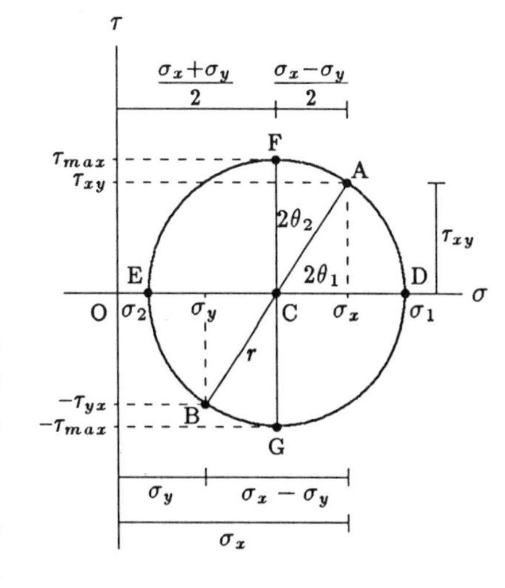

Figure 16.7 *Mohr's circle.*

- Set up a rectangular coordinate system in which the horizontal axis represents the normal stresses and the vertical axis represents the shear stresses. On the τ versus σ diagram, positive normal stresses are plotted to the right of the origin O whereas the negative normal stresses are plotted to the left of O.

- Let A represent a point on the τ-σ diagram whose coordinates are set by the normal and shear stresses acting on the right-hand face of the material element. That is, A has the coordinates σ_x and τ_{xy}. Similarly, B is a point whose coordinates are set by the stress components σ_y and $-\tau_{yx}$ acting on the upper face of the element.

- Connect points A and B with a straight line. Label the point of intersection of line AB and the horizontal axis as C. Point C is the center of the Mohr's circle, and is located at a distance $(\sigma_x + \sigma_y)/2$ from O. Therefore, the stress at C is:

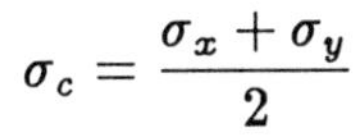

$$\sigma_c = \frac{\sigma_x + \sigma_y}{2}$$

• The distance between A and C (or B and C) is the radius r of the Mohr's circle, which can be calculated as:

$$r = \sqrt{\left(\frac{\sigma_x - \sigma_y}{2}\right)^2 + {\tau_{xy}}^2}$$

• Draw a circle with radius r and center at C. Intersections of this circle with the horizontal axis (where τ=0) correspond to the maximum and minimum (principal) normal stresses, which can be calculated as:

$$\sigma_1 = \sigma_c + r = \frac{\sigma_x + \sigma_y}{2} + \sqrt{\left(\frac{\sigma_x - \sigma_y}{2}\right)^2 + {\tau_{xy}}^2}$$

$$\sigma_2 = \sigma_c - r = \frac{\sigma_x + \sigma_y}{2} - \sqrt{\left(\frac{\sigma_x - \sigma_y}{2}\right)^2 + {\tau_{xy}}^2}$$

If the Mohr's circle is drawn carefully, then σ_1 and σ_2 can also be measured directly from the τ-σ diagram. Notice that the above equations are essentially Eqs. (16.5) and (16.6).

• On the τ-σ diagram, F and G are the points of intersection of the Mohr's circle and a vertical line passing through C. At F and G, normal stresses are both equal to σ_c and the magnitude of the shear stress is maximum. Therefore, points F and G on the Mohr's circle correspond to the state of maximum shear stress. The maximum shear stress is simply equal to the radius of the Mohr's circle:

$$\tau_{max} = \sqrt{\left(\frac{\sigma_x - \sigma_y}{2}\right)^2 + {\tau_{xy}}^2}$$

Again, this is the same equation provided in the previous section for τ_{max}.

• Mohr's circle has its own way of interpreting angles. On the plane stress element (Figure 16.6), the normals of faces A and B (i.e., the x and y axis) are at right angles. On the τ-σ diagram, A and B make an angle of 180°. Therefore, a rotation of θ degrees corresponds to an angle of 2θ on the Mohr's circle. Point D on the τ-σ diagram is related to the maximum normal stress, and therefore, to one of the principal directions. On the τ-σ diagram, point D is located at an angle of $2\theta_1$ clockwise from point A. The direction normal to the principal plane (direction of σ_1) can be obtained by rotating the x axis through an angle of θ_1 counterclockwise. A similar procedure is valid for finding the orientation of the element for which the shear stress is maximum.

Example 16.1 *Uniaxial tension.*
Consider the bar shown in Figure 16.8(a), which is subjected to tension in the x direction. The state of uniaxial stress is shown in Figure 16.8(b) on a plane material element whose sides have normals in the x and y directions.

Using Mohr's circle, determine the maximum shear stress occurring in the bar and the plane of maximum shear stress.

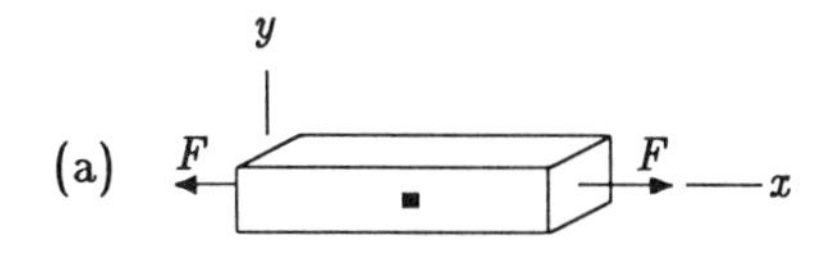

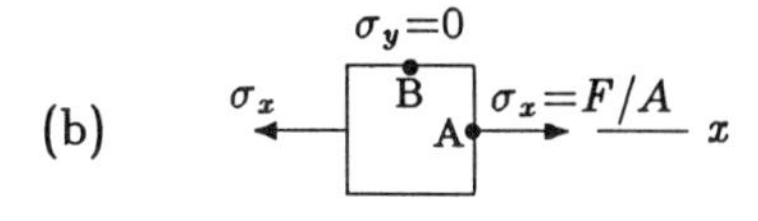

Figure 16.8 *Uniaxial tension.*

Solution: For given magnitude F of the externally applied force and the cross-sectional area A of the bar, the normal stress induced in the bar in the x direction can be determined as $\sigma_x = F/A$. As illustrated in Figure 16.8(b), σ_x is the only component of stress on a material element with sides parallel to the x and y directions.

Based on the plane stress element of Figure 16.8(b), Mohr's circle is drawn in Figure 16.9(a). Notice that there is only a tensile stress of magnitude σ_x on face A, and there is no stress on face B. Therefore, on the τ-σ diagram, point A is located along the σ-axis at a distance σ_x from the origin, and point B is essentially the origin of the τ-σ diagram. The center C of the Mohr's circle lies along the σ-axis between B and A, at a distance $\sigma_x/2$ from both A and B. Therefore, the radius of the Mohr's circle is $\sigma_x/2$.

Point F on the Mohr's circle represents the orientation of the material element for which the shear stress is maximum. The magnitude of the maximum shear stress is equal to the radius of the Mohr's circle; $\tau_{max} = \sigma_x/2$. On the Mohr's circle, point F is located 90° counterclockwise from A. Therefore, as illustrated in Figure 16.9(b), the material element for which the shear stress is maximum can be obtained by rotating the material element given in Figure 16.8(b) in the clockwise direction through an angle $\theta_2 = 45°$.

Note that σ_x is the maximum normal stress on the Mohr's circle. Therefore, point A on the Mohr's circle represents the orientation of the material element for which the normal stresses are maximum and minimum. The material element in Figure 16.8(b) represents the state of principal stresses.

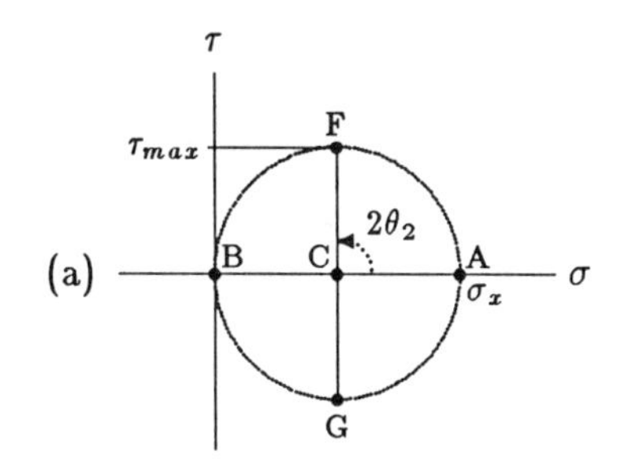

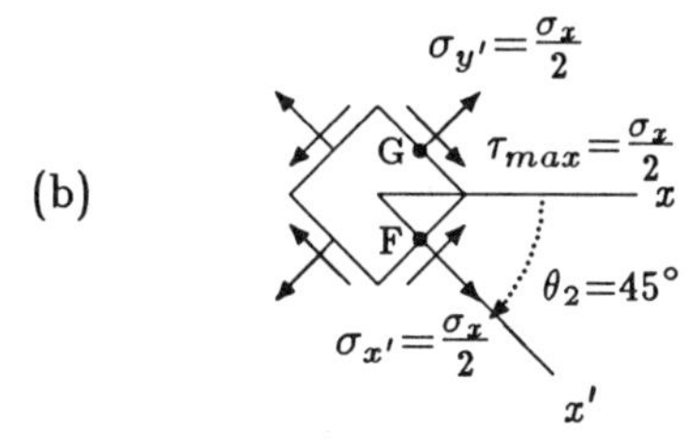

Figure 16.9 *Analysis of the element in Figure 16.8(b).*

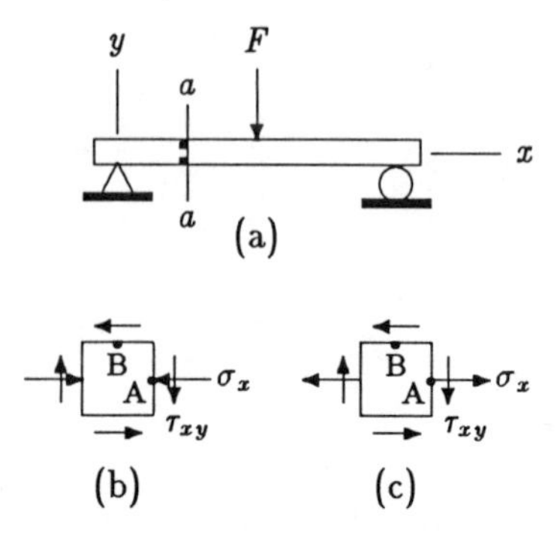

Figure 16.10 *Bending.*

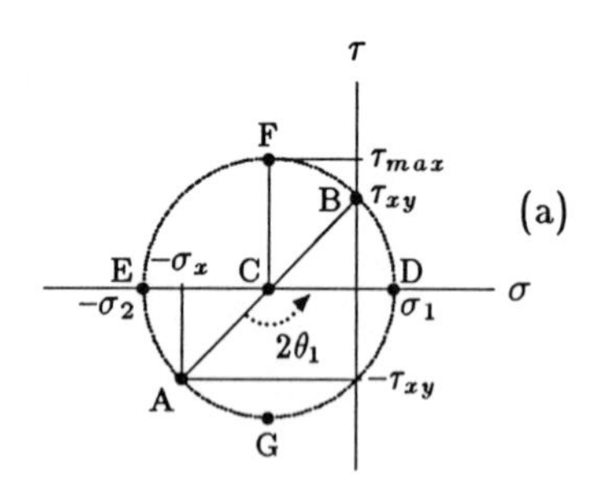

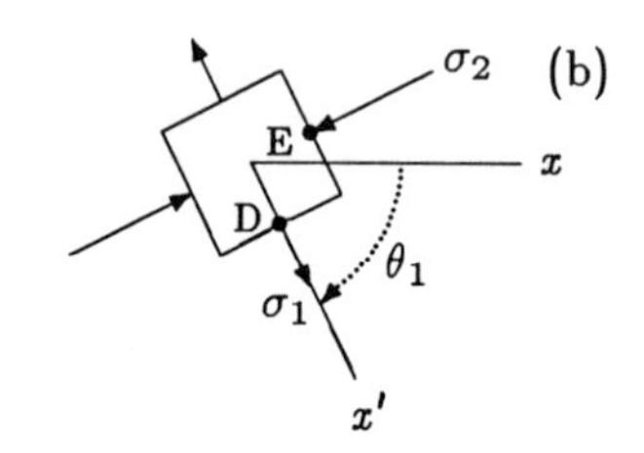

Figure 16.11 *Analysis of the element in Figure 16.10(b).*

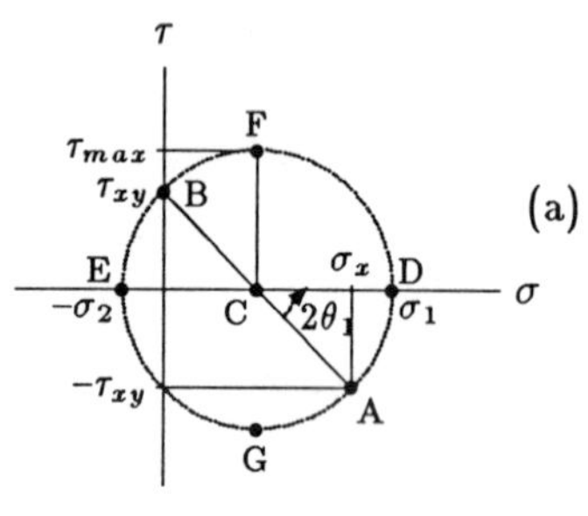

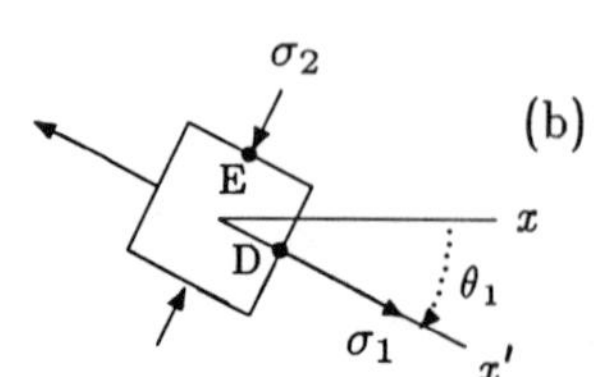

Figure 16.12 *Analysis of the element in Figure 16.10(c).*

Example 16.2 *Bending.*

Consider the beam shown in Figure 16.10(a). The externally applied force and the reactions at the supports bend the beam (three-force bending), subjecting it to shear stresses. In addition to shear, the upper layers of the beam are subjected to compression and the lower layers to tension. Figure 16.10(b) shows the state of stress occurring at a material point along a section (section aa) on the left-hand side of the applied force F and above the neutral plane of the beam. Figure 16.10(c) illustrates the state of stress at the same section below the neutral plane.

Using Mohr's circle, determine the principal stresses and maximum shear stresses occurring in the beam for the states of stress shown in Figures 16.10(b) and 16.10(c).

Solution: At a given section, magnitude F of the externally applied force and parameters defining the geometry of the beam, the normal (flexural) stress σ_x and the shear stress τ_{xy} distributions can be determined using the formulas provided in Chapter 15. Once the state of stress at a material point is known, Mohr's circle can be used for further analyses of the stresses involved.

Mohr's circle in Figure 16.11(a) is drawn by using the plane stress element of Figure 16.10(b). There is a negative (compressive) normal force with magnitude σ_x and a negative shear stress with magnitude τ_{xy} on face A of the stress element in Figure 16.10(b). Therefore, point A on Mohr's circle has coordinates $-\sigma_x$ and $-\tau_{xy}$. Face B on the stress element has only a negative shear stress with magnitude τ_{xy}. Therefore, point B on the τ-σ diagram lies along the τ-axis, τ_{xy} distance above the origin (positive). The center C of Mohr's circle can be determined as the point of intersection of the line connecting A and B with the σ-axis. The radius of the Mohr's circle can also be determined by utilizing the properties of right triangles. In this case, $r=\sqrt{(\sigma_x/2)^2+\tau_{xy}^2}$. Once the radius of the Mohr's circle is established, it is easy to find the principal stresses σ_1 and σ_2 and the maximum shear stress τ_{max}. For the case shown in Figure 16.11(a), $\sigma_1=r-\sigma_x/2$ (tensile), $\sigma_2=r+\sigma_x/2$ (compressive), and $\tau_{max}=r$. To determine the angle of orientation (θ_1) of the plane for which the stresses are maximum and minimum, we need to read the angle between lines CA and CD (in this case, it is $2\theta_1$ counterclockwise), divide it by two, and rotate the stress element in Figure 16.10(b) in the clockwise direction through an angle θ_1. This is illustrated in Figure 16.11(b).

The analysis of the stresses shown in Figure 16.10(c) is similar to that described above. It is illustrated graphically in Figure 16.12.

Example 16.3 *Pure torsion.*
Consider the solid circular cylinder shown in Figure 16.13(a), which is subjected to pure torsion by an externally applied torque M. As illustrated in Figure 16.13(b), the state of stress on a material element with sides parallel to the longitudinal and transverse planes of the cylinder is pure shear. For given M and the parameters defining the geometry of the cylinder, the magnitude τ_{xy} of the torsional shear stress can be calculated using the torsion formula provided in the previous chapter (Eq. 15.13).

Using Mohr's circle, investigate the state of stress in the cylinder.

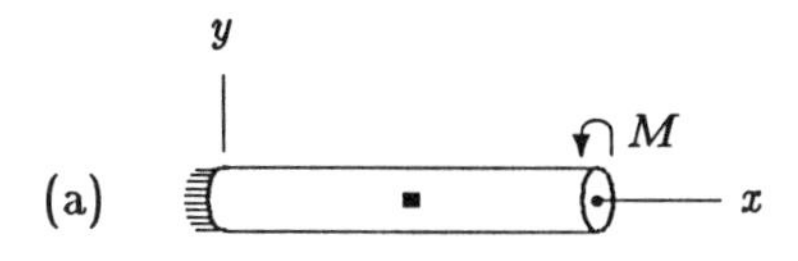

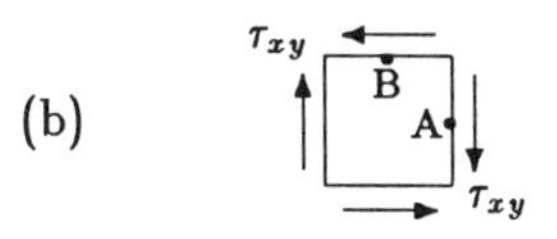

Figure 16.13 *Pure torsion.*

Solution: Mohr's circle in Figure 16.14(a) is drawn by using the stress element of Figure 16.13(b). On faces A and B of the stress element shown in Figure 16.13(b), there is only a negative shear stress with magnitude τ_{xy}. Therefore, both A and B on the τ-σ diagram lie along the τ-axis where σ=0. Furthermore, the origin of the τ-σ diagram constitutes the midpoint between A and B, and hence, the center C of the Mohr's circle. The distance between C and A is equal to τ_{xy}, which is the radius of the Mohr's circle. Mohr's circle cuts the horizontal axis at two locations, both at a distance τ_{xy} from the origin. Therefore, the principal stresses are such that σ_1=τ_{xy} (tensile) and σ_2=τ_{xy} (compressive). Furthermore, τ_{xy} is also the maximum shear stress.

The point where σ=σ_1 on the τ-σ diagram in Figure 16.14(a) is labeled D. The angle between lines CA and CD is 90°, and it is equal to half of the angle of orientation of the plane whose normal is one of the principal directions. Therefore, the planes of maximum and minimum normal stresses can be obtained by rotating the element in Figure 16.13(b) 45° (clockwise). This is illustrated in Figure 16.14(b).

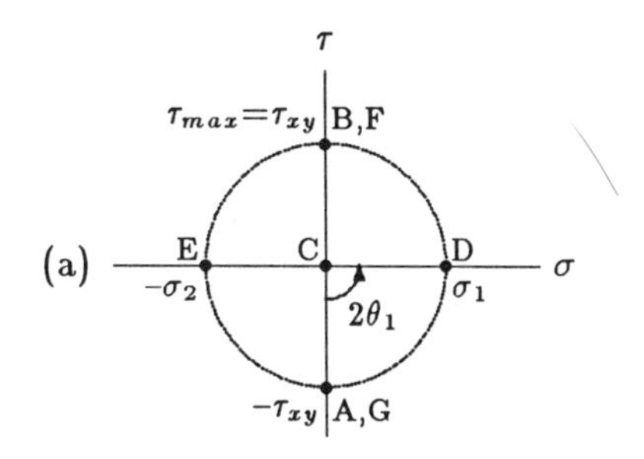

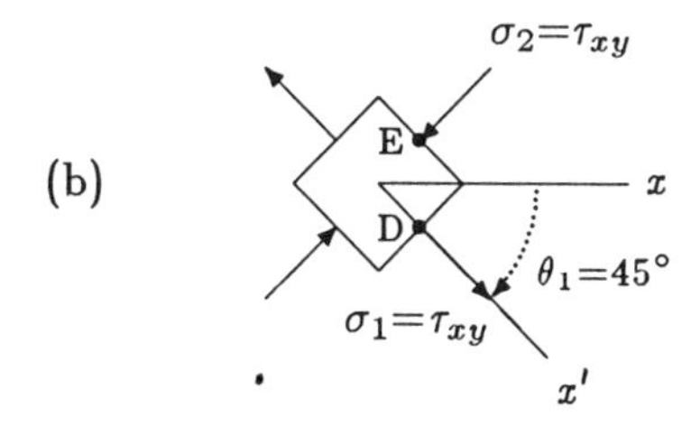

Figure 16.14 *Analysis of the element in Figure 16.13(b).*

The lines that follow the directions of principal stresses are called the *stress trajectories.* As illustrated in Figure 16.15 for a circular cylinder subjected to pure torsion, the stress trajectories are in the form of helices making an angle 45° (clockwise and counterclockwise) with the longitudinal axis of the cylinder. As discussed in Section 15.5, the significance of these stress trajectories is such that if the material is weakest in tension, the failure occurs along a helix such as *bb* in Figure 16.15, where the tensile stresses are at a maximum.

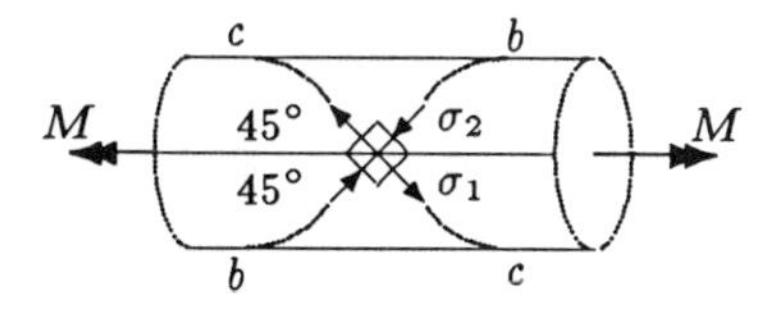

Figure 16.15 *Stress trajectories.*

16.4 Failure Theories

To assure both safety and reliability, a structure must be designed and a proper material must be selected so that the strength of the structure is considerably greater than the stresses to which it will be subjected when it is put in service. For example, if a material is subjected to tensile loads only, then the strength of the material must be judged by its ultimate strength (or yield strength). As discussed in Chapter 14, there are well established normal stress-normal strain and shear stress-shear strain diagrams for most of the common materials. Therefore, it is a relatively straightforward task to predict the response of a material subjected to uniaxial stress or pure shearing stress. However, such a direct approach is not available for a complex state of stress occurring under a combined loading.

Several criteria have been established to predict the conditions under which material failure occurs when it is subjected to combined loading. By material or structural failure, it is meant that the material either ruptures so that it can no longer support any load or undergoes excessive permanent deformation (yielding). Unfortunately, there is no single complete failure criteria that can be used to predict the material response under any type of loading. The purpose of these theories is to relate the stresses to the strength of the material. However, the available data is usually expressed in terms of the yield and ultimate strengths of the material as established in simple tension and pure shear experiments. Therefore, the idea is to utilize this limited information to analyze complex situations in which there may be more than one stress component. A few of these failure criteria will be reviewed next.

16.4.1 Maximum shear stress theory

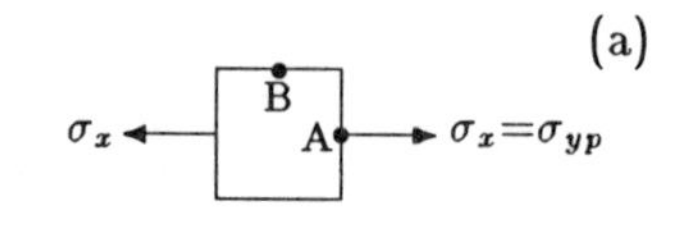

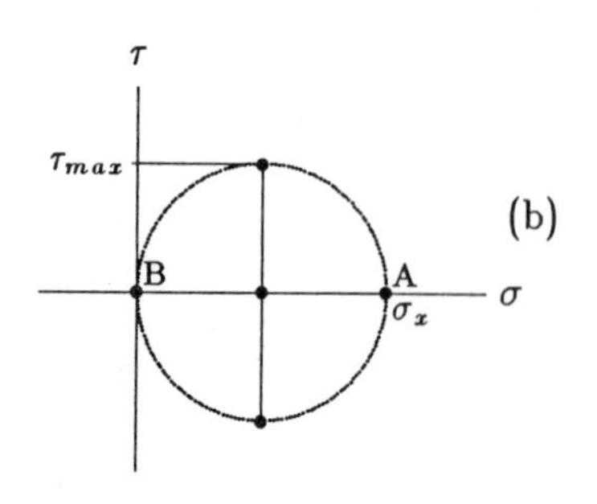

Figure 16.16 *Explaining the maximum shear stress theory.*

The maximum shear stress theory is used to predict yielding and therefore is applicable to ductile materials. It is also known as *Coulomb theory* or *Tresca theory.* This theory assumes that yielding occurs when the maximum shear stress in a material element reaches the value of the maximum shear stress which would be observed at the instant when yielding occurred if the material were subjected to a simple tension test.

Assume that a material is subjected to a simple (uniaxial) tension test until yielding. The stress level at yielding is recorded as σ_{yp}. The maximum shear stress to which this material is subjected can be determined by constructing a Mohr's circle (see Example 16.17) which is illustrated in Figure 16.16. It is clear that the maximum shear stress in simple tension is equal to half of the normal stress. In this case, the normal stress is also the yield strength of the material in tension, and therefore:

$$\tau_{max} = \frac{\sigma_x}{2} = \frac{\sigma_{yp}}{2} \tag{16.9}$$

The maximum stress theory states that if the same material is subjected to any combination of normal and shear stresses and the maximum shear stress is calculated, the yielding will start when the maximum shear stress is equal to τ_{max}. For example, consider that the material is subjected to a combination of normal and shear stresses in the xy plane as illustrated in Figure 16.17(a). For this state of stress, the maximum shear stress τ_{max} can be determined by constructing a Mohr's circle (Figure 16.17b) or using Eq. (16.8). Yielding will occur if τ_{max} is equal to or greater than $\sigma_{yp}/2$.

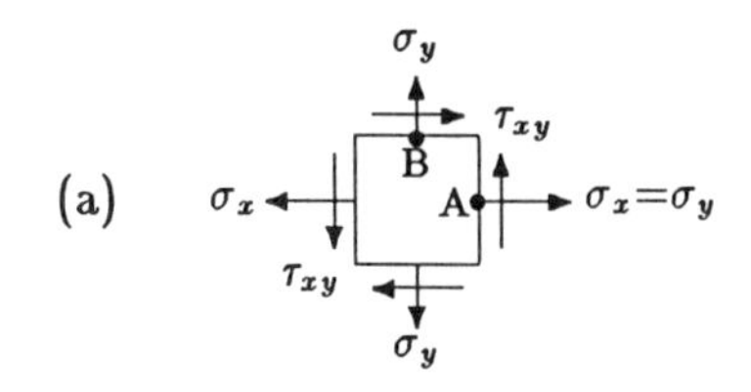

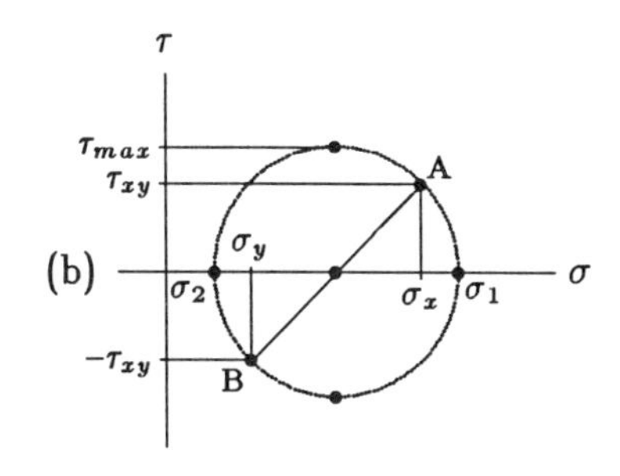

Figure 16.17 *Combined state of stress and its Mohr's circle.*

16.4.2 Maximum distortion energy theory

The maximum distortion energy theory is a widely accepted criterion that states the conditions for yielding of ductile materials. It is also known as the *von Mises yield theory* or the *Mises-Hencky theory*. This theory assumes that yielding can occur when the root mean square of the differences between the principal stresses is equal to the yield strength of the material established in a simple tension test.

Let σ_1 and σ_2 be the principal stresses in a plane state of stress, and σ_{yp} be the yield strength of the material. According to the distortion energy theory, the failure by yielding is predicted by the condition:

$$\sqrt{{\sigma_1}^2 - \sigma_1\sigma_2 + {\sigma_2}^2} = \sigma_{yp} \qquad (16.10)$$

Note that the left-hand side of Eq. (16.10) is known as the *von Mises stress.*

16.4.3 Maximum normal stress theory

The maximum normal stress theory is based on the assumption that failure by yielding occurs whenever the largest principal stress is equal to the yield strength σ_{yp}, or failure by rupture occurs whenever the largest principal stress is equal to the ultimate strength σ_u of the material. This is a relatively simple theory to implement, and it applies both for ductile and brittle materials.

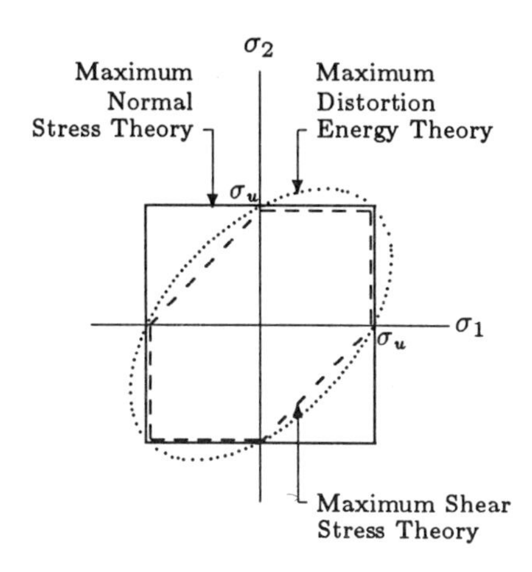

Figure 16.18 *Comparison of the failure theories.*

16.4.4 Comparison of theories

The failure theories reviewed above can be compared by representing them on a common σ_1 versus σ_2 graph, as illustrated in Figure 16.18. For a given theory, failure occurs if the stress level falls on or outside the closed boundary representing that theory. Among these three failure theories, the maximum distortion energy theory predicts yielding with highest accuracy and shows the best agreement with the experimental results when it is applied to ductile materials. For brittle materials, the maximum normal stress theory is more suitable. If the distortion energy theory is accepted as the basis of comparison, then the maximum shear stress theory is always (in all four quadrants) more conservative and safer, whereas the maximum normal stress theory is conservative whenever the signs of the principal stresses are alike (i.e., when the stress level falls in the first or the third quadrants).

The failure theories reviewed here are valid for static loading conditions. These theories have to be modified to account for dynamic or repeatedly applied (fatigue) loads.

16.5 Allowable Stress and Factor of Safety

Stress and failure analyses constitute the fundamental components in the design of structures. A structure must be designed to withstand the maximum possible stress level, called *working stress*, to which it will be subjected when it is put in service. However, the exact magnitudes of the loads that will be acting upon the structure may not be known. The structure may be subjected to unexpectedly high loads, dynamic loading conditions, or a corrosive environment which can alter the physical properties of the structural material. To account for the effects of uncertainties, a stress level called the *allowable stress* must be set considerably lower than the ultimate strength of the material. The allowable stress must be low enough to provide a margin of safety. It must also allow for an efficient use of the material.

Safety against unpredictable conditions can be achieved by considering a *factor of safety*. The factor of safety n is usually determined as the ratio of the ultimate strength of the material to the allowable stress. The factor of safety is a number greater than one, and may vary depending on whether the material is loaded under tension, compression, or shear. Instead of the ultimate strength, the factor of safety can also be based on the yield strength of the material. This is particularly important for operational conditions in which excessive yielding or plastic deformations cannot be tolerated. In the case of fatigue loading, endurance limit or the fatigue strength of the material must be used. Once the factor of safety is established, the allowable stress σ_{all} can be determined. For example, based on the ultimate strength σ_u criteria:

$$\sigma_{all} = \frac{\sigma_u}{n} \tag{16.11}$$

16.6 Factors Affecting the Strength of Materials

There are many physical and enviromental factors that may influence the properties of a material. For example, temperature can alter the strength of a material by altering its physical properties. Common materials expand when heated and contract when cooled. An increase in temperature will lower the ultimate strength of a material. The stresses in structures caused by temperature changes can be quite considerable. A good surface finish can improve the response of a structure to certain types of loads. On the other hand, friction, wear, corrosion, and the presence of discontinuities in the material can lower its strength. The response of a material can be very different to a dynamic, repeated load than to a statically applied load. The effects of some of these factors will be discussed next.

16.7 Fatigue and Endurance

The failure theories reviewed in Section 16.4 are based on the material properties established under static loading configurations. They attempt to predict the response of a material to a loading configuration which is applied once in a specific direction. Many structures, including machine parts and muscles and bones in the human body, are subjected to repeated loading and unloading. Loads that may not cause the failure of a structure in a single application may cause fracture when applied repeatedly. Failure may occur after a few cycles of loading and unloading, or after millions of cycles, depending on the amplitude of the applied load, the physical properties of the material, the size of the structure, the surface quality of the structure, and the operational conditions. Fracture due to repeated loading is called *fatigue*, and in mechanics, fatigue implies a condition of complete structural failure.

Fatigue analysis of structures is quite complicated. There are several experimental techniques developed to understand the fatigue behavior of materials. The fatigue behavior of a material can be determined in a fatigue test using tensile, compressive, bending, or torsional forces. Here, fatigue due to a combination of tension and compression will be explained.

Consider the bar shown in Figure 16.19. Assume that the bar is made of a material whose ultimate strength is σ_u. This bar is first stressed to a level σ_m (a mean stress) which is considerably lower than σ_u. The bar is then subjected to a stress fluctuating over time, sometimes tensile and other times compressive. The amplitude σ_a of the stress is such that the bar is subjected to a maximum tensile stress of $\sigma_{max}=\sigma_m+\sigma_a$, which is less than the ultimate strength σ_u of the material. This reversible and periodic stress is applied until the bar fractures and the number of cycles (N) to fracture is recorded. This experiment is repeated on specimens having the same geometric and material properties by applying sinusoidal stresses of varying amplitude. The results show that the number of cycles to failure depends on the stress amplitude σ_a. The higher the σ_a, the lower the N.

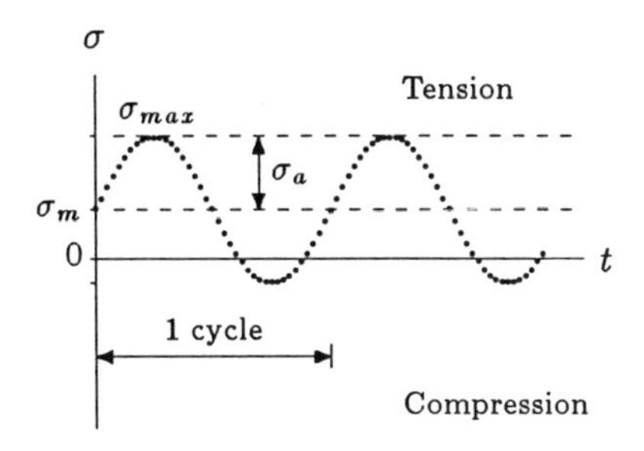

Figure 16.19 *Uniaxial fatigue test.*

A typical result of a fatigue test is plotted in Figure 16.20 on a diagram showing stress amplitude versus number of cycles to

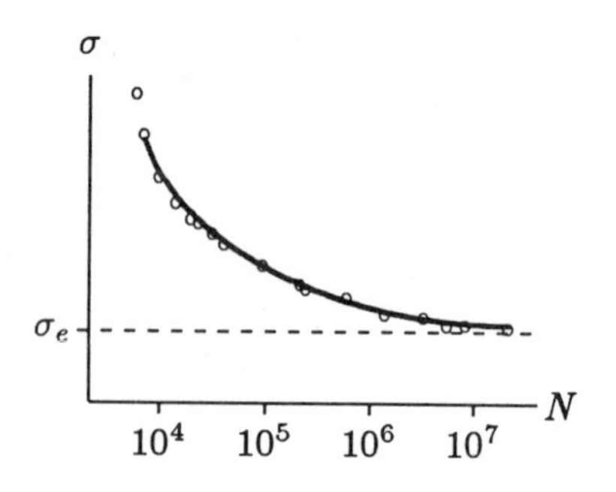

Figure 16.20 *Stress amplitude versus number of cycles to failure.*

failure (σ-N). For a given N, the corresponding stress value is called the ***fatigue strength*** of the material at that number of cycles. For a given stress level, N represents the fatigue life of the material, which increases rapidly with decreasing stress amplitude. In Figure 16.20, the experimental data is represented by a single curve. For some materials, the σ-N curve levels off and becomes essentially a horizontal line. The stress at which the fatigue curve levels off is called the ***endurance limit*** of the material, which is denoted by σ_e in Figure 16.20. Below the endurance limit, the material has a high probability of not failing in fatigue no matter how many cycles of stress are imposed upon the material.

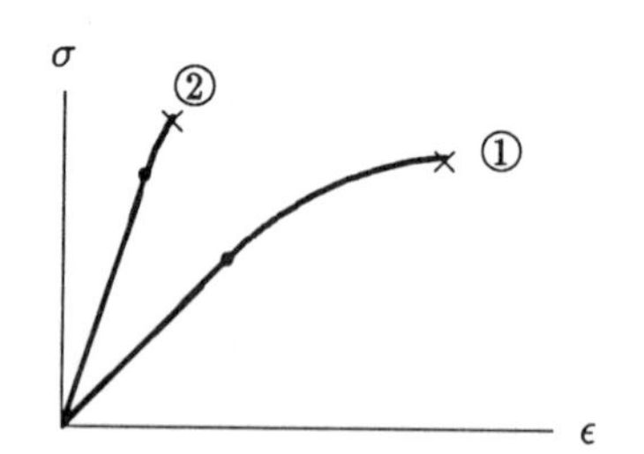

Figure 16.21 *Comparing ductile (1) and brittle (2) fractures.*

A brittle material such as glass or ceramic will undergo elastic deformation when it is subjected to a gradually increasing load. ***Brittle fracture*** occurs suddenly without exhibiting considerable plastic deformation (Figure 16.21). ***Ductile fracture***, on the other hand, is characterized by failure accompanied by considerable elastic and plastic deformations. When a ductile material is subjected to fatigue loading, the failure occurs suddenly without showing considerable plastic deformation (yielding). The fatigue failure of a ductile material occurs in a manner similar to the static failure of a brittle material.

The fatigue behavior of a material depends upon several factors. A good surface finish can improve the fatigue life of a material. The higher the temperature, the lower the fatigue strength. The fatigue behavior is very sensitive to surface imperfections and presence of discontinuities within the material that cause stress concentrations. The fatigue failure starts with the creation of a small crack on the surface of the material, which propagates under the effect of repeated loads, resulting in the rupture of the material.

Orthopaedic devices undergo repeated loading and unloading due to the activities of the patients and the actions of their muscles. Over a period of years, a weight-bearing prosthetic device or a fixation device can be subjected to a considerable number of cycles of stress reversals due to normal daily activity. This cyclic loading and unloading can cause fatigue failure of the components of a prosthetic device.

16.8 Stress Concentration

Consider the rectangular bar shown in Figure 16.22. The bar has a cross-sectional area A and is subjected to a tensile force F. The internal reaction force per unit cross-sectional area is defined as the stress, and in this case:

$$\sigma = \frac{F}{A} \tag{16.12}$$

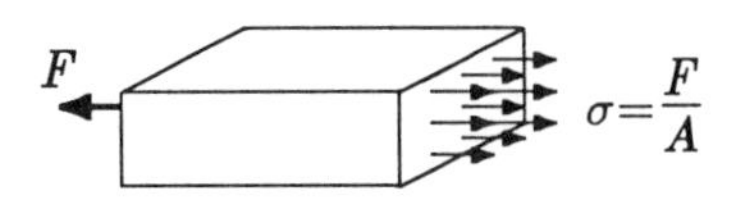

Figure 16.22 *Uniform stress distribution.*

(a)

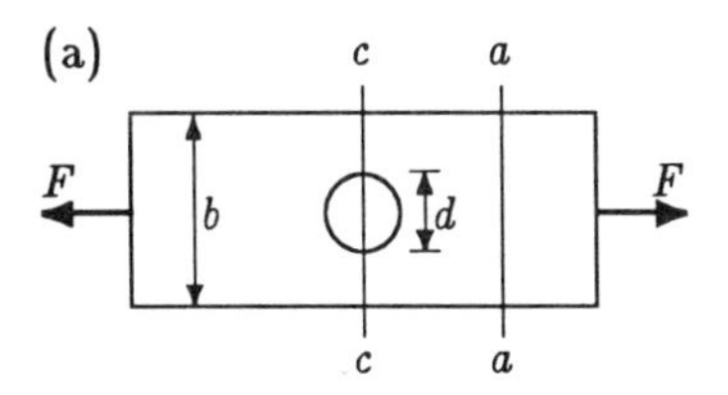

(b)

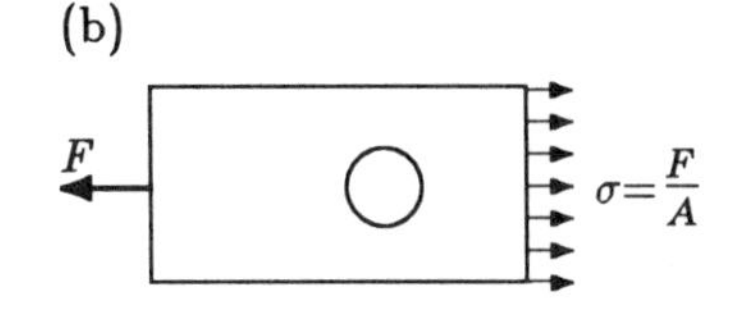

(c)

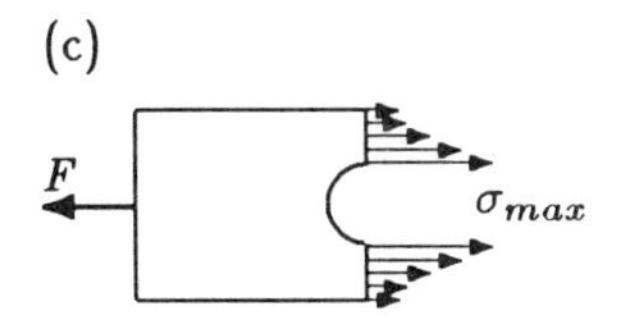

Figure 16.23 *Effects of stress concentration.*

The classic definition of stress is based on the assumption that the external force F is applied over a relatively large area rather than at a single point, and that the cross-sectional area of the bar is constant throughout the length of the bar. Consequently, as illustrated in Figure 16.22, the stress distribution is uniform over the cross-sectional area of the bar throughout the length of the bar. If the uniformity of the cross-sectional area of the bar is disturbed by the presence of holes, cracks, fillets, scratches, or notches, or if the force is applied over a very small area, then the stress distribution will no longer be uniform at the section where the discontinuity is present, or around the region where the force is applied.

Consider the plate with a circular hole of diameter d shown in Figure 16.23(a). The plate is subjected to a tensile load of F. The equilibrium considerations of the plate require that the resultant internal reaction force is equal to F at any section of the plate. As shown in Figure 16.23(b), at a section away from the hole (for example at section aa), the stress distribution is assumed to be uniform. If the cross-sectional area of the plate is A, the magnitude of this uniform stress can be calculated as $\sigma=F/A$. Now, consider the section cc passing through the center of the hole. The magnitude of the average stress at section cc is $\bar{\sigma}=F/(A-A_h)$ where A_h is the hollow area of the cross-section of the plate at section cc. Since $A-A_h$ is always less than A, the magnitude of the average stress at section cc is greater than the magnitude of the uniform stress at section aa. Futhermore, the distribution of stress at section cc is not uniform, but the stress is maximum along the edges of the hole (Figure 16.23c). That is, the stress is concentrated around the hole. This phenomena is known as the *stress concentration.*

Based on experimental observations, there are empirical formulas established to calculate the maximum stresses developed due to the presence of stress concentrators. The general relationship between the maximum stress σ_{max} (or τ_{max}) and the average stress $\bar{\sigma}$ (or $\bar{\tau}$) is such that:

$$\sigma_{max} = k\,\bar{\sigma} \tag{16.13}$$

In Eq. (16.13), k is known as the *stress concentration factor.* The value of the stress concentration factor is greater than one and varies depending on many factors such as the size of the stress concentrator relative to the size of the structure (for example, the ratio of the diameter of the hole and the width of the structure), the type of applied load (tension, compression, shear, bending, torsion, or combined), and the physical properties of the material (ductility, brittleness, hardness).

Although the stress levels measured by considering the uniform cross-sectional area of the structure may be below the fracture strength of the material, the structure may fail unexpectedly due to stress concentration effects. The fracture or ultimate strength of a material may be exceeded locally due to the presence of a stress concentrator. Note here that the fatigue failure of structures is explained by the localized stress theory due to the stress concentration effects. There may be very small imperfections or discontinuities inside or on the surface of a structure. These small holes or notches may not cause any serious problem when the structure is subjected to static loading configurations. However, repeatedly applied loads can start minute cracks in the material at the locations of discontinuities. With each application of the load these cracks may propagate, and eventually cause the material to rupture.

The effect of stress concentrations on the lives of bones and orthopaedic devices is very important. It is noted that after the removal of orthopaedic screws from a bone, the screw holes remain in the bone for many months. During the first few months after the removal of screws, the bones may fracture through the sections of one of the screw holes. A screw hole in the bone causes stress concentration effects and makes the bone weaker, particularly in bending and torsion.

The effects of stress concentrations can be reduced by good surface finish and by avoiding unnecessary holes or any other sudden shape changes in the structure.

16.9 Wear and Corrosion

As discussed in Section 3.16, *friction* occurs when two surfaces roll or slide over one another. Friction dissipates energy primarily as heat. Another consequence of the sliding action of two surfaces is the removal of material from the surfaces, which is called *wear*. Wear can alter the surface quality of structures, expose them to corrosive environments, and consequently reduce their mechanical strength. Friction and wear can be reduced by the use of a lubricant.

Corrosion is one of the primary causes of mechanical failure. Failure by corrosion can be accelerated by the presence of stresses. Corrosion can cause the development of minute cracks in the material, which can propagate in a stressed environment. As in fatigue, failure may occur when stresses are below the yield strength of the material.

Chapter 17

VISCOELASTICITY AND BIOLOGICAL TISSUES

17.1 An Overview

The material response discussed in the previous chapters was limited to response of elastic materials – in particular to linearly elastic materials. Most metals, for example, exhibit linearly elastic behavior when they are subjected to relatively low stresses at room temperature. They undergo plastic deformations at high stress levels. For an elastic material, the relationship between stress and strain can be expressed in the following general form:

$$\sigma = \sigma(\epsilon) \tag{17.1}$$

Equation (17.1) states that the normal stress σ is a function of normal strain ϵ only. The relationship between the shear stress τ and shear strain γ can be expressed in a similar manner. For a linearly elastic material, stress is linearly proportional to strain, and in the case of normal stress and strain, the constant of proportionality is the elastic modulus E of the material (Figure 17.1):

$$\sigma = E\,\epsilon \tag{17.2}$$

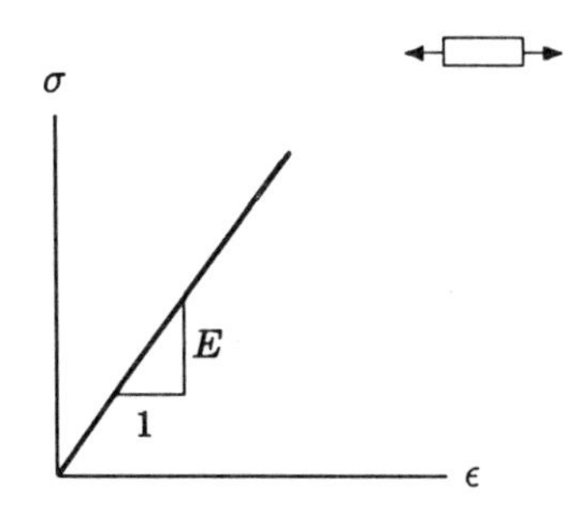

Figure 17.1 *Linearly elastic material behavior.*

While investigating the response of an elastic material, the concept of time does not enter into the discussions. Elastic materials show time-independent material behavior. Elastic materials deform instantaneously when they are subjected to externally applied loads. They resume their original (unstressed) shapes almost instantly when the applied loads are removed.

There is a different group of materials – such as polymer plastics, almost all biological materials, and metals at high temperatures – that exhibits gradual deformation and recovery when they are subjected to loading and unloading. The response of such materials is dependent upon how quickly the load is applied or removed, the extent of deformation being dependent upon the rate at which the deformation-causing loads are applied. This time-dependent material behavior is called ***viscoelasticity***. Viscoelasticity is made up of two words: viscosity and elasticity. ***Viscosity*** is a fluid property and is a measure of resistance to flow. ***Elasticity***, on the other hand, is a solid material property. Therefore, a viscoelastic material is one that possesses both fluid and solid properties.

For viscoelastic materials, the relationship between the stress and strain can be expressed as:

$$\sigma = \sigma(\epsilon, \dot{\epsilon}) \tag{17.3}$$

Equation (17.3) states that stress σ is not only a function of strain ϵ but is also a function of the ***strain rate*** $\dot{\epsilon}=d\epsilon/dt$, where t is time. A more general form of Eq. (17.3) can be obtained by including higher order time derivatives of strain. Equation (17.3) indicates that the stress-strain diagrams of viscoelastic materials are not unique but dependent upon the rate at which the strain is developed in the material (Figure 17.2).

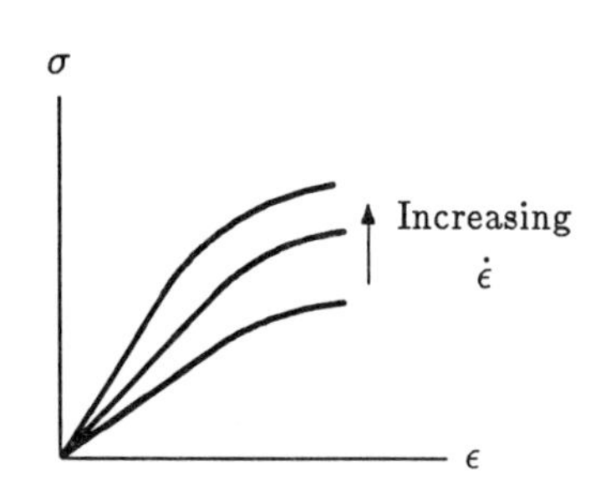

Figure 17.2 *Strain rate dependent viscoelastic behavior.*

17.2 Analogies Based on Springs and Dashpots

In Section 14.8, an analogy was made between linearly elastic materials and linear springs. An elastic material deforms, stores potential energy, and recovers deformations in a manner similar to that of a linear spring. The elastic modulus E for a linearly elastic material relates stresses and strains, whereas the constant k for a linear spring relates applied forces and corresponding deformations (Figure 17.3). Both E and k are measures of stiffness. The similarities between elastic materials and springs suggest that springs can be used to represent elastic material behavior.

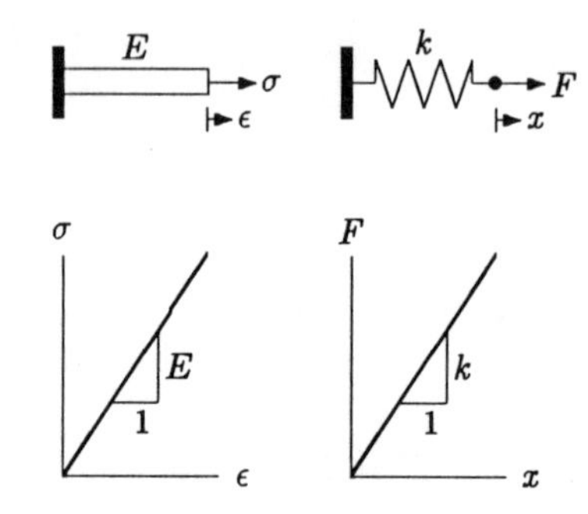

Figure 17.3 *Analogy between a linear spring and an elastic solid.*

When subjected to external loads, fluids deform as well. Fluids deform continuously, or *flow*. For fluids, stresses are not dependent upon the strains but on the strain rates. If the stresses and strain rates in a fluid are linearly proportional, then the fluid is called a ***linearly viscous fluid*** or a *Newtonian fluid*. Examples of linearly viscous fluids include water and blood plasma. For a linearly viscous fluid:

$$\sigma = \eta\, \dot{\epsilon} \tag{17.4}$$

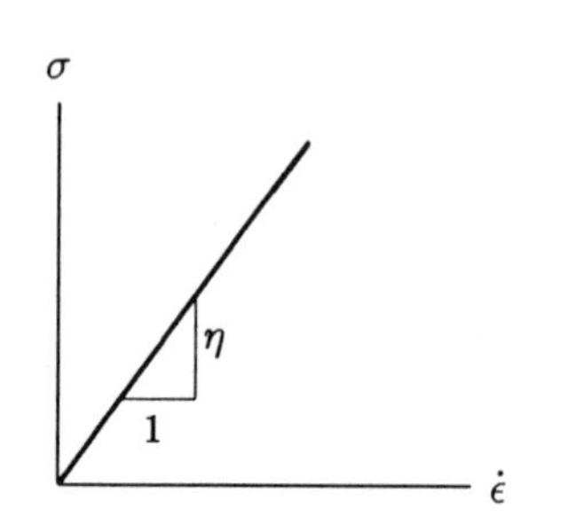

Figure 17.4 *Stress-strain rate diagram for a linearly viscous fluid.*

In Eq. (17.4), η (eta) is the constant of proportionality between the stress σ and the strain rate $\dot{\epsilon}$ and is called the *coefficient of viscosity* of the fluid. As illustrated in Figure 17.4, the coefficient of viscosity is the slope of the σ-$\dot{\epsilon}$ graph of the Newtonian fluid. The physical significance of this coefficient is similar to that of the coefficient of friction between the contact surfaces of solid bodies. The higher the coefficient of viscosity, the "thicker" the fluid and the slower the flow of fluid (i.e., the more difficult to deform it). The coefficient of viscosity for water is about 1 centipoise at room temperature, while it is about 1.2 centipoise for the blood plasma.

The spring is one of the two basic mechanical elements used to simulate the mechanical behavior of materials. The second basic mechanical element is called the ***dashpot***, which is used to simulate fluid behavior. As illustrated in Figure 17.5, a dashpot is a simple piston-cylinder or a syringe type of arrangement. A force applied on the piston will advance the piston in the direction of the applied force. The speed of the piston is dependent upon the magnitude of the applied force and the friction occurring between the contact surfaces of the piston and cylinder. For a linear dashpot, the applied force and speed (rate of displacement or deformation) are linearly proportional, the *coefficient of friction* μ (mu) being the constant of proportionality. If the applied force and the displacement are both in the x direction, then:

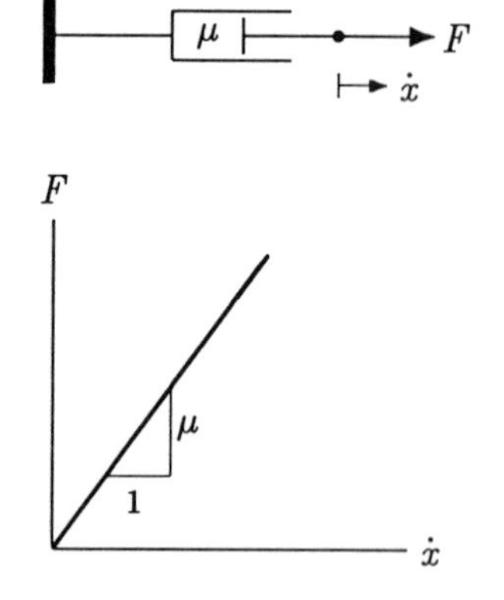

Figure 17.5 *A linear dashpot and its force-displacement rate diagram.*

$$F = \mu\, \dot{x} \tag{17.5}$$

In Eq. (17.5), $\dot{x}=dx/dt$ is the time rate of change of displacement or the speed.

By comparing Eqs. (17.4) and (17.5), an analogy can be made between linearly viscous fluids and linear dashpots. The stress and the strain rate for a linearly viscous fluid are respectively analogous to the force and the displacement rate for a dashpot, and the coefficient of viscosity is analogous to the coefficient of viscous friction for a dashpot. These analogies suggest that dashpots can be used to represent fluid behavior.

17.3 Empirical Models of Viscoelasticity

Springs and dashpots constitute the building blocks of model analyses in viscoelasticity. Springs and dashpots connected to one another in various forms are used to construct empirical viscoelastic models. Springs are used to account for the elastic solid behavior and dashpots are used to describe the viscous fluid behavior (Figure 17.6). It is assumed that a constantly applied force (stress) produces a constant deformation (strain) in a spring and a constant rate of deformation (strain rate) in a dashpot. The deformation in a spring is completely recoverable upon release of applied forces, whereas the deformation that the dashpot undergoes is permanent.

SPRING: ELASTIC SOLID

$\sigma = E\,\epsilon$

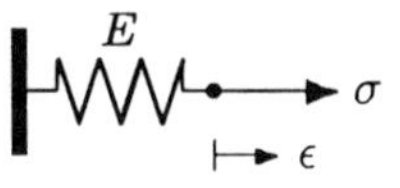

DASHPOT: VISCOUS FLUID

$\sigma = \eta\,\dot{\epsilon}$

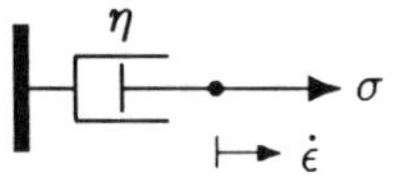

Figure 17.6 *Spring represents elastic and dashpot represents viscous material behaviors.*

17.3.1 Kelvin-Voight model

The simplest forms of empirical models are obtained by connecting a spring and a dashpot together in parallel and in series configurations. As illustrated in Figure 17.7, the Kelvin-Voight model is a system consisting of a spring and a dashpot connected in a parallel arrangement. If subscripts "s" and "d" denote the spring and dashpot, respectively, then a stress σ applied to the entire system will produce stresses σ_s and σ_d in the spring and the dashpot. The total stress applied to the system will be shared by the spring and the dashpot such that:

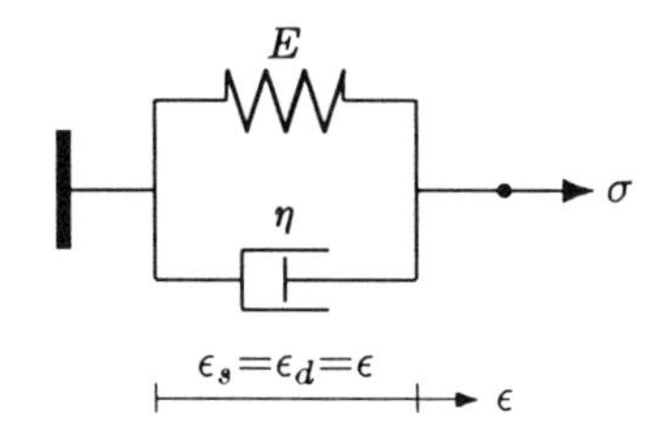

Figure 17.7 *Kelvin-Voight model.*

$$\sigma = \sigma_s + \sigma_d \tag{17.6}$$

As the stress σ is applied, the spring and dashpot will deform by an equal amount because of their parallel arrangement. Therefore, the strain ϵ of the system will be equal to the strains ϵ_s and ϵ_d occurring in the spring and the dashpot:

$$\epsilon = \epsilon_s = \epsilon_d \tag{17.7}$$

The stress-strain relationship for the spring and the stress-strain rate relationship for the dashpot are:

$$\sigma_s = E\,\epsilon_s \tag{17.8}$$

$$\sigma_d = \eta\,\dot{\epsilon}_d \tag{17.9}$$

Substituting Eqs. (17.8) and (17.9) into Eq. (17.6) will yield:

$$\sigma = E\,\epsilon_s + \eta\,\dot{\epsilon}_d \tag{17.10}$$

From Eq. (17.7), $\epsilon_s=\epsilon_d=\epsilon$. Therefore:

$$\sigma = E\,\epsilon + \eta\,\dot{\epsilon} \tag{17.11}$$

Note that the strain rate $\dot{\epsilon}$ can alternatively be written as $d\epsilon/dt$. Consequently:

$$\sigma = E\,\epsilon + \eta\,\frac{d\epsilon}{dt} \tag{17.12}$$

Equation (17.12) relates stress to strain and to the strain rate for the Kelvin-Voight model. It is a two-parameter (E and η) viscoelastic model. Equation (17.12) is an ***ordinary differential equation***. More specifically, it is a first order, linear ordinary differential equation. For a given stress σ, Eq. (17.12) can be solved for the corresponding strain ϵ. For prescribed strain ϵ, it can be solved for stress σ.

17.3.2 Maxwell model

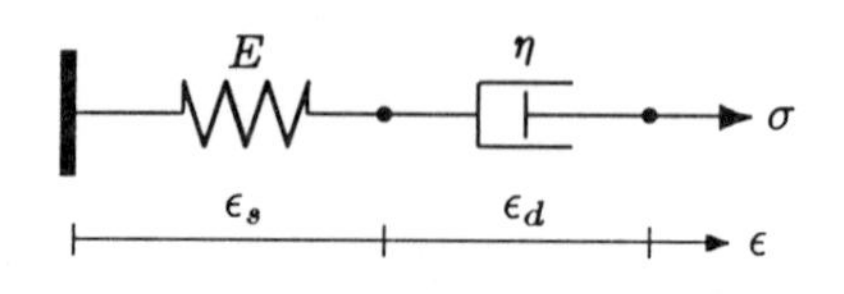

Figure 17.8 *Maxwell model.*

As shown in Figure 17.8, the Maxwell model is constructed by connecting a spring and a dashpot in series. In this case, a stress σ applied to the entire system is applied equally on the spring and the dashpot ($\sigma=\sigma_s=\sigma_d$), and the resulting strain ϵ is the sum of the strains in the spring and the dashpot ($\epsilon=\epsilon_s+\epsilon_d$). Through stress-strain analyses similar to those carried out for the Kelvin-Voight model, a differential equation relating stresses and strains for the Maxwell model can be derived in the following form:

$$\eta\,\dot{\sigma} + E\,\sigma = E\eta\,\dot{\epsilon} \tag{17.13}$$

This is also a first order, linear, ordinary differential equation representing a two-parameter (E and η) viscoelastic behavior. For a given stress (or strain), Eq. (17.13) can be solved for the corresponding strain (or stress).

Notice that springs are used to represent the elastic solid behavior, and there is a limit to how much a spring can deform. On the other hand, dashpots are used to represent fluid behavior and are assumed to deform continuously (flow) as long as there is a force present which will deform them. For example, in the case of a Maxwell model, a force applied will cause both the spring and the dashpot to deform. The deformation of the spring will be finite. The dashpot will keep deforming as long as the force is maintained. Therefore, the overall behavior of the Maxwell model is more like a fluid than a solid, and is known to be a ***viscoelastic fluid*** model. The deformation of a dashpot connected in parallel to a spring, as in the Kelvin-Voight model, is restricted by the response of the spring to the applied loads. The dashpot in the Kelvin-Voight model cannot undergo continuous deformations. Therefore, the Kelvin-Voight model represents a ***viscoelastic solid*** behavior.

17.3.3 Standard solid model

The Kelvin-Voight solid and Maxwell fluid are the basic viscoelastic models constructed by connecting a spring and a dashpot together. They do not represent any known real material. However, in addition to springs and dashpots, they can be used to construct more complex viscoelastic models, such as the standard solid model. As illustrated in Figure 17.9, the standard solid model is composed of a spring and a Kelvin-Voight solid connected in series. The standard solid model is a three-parameter model with elastic moduli E_1 and E_2, and the coefficient of viscosity η. The material function relating the stress, strain, and their rates for this model is:

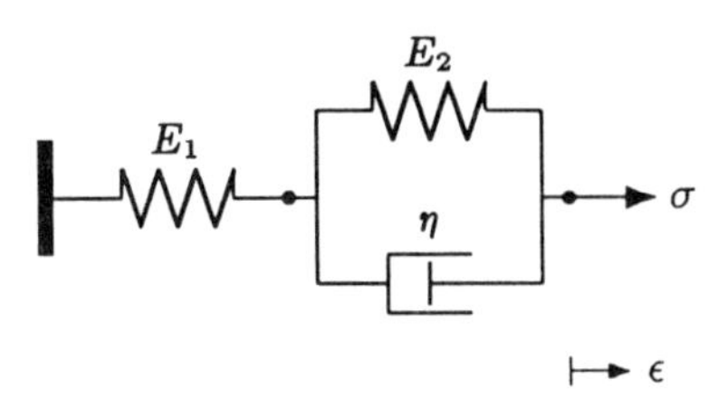

Figure 17.9 *Standard solid model.*

$$E_1 E_2\,\epsilon + E_1\eta\,\dot{\epsilon} = (E_1 + E_2)\,\sigma + \eta\,\dot{\sigma} \tag{17.14}$$

In Eq. (17.14), $\dot{\sigma}=d\sigma/dt$ is the stress rate. The standard solid model is used to describe the viscoelastic behavior of a number of biological materials such as the cartilage and the white blood cell membrane.

17.4 Time-Dependent Material Response

An empirical model for a given viscoelastic material can be established through a series of experiments. There are several experimental techniques designed to analyze the time-dependent aspects of material behavior. As illustrated in Figure 17.10(a), a *creep and recovery (recoil)* test is conducted by applying a load (stress σ_o) on the material at time t_0, maintaining the load at a constant level until time t_1, suddenly removing the load at t_1, and observing the material response. As illustrated in Figure 17.10(b), the *stress relaxation* experiment is done by straining the material to a level ϵ_o and maintaining the constant strain while observing the stress response of the material. In an *oscillatory response* test, a harmonic stress is applied and the strain response of the material is measured (Figure 17.10c).

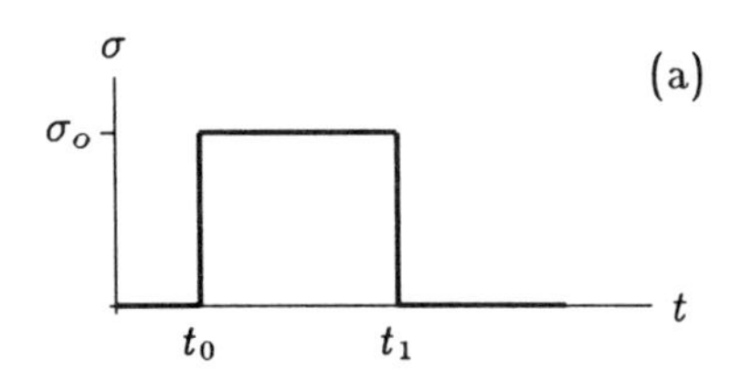

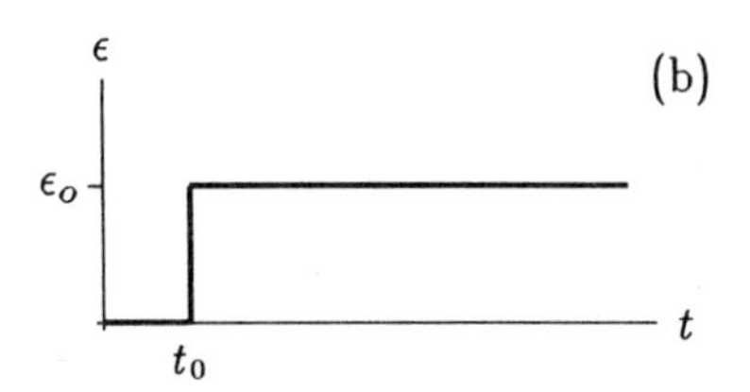

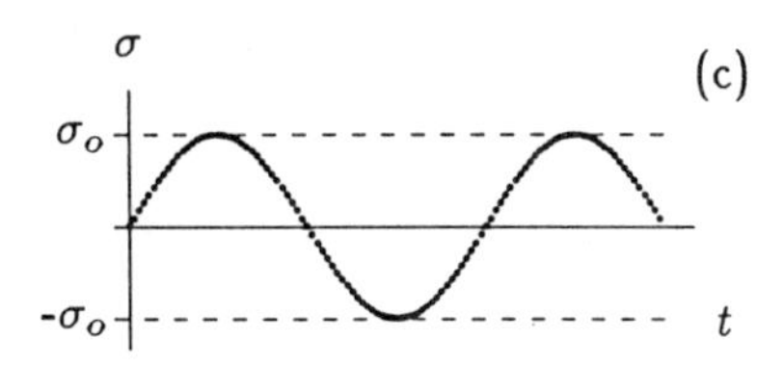

Figure 17.10 *(a) Creep and recovery, (b) stress relaxation, and (c) oscillatory response tests.*

Consider a viscoelastic material. Assume that the material is subjected to a creep test. The results of the creep test can be represented by plotting the measured strain as a function of time. An empirical viscoelastic model for the material behavior can be established through a series of trials. For this purpose, an empirical model is constructed by connecting a number of springs and dashpots together. A differential equation relating stress, strain, and their rates is derived through the procedure outlined in Section 17.3.1 for the Kelvin-Voight model. The imposed condition in a creep test is $\sigma=\sigma_o$. This condition of constant stress is substituted into the differential equation, which is then solved (integrated) for strain ϵ. The result obtained is another equation relating strain to stress σ_o, the elastic moduli and coefficients of viscosity of the empirical model, and time. For a given σ_o and assigned elastic and viscous moduli, this equation is reduced to a function relating strain to time. This function is then used to plot a strain versus time graph and is compared to the experimentally obtained graph. If the general characteristics of the two (experimental and analytical) curves match, the analyses are furthered to establish the elastic and viscous moduli (material constants) of the material. This is achieved by varying the values of the elastic and viscous moduli until the analytical curve matches the experimental curve as closely as possible. In general, this procedure is called ***curve fitting***. If there is no general match between the two curves, the model is abandoned and a new model is constructed and checked.

The result of these mathematical model analyses is an empirical model and a differential equation relating stresses and strains. The stress-strain relationship for the material can be used in conjunction with the fundamental laws of mechanics to analyze the response of the material to different loading conditions.

Note that the deformation processes occurring in viscoelastic materials are quite complex, and it is sometimes necessary to use an array of empirical models to describe the response of a viscoelastic material to different loading conditions. For example, the shear response of a viscoelastic material may be explained with one model and a different model may be needed to explain its response to normal loading. Different models may also be needed to describe the response of a viscoelastic material at low and high strain rates.

17.5 Comparison of Elasticity and Viscoelasticity

There are various criteria with which elastic and viscoelastic behavior of materials can be compared. Some of these criteria will be discussed in this section.

17.5.1 Stress-strain diagrams

An elastic material has a unique stress-strain relationship which is independent of the time or strain rate. For elastic materials, normal and shear stresses can be expressed as functions of normal and shear strains:

$$\sigma = \sigma(\epsilon) \qquad \text{and} \qquad \tau = \tau(\gamma)$$

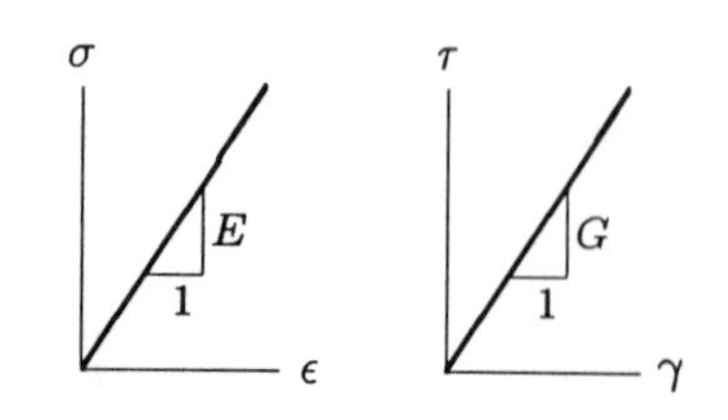

Figure 17.11 *An elastic material has unique normal and shear stress-strain diagrams.*

For example, the stress-strain relationships for a linearly elastic solid are $\sigma = E\epsilon$ and $\tau = G\gamma$, where E and G are constant elastic moduli of the material. As illustrated in Figure 17.11, a linearly elastic material has a unique normal stress-strain diagram and a unique shear stress-strain diagram.

Viscoelastic materials exhibit time-dependent material behavior. The response of a viscoelastic material to an applied stress not only depends upon the magnitude of the stress but also on how fast the stress is applied to or removed from the material. Therefore, the stress-strain relationship for a viscoelastic material is not unique but is a function of the time or rate at which the stresses and strains are developed in the material:

$$\sigma = \sigma(\epsilon, \dot{\epsilon}, .., t) \qquad \text{and} \qquad \tau = \tau(\gamma, \dot{\gamma}, .., t)$$

Consequently, as illustrated in Figure 17.12, a viscoelastic material does not have a unique stress-strain diagram.

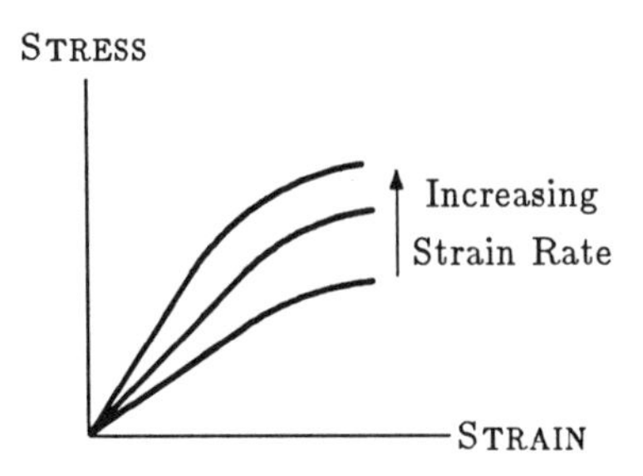

Figure 17.12 *Stress-strain diagram for a viscoelastic material is not unique.*

17.5.2 Loading and unloading paths

For an elastic body, the energy supplied to deform the body (strain energy) is stored in the body as potential energy. This energy is available to return the body to its original (unstressed) size and shape once the applied stress is removed. As illustrated in Figure 17.13, the loading and unloading paths for an elastic material coincide. This indicates that there is no loss of energy during loading and unloading.

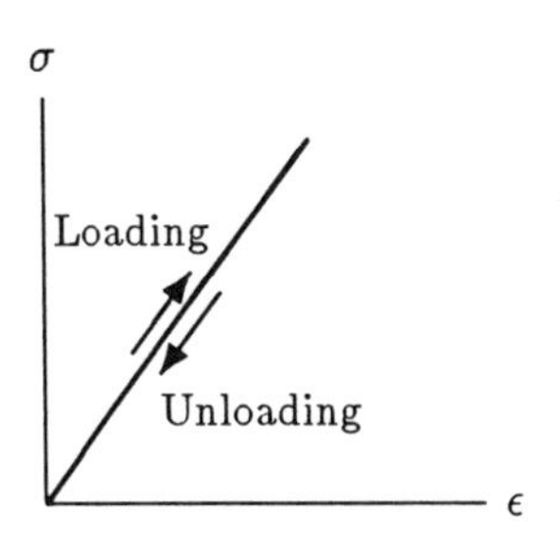

Figure 17.13 *For an elastic material, loading and unloading paths coincide.*

For a viscoelastic body, some of the strain energy is stored in the body as potential energy and some of it is dissipated as heat. For example, consider the Maxwell model. The energy provided to stretch the spring is stored in the spring while the energy supplied to deform the dashpot is dissipated as heat due to the friction between the moving parts of the dashpot. Once the applied load is removed, the potential energy stored in the spring is released to recover the deformation of the spring, but there is no energy available in the dashpot to regain its original configuration.

Consider the three-parameter standard solid model shown in Figure 17.9. A typical loading and unloading diagram for this model is shown in Figure 17.14. The area enclosed by the loading and unloading paths is called the ***hysteresis loop***, which represents the energy dissipated as heat during the deformation and recovery phases. This area, and consequently the amount of energy dissipated as heat, is dependent upon the rate of strain employed to deform the body. The presence of the hysteresis loop in the stress-strain diagram for a viscoelastic material indicates that continuous loading and unloading would result in an increase in temperature of the material.

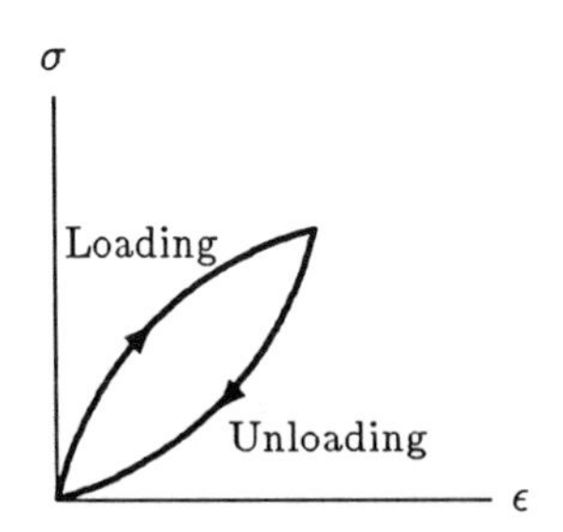

Figure 17.14 *Hysteresis loop.*

Note here that most of the elastic materials exhibit plastic behavior at stress levels beyond the yield point. For elastic-plastic materials, some of the strain energy is dissipated as heat during plastic deformations. This is indicated with the presence of a hysteresis loop in their loading and unloading diagrams (Figure 17.15). For such materials, energy is dissipated as heat only if the

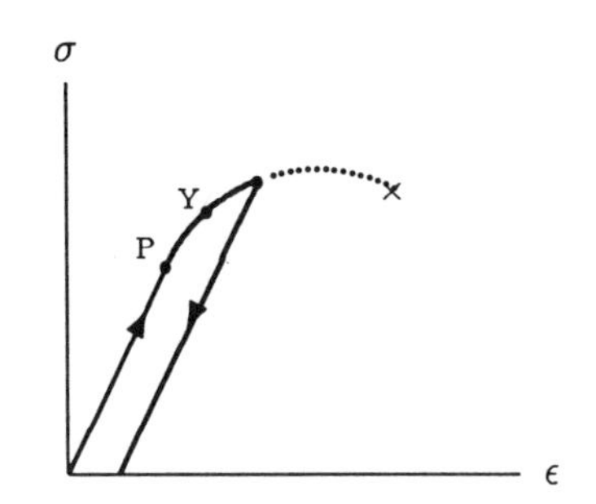

Figure 17.15 *Hysteresis loop for an elastic-plastic material.*

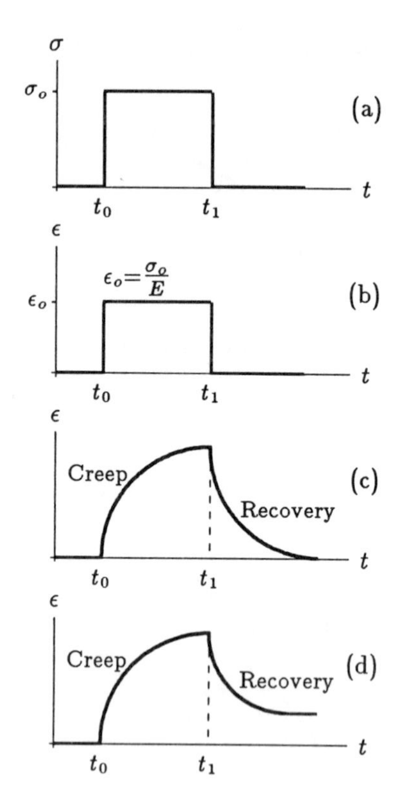

Figure 17.16 *Creep and recovery.*

plastic region is entered. Viscoelastic materials dissipate energy regardless of whether the strains or stresses are small or large.

17.5.3 Time-dependent responses

Since viscoelastic materials exhibit time-dependent material behavior, the differences between elastic and viscoelastic material responses are most evident under time-dependent loading conditions, such as during the creep and stress relaxation experiments.

As discussed earlier in this chapter, a creep and recovery test is conducted by observing the response of a material to a constant stress σ_o applied at time t_0 and removed at a later time t_1 (Figure 17.16a). As illustrated in Figure 17.16(b), such a load will cause a strain $\epsilon_o = \sigma_o / E$ in a linearly elastic material instantly at time t_0. This constant strain will remain in the material until time t_1. At time t_1, the material will instantly and completely recover the deformation. To the same constant loading condition, a viscoelastic material will respond with a strain increasing between times t_0 and t_1. At time t_1, gradual recovery will start. For a viscoelastic solid material, the recovery will eventually be complete (Figure 17.16c). For a viscoelastic fluid, complete recovery will never be achieved and there will be a residue of deformation left in the material (Figure 17.16d).

As illustrated in Figure 17.17(a), the stress relaxation test is performed by straining a material instantaneously, maintaining the constant strain level ϵ_o in the material and observing the response of the material. A linearly elastic material response is illustrated in Figure 17.17(b). The constant stress $\sigma_o = E\epsilon_o$ developed in the material will remain as long as the strain ϵ_o is maintained. In other words, an elastic material will not exhibit a stress relaxation behavior. A viscoelastic material, on the other hand, will respond with an initial high stress which will decrease over time. If the material is a viscoelastic solid, the stress level will never reduce to zero (Figure 17.17c). As illustrated in Figure 17.17(d), the stress will eventually reduce to zero for a viscoelastic fluid.

Almost all biological materials exhibit viscoelastic properties, and the rest of this chapter is devoted to the discussion and review of the mechanical properties of biological tissues including bone, tendons, ligaments, muscles, and articular cartilage.

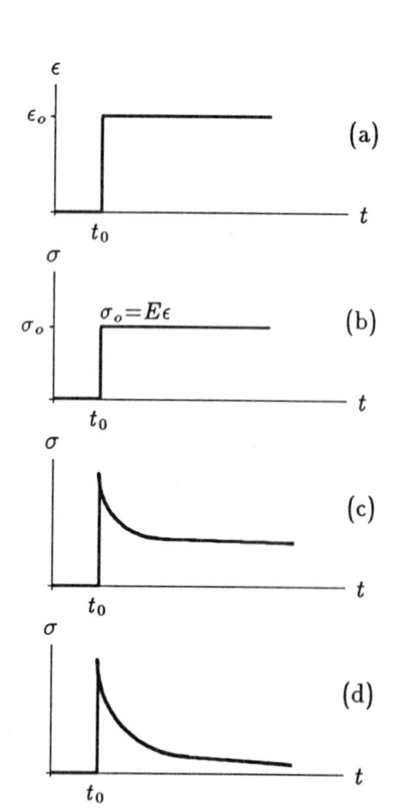

Figure 17.17 *Stress relaxation.*

17.6 Biomechanics of Bone

A selected summary of the biomechanical properties of bone and the methods commonly used to determine these properties will be provided in the following sections. For more information the interested reader should refer to Carter (1985); Cowin, Van Buskirk, and Ashman (1987); Hayes (1986); Melvin and Evans (1985); Nordin and Frankel (1989); and Reilly and Burstein (1974).

Bone is the primary structural element of the human body. Bones form the building blocks of the skeletal system which protects the internal organs, provides kinematic links, provides muscle attachment sites, and facilitates muscle actions and body movements. Bone has unique structural and mechanical properties that allow it to carry out these functions. As compared to other structural materials, bone is also unique in that it is self-repairing. Bone can also alter its shape, mechanical behavior, and mechanical properties to adapt to the changes in mechanical demand. The major factors that influence the mechanical behavior of bone are: the composition of bone, the mechanical properties of the tissues comprising the bone, the size and geometry of the bone, and the direction, magnitude, and the rate of applied loads.

17.6.1 Composition of bone

In biological terms, bone is a *connective tissue* which binds together various structures of the body. In mechanical terms, bone is a *composite material* with various solid and fluid phases. Bone consists of cells and an organic mineral matrix of fibers and a ground substance. Bone also contains inorganic substances in the form of mineral salts. The inorganic component of bone makes it hard and relatively rigid, and its organic component provides flexibility and resilience. The composition of bone varies with species, age, sex, type of bone, type of bone tissue, and the presence of bone disease.

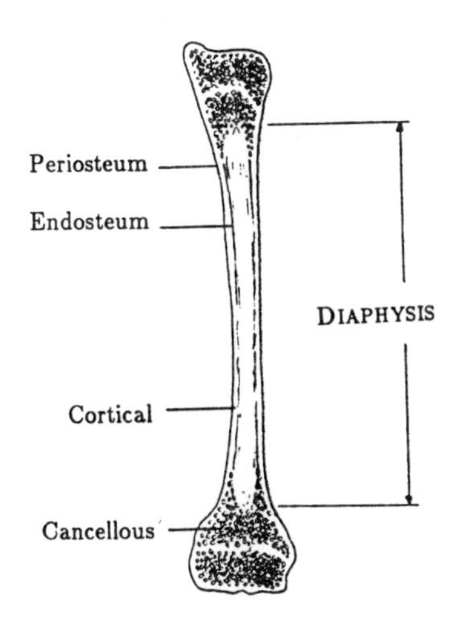

Figure 17.18 *Sectional view of bone.*

At the macroscopic level, all bones consist of two types of tissues (Figure 17.18). The *cortical*, or *compact*, bone tissue is a dense material forming the outer shell (cortex) of bones and the diaphysial region of long bones. The *cancellous*, *trabecular*, or *spongy*, bone tissue consists of thin plates (trabeculae) in a loose mesh structure which is enclosed by the cortical bone. Bones are surrounded by a dense fibrous membrane called the *periosteum*. The periosteum covers the entire bone except for the joint surfaces which are covered with articular cartilage.

17.6.2 Mechanical properties of bone

Bone is a nonhomogeneous material because it consists of various cells, organic and inorganic substances with different material properties. Bone is an anisotropic material because its mechanical properties in different directions are different. That is, the mechanical response of bone is dependent upon the direction as well as the magnitude of the applied load. For example, compressive strength of bone is greater than its tensile strength. Bone possesses viscoelastic (time-dependent) material properties. The mechanical response of bone is dependent on the rate at which the loads are applied. Bone can resist rapidly applied loads much better than slowly applied loads. In other words, bone is stiffer and stronger at higher strain rates.

Bone is a complex structural material, and it is subject to the same engineering laws that govern the behavior of engineering materials such as metals, concrete, and polymers. The mechanical response of bone can be observed by subjecting it to tension, compression, bending, and torsion. Various tests to implement these conditions were discussed in the previous chapters. These tests can be performed using uniform bone specimens or whole bones. If the purpose is to investigate the mechanical response of a specific bone tissue (cortical or cancellous), then the tests are performed using bone specimens. Testing a whole bone, on the other hand, attempts to determine the "bulk" properties of that bone.

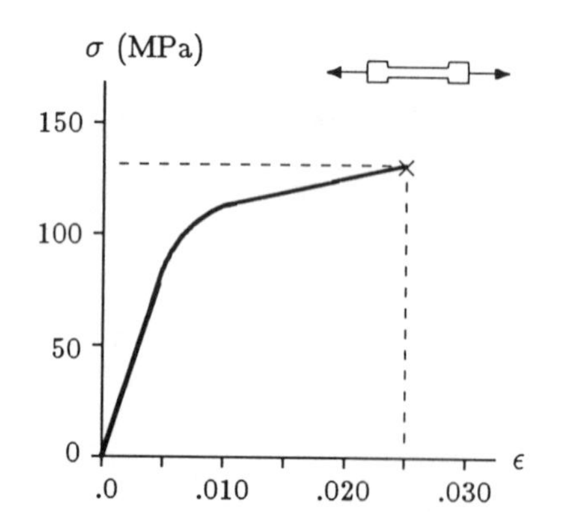

Figure 17.19 *Tensile stress-strain diagram for human cortical bone loaded in the longitudinal direction (strain rate $\dot{\epsilon}$=0.05 s^{-1}).*

The tensile stress-strain diagram for the cortical bone is shown in Figure 17.19. This σ-ϵ curve is drawn using the averages of the elastic modulus, strain hardening modulus, ultimate stress, and ultimate strain values determined for the human femoral cortical bone by Reilly, Burstein, and Frankel (1974). Reilly *et al* tested specimens of bone tissues (human and bovine) under tensile and compressive loads applied in the longitudinal direction at a moderate strain rate ($\dot{\epsilon}$=0.05 s^{-1}). The σ-ϵ curve in Figure 17.19 has three distinct regions. In the initial linearly elastic region, the σ-ϵ curve is nearly a straight line and the slope of this line is equal to the elastic modulus (E) of the bone which is about 17 GPa. In the intermediate region, the bone exhibits nonlinear elasto-plastic material behavior. Material yielding also occurs in this region. By the offset method described in Chapter 14, the yield strength of the cortical bone for the σ-ϵ diagram shown in Figure 17.19 can be determined to be about 110 MPa. In the final region, the bone exhibits a linearly plastic material behavior and the σ-ϵ diagram is another straight line. The slope of this line is the strain hardening modulus (E') of bone tissue which is about 0.9 GPa. The bone fractures when the tensile stress is about 128 MPa, for which the tensile strain is about 0.026. Therefore, based on the data which is presented in Figure 17.19, the tensile ultimate strength of the human cortical bone is about 128 MPa.

The elastic moduli and strength values for bone are dependent upon many factors including the test conditions such as the rate

at which the loads are applied (see Carter and Spengler, 1982). This viscoelastic nature of bone tissue is demonstrated in Figure 17.20. The stress-strain diagrams in Figure 17.20 for different strain rates indicate that a specimen of bone tissue which is subjected to rapid loading (high $\dot{\epsilon}$) have a greater elastic modulus and ultimate strength than a specimen which is loaded more slowly (low $\dot{\epsilon}$). Figure 17.20 also demonstrates that the energy absorbed (which is proportional to the area under the σ-ϵ curve) by the bone tissue increases with an increasing strain rate. Note that during normal daily activities, bone tissues are subjected to a strain rate of about 0.01 s^{-1}.

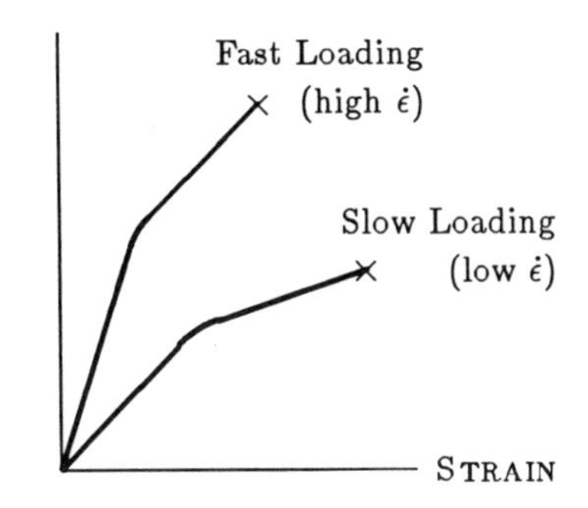

Figure 17.20 *The strain rate dependent stress-strain curves for cortical bone tissue.*

The stress-strain behavior of bone is also dependent upon the orientation of bone with respect to the direction of loading. This anisotropic material behavior of bone is demonstrated in Figure 17.21. Notice that the cortical bone has a larger ultimate strength (i.e., stronger) and a larger elastic modulus (i.e., stiffer) in the longitudinal direction than the transverse direction. Furthermore, bone specimens loaded in the transverse direction fail in a more brittle manner (without showing considerable yielding) as compared to bone specimens loaded in the longitudinal direction. The ultimate strength values for adult femoral cortical bone under various modes of loading, and its elastic and shear moduli are listed in Table 17.1, which is adopted from Reilly and Burstein (1974). Ultimate strength values in Table 17.1 indicate that the bone strength is highest under compressive loading in the longitudinal direction (direction of osteon orientation) and lowest under tensile loading in the transverse direction (direction perpendicular to the longitudinal direction). The elastic modulus of cortical bone in the longitudinal direction is higher than its elastic modulus in the transverse direction. Therefore, cortical bone is stiffer in the longitudinal direction than in the transverse direction.

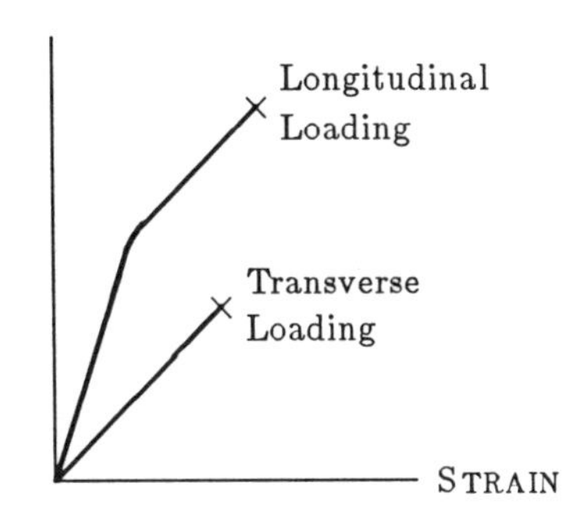

Figure 17.21 *The direction dependent stress-strain curves for bone tissue.*

LOADING MODE	ULTIMATE STRENGTH
LONGITUDINAL	
Tension	133 MPa
Compression	193 MPa
Shear	68 MPa
TRANSVERSE	
Tension	51 MPa
Compression	133 MPa
ELASTIC MODULI, E	
Longitudinal	17.0 GPa
Transverse	11.5 GPa
SHEAR MODULUS, G	3.3 GPa

Table 17.1 *Ultimate strength, and elastic and shear moduli for human femoral cortical bone. (1 GPa=10^9 Pa, 1 MPa=10^6 Pa)*

The chemical compositions of cortical and cancellous bone tissues are similar. The distinguishing characteristic of the cancellous bone is its porosity. This physical difference between the two bone tissues is quantified in terms of the ***apparent density*** of bone, which is defined as the mass of bone tissue present in a unit volume of bone. To a certain degree, both cortical and cancellous bone tissues can be regarded as a single material of variable density. The material properties (for example, strength and stiffness) and the stress-strain characteristics of cancellous bone depend not only on the apparent density of bone which may be different for different bone types or at different parts of a single bone, but also on the mode of loading. The compressive stress-strain curves (Figure 17.22) of cancellous bone contain an initial linearly elastic region up to a strain of about 0.05. The material yielding occurs as the trabeculae begin to fracture. This initial elastic region is followed by a plateau region of almost

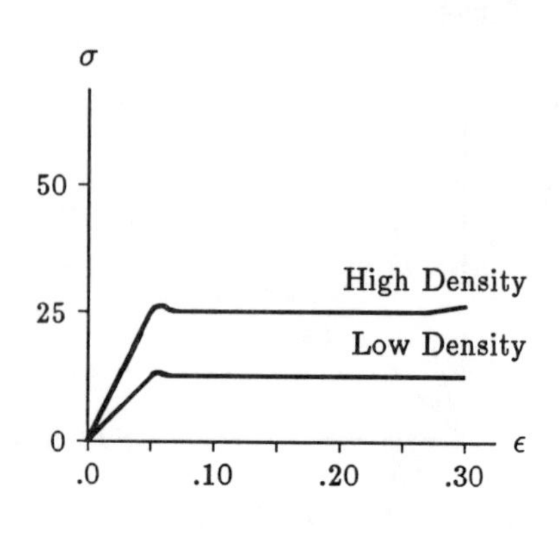

Figure 17.22 *Apparent density dependent stress-strain curves for cancellous bone tissue.*

constant stress until fracture, exhibiting a ductile material behavior. By contrast, cancellous bone fractures abruptly under tensile forces, showing a brittle material behavior. The energy absorption capacity of cancellous bone is considerably higher under compressive loads than under tensile loads.

17.6.3 Structural integrity of bone

There are several factors which may affect the structural integrity of bones. For example, the size and geometry of a bone determine the distribution of the internal forces throughout the bone, thereby influencing its response to externally applied loads. The larger the bone, the larger the area upon which the internal forces are distributed and the smaller the intensity (stress) of these forces. Consequently, the larger the bone, the more resistant the bone to applied loads.

A common characteristic of long bones is their tubular structure in the diaphysial region, which has considerable mechanical advantage over solid circular structures of the same mass. Recall from Chapter 15 that the normal stresses in a structure subjected to torsion are inversely proportional with polar moment of inertia (J) of the cross-sectional area of the structure, and the normal stresses in a structure subjected to bending are inversely proportional with the area moment of inertia (I) of the cross-section of the structure. The larger the polar and area moments of inertia of a structure, the lower the maximum normal stresses due to torsion and bending. Since tubular structures have larger polar and area moments of inertia as compared to solid cylindrical structures of the same volume, tubular structures are more resistant to torsional and bending loads as compared to solid cylindrical structures. Furthermore, a tubular structure can distribute the internal forces more evenly over its cross-section as compared to a solid cylindrical structure of the same cross-sectional area.

Certain skeletal conditions such as osteoporosis can reduce the structural integrity of bone by reducing its apparent density. Small decreases in bone density can generate large reductions in bone strength and stiffness. As compared to a normal bone with the same geometry, an osteoporotic bone will deform easier and fracture at lower loads. The density of bone can also change with aging, after periods of disuse, or after chronic exercise, thereby changing its overall strength. Certain surgical procedures that alter the normal bone geometry may also reduce the strength of bone. Bone defects such as screw holes reduce the load-bearing ability of bone by causing stress concentrations around the defects.

17.6.4 Bone fractures

When bones are subjected to moderate loading conditions, they respond by small deformations that are only present while the loads are applied. When the loads are removed, bones exhibit elastic material behavior by resuming their original (unstressed) shapes and positions. Large deformations occur when the applied loads are high. Bone fractures when the stresses generated in any region of bone are larger than the ultimate strength of bone.

Fractures caused by pure tensile forces are observed in bones with a large proportion of cancellous bone tissue. Fractures due to compressive loads are commonly encountered in the vertebrae of the elderly, whose bones are weakened as a result of aging. Bone fractures caused by compression occur in the diaphysial regions of long bones. Compressive fractures are identified by their oblique fracture pattern. Long bone fractures are usually caused by torsion and bending. Torsional fractures are identified by their spiral oblique pattern, whereas bending fractures are usually identified by the formation of "butterfly" fragments. Fatigue fracture of bone occurs when the damage caused by repeated mechanical stress outpaces the bone's ability to repair to prevent failure. Bone fractures caused by fatigue are common among professional athletes and dedicated joggers. Clinically, most bone fractures occur as a result of complex, combined loading situations rather than simple loading mechanisms.

17.7 Biomechanics of Soft Tissues

Examples of soft tissues include skin, cardiovascular tissues, articular cartilage, muscles, tendons, and ligaments. The mechanical properties and structural behavior of some of these biological tissues will be summarized in the following sections. As a first step towards more complete information, the interested reader can refer to the review articles related to the mechanical properties of tendons and ligaments (Aspen and Hukins, 1989; Viidik, 1987; Woo 1986; and Woo, Gomez, and Akeson, 1985), muscles (McMahon, 1987; and Winters and Woo, 1990), and articular cartilage (Woo, Mow, and Lai, 1987; and Armstrong, Beverland, and McCoy, 1988).

From a mechanical point of view, all soft tissues are composite materials. Among the common components of soft tissues, collagen and elastin fibers have the most important mechanical properties affecting the overall mechanical behavior of the tissues in which they appear. *Collagen* is a protein made of crimped fibrils which aggregate into fibers. The mechanical properties of collagen fibrils are such that each fibril can be considered a mechanical spring and each fiber an assemblage of springs. The primary mechanical function of collagen fibers is to withstand axial tension. Because of their high length-to-diameter ratios (aspect ratio), collagen fibers are not effective under compressive loads. Whenever a fiber is pulled, its crimp straightens, and its length increases. Like a mechanical spring, the energy supplied to stretch the fiber is stored in the fiber, and it is the release of this energy that returns the fiber to its unstretched configuration when the applied load is removed. The individual fibrils of the collagen fibers are surrounded by a gel-like *ground substance*, which consists largely of water. The collagen fibers possess a two-phase, solid-fluid, or viscoelastic material behavior.

The geometric configuration of collagen fibers and their interaction with the noncollagenous tissue components form the basis of the mechanical properties of soft tissues. Among the noncollagenous tissue components, *elastin* is another fibrous protein whose material properties resemble the properties of rubbers. Elastin and microfibrils form the elastic fibers that are highly extensible, and their extension is reversible even at high strains. In summary, elastin fibers possess a low-modulus elastic material property, while collagen fibers show a higher-modulus viscoelastic material behavior.

17.7.1 Tendons and ligaments

Tendons and ligaments are fibrous connective tissues. Tendons help execute joint motion by transmitting mechanical forces (tensions) from muscles to bones. Ligaments join bones and provide stability to the joints. Unlike muscles, which are active tissues and can produce mechanical forces, tendons and ligaments are passive tissues and cannot actively contract to generate forces.

Around many joints of the human body, there is insufficient space to attach more than one or a few muscles. This requires that to accomplish a certain task, one or a few muscles must share the burden of generating and withstanding large loads whose intensity (stress) is even larger at regions closer to the bone attachments where the cross-sectional areas of the muscles are small. As compared to muscles, tendons are stiffer, have higher tensile strengths, and can endure larger stresses. Therefore, around the joints where the space is limited, muscle attachments to bones are made by tendons. Tendons are capable of supporting very large loads with very small deformations. This property of tendons enables the muscles to transmit forces to bones without wasting energy to stretch tendons.

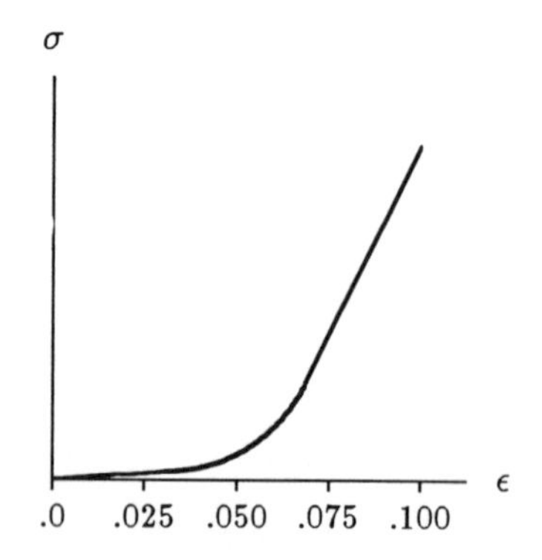

Figure 17.23 *Tensile stress-strain diagram for tendon.*

The mechanical properties of tendons and ligaments depend upon their composition which can vary considerably. The most common means of evaluating the mechanical response of tendons and ligaments is the uniaxial tension test. Figure 17.23 shows a tensile stress-strain diagram for a typical tendon. The shape of this curve is the result of the interaction between elastic elastin fibers and the viscoelastic collagen fibers. At low strains (up to about 0.05), less stiff elastic fibers dominate and the crimp of the collagen fibers straightens, requiring very little force to stretch the tendon. The tendon becomes stiffer when the crimp is straightened. At the same time, the fluid-like ground substance in the collagen fibers tends to flow. At higher strains, therefore, the stiff and viscoelastic nature of the collagen fibers begins to take an increasing portion of the applied load. Tendons are believed to function in the body at strains of up to about 0.04, which is believed to be their yield strain (ϵ_y). Tendons rupture at strains of about 0.1 (ultimate strain, ϵ_u), or stresses of about 60 MPa (ultimate stress, σ_u).

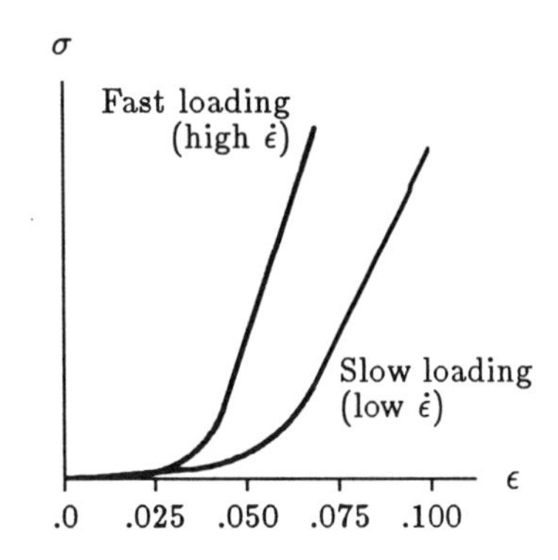

Figure 17.24 *The strain rate dependent stress-strain curves for tendon.*

Note that the shape of the stress-strain curve in Figure 17.23 is such that the area under the curve is considerably small. In other words, the energy stored in the tendon to stretch the tendon to a stress level is much smaller than the energy stored to stretch a linearly elastic material (whose stress-strain diagram would be a straight line) to the same stress level. Therefore, the tendon has higher resilience than linearly elastic materials.

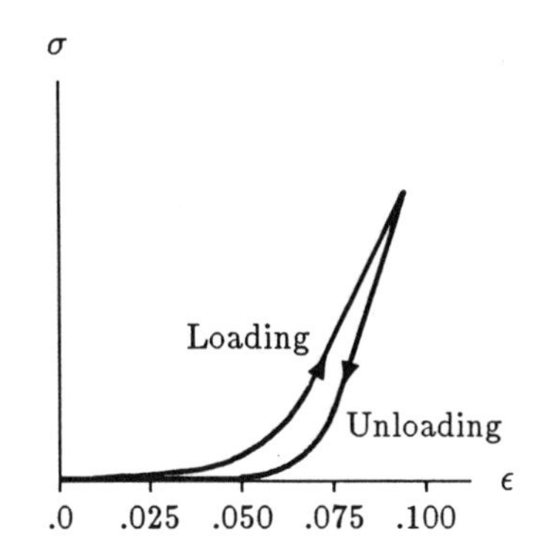

Figure 17.25 *The hysteresis loop of stretching and relaxing modes of the tendon.*

The time-dependent, viscoelastic nature of the tendon is illustrated in Figures 17.24 and 17.25. When the tendon is stretched rapidly, there is less chance for the ground substance to flow, and consequently, the tendon becomes stiffer. The hysteresis loop shown in Figure 17.25 demonstrates the time-dependent loading and unloading behavior of the tendon. Note that more work is done in stretching the tendon than is recovered when the tendon is allowed to relax, and therefore, some of the energy is dissipated in the process.

The mechanical role of ligaments is to transmit forces from one bone to another. Ligaments also have a stabilizing role for the skeletal joints. The composition and structure of ligaments depend upon their function and position within the body. Like tendons they are composite materials containing crimped collagen fibers surrounded by ground substance. As compared to tendons, they often contain a greater proportion of elastic fibers which accounts for their higher extensibility but lower strength and stiffness. The mechanical properties of ligaments are qualitatively similar to those of tendons. Like tendons, they are viscoelastic and exhibit hysteresis, but deform elastically up to strains of about ϵ_y=0.25 (about five times as much as the yield strain of tendons) and stresses of about σ_y=5 MPa. They rupture at a stress of about 20 MPa.

Since tendons and ligaments are viscoelastic, some of the energy supplied to stretch them is dissipated by causing the flow of the fluid within the ground substance, and the rest of the energy is stored in the stretched tissue. This stored energy can disrupt structures within the materials that can cause damage. Tendons and ligaments are tough materials and do not rupture easily. Most common damages to tendons and ligaments occur at their junctions with bones.

17.7.2 Skeletal muscles

There are three types of muscles: skeletal, smooth, and cardiac. Smooth muscles line the internal organs, and cardiac muscles form the heart. Here, we are concerned with the characteristics of the skeletal muscles, each of which is attached, via aponeuroses and/or tendons, to at least two bones causing and/or controlling the relative movement of one bone with respect to the other. When its fibers contract under the stimulation of a nerve, the muscle exerts a pulling effect on the bones to which it is attached. *Contraction* is a unique ability of the muscle tissue, which is defined as the development of tension in the muscle. Muscle contraction can occur as a result of muscle shortening (concentric contraction) or muscle lengthening (eccentric contraction), or it can occur without any change in the muscle length (static or isometric contraction).

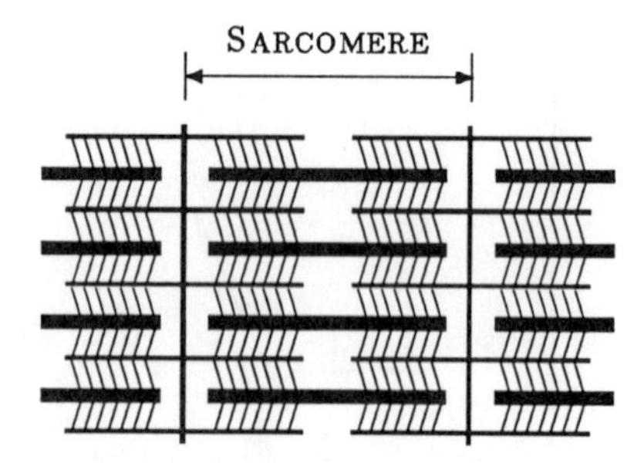

Figure 17.26 *Basic structure of the contractile element of muscle. (Thick lines represent myosin filaments, thin horizontal lines are actin filaments, and cross-hatched lines are cross-bridges.)*

The skeletal muscle is composed of muscle fibers and myofibrils. Myofibrils in turn are made of contractile elements: *actin* and *myosin* proteins. Actin and myosin appear in bands or filaments. Several relatively thick myosin filaments interact across cross-bridges with relatively thin actin filaments to form the basic structure of the contractile element of the muscle, called the *sarcomere* (Figure 17.26). Many sarcomere elements connected in a series arrangement form the contractile element (motor unit) of the muscle. It is within the sarcomere that the muscle force (tension) is generated, and where muscle shortening and lengthening takes place. The active contractile elements of the muscle are contained within a fibrous passive connective tissue, called *fascia*. Fascia encloses the muscles, separates them into layers, and connects them to tendons.

The force and torque developed by a muscle is dependent on many factors, including the number of motor units within the muscle, the number of motor units recruited, the manner in which the muscle changes its length, the velocity of muscle contraction, and the length of the lever arm of the muscle force. For muscles, two different forces can be distinguished. *Active tension* is the force produced by the contractile elements of the muscle and is a result of voluntary muscle contraction. *Passive tension*, on the other hand, is the force developed within the connective muscle tissue when the muscle length surpasses its resting length. The net tensile force in a muscle is dependent on the force-length characteristics of both the active and passive components of the muscle. A typical tension versus muscle length diagram is shown in Figure 17.27. The number of cross-bridges between the filaments is maximum, and therefore, the active tension (T_a) is maximum at the resting length (ℓ_o) of the muscle. As the muscle lengthens, the filaments are pulled apart, the number of cross-bridges is reduced and the active tension is decreased. At full length, there are no cross-bridges and the active tension

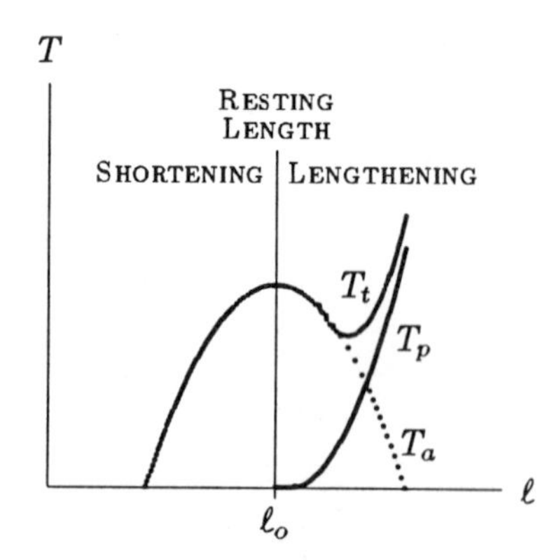

Figure 17.27 *Muscle force (T) versus muscle length (ℓ).*

reduces to zero. As the muscle shortens, the cross-bridges overlap and the active tension is again reduced. When the muscle is at its resting length or less, the passive (connective) component of the muscle is in a loose state with no tension. As the muscle lengthens, a passive tensile force (T_p) starts building up in the connective tissues. The force-length characteristic of this passive component resembles that of a non-linear spring. Passive tensile force increases at an increasing rate as the length of the muscle increases. The overall, total, or net muscle force (T_t) which is transmitted via tendons is the sum of the forces in the active and passive elements of the muscle. Note here that for a given muscle, the tension-length diagram is not unique but dependent on the number of motor units recruited. The magnitude of the active component of the muscle force can vary depending on how the muscle is excited, and usually expressed as the percentage of the maximum voluntary contraction. There are several mechanical models that have been suggested to describe and predict tension in the muscles. Some of these models are reviewed in McMahon (1987).

The force generated by a contracting muscle is usually transmitted to a bone through a tendon. There is a functional reason for tendons to make the transfer of forces from muscles to bones. As compared to tendons, muscles have lower tensile strengths. The relatively low ultimate strength requires muscles to have relatively large cross-sectional areas in order to transmit sufficiently high forces without tearing. Tendons are better designed to perform this function.

17.7.3 Articular cartilage

Cartilage covers the articulating surfaces of bones at the diarthrodial (synovial) joints. The primary function of cartilage is to facilitate the relative movement of articulating bones. Cartilage reduces stresses applied to bones by increasing the area of contact between the articulating surfaces and reduces bone wear by reducing the effects of friction.

Cartilage is a two-phase material consisting of about 75% water and 25% organic solid. A large portion of the solid phase of the cartilage material is made up of collagen fibers. The remaining ground substance is mainly proteoglycan (hydrophilic molecules). Collagen fibers are relatively strong and stiff in tension, while proteoglycans are strong in compression. The solid-fluid composition of cartilage makes it a viscoelastic material.

The mechanical properties of cartilage under various loading conditions have been investigated using a number of different techniques. For example, Sokoloff (1966) investigated the response of human patella to compressive loads using an *indentation test*. As illustrated in Figure 17.28(a), a small cylindrical or hemispherical

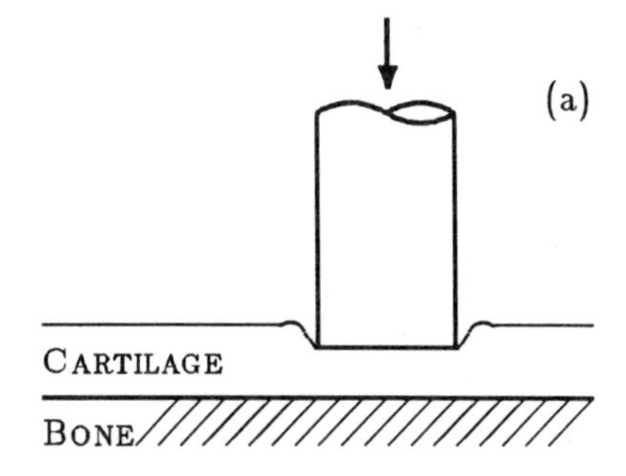

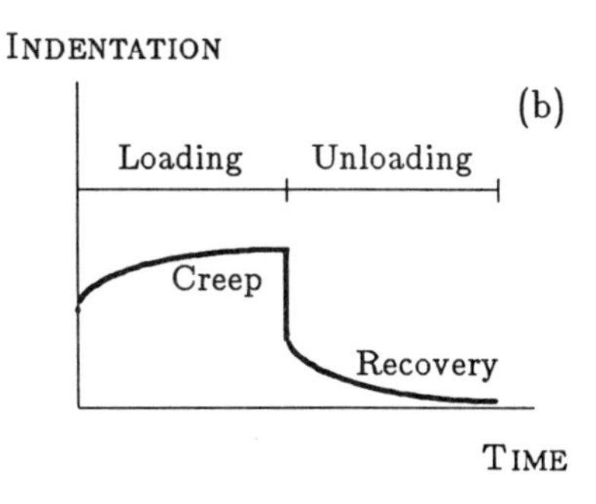

Figure 17.28 *Indentation test.*

indentor is pressed into the articulating surface, and the resulting deformation is recorded. A typical result of an indentation test is shown in Figure 17.28(b). When a constant magnitude load is applied, the material initially responds with a relatively large elastic deformation. The applied load causes pressure gradients to occur in the interstitial fluid, and the variations in pressure cause the fluid to flow through and out of the cartilage matrix. As the load is maintained, the amount of deformation increases at a decreasing rate. The deformation tends towards an equilibrium state as the pressure variations within the fluid are dissipated. When the applied load is removed (unloading phase), there is an instantaneous elastic recovery (recoil) which is followed by a more gradual recovery leading to complete recovery. This creep-recovery response of cartilage may be qualitatively represented by the three-parameter viscoelastic solid model (Figure 17.29), which consists of a linear spring and a Kelvin-Voight unit connected in series.

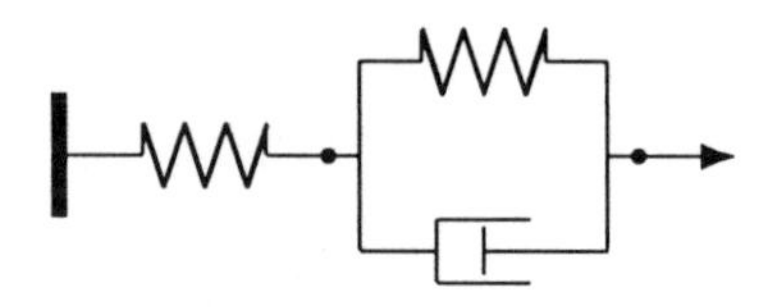

Figure 17.29 *Standard solid model can be used to represent the creep-recovery behavior of cartilage.*

Another experiment designed to investigate the response of cartilage to compressive loading conditions is the *confined compression test* (see Woo, Mow, and Lai, 1987) illustrated in Figure 17.30. In this test, the specimen is confined in a rigid cylindrical die and loaded with a rigid permeable block. The compressive load causes pressure variations in the interstitial fluid and consequent fluid exudation. Eventually, the pressure variations dissipate and an equilibrium is reached. The state at which the equilibrium is reached is indicative of the compressive stiffness of the cartilage. The compressive stiffness and resistance of cartilage depend upon the water and proteoglycan content of the tissue. The higher the proteoglycan content, the higher the compressive resistance of the tissue.

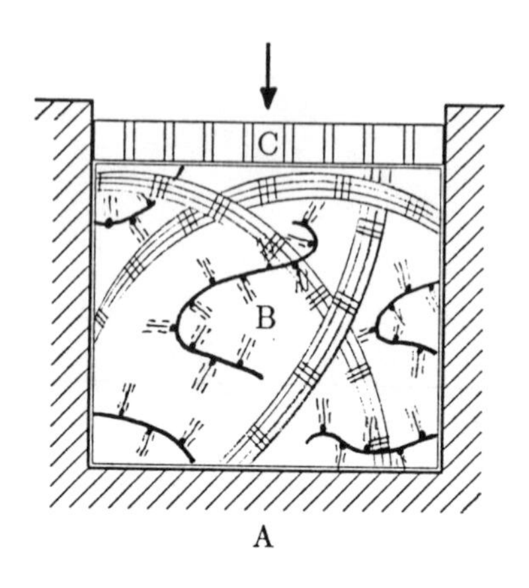

Figure 17.30 *Confined compression test with rigid die (A), specimen (B), and permeable block (C).*

During daily activities, the articular cartilage is subjected to tensile and shear stresses as well as compressive stresses. Under tension, cartilage responds by realigning the collagen fibers which carry the tensile loads applied to the tissue. The tensile stiffness and strength of cartilage depend on the collagen content of the tissue. The higher the collagen content, the higher the tensile strength of cartilage. Shear stresses on the articular cartilage are due to the frictional forces between the relative movement of articulating surfaces. However, the coefficient of friction for synovial joints is so low (of the order 0.001-0.06) that friction has an insignificant effect on the stress resultants acting on the cartilage.

Both structural (such as intraarticular fracture) and anatomical abnormalities (such as rheumatoid arthritis and acetabular dysplasia) can cause cartilage damage, degeneration, wear, and failure. These abnormalities can change the load-bearing ability of the joint by altering its mechanical properties. The importance of the load-bearing ability of the cartilage and maintaining its mechanical integrity may become clear if we consider that the magnitude of the forces involved at the human hip joint is about five times body weight during ordinary walking (much higher during running or jumping). The hip contact area over which these forces are applied is about 15 cm^2 (0.0015 m^2). Therefore, the compressive stresses (pressures) involved are of the order 3 MPa for an 85 kg person.

17.8 Conclusion

Here we have covered, very briefly, the mechanical properties of selected biological tissues. We believe that the knowledge of the mechanical properties and structural behavior of biological tissues is an essential prerequisite for any experimental or theoretical analysis of their physiological function in the body. We are aware of the fact that the proper coverage of each of these topics deserves at least a full chapter. Our purpose here was to provide a summary, illustrate how biological phenomena can be described in terms of the mechanical concepts that we introduced earlier, and hope that the interested reader would refer to more complete sources of information to improve his or her knowledge of the subject matter.

Finally, by preparing this text, we hope to contribute to an improved dialogue between those who are interested in the biological and physiological aspects of the human body and those who are interested in the structural behavior of the human body through an engineering approach.

17.9 References Cited

Armstrong, C.G., Beverland, D.E., and McCoy, G.F. 1988. Properties of cartilage and menisci. In *Encyclopedia of Medical Devices and Instrumentation*, ed. J.G. Webster, 1:616-624. New York: John Wiley & Sons.

Aspen, R.M., and Hukins, D.W.L. 1988. Properties of ligament and tendon. In *Encyclopedia of Medical Devices and Instrumentation*, ed. J.G. Webster, 2:1764-1772. New York: John Wiley & Sons.

Carter, D.R. 1985. Biomechanics of bone. In *The Biomechanics of Trauma*, ed. A.M. Nahum and J. Melvin, pp. 135-165. Norwalk, CT: Appleton-Century-Crofts.

Carter, D.R., and Spengler, D.M. 1982. Biomechanics of fracture. In *Bone and Clinical Orthopaedics*, ed. G. Summer-Smith, pp. 305-334. Philadelphia: Saunders.

Cowin, S.C., Van Buskirk, W.C., and Ashman, R.B. 1987. Properties of bone. In *Handbook of Bioengineering*, ed. R. Skalak and S. Chien, pp. 2.1-2.27. New York: McGraw-Hill.

Hayes, W.C. 1986. Bone mechanics: From tissue mechanical properties to an assessment of structural behavior. In *Frontiers in Biomechanics*, ed. G.W. Schmid-Schönbein, S.L-Y. Woo, and B.W. Zweifach, pp. 196-207. New York: Springer Verlag.

McMahon, T.A. 1987. Muscle mechanics. In *Handbook of Bioengineering*, ed. R. Skalak and S. Chien, pp. 7.1-7.26. New York: McGraw-Hill.

Melvin, J.W., and Evans, F.G. 1985. Extremities: Experimental aspects. In *The Biomechanics of Trauma*, ed. A.M. Nahum and J. Melvin, pp. 135-165. Norwalk, CT: Appleton-Century-Crofts.

Nordin, M., and Frankel, V.H. 1989. Biomechanics of bone. In *Basic Biomechanics of the Musculoskeletal System*, ed. M. Nordin and V.H. Frankel, pp. 3-29. Philadelphia: Lea & Febiger.

Reilly, D.T., and Burstein, A.H. 1974. The mechanical properties of cortical bone. *J. Bone and Joint Surgery* 56A:1001-1022.

Reilly, D.T., and Burstein, A.H. 1975. The elastic and ultimate properties of compact bone tissue. *J. Biomechanics* 8:393-405.

Reilly, D.T., Burstein, A.H., and Frankel, V.H. 1974. The elastic modulus for bone. *J. Biomechanics* 7:271-275.

Sokoloff, L. 1966. Elasticity of ageing cartilage, *Fed. Proc., Fed. Am. Soc. Exp. Biol.* 25:1089-1095.

Viidik, A. 1987. Properties of tendons and ligaments. In *Handbook of Bioengineering*, ed. R. Skalak and S. Chien, pp. 2.1-2.27. New York: McGraw-Hill.

Winters, J.M., and Woo, S.L-Y. (Eds.) 1990. *Multiple Muscle Systems*. New York: Springer Verlag.

Woo, S.L-Y. 1986. Biomechanics of tendons and ligaments. In *Frontiers in Biomechanics*, ed. G.W. Schmid-Schönbein, S.L-Y. Woo, and B.W. Zweifach, pp. 180-195. New York: Springer Verlag.

Woo, S.L-Y., Gomez, M.A., and Akeson, W.H. 1985. Mechanical behaviors of soft tissues: Measurements, modifications, injuries, and treatment. In *The Biomechanics of Trauma*, ed. A.M. Nahum and J. Melvin, pp. 109-133. Norwalk, CT: Appleton-Century-Crofts.

Woo, S.L-Y., Mow, V.C., and Lai, W.M. 1987. Biomechanical properties of articular cartilage. In *Handbook of Bioengineering*, ed. R. Skalak and S. Chien, pp. 4.1-4.44. New York: McGraw-Hill.

Appendix A

PLANE GEOMETRY

A.1 Angles

An *angle* is formed by two intersecting straight lines. Angles designated as θ in Figure A.1 which are formed by two intersecting straight lines aa and bb are equal and called *opposite angles.* If the two lines are perpendicular to one another, then the angles formed are called *right angles.* A right angle is equal to 90°. Graphically, right angles are usually denoted by small square boxes. An angle is called *acute* if it is smaller than 90°, and is called *obtuse* if it is greater than 90°.

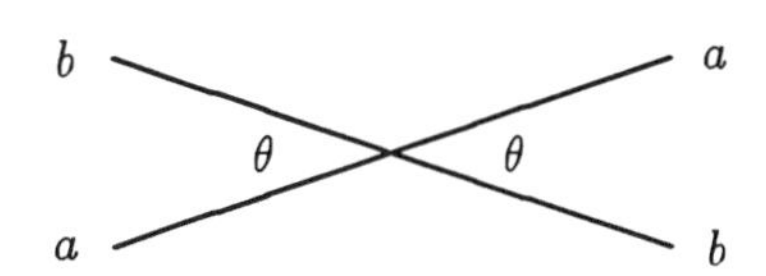

Figure A.1 *Opposite angles.*

Angles designated as θ in Figure A.2 are equal and called *alternate angles.* These angles are formed by a straight line cc intersecting two parallel straight lines aa and bb.

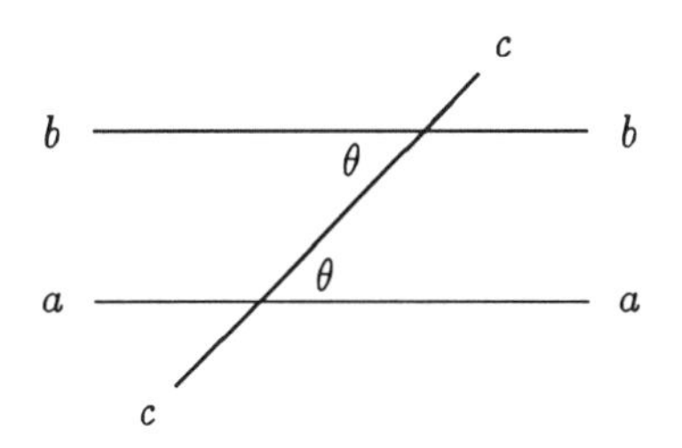

Figure A.2 *Alternate angles.*

Angles designated as θ in Figure A.3 are equal. In this case, straight line cc is perpendicular to bb, and dd is perpendicular to aa. The geometry illustrated in Figure A.3 is utilized extensively in physics and mechanics, for example while analyzing motions on inclined surfaces. For such problems aa represents the horizontal, bb represent the inclined surface which makes an angle θ with the horizontal, cc is perpendicular to the inclined surface, and dd is a vertical straight line.

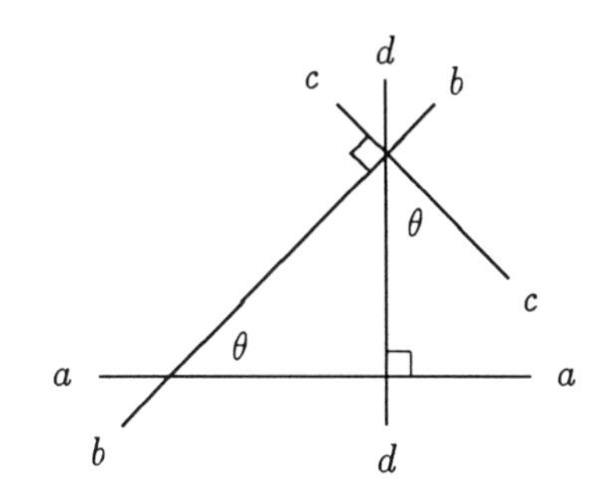

Figure A.3 *Lines bb and cc, and aa and dd are perpendicular.*

A.2 Triangles

A *triangle* is a geometric shape with three sides and three angles (Figure A.4). For any triangle, the sum of the three angles is equal to 180°:

$$\alpha + \beta + \theta = 180°$$

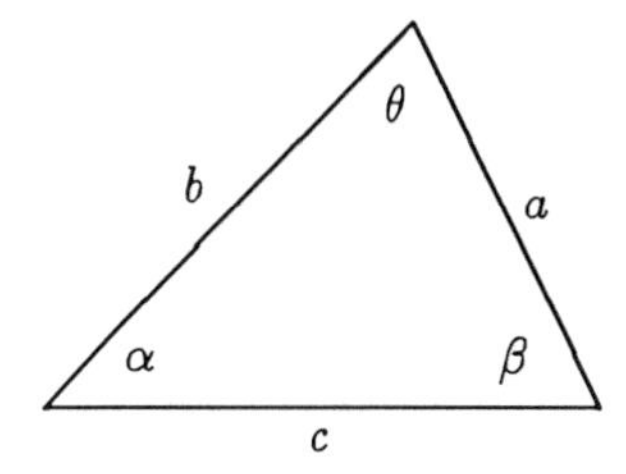

Figure A.4 *A triangle.*

A.3 Law of Sines

For any triangle, such as the one in Figure A.4, the angles and sides of the triangle are related through the *law of sines* which states that:

$$\frac{\sin\alpha}{a} = \frac{\sin\beta}{b} = \frac{\sin\theta}{c} \tag{A.1}$$

The definition of sine (abbreviated as sin) is given in Section A.6.

A.4 The Right-Triangle

If one of the three angles of a triangle is equal to 90°, then it is called a *right-triangle.* For the right-triangle shown in Figure A.5, angle θ is equal to 90°, and the sum of the remaining angles is equal to 90°:

$$\theta = \alpha + \beta = 90°$$

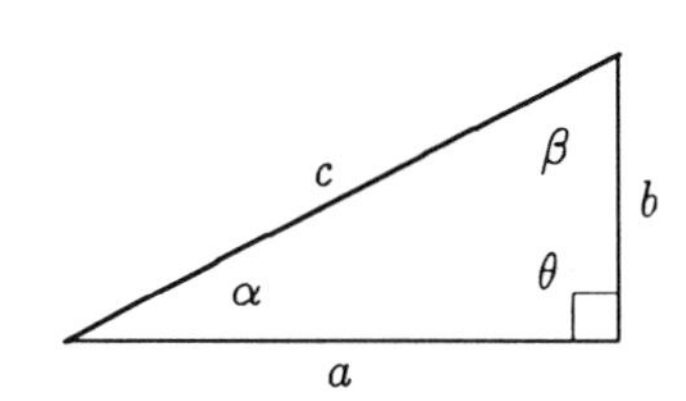

Figure A.5 *The right-triangle.*

In Figure A.5, side c of the triangle opposite the right angle (angle θ) is called the *hypotenuse,* and it is the longest side of the triangle. With respect to angle α, a is called the *adjacent* side and b the *opposite* side.

A.5 Pythagorean Theorem

The *Pythagorean theorem* states that the square of the length of the hypotenuse of a right-triangle is equal to the sum of the squares of the lengths of the other sides of the right-triangle:

$$c^2 = a^2 + b^2 \tag{A.2}$$

Considering the square-root of both sides:

$$c = \sqrt{a^2 + b^2}$$

A.6 Sine, Cosine, and Tangent

The *sine* of an acute angle (angles other than the right angle) in a right-triangle is equal to the ratio of the lengths of the opposite side and the hypotenuse:

$$\sin\alpha = \frac{b}{c} \qquad \sin\beta = \frac{a}{c} \tag{A.3}$$

The *cosine* of an acute angle in a right-triangle is the ratio of the lengths of the adjacent side and the hypotenuse:

$$\cos\alpha = \frac{a}{c} \qquad \cos\beta = \frac{b}{c} \tag{A.4}$$

Note that the sine of angle α in Figure A.5 is equal to the cosine of angle β.

The *tangent* of an acute angle in a right-triangle is the ratio of the lengths of the opposite side and the adjacent side:

$$\tan\alpha = \frac{b}{a} \qquad \tan\beta = \frac{a}{b} \tag{A.5}$$

The sine, cosine, and tangent of a few selected angles are listed in Table A.1. Note that the tangent of an angle is also equal to the ratio of the sine and the cosine of that angle:

$$\tan\alpha = \frac{\sin\alpha}{\cos\alpha} \qquad \tan\beta = \frac{\sin\beta}{\cos\beta}$$

Also note that Eqs. (A.3) and (A.4) can alternatively be written as follows:

$$\begin{aligned} a &= c\ \cos\alpha \\ b &= c\ \sin\alpha \end{aligned} \tag{A.6}$$

Angle	sin	cos	tan
0°	0.000	1.000	0.000
30°	0.500	0.866	0.577
45°	0.707	0.707	1.000
60°	0.866	0.500	1.732
90°	1.000	0.000	

Table A.1 *Sine, cosine, and tangent of selected angles.*

We can take the squares of a and b in Eq. (A.6) and substitute them in Eq. (A.2) to get:

$$c^2 = a^2 + b^2$$
$$c^2 = (c \cos \alpha)^2 + (c \sin \alpha)^2$$
$$c^2 = c^2(\cos^2 \alpha + \sin^2 \alpha)$$

Dividing both sides of the above equation by c^2 will yield an important property of trigonometric functions:

$$\cos^2 \alpha + \sin^2 \alpha = 1 \tag{A.7}$$

A right-triangle can uniquely be defined if the lengths of two sides of the triangle are given, or one of the acute angles (α or β) is specified along with one of the sides of the triangle. The remaining sides and angles can be determined using the formulas given above.

A.7 Inverse Sine, Cosine, and Tangent

Sometimes the sine, cosine, or the tangent of an angle is known and the task is to determine the angle itself. For this purpose, *inverse* trigonometric functions are defined such that:

$$\begin{array}{lll} \text{If } \sin\alpha = A & \text{then} & \alpha = \arcsin(A) \\ \text{If } \cos\alpha = B & \text{then} & \alpha = \arccos(B) \\ \text{If } \tan\alpha = C & \text{then} & \alpha = \arctan(C) \end{array}$$

Arcsin, arccos, and arctan are abbreviated forms of arcsine, arccosine, and arctangent, respectively. They are alternatively referred to with $\sin^{-1}$, $\cos^{-1}$, and $\tan^{-1}$.

Example A.1 For the right-triangle shown in Figure A.6, determine angles α and β, and the length c of the hypotenuse.

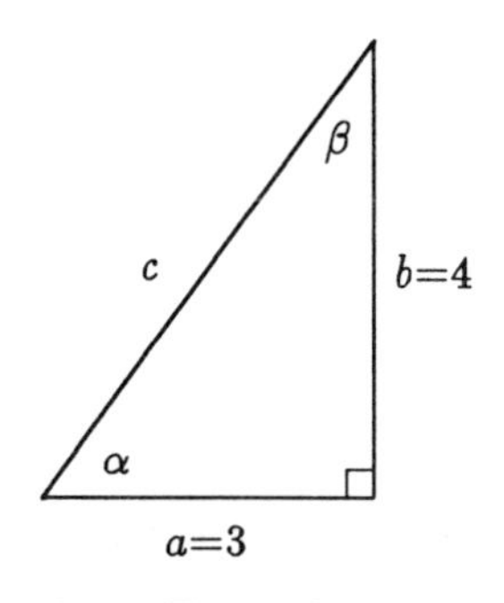

Figure A.6 *Example A.1.*

Solution: From Eq. (A.2):

$$c^2 = a^2 + b^2$$

Substitute the numerical values of a and b:

$$c^2 = 3^2 + 4^2 = 9 + 16 = 25$$

Take the square-root of both sides:

$$c = 5$$

The cosine of angle α is:

$$\cos \alpha = \frac{a}{c} = \frac{3}{5} = 0.6$$

Take the inverse cosine of both sides:

$$\alpha = \cos^{-1}(0.6) = 53.13°$$

The cosine of angle β is:

$$\cos\beta = \frac{b}{c} = \frac{4}{5} = 0.8$$

Take the inverse cosine of both sides:

$$\beta = \cos^{-1}(0.8) = 36.87°$$

Check whether the results are correct:

$$\alpha + \beta = 53.13° + 36.87° = 90° \qquad (\surd)$$

Example A.2 For the right-triangle shown in Figure A.7, determine angle β, and sides b and c.

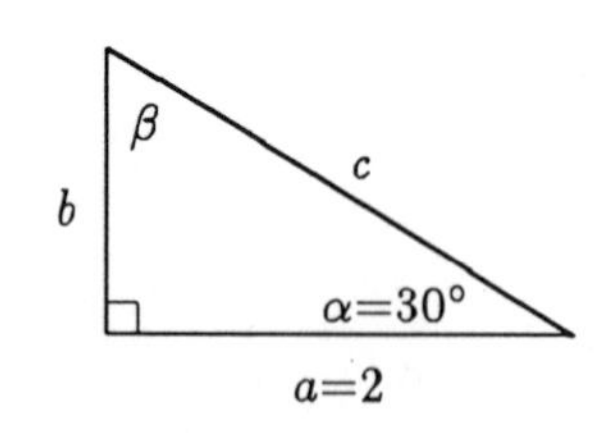

Figure A.7 *Example A.2.*

Solution: The sum of acute angles of a right-triangle must be equal to 90°. Therefore:

$$\beta = 90° - \alpha = 90° - 30° = 60°$$

Consider the cosine of angle α:

$$\cos\alpha = \frac{a}{c}$$

Multiply both sides by c and divide by $\cos\alpha$:

$$c = \frac{a}{\cos\alpha}$$

Therefore:

$$c = \frac{2}{\cos 30°} = \frac{2}{0.866} = 2.31$$

Consider the tangent of angle α:

$$\tan\alpha = \frac{b}{a}$$

Multiply both sides by a:

$$b = a \tan\alpha$$

Therefore:

$$b = (2)(\tan 30°) = (2)(0.577) = 1.15$$

Check if the results are correct:

$$a^2 + b^2 = c^2$$

$$(2)^2 + (1.15)^2 \stackrel{?}{=} (2.31)^2$$

$$5.3 \stackrel{\surd}{=} 5.3$$

Appendix B

ALGEBRA

An equation, such as $2+5=7$, which involves only numbers is called an *arithmetic equation*. Equations may also contain combinations of numbers and symbols which are usually the letters of the alphabet. Equations such as $3z+7=16$ and $x^2+y^2=25$ are called *algebraic equations*. Manipulation of algebraic equations often requires solving one or more equations for one or more unknowns. For example, in the equation $3z+7=16$, the unknown is z and this equation is true if $z=3$.

B.1 Equations with One Unknown

The following examples illustrate the solution of algebraic equations with one unknown A.

Example B.1 $2\times A+3\times 4=18$

$$\begin{aligned}2A+12&=18\\2A&=18-12\\2A&=6\\A&=6/2\\A&=3\end{aligned}$$

Example B.2 $A^2-3^2=16$

$$\begin{aligned}A^2-3\times 3&=16\\A^2-9&=16\\A^2&=16+9\\A^2&=25\\\sqrt{A^2}&=\sqrt{25}\\A&=5\end{aligned}$$

Example B.3 $2(A+9)-5=7$

$$\begin{aligned}2(A+9)&=7+5\\2(A+9)&=12\\A+9&=12/2\\A+9&=6\\A&=6-9\\A&=-3\end{aligned}$$

Example B.4 $(A/5) - (2/3) = (1/2)$

$$\frac{A}{5} - \frac{2}{3} = \frac{1}{2}$$
$$\frac{A}{5} = \frac{1}{2} + \frac{2}{3}$$
$$\frac{A}{5} = \frac{1 \times 3 + 2 \times 2}{2 \times 3}$$
$$\frac{A}{5} = \frac{3 + 4}{6}$$
$$\frac{A}{5} = \frac{7}{6}$$
$$A = 5 \times \frac{7}{6}$$
$$A = \frac{35}{6}$$
$$A = 5.8333$$

B.2 Equations with Two Unknowns

The following examples illustrate the solutions of systems of two algebraic equations for two unknowns A and B.

Example B.5 $A + B = 10$
$A + 2B = 5$

$$A = 10 - B \qquad (i)$$
$$A + 2B = 5 \qquad (ii)$$

Substitute (i) into (ii):

$$(10 - B) + 2B = 5$$
$$10 - B + 2B = 5$$
$$B = 5 - 10$$
$$B = -5 \qquad (iii)$$

Substitute (iii) into (i):

$$A = 10 - (-5)$$
$$A = 10 + 5$$
$$A = 15$$

Example B.6 $2A + 3B = 12$
$3A - 2B = 5$

$$2A + 3B = 12 \qquad (i)$$
$$3A - 2B = 5 \qquad (ii)$$

Multiply (i) by 3 and (ii) by 2:

$$6A + 9B = 36$$
$$6A - 4B = 10$$

$$6A = 36 - 9B \qquad (iii)$$
$$6A - 4B = 10 \qquad (iv)$$

Substitute (iii) into (iv):

$$(36 - 9B) - 4B = 10$$
$$36 - 9B - 4B = 10$$
$$-13B = 10 - 36$$
$$-13B = -26$$
$$B = (-26)/(-13)$$
$$B = 2 \qquad (v)$$

Substitute (v) into (iii)

$$6A = 36 - 9(2)$$
$$6A = 36 - 18$$
$$6A = 18$$
$$A = 18/6$$
$$A = 3$$

B.3 Equations with Three Unknowns

The following examples illustrate the solutions of systems of three algebraic equations for three unknowns A, B, and C.

Example B.7 $2A + B = 10$
$A - C = 5$
$A - B + 2C = 20$

These equations can also be written as:

$$B = 10 - 2A \qquad (i)$$
$$C = -(5 - A) \qquad (ii)$$
$$A - B + 2C = 20 \qquad (iii)$$

Substitute (i) and (ii) into (iii):

$$\begin{aligned} A-(10-2A)+2[-(5-A)] &= 20 \\ A-10+2A-2(5-A) &= 20 \\ A-10+2A-10+2A &= 20 \\ 5A-20 &= 20 \\ 5A &= 40 \\ A &= 40/5 \\ A &= 8 \end{aligned} \qquad (iv)$$

Substitute (iv) into (i):

$$\begin{aligned} B &= 10-2(8) \\ B &= 10-16 \\ B &= -6 \end{aligned}$$

Substitute (iv) into (ii):

$$\begin{aligned} C &= -(5-8) \\ C &= -(-3) \\ C &= 3 \end{aligned}$$

Example B.8

$$A+B+C=6 \qquad (i)$$
$$A-B+C=2 \qquad (ii)$$
$$A+B-C=0 \qquad (iii)$$

Add both sides of (ii) and (iii):

$$\begin{aligned} 2A &= 2 \\ A &= 1 \end{aligned} \qquad (iv)$$

Add both sides of (i) and (ii):

$$\begin{aligned} 2A+2C &= 8 \\ 2C &= 8-2A \end{aligned} \qquad (v)$$

Substitute (iv) into (v):

$$\begin{aligned} 2C &= 8-2(1) \\ 2C &= 6 \\ C &= 3 \end{aligned} \qquad (vi)$$

Substitute (iv) and (vi) into either (i), (ii), or (iii):

$$\begin{aligned} A+B+C &= 6 \\ 1+B+3 &= 6 \\ B+4 &= 6 \\ B &= 2 \end{aligned}$$

B.4 General Approach and Cautions

Any set of two algebraic equations can be written in the following general form:

$$a_1\,A + b_1\,B = d_1$$
$$a_2\,A + b_2\,B = d_2 \qquad \text{(B.1)}$$

where a_1, a_2, b_1, b_2, d_1, and d_2 are some known numbers, and A and B are the unknowns to be determined. The simultaneous solution of Eqs. (B.1) for A and B will give:

$$A = \frac{d_1 b_2 - d_2 b_1}{a_1 b_2 - a_2 b_1}$$
$$B = \frac{a_1 d_2 - a_2 d_1}{a_1 b_2 - a_2 b_1} \qquad \text{(B.2)}$$

Caution. A condition must be checked before applying Eqs. (B.2) to evaluate A and B. For a solution to exist, the denominator in Eqs. (B.2) must be different than zero. That is:

$$a_1 b_2 - a_2 b_1 \neq 0$$

Example B.9 $2A + 3B = -5$
$7A + B = 11$

Compare the given equations with Eqs. (B.1):

$$a_1 = 2 \qquad b_1 = 3 \qquad d_1 = -5$$
$$a_2 = 7 \qquad b_2 = 1 \qquad d_2 = 11$$

Check whether a solution exists:

$$a_1 b_2 - a_2 b_1 = (2)(1) - (7)(3) = 2 - 21 = -19 \qquad (\surd)$$

Apply Eqs. (B.2):

$$A = \frac{(-5)(1) - (11)(3)}{-19} = \frac{-38}{-19} = 2$$

$$B = \frac{(2)(11) - (7)(-5)}{-19} = \frac{57}{-19} = -3$$

Check the validity of results by substituting A=2 and B=–3 into one of the given algebraic equations:

$$7A + B \stackrel{?}{=} 11$$
$$7(2) + (-3) \stackrel{?}{=} 11$$
$$11 \stackrel{\surd}{=} 11$$

Any set of three algebraic equations can be written in the following general form:

$$
\begin{aligned}
a_1\,A + b_1\,B + c_1\,C &= d_1 \\
a_2\,A + b_2\,B + c_2\,C &= d_2 \\
a_3\,A + b_3\,B + c_3\,C &= d_3
\end{aligned}
\tag{B.3}
$$

where A, B, and C are the unknowns, and a_1, a_2, a_3, b_1, b_2, b_3, c_1, c_2, c_3, d_1, d_2, and d_3 are some known numbers. The simultaneous solution of Eqs. (B.3) for A, B, and C will result in:

$$
\begin{aligned}
C &= \frac{b_{12}d_{23} - b_{23}d_{12}}{b_{12}c_{23} - b_{23}c_{12}} \\
B &= \frac{d_{12} - c_{12}\,C}{b_{12}} \\
A &= \frac{d_1 - b_1\,B - c_1\,C}{a_1}
\end{aligned}
\tag{B.4}
$$

where

$$
\begin{aligned}
b_{12} &= a_1\,b_2 - a_2\,b_1 \\
b_{23} &= a_2\,b_3 - a_3\,b_2 \\
c_{12} &= a_1\,c_2 - a_2\,c_1 \\
c_{23} &= a_2\,c_3 - a_3\,c_2 \\
d_{12} &= a_1\,d_2 - a_2\,d_1 \\
d_{23} &= a_2\,d_3 - a_3\,d_2
\end{aligned}
\tag{B.5}
$$

Caution. Before applying Eqs. (B.4), the following conditions must be satisfied:

$$b_{12} \neq 0$$

$$b_{12}c_{23} - b_{23}c_{12} \neq 0$$

If these conditions are not satisfied, then there exists no real solution for A, B, or C.

Example B.10
$$A + 2B - C = 5$$
$$2A + 3C = 1$$
$$A - 3B + 4C = -5$$

Compare the given algebraic equations with Eqs. (B.3):

$$\begin{array}{llll} a_1 = 1 & b_1 = 2 & c_1 = -1 & d_1 = 5 \\ a_2 = 2 & b_2 = 0 & c_2 = 3 & d_2 = 1 \\ a_3 = 1 & b_3 = -3 & c_3 = 4 & d_3 = -5 \end{array}$$

Evaluate the parameters in Eqs. (B.5):

$$b_{12} = (1)(0) - (2)(2) = -4$$
$$b_{23} = (2)(-3) - (1)(0) = -6$$
$$c_{12} = (1)(3) - (2)(-1) = 5$$
$$c_{23} = (2)(4) - (1)(3) = 5$$
$$d_{12} = (1)(1) - (2)(5) = -9$$
$$d_{23} = (2)(-5) - (1)(1) = -11$$

Check whether a solution exists:

$$b_{12} = -4 \qquad (\surd)$$

$$b_{12}c_{23} - b_{12}c_{23} = (-4)(5) - (-6)(5) = 10 \qquad (\surd)$$

Apply Eqs. (B.4):

$$C = \frac{(-4)(-11) - (-6)(-9)}{(-4)(5) - (-6)(5)} = \frac{-10}{10} = -1$$

$$B = \frac{(-9) - (5)(-1)}{(-4)} = \frac{-4}{-4} = 1$$

$$A = \frac{(5) - (2)(1) - (-1)(-1)}{(1)} = 2$$

By substituting the values calculated for A, B, and C into one of the original algebraic equations, check whether the solution is correct:

$$A - 3B + 4C \stackrel{?}{=} -5$$
$$(2) - 3(1) + 4(-1) \stackrel{?}{=} -5$$
$$2 - 3 - 4 \stackrel{?}{=} -5$$
$$-5 \stackrel{\surd}{=} -5$$

B.5 The Quadratic Formula

If an unknown appears in an algebraic equation to the first power, then the equation is said to be *linear*. For example, $3x = 7$ is a simple linear equation with unknown x. If an unknown appears not only to the first power but squared as well, then the equation is *quadratic*. For example, $x^2 + 3x = 5$ is a quadratic equation. Quadratic equations can be written in the following general form:

$$a\,x^2 + b\,x + c = 0 \tag{B.6}$$

The quantities a, b, and c are some known numbers, and x is the unknown parameter. The general solution of Eq. (B.6) is:

$$x = \frac{-b \pm \sqrt{b^2 - 4ac}}{2a} \tag{B.7}$$

The $\pm$ sign indicates that there are two solutions for x, which can be determined by considering the plus and minus signs.

Caution. Note that in Eqs. (B.7), b^2 must be greater than or equal to $4ac$, so that the square root of $(b^2 - 4ac)$ yields a real number. Otherwise, $(b^2 - 4ac)$ will be a negative number, and the square root of a negative number is an "imaginary" number (as opposed to a "real" number).

Example B.11 Find the solutions for x in the following equation:

$$2x^2 + 7x = 4$$

First, rewrite the given equation in the standard form given in Eq. (B.6) by subtracting 4 from both sides:

$$2x^2 + 7x - 4 = 0$$

Compare this with Eq. (B.6) to observe that:

$$a = 2 \qquad b = 7 \qquad c = -4$$

The solutions for x are:

$$x = \frac{-(7) \pm \sqrt{(7^2) - 4(2)(-4)}}{2(2)} = \frac{-7 \pm \sqrt{81}}{4} = \frac{-7 \pm 9}{4}$$

Consider the plus sign:

$$x = \frac{-7 + 9}{4} = \frac{2}{4} = \frac{1}{2}$$

Consider the minus sign:

$$x = \frac{-7 - 9}{4} = \frac{-16}{4} = -4$$

Appendix C

CALCULUS

C.1 Functions

Almost all problems in calculus involve functions. Consider two quantities X and Y. Assume that these quantities are somewhat related, such that a change in quantity X causes quantity Y to vary. The variations in Y with respect to variations in X can be represented by various schemes, such as a table or a diagram. Another way of representing the relationship between X and Y may be by means of a mathematical equation called a *function*. Functions are usually denoted by single letters, such as f or g. For example, $Y=f(X)$ implies that Y is a function of X. X in $Y=f(X)$ is the "input" or "cause" of an operation or process, while Y is the "output" or "effect." The input of a function is called the *independent variable*, and the output is called the *dependent variable*.

Experiment	X	Y
1	0.5	2.0
2	1.0	2.9
3	1.5	4.1
4	2.0	5.2
5	2.5	6.0
6	3.0	6.9
7	3.5	8.0
8	4.0	8.9
9	4.5	10.0
10	5.0	11.0

Table C.1 *Experimental data.*

Assume that a series of ten experiments are conducted, where by varying quantity X, corresponding Y values are measured. The data collected is represented in Table C.1. In general, information can be extracted more easily from a diagram than from a set of numbers. For this purpose, a rectangular coordinate system can be constructed by assigning the horizontal coordinate (abscissa) to the input X and the vertical coordinate (ordinate) to the output Y (Figure C.1). Each pair of X and Y values recorded in Table C.1 will then correspond to a point on this XY plane. A curve can be obtained by connecting these points, which will represent the graph of the data obtained. The curve obtained can be compared with the graph of known functions, and the function which may represent the experimental data can be established. For example, the straight line in Figure C.1 implies that quantity Y is a "linear" function of X, and that the relationship between X and Y can be represented by the equation $Y=1+2X$. Therefore, the function relating X and Y is $f(X)=1+2X$.

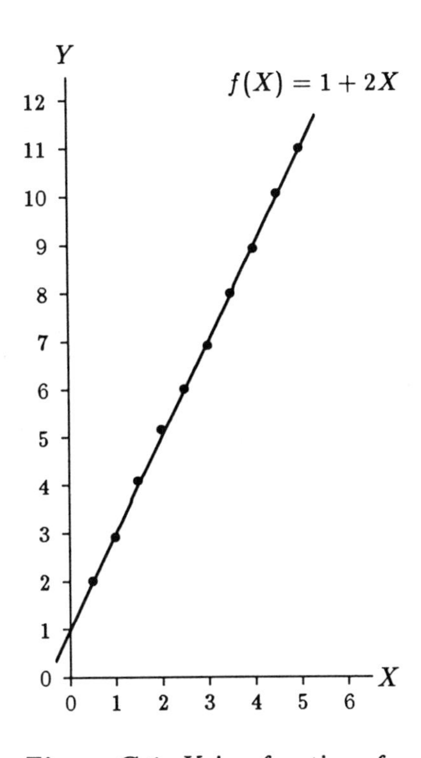

Figure C.1 *Y is a function of X, such that $Y=f(X)=1+2X$.*

Once the function relating two or more quantities is established, physical laws and mathematical tools can be utilized to determine related physical quantities without employing additional experiments. Note however that to be able to find the most suitable function to represent the data collected, we have to be familiar with the characteristics of commonly encountered functions.

There are a few basic functions in calculus that may be sufficient to describe a large number of physical phenomena. These basic functions are constant functions, power functions, trigonometric functions, logarithmic functions, and exponential functions. Basic functions can be combined together in various ways by means of addition, subtraction, multiplication, or division to obtain relatively complex functions.

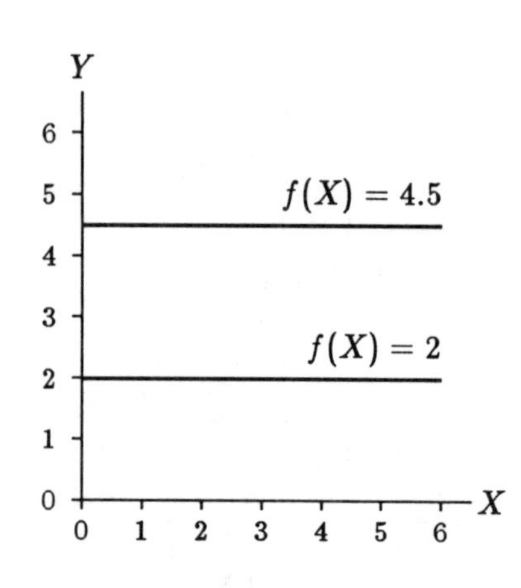

Figure C.2 *Constant functions.*

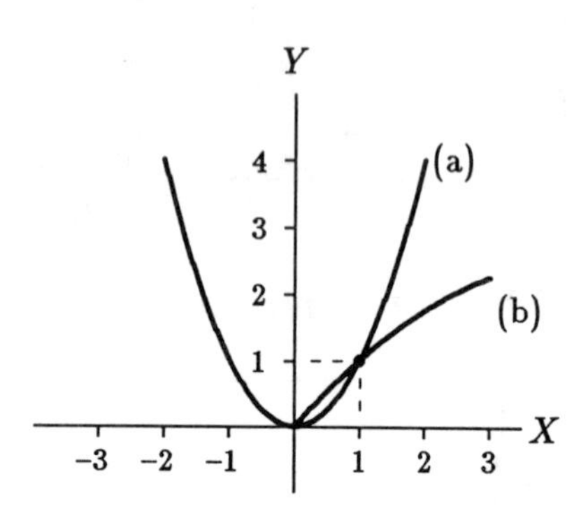

Figure C.3 *Examples of power functions. (a) $f(X)=X^2$ and (b) $f(X)=\sqrt{X}$.*

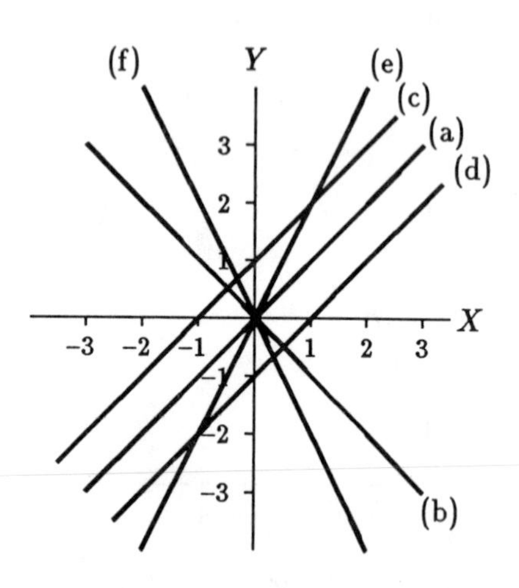

Figure C.4 *Linear functions. (a) $Y=X$, (b) $Y=-X$, (c) $Y=1+X$, (d) $Y=-1+X$, (e) $Y=2X$, and (f) $Y=-2X$.*

C.1.1 Constant functions

In general, *constant functions* can be represented as

$$f(X) = c \tag{C.1}$$

where c is a symbol with a constant numerical value. For example, if $f(X)=2$ for all X, then f is a constant function. The fact that the independent variable X does not appear on the right-hand side of Eq. (C.1) makes it a constant function. The graph of a constant function is a horizontal line, as illustrated in Figure C.2.

C.1.2 Power functions

Power functions can be represented in the following general form:

$$f(X) = X^r \tag{C.2}$$

In Eq. (C.2), r can be any number (integer or real) including zero, or ratio of numbers. Note that a constant function can be considered a power function for which $r=0$. The following are examples of power functions:

$$f(X) = X^1 = X$$
$$f(X) = X^3 = X \cdot X \cdot X$$
$$f(X) = X^{-1} = \frac{1}{X}$$
$$f(X) = X^{-2} = \frac{1}{X^2}$$
$$f(X) = X^{\frac{1}{2}} = \sqrt{X}$$

Graphs of some power functions are shown in Figure C.3.

C.1.3 Linear functions

Linear functions can be represented in the following general form:

$$f(X) = a + bX \tag{C.3}$$

In Eq. (C.3), a and b are some constant coefficients. The graph of a linear function is a straight line with coefficient a representing the point at which the line intersects the Y axis, and coefficient b is the slope of the line. Examples of linear functions include:

$$f(X) = 1 + 2X$$
$$f(X) = 0.5 - 5X$$
$$f(X) = X$$

The graphs of some linear functions are shown in Figure C.4.

C.1.4 Quadratic functions

Quadratic functions can be represented in the following general form:

$$f(X) = a + bX + cX^2 \qquad \text{(C.4)}$$

In Eq. (C.4), a, b, and c are again some constant coefficients. The graph of a quadratic function is a parabola. Examples of quadratic functions include:

$$f(X) = 2 + X - 0.5X^2$$
$$f(X) = 5 + X^2$$
$$f(X) = \ - 3X + 4X^2$$
$$f(X) = X^2$$

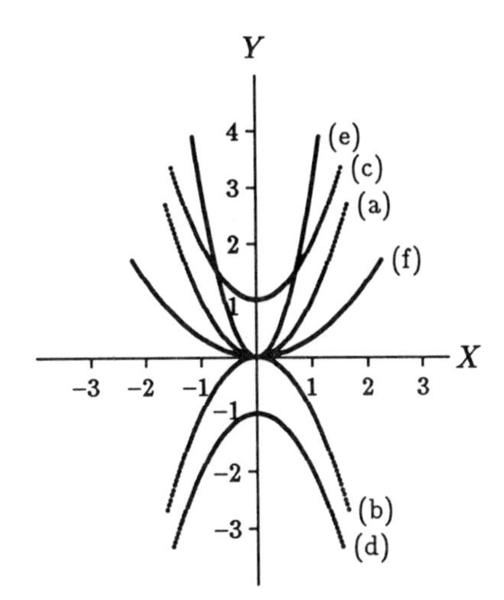

Figure C.5 *Quadratic functions. (a) $Y{=}X^2$, (b) $Y{=}{-}X^2$, (c) $Y{=}1+X^2$, (d) $Y{=}{-}1{-}X^2$, (e) $Y{=}3X^2$ and (f) $Y{=}X^2/3$.*

The graphs of some quadratic functions are shown in Figure C.5. The distinguishing characteristic of these functions is that the highest power of X appearing in these equations is two.

C.1.5 Polynomial functions

A *polynomial function* is one for which

$$f(X) = A_0 + A_1X + A_2X^2 + A_3X^3 + \cdots + A_nX^n \qquad \text{(C.5)}$$

Coefficients A_0, $A_1, \cdots A_n$ in Eq. (C.5) are some constants, and n is a positive integer number corresponding to the highest power of X. n defines the "order" of the polynomial. For example,

$$f(X) = 1 - X - 2X^2 + 5X^3$$

is a polynomial of order 3 with coefficients $A_0{=}1$, $A_1{=}{-}1$, $A_2{=}{-}2$, and $A_3{=}5$.

$$f(X) = 2 + 3X^2$$

is a polynomial of order 2 with coefficients $A_0{=}2$, $A_1{=}0$, and $A_2{=}3$. It is also a quadratic function with coefficients $a{=}2$, $b{=}0$, and $c{=}3$.

$$f(X) = 5 - 4X$$

is a polynomial of order 1 with coefficients $A_0{=}5$ and $A_1{=}{-}4$. It is also a linear function.

Notice that constant, linear, and quadratic functions are special forms of polynomial functions. A constant function is also a zero-order polynomial, a linear function is a first-order polynomial, and a quadratic function is a second-order polynomial.

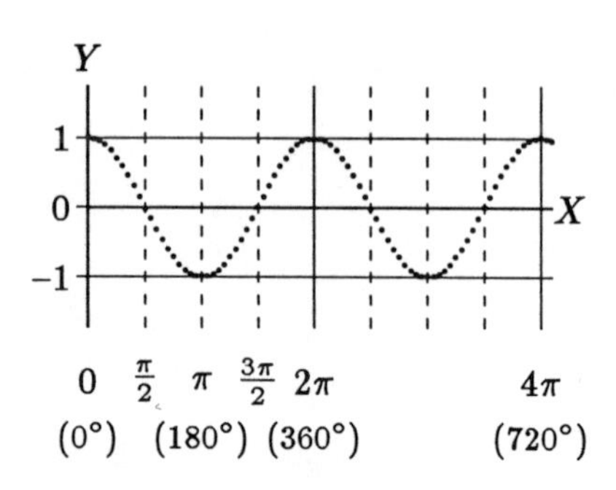

Figure C.6 *Function* $Y = \cos X$.

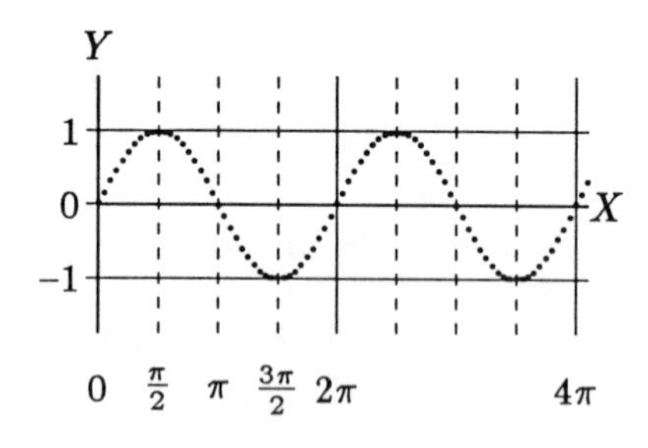

Figure C.7 *Function* $Y = \sin X$.

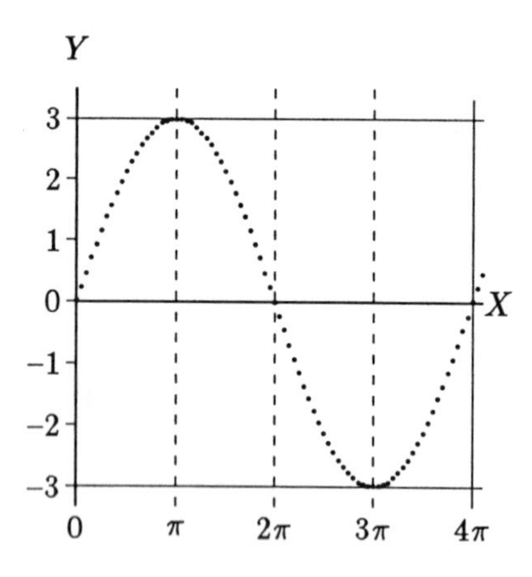

Figure C.8 $Y=3\sin(\frac{X}{2})$.

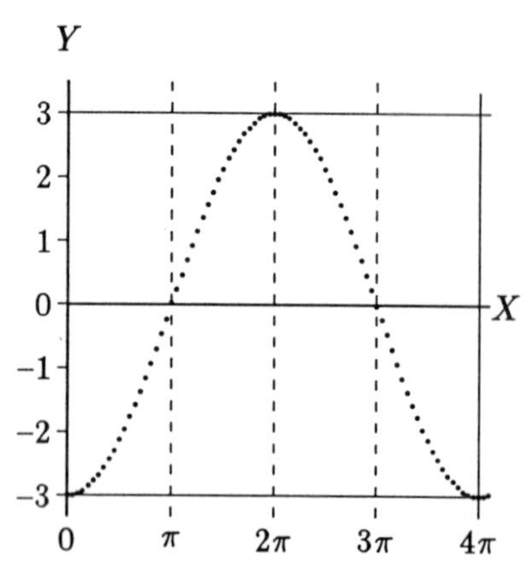

Figure C.9 $Y=3\sin(\frac{X}{2} - \frac{\pi}{2})$.

C.1.6 Trigonometric functions

If a quantity Y depends on another quantity X through some trigonometric relations, such as $Y = \sin X$ or $Y = \cos X$, then Y is said to be a ***trigonometric function*** of X. $Y = \sin X$ and $Y = \cos X$ are called the ***sine*** and ***cosine*** functions, respectively. The graphs of these functions are shown with dotted curves in Figures C.6 and C.7, which are obtained by assigning values to X, calculating corresponding Y values, and plotting them.

X in $Y = \sin X$ and $Y = \cos X$ can be measured either in ***degrees*** or in ***radians***. An angle of 180° is called π radians with π=3.1416. Therefore, degrees can be converted into radians using the following formula:

$$\text{radian} = \frac{\pi}{180} \times \text{degree}$$

For example, 0°=0 rad, 45°=$\pi/4$ rad, 90°=$\pi/2$ rad, 180°=π rad, 270°=$3\pi/2$ rad, 360°=2π rad, and 720°=4π rad.

Trigonometric functions are cyclic or periodic in the sense that their graphs repeat a pattern. The graphs of $Y = \sin X$ and $Y = \cos X$ in Figures C.6 and C.7 repeat after every 2π radians (360°). This means that the ***period*** of $Y = \sin X$ and $Y = \cos X$ is 2π radians. Furthermore, $Y = \sin X$ and $Y = \cos X$ assume values (or vary) between −1 and +1. Therefore, the ***amplitude*** of $Y = \sin X$ and $Y = \cos X$ is 1.

$Y = \sin X$ and $Y = \cos X$ are the simplest forms of trigonometric functions. Sine functions can be expressed in a more general form as follows:

$$f(X) = a\ \sin(bX) \qquad \text{(C.6)}$$

Here, a and b are some constants. The sine function defined by Eq. (C.6) has an amplitude a and a period $2\pi/b$. For example, as illustrated in Figure C.8, the amplitude and period of function $Y=3\sin(\frac{X}{2})$ are 3 and 4π, respectively. The sine function in Eq. (C.6) can further be generalized as:

$$f(X) = a\ \sin(bX + c) \qquad \text{(C.7)}$$

Here, a is the amplitude, $2\pi/b$ is the period, and the graph of this function is shifted by an amount c to the right or the left as compared to the graph of the function $Y = a\sin(bX)$. For example, the graph of $Y=3\sin(\frac{X}{2} - \frac{\pi}{2})$ is as the one shown in Figure C.9.

Note that the graph of a sine function which is shifted by $\pi/2$ is essentially the graph of a negative cosine function. In other words, $\sin(\frac{X}{2} - \frac{\pi}{2}) = -\cos(\frac{X}{2})$. There are a number of other trigonometric identities and formulas which are useful in handling trigonometric functions. Some of these formulas are provided next.

Negative angle formulas:

$$\sin(-X) = -\sin X$$
$$\cos(-X) = \cos X$$

Addition formulas:

$$\sin(X+Y) = \sin X \cos Y + \cos X \sin Y$$
$$\sin(X-Y) = \sin X \cos Y - \cos X \sin Y$$
$$\cos(X+Y) = \cos X \cos Y - \sin X \sin Y$$
$$\cos(X-Y) = \cos X \cos Y + \sin X \sin Y$$

Product formulas:

$$2 \sin X \cos Y = \sin(X+Y) + \sin(X-Y)$$
$$2 \cos X \sin Y = \sin(X+Y) - \sin(X-Y)$$
$$2 \cos X \cos Y = \cos(X+Y) + \cos(X-Y)$$
$$2 \sin X \sin Y = \cos(X-Y) - \cos(X+Y)$$

Double-angle formulas:

$$\sin(2X) = 2 \sin X \cos X$$
$$\cos(2X) = \cos^2 X - \sin^2 X$$

Half-angle formulas:

$$2 \sin^2 \left(\frac{X}{2}\right) = 1 - \cos X$$
$$2 \cos^2 \left(\frac{X}{2}\right) = 1 + \cos X$$

Pythagorean identity:

$$\sin^2 X + \cos^2 X = 1$$

Reduction formulas:

$$\cos\left(\frac{\pi}{2} - X\right) = \sin X \qquad \cos(\pi - X) = -\cos X$$
$$\sin\left(\frac{\pi}{2} - X\right) = \cos X \qquad \sin(\pi - X) = \sin X$$

Sine and cosine are the basic trigonometric functions. Other trigonometric functions, namely the tangent (tan), cotangent (cot), secant (sec), and cosecant (csc), can be derived from sine and cosine using the following definitions:

$$\tan X = \frac{\sin X}{\cos X} \qquad \cot X = \frac{\cos X}{\sin X} = \frac{1}{\tan X}$$
$$\sec X = \frac{1}{\cos X} \qquad \csc X = \frac{1}{\sin X}$$

C.1.7* Exponential and logarithmic functions

Functions such as 3^X and $(\frac{1}{2})^X$ are called *exponential functions.* The general form of exponential functions is b^X, where b is called the *base.* The most popular base in calculus is e, an irrational number between 2.71 and 2.72, and the function e^X or $\exp X$ is often referred to as the exponential function.

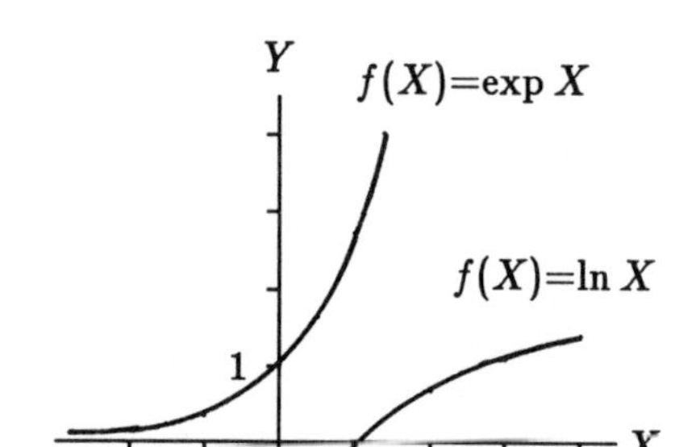

Figure C.10 *Exponential and logarithmic functions.*

The exponential function $\exp X$ has an inverse, called the *natural logarithmic function* denoted by $\ln X$. It can also be written as $\log_e X$ and called the logarithm with base e. The properties of $\exp X$ and $\ln X$ are such that:

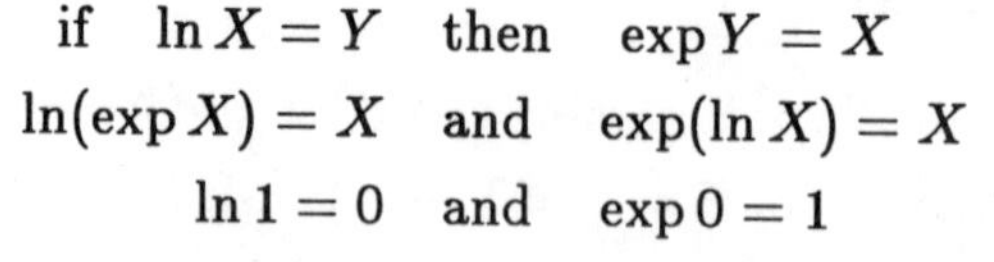

$$\text{if} \quad \ln X = Y \quad \text{then} \quad \exp Y = X$$
$$\ln(\exp X) = X \quad \text{and} \quad \exp(\ln X) = X$$
$$\ln 1 = 0 \quad \text{and} \quad \exp 0 = 1$$

Graphs of $\exp X$ and $\ln X$ are shown in Figure C.10. Additional properties of exponential and logarithmic functions include the following:

$$\exp X \exp Y = \exp(X + Y)$$
$$\frac{\exp X}{\exp Y} = \exp(X - Y)$$
$$\exp(-X) = \frac{1}{\exp X}$$
$$(\exp X)^Y = \exp(XY)$$

$$\ln(XY) = \ln X + \ln Y$$
$$\ln\left(\frac{X}{Y}\right) = \ln X - \ln Y$$
$$\ln\left(\frac{1}{X}\right) = -\ln X$$
$$\ln(X^Y) = Y \ln X$$

Note that it is not allowed to take the "ln" of a negative number, but it is possible for $\ln X$ to come out negative. In other words, $\ln X$ is defined for X greater than zero. On the other hand, $\exp X$ is defined for all X. However, $\exp X$ is always positive.

We have reviewed the definitions and properties of the basic and most commonly used functions in calculus. More complex functions can be constructed by combining these functions in various different ways. Next we shall review the "differentiation" and "integration" of functions.

C.2 The Derivative

The derivative is one of the fundamental mathematical tools used extensively for many purposes. Applications of the derivative include the prediction of graphs of functions, calculation of maximums and minimums, statistical analyses, velocity and acceleration computations. The process of finding the derivative of a function is called *differentiation*. The branch of calculus dealing with the derivative is called *differential calculus*.

The derivative of a function represents the slope of that function. For example, consider the linear function $Y=1+2X$ whose graph is shown in Figure C.11. The slope of the line representing $Y=1+2X$ can be determined by considering any two points 1 and 2 along the line, such as those points with coordinates $X_1=3$, $Y_1=7$, and $X_2=5$, $Y_2=11$. The slope of a line is defined by the tangent of the angle that the line makes with the horizontal. For the line shown in Figure C.11:

$$\text{Slope} = \tan\theta = \frac{Y_2 - Y_1}{X_2 - X_1} = \frac{11-7}{5-2} = 2$$

Therefore, the derivative of the function $Y=1+2X$ must be equal to 2. We shall demonstrate that this 2 is the same 2 in front of the X in $Y=1+2X$.

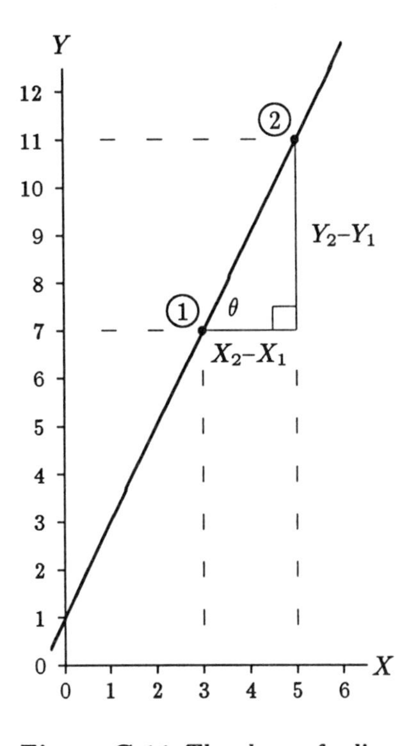

Figure C.11 *The slope of a line.*

The slope of a line is constant. The curve that is not a line will not have a unique slope, and it is usually difficult to predict the varying slope of a curve along the curve. The concept of differentiation makes it easier to determine the slopes of curves. To be able to utilize this mathematical tool efficiently and properly, the principles of differentiation must be known.

In general, the derivative of a function is another function. For a function $Y=f(X)$ there are many symbols used to denote the derivative. For example:

$$f' \qquad f'(X) \qquad \frac{df}{dX} \qquad Y' \qquad \frac{dY}{dX}$$

The notations df/dX and dY/dX look like fractions but represent single symbols.

C.2.1 Derivatives of basic functions

The graph of a constant function $f(X)=c$ is a horizontal line whose slope is zero. Therefore, the derivative of constant functions is always zero:

$$\frac{d}{dX}(c) = 0 \tag{C.8}$$

The graph of linear function $f(X)=X$ is a straight line whose slope is equal to 1. Therefore:

$$\frac{d}{dX}(X) = 1 \tag{C.9}$$

It is easy to find the derivatives of functions such as $f(X)=c$ and $f(X)=X$ using slopes. However, the graphs of power functions such as X^3, $\sqrt{X}$, and X^{-2} have varying slopes, and it is not easy to predict their derivatives. To differentiate a power function X^r, we shall adopt the following rule: multiply the function by its power (r) and lower the power by 1. Mathematically speaking:

$$\frac{d}{dX}(X^r) = r\,X^{r-1} \tag{C.10}$$

The following examples illustrate the use of Eq. (C.10).

Function, $f(X)$	*Derivative, f'*
X^4	$4X^3$
X^{-2}	$-2X^{-3}$
$X^{2.3}$	$2.3X^{1.3}$
$\sqrt{X} = X^{\frac{1}{2}}$	$\frac{1}{2}X^{-\frac{1}{2}} = \frac{1}{2\sqrt{X}}$

If the power r is zero then $X^0=1$, regardless of what X is. Therefore, 1 can be treated as if it is X^0, or a power function with $r=0$, and the ***power rule*** as given in Eq. (C.10) can be applied to differentiate it:

$$\frac{d}{dX}(1) = \frac{d}{dX}(X^0) = 0X^{-1} = 0$$

Note that anything multiplied by zero is equal to zero.

$f(X)=X$ is also a power function with $r=1$. Therefore:

$$\frac{d}{dX}(X) = \frac{d}{dX}(X^1) = 1X^0 = 1$$

The definition of the derivative can be utilized for the differentiation of other basic functions. We shall adopt the following definitions without presenting any proofs:

$$\frac{d}{dX}(\sin X) = \cos X \tag{C.11}$$

$$\frac{d}{dX}(\cos X) = -\sin X \tag{C.12}$$

$$\frac{d}{dX}(\exp X) = \exp X \tag{C.13}$$

$$\frac{d}{dX}(\ln X) = \frac{1}{X} \tag{C.14}$$

C.2.2 The constant multiple rule

To differentiate the product of a constant c and a function $f(X)$, take the derivative of the function and then multiply it with the constant:

$$\frac{d}{dX}[c\, f(X)] = c\,\frac{df}{dX} = c\, f' \tag{C.15}$$

For example:

Function, $f(X)$	*Derivative, f'*
$2X^2$	$4X$
$-1.25X^{-3}$	$3.75X^{-4}$
$0.5\cos X$	$-0.5\sin X$
$-3\exp X$	$-3\exp X$

C.2.3 The sum rule

The derivative of the sum of two functions is equal to the sum of the derivatives of the functions. If $f_1(X)$ and $f_2(X)$ represent two functions, then the derivative of the function $f(X)=f_1+f_2$ is:

$$\frac{d}{dX}[f_1(X) + f_2(X)] = f_1' + f_2' \tag{C.16}$$

For example:

$$\frac{d}{dX}(3X - 2\sin X) = \frac{d}{dX}(3X) + \frac{d}{dX}(-2\sin X) = 3 - 2\cos X$$

The sum rule can also be applied to take the derivative of linear and quadratic functions. If a, b, and c are constants, then:

$$\frac{d}{dX}(a + bX) = b \tag{C.17}$$

$$\frac{d}{dX}(a + bX + cX^2) = b + 2cX \tag{C.18}$$

Recall that linear and quadratic functions are special forms of polynomial functions. If A_0, A_1, $A_2, \cdots$, and A_n are some constants, then the derivative of a polynomial is:

$$\frac{d}{dX}(A_0 + A_1X + A_2X^2 + \cdots + A_nX^n) = A_1 + 2A_2X + \cdots + nA_nX^{n-1} \tag{C.19}$$

Note that differentiation reduces the order of the polynomial by one. For example, the derivative of a quadratic function (second-order polynomial) is a linear function (first-order polynomial), and the derivative of a linear function is a constant (zero-order polynomial). Similarly, the derivative of a third-order polynomial is a quadratic function. For example:

$$\frac{d}{dX}(3 - 2X + 5X^2 - X^3) = -2 + 10X - 3X^2$$

C.2.4 The product rule

The derivative of a function in the form of a product of two functions is equal to the first function multiplied by the derivative of the second function, plus the second function itself multiplied by the derivative of the first function. If $f_1(X)$ and $f_2(X)$ represent two functions, then the derivative of the function $f(X)=f_1 f_2$ is:

$$\frac{d}{dX}[f_1(X)f_2(X)] = f_1 f_2' + f_1' f_2 \qquad \text{(C.20)}$$

For example:

$$\begin{aligned}\frac{d}{dX}(X^2 \cos X) &= (X^2)\frac{d}{dX}(\cos X) + (\cos X)\frac{d}{dX}(X^2)\\ &= (X^2)(-\sin X) + (\cos X)(2X)\\ &= 2X \cos X - X^2 \sin X\end{aligned}$$

Note that the application of the product rule can be expanded to include functions in the form of the product of more than two functions. For example, for three functions:

$$\frac{d}{dX}[f_1(X)f_2(X)f_3(X)] = f_1 f_2 f_3' + f_1 f_2' f_3 + f_1' f_2 f_3$$

C.2.5 The quotient rule

The derivative of a function in the form of a ratio of two functions is equal to the derivative of the numerator times the denominator, minus the numerator times the derivative of the denominator, all divided by the square of the denominator. If $f_1(X)$ and $f_2(X)$ represent two functions, then the derivative of the function $f(X)=f_1/f_2$ is:

$$\frac{d}{dX}\left[\frac{f_1(X)}{f_2(X)}\right] = \frac{f_1' f_2 - f_1 f_2'}{{f_2}^2} \qquad \text{(C.21)}$$

For example:

$$\begin{aligned}\frac{d}{dX}\left(\frac{2+3X^2}{X^2}\right) &= \frac{(X^2)\frac{d}{dX}(2+3X^2) - (2+3X^2)\frac{d}{dX}(X^2)}{(X^2)^2}\\ &= \frac{(X^2)(6X) - (2+3X^2)(2X)}{X^4}\\ &= \frac{6X^3 - 4X - 6X^3}{X^4}\\ &= -\frac{4}{X^3}\end{aligned}$$

Alternatively:

$$\begin{aligned}
\frac{d}{dX}\left(\frac{2+3X^2}{X^2}\right) &= \frac{d}{dX}\left(\frac{2}{X^2}+3\right) \\
&= \frac{(X^2)\frac{d}{dX}(2)-(2)\frac{d}{dX}(X^2)}{(X^2)^2}+\frac{d}{dX}(3) \\
&= \frac{(X^2)(0)-(2)(2X)}{X^4}+(0) \\
&= -\frac{4}{X^3}
\end{aligned}$$

Note that trigonometric functions $\tan X$, $\cot X$, $\sec X$, and $\csc X$ are various quotients of $\sin X$ and $\cos X$. Therefore, the quotient rule can be applied to determine their derivatives. For example:

$$\begin{aligned}
\frac{d}{dX}(\tan X) &= \frac{d}{dX}\left(\frac{\sin X}{\cos X}\right) \\
&= \frac{(\cos X)\frac{d}{dX}(\sin X)-(\sin X)\frac{d}{dX}(\cos X)}{(\cos X)^2} \\
&= \frac{(\cos X)(\cos X)-(\sin X)(-\sin X)}{\cos^2 X} \\
&= \frac{\cos^2 X+\sin^2 X}{\cos^2 X}
\end{aligned}$$

Since $\cos^2 X+\sin^2 X=1$:

$$\frac{d}{dX}(\tan X)=\frac{1}{\cos^2 X}=\sec^2 X \tag{C.22}$$

Similarly:

$$\frac{d}{dX}(\cot X)=-\frac{1}{\sin^2 X}=-\csc^2 X \tag{C.23}$$

$$\frac{d}{dX}(\sec X)=\frac{\sin X}{\cos^2 X}=\sec X\tan X \tag{C.24}$$

$$\frac{d}{dX}(\csc X)=-\frac{\cos X}{\sin^2 X}=-\csc X\cot X \tag{C.25}$$

C.2.6 The chain rule

Sometimes functions appear in forms other than those analyzed in previous sections. None of the above rules can be applied to differentiate functions such as $\cos(X^3)$, $\exp(2-5X)$, and $\ln(4X)$. For example, consider the function $f(X)=\exp(2-5X)$. We have already seen the derivatives of $\exp X$ and $2-5X$, but the derivative of $\exp(2-5X)$ follows a different rule. To take the derivative of $f(X)=\exp(2-5X)$, define the terms in the parentheses as $Z=2-5X$, so that the original function can be reduced to a form $f(Z)=\exp Z$. The ***chain rule*** states that:

$$\frac{df}{dX}=\frac{df}{dZ}\frac{dZ}{dX} \tag{C.26}$$

Now, we can take the derivative of $f(X)$=exp$(2-5X)$.

$$\begin{aligned}\frac{d}{dX}[\exp(2-5X)] &= \frac{d}{dZ}(\exp Z)\frac{d}{dX}(Z)\\ &= (\exp Z)(-5)\\ &= -5\exp(2-5X)\end{aligned}$$

The chain rule has a vast number of applications. For example, to take the derivative of $Y=\sin(X^2)$, let $Z=X^2$ and $Y=\sin Z$. The derivative of Y with respect to Z is $\cos X$, and the derivative of Z with respect to X is $2X$. Therefore:

$$\begin{aligned}\frac{d}{dX}[\sin(X^2)] &= \frac{d}{dZ}(\sin Z)\frac{d}{dX}(Z)\\ &= (\cos Z)(2X)\\ &= 2X\cos(X^2)\end{aligned}$$

For another example, consider $Y=(3+X^2)^2$. There are two ways to take the derivative of this function. One way is by expanding the parentheses, writing the function as $Y=9+6X^2+X^4$, and then taking the derivative:

$$\begin{aligned}\frac{d}{dX}[(3+X^2)^2] &= \frac{d}{dX}(9+6X^2+X^4)\\ &= 0+12X+4X^3\\ &= 12X+4X^3\end{aligned}$$

The second way is to let $Z=3+X^2$ so that $Y=Z^2$, taking individual derivatives, and then applying the chain rule:

$$\begin{aligned}\frac{d}{dX}[(3+X^2)^2] &= \frac{d}{dZ}(Z^2)\frac{d}{dX}(3+X^2)\\ &= (2Z)(2X)\\ &= 12X+4X^3\end{aligned}$$

For any Z which is a function of X, applications of the chain rule can be summarized in the following form:

$$\frac{d}{dX}(Z^n) = nZ^{n-1}\frac{dZ}{dX} \tag{C.27}$$

$$\frac{d}{dX}(\sin Z) = \cos Z\frac{dZ}{dX} \tag{C.28}$$

$$\frac{d}{dX}(\cos Z) = -\sin Z\frac{dZ}{dX} \tag{C.29}$$

$$\frac{d}{dX}(\exp Z) = \exp Z\frac{dZ}{dX} \tag{C.30}$$

$$\frac{d}{dX}(\ln Z) = \frac{1}{Z}\frac{dZ}{dX} \tag{C.31}$$

C.2.7* Implicit differentiation

Sometimes a quantity may be an *implicit* function of another quantity. For example, in

$$Y^2 - X^2 = 4$$

Y is an implicit function of X. The same equation can be rewritten in an *explicit* form as:

$$Y = (4 + X^2)^{\frac{1}{2}}$$

The derivative of Y with respect to X can now be determined by applying the chain rule:

$$\frac{dY}{dX} = \frac{1}{2}(4 + X^2)^{-\frac{1}{2}}(2X) = \frac{X}{(4 + X^2)^{\frac{1}{2}}}$$

Note that the derivative of Y with respect to X can also be determined directly from the implicit expression, by taking the derivative of both sides of the equation with respect to X:

$$2Y\frac{dY}{dX} - 2X = 0$$

$$\frac{dY}{dX} = \frac{X}{Y} = \frac{X}{(4 + X^2)^{\frac{1}{2}}}$$

C.2.8* Higher derivatives

The derivative $f'(X)$ of a function $f(X)$ is also a function. The derivative of $f'(X)$ is yet another function, called the *second derivative* of $f(X)$ and denoted by $f''(X)$. Some of the notations used for the second derivative are:

$$f'' \qquad f''(X) \qquad \frac{d^2 f}{dX^2} \qquad Y'' \qquad \frac{d^2Y}{dX^2}$$

Similarly, $f'''(X)$ refers to the first derivative of $f''(X)$, the second derivative of $f'(X)$, and the third derivative of $f(X)$.

Consider the following examples:

$$\begin{aligned} Y &= 1 + 3X - X^2 + 2X^3 \\ Y' &= 3 - 2X + 6X^2 \\ Y'' &= -2 + 12X \\ Y''' &= 12 \end{aligned}$$

$$\begin{aligned} Y &= X\sin(2X) \\ Y' &= \sin(2X) + 2X\cos(2X) \\ Y'' &= 2\cos(2X) + 2\cos(2X) - 4X\sin(2X) \\ &= 4\cos(2X) - 4X\sin(2X) \\ Y''' &= -8\sin(2X) - 4\sin(2X) - 8X\cos(2X) \\ &= -12\sin(2X) - 8X\cos(2X) \end{aligned}$$

C.3 The Integral

In Section C.2 we concentrated on finding the derivative $f'(X)$ of a given function $f(X)$. We have seen that the differentiation of basic functions is relatively simple and straightforward, and the differentiation of relatively complex functions is possible using the derivatives of basic functions along with the rules for the derivatives of combinations (sums, products, quotients) of functions.

Next we want to determine the functions whose derivatives are known. The reversed operation of differentiation is called *integration*, or *antidifferentiation*. As compared to differentiation, integration is more difficult. There are no standard product, quotient, or chain rules for integration. In the absence of sufficient rules to integrate combinations of functions, it is common practice to use integral tables.

Integration has many applications, such as area and volume calculations, work and energy computations.

The integral of a function $Y{=}f(X)$ with respect to X can be expressed in two ways:

$$\int f(X)\,dX \tag{C.32}$$

$$\int_a^b f(X)\,dX \tag{C.33}$$

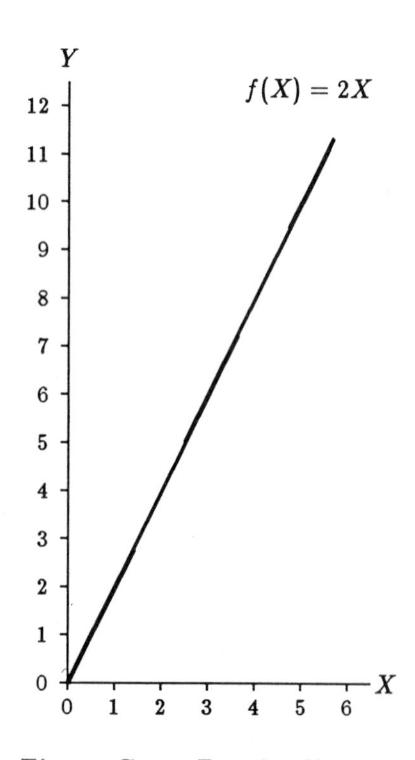

Figure C.12 *Function $Y{=}2X$.*

Eq. (C.32) is called the *indefinite* integral, and Eq. (C.33) is called the *definite* integral. The integral symbol $\int$ is an elongated "S" which stands for summation, the function $f(X)$ to be integrated is called the *integrand*, dX is an increment in X, and a and b in Eq. (C.33) are called the *lower* and *upper limits* of integration.

Consider the function $Y{=}f(X){=}2X$ whose graph is shown in Figure C.12. Also consider $F_1(X){=}X^2$. The derivative of $F_1(X)$ is essentially equal to $f(X)$, and therefore, $F_1(X)$ can be the integral of $f(X){=}2X$. Now consider another function $F_2(X){=}X^2{+}c_0$ where c_0 is a constant. The derivative of $F_2(X)$ is again equal to $f(X){=}2X$ because the derivative of a constant is zero. Note that F_1 is in fact a special form of F_2 for which $c_0{=}0$. Therefore, the indefinite integral of $f(X){=}2X$ is:

$$\int (2X)\,dX = X^2 + c_0$$

Here, c_0 is called the *constant of integration*. Note that the indefinite integral of a function is another function which is not unique. There are different solutions for different values of c_0. These different solutions have parallel graphs (Figure C.13), all of which have slope $2X$.

The definite integral of a given function is unique. To evaluate the integral of a function $f(X)$ between a and b, first take the integral of the given function. For definite integrals, the constant of integration cancels out during the course of integration. Therefore, it can be assumed that the constant of integration c_0 is zero. If $F(X)$ represents the integral of $f(X)$, then evaluate the values of $F(X)$ at $X{=}a$ and $X{=}b$ by subsequently substituting the numerical values of a and b wherever you see X in $F(X)$. In other words, evaluate $F(a)$ and $F(b)$. The integral of $f(X)$ between a and b is equal to $F(b)$ minus $F(a)$:

$$\int_a^b f(X)\,dX = F(b) - F(a)$$

For example, consider the function $Y{=}f(X){=}2X$. Assume that the integral of $f(X)$ between $X{=}0$ and $X{=}3$ (i.e., $a{=}0$ and $b{=}3$) is to be evaluated. Recall that the integral of $f(X){=}2X$ is $F(X){=}X^2$ (with $c_0{=}0$). The steps to be followed while evaluating the definite integral are as follows:

Step 1: $$\int_0^3 (2X)\,dX = \left[X^2\right]_0^3$$

Step 2: $$= \left[(3^2) - (0^2)\right]$$

Step 3: $$= (9 - 0) = 9$$

That is, determine $F(X)$ by taking the integral of the given function (Step 1), evaluate $F(b)$ and $F(a)$ by substituting the upper and lower limits of integration (Step 2), and evaluate the difference $F(b)$–$F(a)$ (Step 3).

The physical meaning of the definite integral is such that it represents the area bounded by the given function $f(X)$, the X axis, and vertical lines passing through $X{=}a$ and $X{=}b$. For example, the integral of function $f(X){=}2X$ between $X{=}0$ and $X{=}3$ is equal to the area of the triangle ABC (shaded area) illustrated in Figure C.14. Note that the area of the right triangle ABC is half of the area of the rectangle ABCD:

$$\text{area ABC} = \frac{\text{area ABCD}}{2} = \frac{(3)(6)}{2} = 9$$

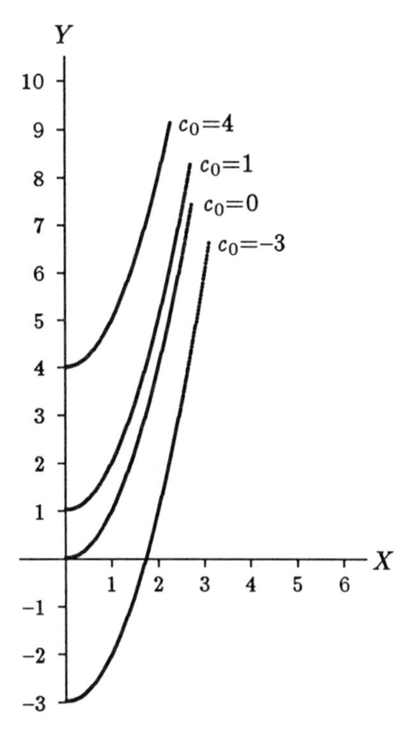

Figure C.13 *Slope of functions $Y{=}X^2 + c_0$ (c_0 is a constant) is equal to $2X$.*

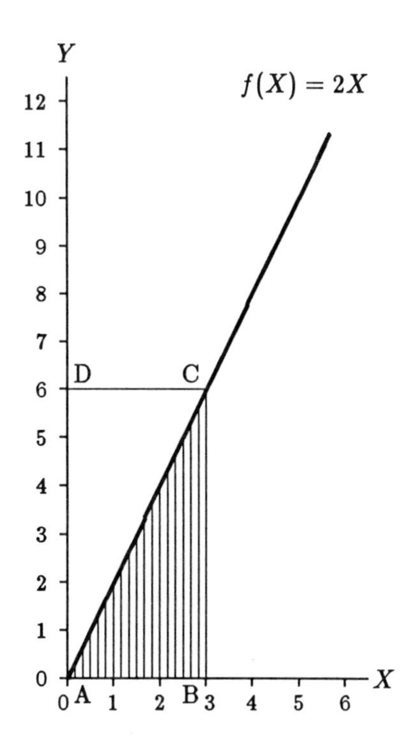

Figure C.14 *Integral represents the area under the curve.*

C.3.1 Properties of the indefinite integrals

Integrals of basic functions:

$$\int dX = X + c_0$$

$$\int c\, dX = cX + c_0 \qquad \text{(constant } c\text{)} \tag{C.34}$$

$$\int X\, dX = \frac{X^2}{2} + c_0$$

$$\int X^2\, dX = \frac{X^3}{3} + c_0$$

$$\int X^r\, dX = \frac{X^{r+1}}{r+1} + c_0 \qquad (r \neq -1) \tag{C.35}$$

$$\int \sin X\, dX = -\cos X + c_0 \tag{C.36}$$

$$\int \cos X\, dX = \sin X + c_0 \tag{C.37}$$

$$\int \exp X\, dX = \exp X + c_0 \tag{C.38}$$

$$\int \frac{1}{X}\, dX = \ln X + c_0 \qquad (X > 0) \tag{C.39}$$

Constant multiple and sum rules:

$$\int [cf(X)]\, dX = c \int f(X)\, dX \tag{C.40}$$

$$\int [f_1(X) + f_2(X)]\, dX = \int f_1(X)\, dX + \int f_2(X)\, dX \tag{C.41}$$

Examples:

$$\int 2\cos X\, dX = 2 \int \cos X\, dX$$
$$= 2\sin X + c_0$$

$$\int \left(5 - \frac{1}{2}X^3\right) dX = 5 \int dX - \frac{1}{2} \int X^3\, dX$$
$$= 5(X) - \frac{1}{2}\left(\frac{X^4}{4}\right) + c_0$$
$$= 5X - \frac{1}{8}X^4 + c_0$$

$$\int \left(\frac{1}{X^2} + \sin X\right) dX = \int X^{-2}\, dX + \int \sin X\, dX$$
$$= \left(\frac{X^{-1}}{-1}\right) + (-\cos X) + c_0$$
$$= -\frac{1}{X} - \cos X + c_0$$

C.3.2 Properties of the definite integrals

Let $F(X)$ represent the integral of a function $f(X)$ with respect to X. That is:

$$F(X) = \int f(X)\, dX$$

Also let $f_1(X)$ and $f_2(X)$ be two other functions. Then:

$$\int_a^b f(X)\, dX = F(b) - F(a) \tag{C.42}$$

$$\int_a^b [cf(X)]\, dX = c\int_a^b f(X)\, dX = c[F(b) - F(a)] \tag{C.43}$$

$$\int_a^b [f_1(X) + f_2(X)]\, dX = \int_a^b f_1(X)\, dX + \int_a^b f_2(X)\, dX \tag{C.44}$$

$$\int_a^b f(X)\, dX + \int_b^c f(X)\, dX = \int_a^c f(X)\, dX \tag{C.45}$$

$$\int_a^b f(X)\, dX = -\int_b^a f(X)\, dX \tag{C.46}$$

Definite integrals of basic functions:

$$\int_a^b dX = [X]_a^b = b - a$$

$$\int_a^b X\, dX = \left[\frac{X^2}{2}\right]_a^b = \left[\frac{b^2}{2} - \frac{a^2}{2}\right] = \frac{1}{2}(b^2 - a^2)$$

$$\int_a^b X^2\, dX = \left[\frac{X^3}{3}\right]_a^b = \left[\frac{b^3}{3} - \frac{a^3}{3}\right] = \frac{1}{3}(b^3 - a^3)$$

$$\int_a^b \sin X\, dX = [-\cos X]_a^b = -\cos b + \cos a$$

$$\int_a^b \cos X\, dX = [\sin X]_a^b = \sin b - \sin a$$

$$\int_a^b \exp X\, dX = [\exp X]_a^b = \exp b - \exp a$$

$$\int_a^b \frac{1}{X}\, dX = [\ln X]_a^b = \ln b - \ln a$$

Examples:

$$\begin{aligned}\int_{45^\circ}^{90^\circ} \cos X\, dX &= \Big[\sin X\Big]_{45^\circ}^{90^\circ} \\ &= \sin 90^\circ - \sin 45^\circ \\ &= 1.0 - 0.7 = 0.3\end{aligned}$$

$$\int_1^2 (4X + 9X^2)\, dX = \int_1^2 (4X)\, dX + \int_1^2 (9X^2)\, dX$$
$$= 4\left[\frac{X^2}{2}\right]_1^2 + 9\left[\frac{X^3}{3}\right]_1^2$$
$$= 2\left[X^2\right]_1^2 + 3\left[X^3\right]_1^2$$
$$= 2(2^2 - 1^2) + 3(2^3 - 1^3)$$
$$= 2(4 - 1) + 3(8 - 1)$$
$$= 2(3) + 3(7) = 27$$

C.3.3 Methods of integration

There is no set rule for integrating functions in the form of the product and quotient of other functions whose integrals are known. For such functions there are integral tables and a number of methods of integration, including:

- Integration by Substitution
- Integration by Parts
- Integration by Trigonometric Substitution
- Integration by Partial Fraction Decomposition
- Numerical Integration

For example, assume that the following integral has to be evaluated:

$$\int 2X \sin(X^2)\, dX$$

The function $2X \sin(X^2)$ can be integrated by observing that $2X$ is the derivative of X^2. If we let $Z{=}X^2$, then $dZ{=}2X\, dX$. By substituting Z and dZ into the given integral we can write:

Substitution: $\int 2X \sin(X^2)\, dX = \int \sin Z\, dZ$

Integration: $= -\cos Z + c_0$

Back substitution: $= -\cos(X^2) + c_0$

The method used here is called "integration by substitution."

Detailed descriptions of these methods are beyond the scope of this text. For further information, the interested reader should refer to calculus textbooks, such as:

Ash, C., and Ash, R.B. 1986. *The Calculus Tutoring Handbook.* New York: IEEE Press.

Ayres, F. 1972. *Schaum's Outline Series: Theory and Problems of Differential and Integral Calculus.* 2nd ed. New York: McGraw-Hill.

INDEX